# Mathematics: A Universal L

Sir Ronald Aylmer Fisher

John Maynard Keynes

Richard Courant

John von Neumann

Ol'ga Arsen'evna Oleinik

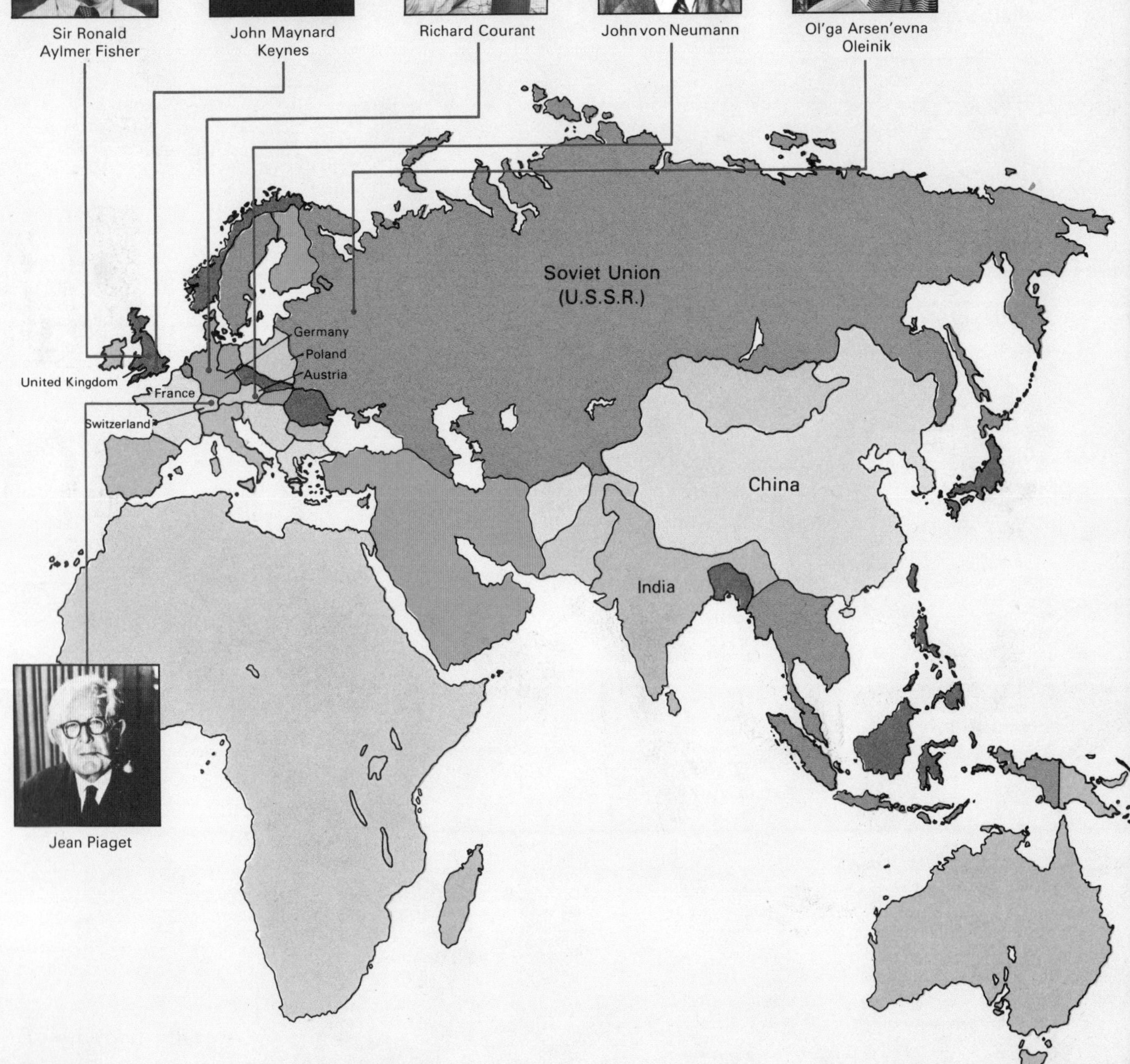


Jean Piaget

# FORMULAS FROM GEOMETRY

**Triangle**

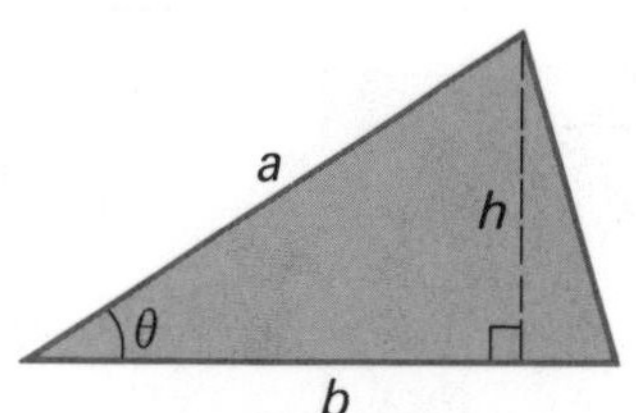


Area $= \frac{1}{2} bh$

Area $= \frac{1}{2} ab \sin \theta$

**Parallelogram**

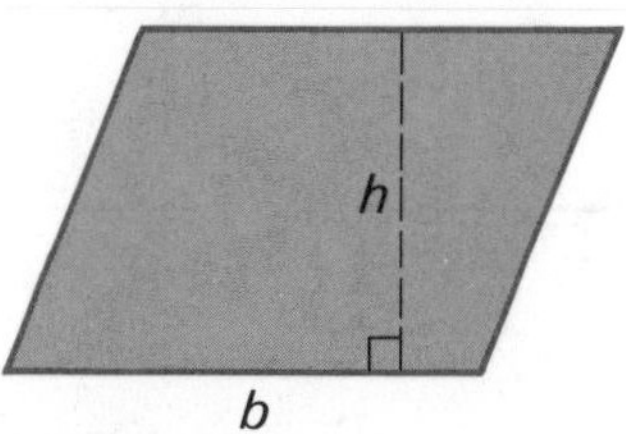


Area $= bh$

**Trapezoid**

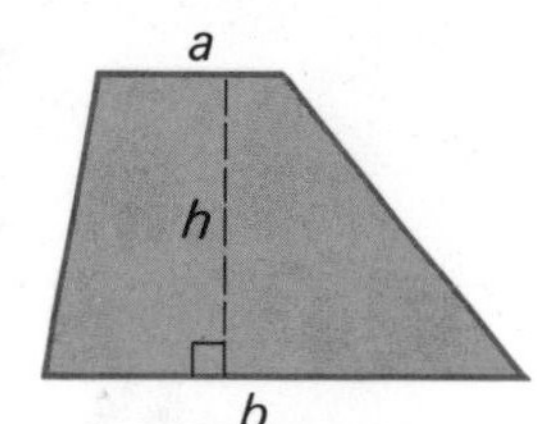


Area $= \frac{a+b}{2} h$

**Circle**

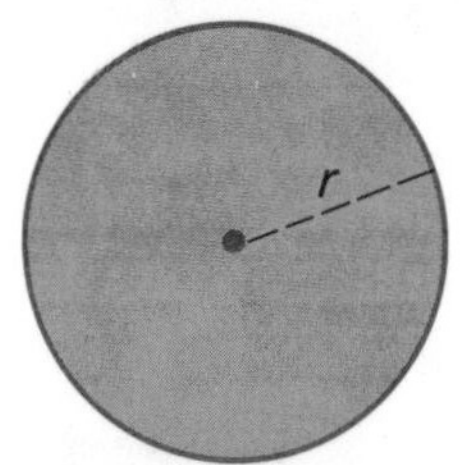


Circumference $= 2\pi r$

Area $= \pi r^2$

**Sector of Circle**

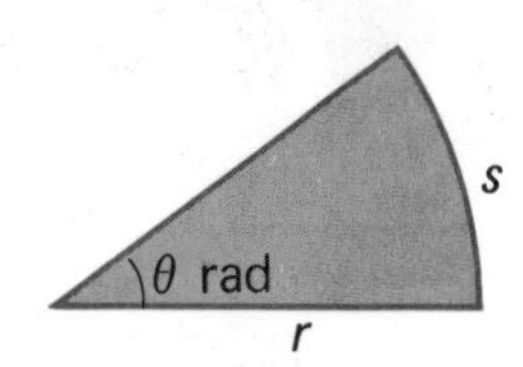


Arc length $s = r\theta$

Area $= \frac{1}{2} r^2 \theta$

**Polar Rectangle**

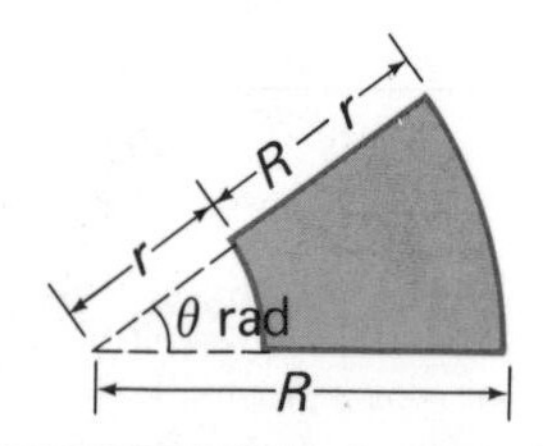


Area $= \frac{R+r}{2} (R-r)\theta$

**Right Circular Cylinder**

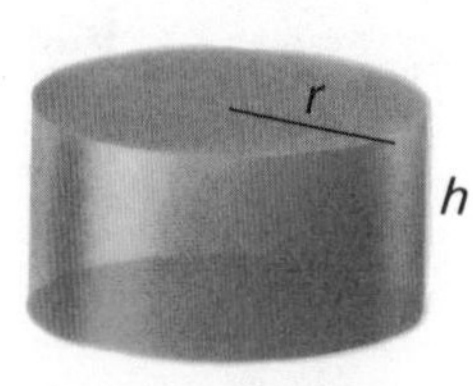


Lateral area $= 2\pi rh$

Volume $= \pi r^2 h$

**Ball**

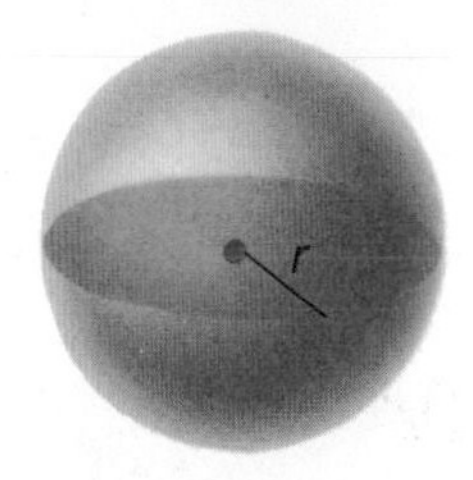


Area $= 4\pi r^2$

Volume $= \frac{4}{3} \pi r^3$

**Right Circular Cone**

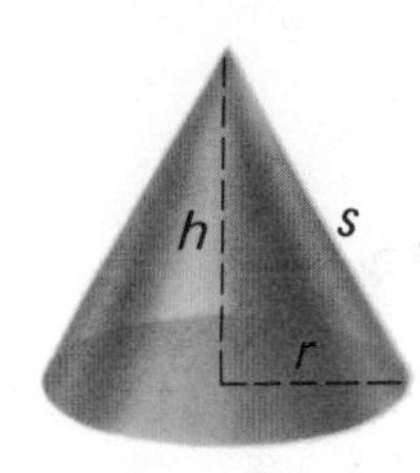


Lateral area $= \pi rs$

Volume $= \frac{1}{3} \pi r^2 h$

**Frustum of Right Circular Cone**

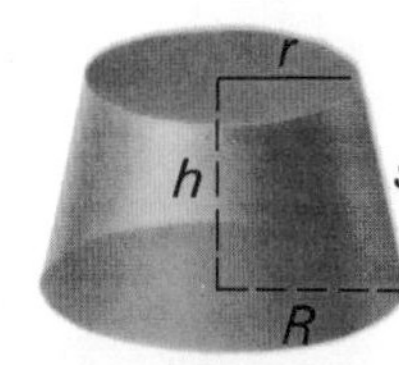


Lateral area $= \pi s(r + R)$

Volume $=$

$\frac{1}{3} \pi (r^2 + rR + R^2)h$

**General Cone**

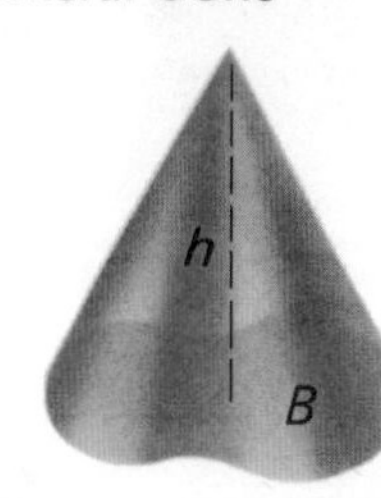


Volume $= \frac{1}{3}$ (area $B$)$h$

**Wedge**

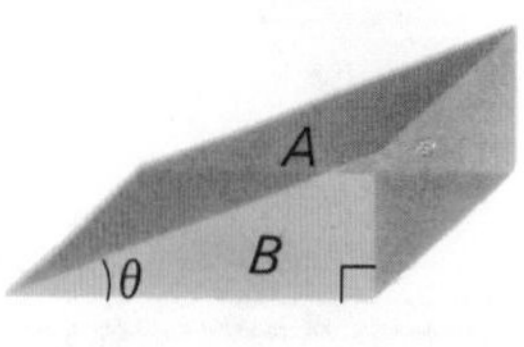


Area $A =$ (area $B$)$\sec \theta$

# Applied Calculus for Management, Social, and Life Sciences

# Applied Calculus for Management, Social, and Life Sciences

**Dale Varberg**
*Hamline University*
**Walter Fleming**
*Hamline University*

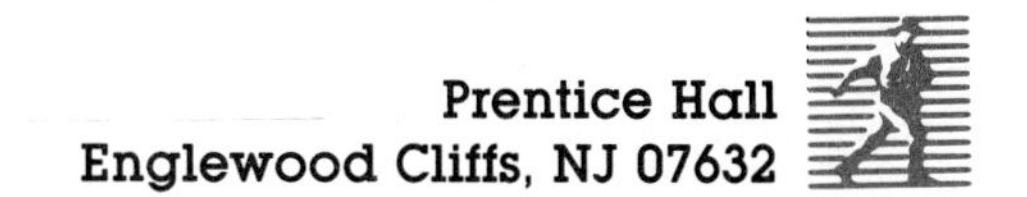
Prentice Hall
Englewood Cliffs, NJ 07632

Library of Congress Cataloging-in-Publication Data

Varberg, Dale E.
Applied calculus : for management, social, and life sciences /
Dale E. Varberg, Walter Fleming.
p. cm.
ISBN 0-13-039769-5 : $35.00
1. Calculus. 2. Social sciences—Mathematics. 3. Biomathematics.
I. Fleming, Walter II. Title.
QA303.F5836 1990
515—dc20 90-32111
CIP

Interior design: *Lorraine Mullaney*
Cover design: *Bruce Kenselaar*
Cover art: *Salem Krieger*
Photo editor: *Lorinda Morris-Nantz*
Photo research: *Barbara Scott*
Pre-press buyer: *Paula Massenaro*
Manufacturing buyer: *Lori Bulwin*

*Photo credits:* **Chapter 1**, Laima Druskis; **p. 1**, Historical Pictures Service, Chicago; **46**, Ed Lettau/Photo Researchers; **47**, Historical Pictures Service, Chicago; **84**, Tcherevkoff Studio/The Image Bank; **85**, Historical Pictures Service, Chicago; **120**, FORTUNE, © 1989 The Time Inc. Magazine Company. All rights reserved; **121**, Massachusetts Institute of Technology; **160**, *Fundamentals of Anatomy and Physiology* (Frederic Martini), Englewood Cliffs, N.J.: Prentice Hall, 1989, p. 40; **161**, University of Cambridge; **188**, NASA; **189**, S. Preobrazhensky/Sovfoto; **216**, Mel Digiacomo/The Image Bank; **217**, Historical Pictures Service, Chicago; **244**, from "Mathematical Games," Martin Gardner. Copyright © 1986 by SCIENTIFIC AMERICAN, Inc. All rights reserved; **245**, UPI/Bettmann Newsphotos; **274**, from "Climate Modeling," Stephen H. Schneider, Copyright © 1987 by SCIENTIFIC AMERICAN, Inc. All rights reserved; **275**, University of California, Berkeley; **310**, from "The Allocation of Resources by Linear Programming," by Robert G. Bland. Copyright © 1981 by SCIENTIFIC AMERICAN, Inc. All rights reserved; **311**, Stanford University.

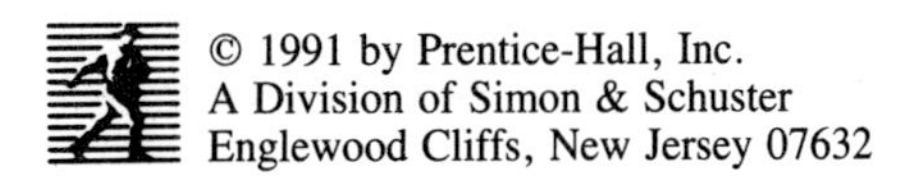



Printed in the United States of America
10 9 8 7 6 5 4 3 2 1

ISBN 0-13-039769-5

Prentice-Hall International (UK) Limited, *London*
Prentice-Hall of Australia Pty. Limited, *Sydney*
Prentice-Hall Canada Inc., *Toronto*
Prentice-Hall Hispanoamericana, S.A., *Mexico*
Prentice-Hall of India Private Limited, *New Delhi*
Prentice-Hall of Japan, Inc., *Tokyo*
Simon & Schuster Asia Pte. Ltd., *Singapore*
Editora Prentice-Hall do Brasil, Ltda., *Rio de Janeiro*

# Contents

## 10 Multivariate Calculus 311



## Appendix 353



*Sections marked with stars are optional and can be omitted without loss of continuity.

# Preface

## Audience

This is a book for those people who want to see the power and usefulness of the calculus but who feel they do not have time to work through the massive tomes that are common in engineering courses. In particular, it was written for students in the management, social, and life sciences; however, we believe it can serve as an introduction to calculus for students in all disciplines. Our goal was to make the book *lean and lively*; readers will judge whether we have succeeded.

## Algebra Review

Our book has a number of novel features that we want to describe. Since we assumed that our readers would want to get to the meat of the subject quickly, we have foregone the general review of algebra one finds at the beginning of many books. Of course, algebra is foundational to calculus. Therefore, we review essential topics as they are needed at various points throughout the book. A further discussion of algebraic concepts can be found in the Appendix. But calculus is our subject, and we introduce the notions of slope and rate of change in the first chapter.

## Pacing

Our general procedure is to postpone discussion of a topic until it will make sense to the student. Thus, the concept of continuity waits until it can be discussed along with differentiability and other properties of curves (Chapter 3). More significant is our decision to leave the Chain Rule until Chapter 5 where the exponential and logarithmic functions are available for examples. Of course, the special case that we call the General Power Rule occurs early (Chapter 2). We have always felt that the Chain Rule falls flat if it can be illustrated only with power functions.

## Conceptual Emphasis

Like most good authors, we attempt to be clear and informative rather than encyclopedic. We give only those proofs that are easy to understand and which provide genuine insight. We try to make all of our descriptions conceptual in nature and, if a diagram will help, we display it. A notable example is our treatment of the definite integral. We think that a rigorous definition based on upper and lower sums is too abstract for our audience. On the other hand, the common practice of defining the definite integral as $F(b) - F(a)$ gives no insight into the real meaning of the integral and obscures the significance of the Fundamental Theorem of Calculus. Therefore, we define the definite integral geometrically, taking area to be an intuitively obvious notion. Later, we show that the definite integral is the limit of Riemann sums.

## Road Signs

We think of our book as providing a road map throught difficult and sometimes confusing terrain. In line with this metaphor, our margins have road signs of four kinds. There are CAUTION signs warning readers of dangerous mistakes. There are RECALL signs reminding students of facts that may have been forgotten. A few signs carry the words SHARP CURVE, suggesting that the argument being made has a subtle turn that requires careful navigation. Finally, there are many signs with the message POINT OF INTEREST. When you see one of these signs, it is time to stop and enjoy a historical note or a bit of mathematical lore.

## Applications

Our book is called *Applied Calculus* for good reason. Four of our chapters are devoted in toto to applications of the calculus. In addition, most sections contain an application or two, and the problem sets introduce many more. These applications come from everywhere, but we tend to emphasize those that are important in the management, social, and life sciences. Economic and business applications occur the most frequently.

## Biographies

To support our applied calculus emphasis, we begin each chapter with a picture and brief biography of an applied mathematician. These people come from a wide variety of disciplines but have several characteristics in common. All have made notable contributions to their fields and all have used mathematics in a significant way. They come from many different countries, demonstrating that mathematics (including calculus) is a universal language. They did their major work in this century, suggesting that calculus is not dead. There are ten of them, corresponding to the ten chapters of this book. The front endpaper with the map of the world indicates their diverse national backgrounds.

## Examples and Problems

Our book is notable for the wealth of examples and problems. We chose our examples with two criteria in mind: They must illustrate the major ideas of calculus, and they should be realistic. The problem sets that conclude each section are carefully graded in difficulty thereby giving students a chance to build their confidence before hitting the hard nuts. First, there is a collection of relatively easy problems, all clearly related to the examples of the section. Then (following a horizontal bar), there is a set of miscellaneous problems. These latter problems are designed to integrate all the concepts of the section and usually culminate in 3 or 4 substantial applications. Each chapter ends with a lengthy set of review problems. The symbol [C] attached to a problem indicates that a calculator will be helpful. A very few problems are identified with the label [PC] suggesting that their solution requires a programmable calculator or a computer. Such problems can be omitted without loss of continuity.

## Supplements

*For students* using this text, the publisher has prepared a **Student Solutions Manual** that includes complete solutions to every third problem.

*For adopters* of this text, the following free supplements are available:

- An **Instructor's Edition** contains the complete text as in the student's edition, along with Teaching Outlines at the beginning of the text. These outlines lay out in terse form the concepts we think should be emphasized and suggest examples to use in illustrating them. Answers to all exercises are provided at the back.
- An **Instructor's Solutions Manual** contains complete solutions to every exercise.
- Included in the **Transparency Pack** are 35 color transparencies of important illustrations from the text along with 20 black and white transparency masters. A sample transparency has been inserted in the Instructor's Edition.
- The **Test Item File** contains 1500 multiple choice test questions that are referenced by chapter number and section.
- The **Prentice Hall Datamanager** allows the instructor to access questions from the **Test Item File** and personally select questions and print out tests on IBM personal computers. The program allows the questions to be edited. It is available on both $5\frac{1}{4}''$ and $3\frac{1}{2}''$ floppy disks. A demonstration disk is available from your Prentice Hall representative.
- A set of eight **Videotaped Lectures** are designed to eliminate the lines of students waiting outside professors' offices. Referenced to the text by chapter and section, the lectures serve as a tutoring aid for difficult material or can be used in a telecourse.

- ***How to Study Calculus*** is a booklet for students that contains strategies, suggestions, and hints for learning and achieving success in calculus courses. One copy of the supplement is provided free for every *new* student text purchased.

- ***EPIC—Exploration Programs in Calculus.*** This interactive software for the IBM PC covers all basic topics of the calculus course and is referenced to the text by chapter and section. Free upon adoption.

- ***Interactive Experiments in Calculus*** is a fully interactive software product for the Macintosh computer, featuring approximately twenty interactive experiments that cover the major calculus topics. Free upon adoption.

## Acknowledgements

We acknowledge with thanks the generous help of many people. Louis Guillou (St. Mary's College, Winona, MN) critiqued the whole manuscript, checked our answers to all the problems, and prepared the instructor's and student's solution manuals. Stephen Mondy (Normandale Community College, Bloomington, MN) read the entire manuscript, checked all the examples, and helped correct the galley proofs. A team of four mathematicians (Frisinger, Montgomery, Davitt, and Kasube—identified by school below) met with authors over a two-day period, offering suggestions for improvements in the manuscript. Finally, each of the following people served as a reviewer for all or part of the book.

Philip Crooke—Vanderbilt University, Nashville, TN
Richard Davitt—University of Louisville, Louisville, KY
Howard Frisinger—Colorado State University, Fort Collins, CO
Chaitan P. Gupta—Northern Illinois University, DeKalb, IL
Herbert E. Kasube—Bradley University, Peoria, IL
Walter Thomas Kyner—University of New Mexico, Albuquerque, NM
Philip R. Montgomery—University of Kansas, Lawrence, KA
Robert Moreland—Texas Technical University, Lubbock, TX
S.F. Neustadter—California State University, San Francisco, CA
James W. Newsom—Tidewater Community College, Portsmouth, VA
David Nilsson—University of Texas, Austin, TX
Clifton Thomas Wyburn—University of Houston, Houston, TX
Melvin R. Woodard—Indiana University of Pennsylvania, Indiana, PA

We express our deep appreciation to Prentice Hall, to mathematics editor Robert Sickles, designer Lorraine Mullaney, project editor John Morgan, and to all others who helped produce this book.

Prentice Hall has taken
every possible step to ensure the precision and accuracy of this text. Experts reviewed content, checked exercises, and proofread technical material. They assisted the authors and Prentice Hall in producing an outstanding text of the highest quality.

You can help Prentice Hall maintain these high standards by perusing this text and relaying pertinent information found, including errors, to Mathematics Editor, Prentice Hall, Englewood Cliffs, NJ 07632 or to your local Prentice Hall representative. We look forward to serving your text needs now and in the future.
Thank you very much for your support.

*It is said that to do well in the stock market, a person must have the nerves of a gambler, the sixth sense of a clairvoyant, and the courage of a lion. But one must also have a deep understanding of economic principles. John Maynard Keynes had all the qualifications.*

# 1

# Slopes and Rates of Change



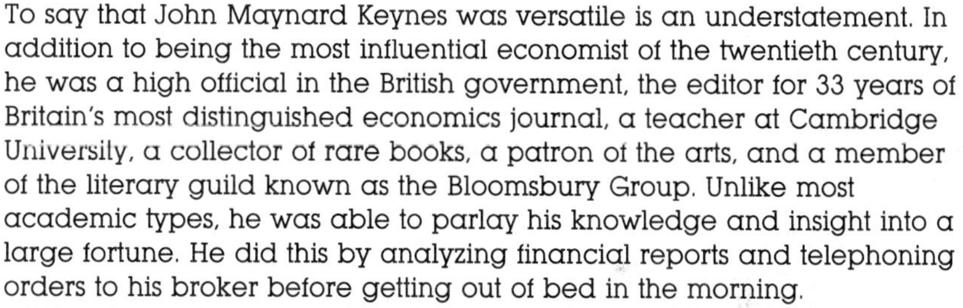

To say that John Maynard Keynes was versatile is an understatement. In addition to being the most influential economist of the twentieth century, he was a high official in the British government, the editor for 33 years of Britain's most distinguished economics journal, a teacher at Cambridge University, a collector of rare books, a patron of the arts, and a member of the literary guild known as the Bloomsbury Group. Unlike most academic types, he was able to parlay his knowledge and insight into a large fortune. He did this by analyzing financial reports and telephoning orders to his broker before getting out of bed in the morning.

Like his countryman and contemporary, Bertrand Russell, Keynes got his early training in mathematics. In fact, he mastered the subject so well that he was able to write a treatise on the foundations of probability, a book still read with profit by students of mathematics and philosophy. It is in mathematics, perhaps more than any other subject, that one learns to think about assumptions, to construct models, to analyze arguments, and to draw rigorous conclusions. Dare we suggest that it was this training that prepared Keynes to make even greater contributions when he turned his attention to the field of economics?

John Maynard Keynes
(1883–1946)

In the middle of the great world depression, Keynes published his masterpiece: *The General Theory of Employment, Interest and Money* (1936). In it, he argued that government could and should influence the economic welfare of the state by reducing interest rates, by equalizing incomes (through progressive taxation), and by stimulating the economy with such "pump-priming" measures as public work projects. These ideas had a profound impact on Franklin Delano Roosevelt and inspired many of the programs of the New Deal. It is safe to say that the United States has been permanently affected by Keynes, and even the conservative economist, Milton Friedman, has been quoted as saying, "We are all Keynesians now."

## 1.1
# FUNCTIONS

Much of science involves studying the relationship between two variables. An economist believes that the number of housing starts in a year depends in a definite way on the magnitude of the prime interest rate. A business executive knows that the price charged for a certain product must be adjusted to the supply available. A physiologist observes that the response time of an automobile driver increases with alcohol intake. In each of these examples, the size of one variable determines the size of another. Mathematicians use the term *function* to describe such relationships.

> **DEFINITION**
>
> A **function** is a rule that associates with each number $x$ in one set a definite number $y$ from another set.

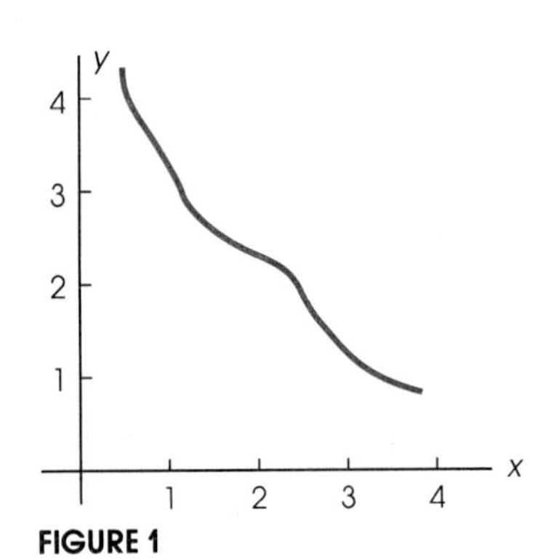

FIGURE 1

In newspapers, magazines, and many social science textbooks, functions are specified by drawing graphs (as in Figure 1). However, in mathematics, our normal starting point is a formula or equation. Consider, for example, the equation

$$y = x^2$$

which readers will recognize as the formula for the area of a square. As soon as the side length $x$ is given, the area $y$ is determined. If $x = 5$, then $y = 5^2 = 25$; if $x = 3.12$, then $y = (3.12)^2 = 9.7344$.

There is nothing sacred about the use of $x$ and $y$ as variables. In geometry, we learned that the volume $V$ of a sphere of radius $r$ (Figure 2) is given by

$$V = \tfrac{4}{3}\pi r^3$$

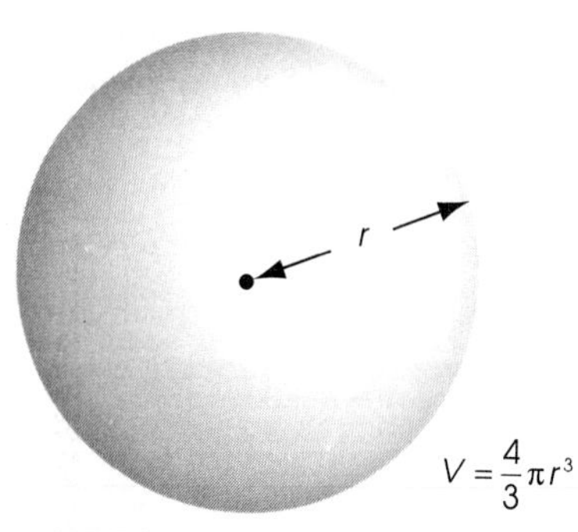

FIGURE 2

Here $\pi$ is that often appearing constant with approximate value 3.1416 (we write $\pi \approx 3.1416$). Knowing a value for $r$, we can easily determine the corresponding value of $V$. If $r = 3$, then $V = \frac{4}{3}\pi 3^3 = 36\pi \approx 113.0973$. Since $r$ can be given any (positive) value, it is called the **independent variable.** $V$ depends on $r$; it is called the **dependent variable.**

### Functional Notation

A single letter like $f$ (or $g$ or $F$) will be used to name a function. Then $f(x)$, read "$f$ of $x$" or "$f$ at $x$", denotes the value that $f$ assigns to $x$. Thus, if $f$ names the "area of a square" function mentioned earlier, it is clear that

$$f(x) = x^2$$

and

$$f(2) = 2^2 = 4$$
$$f(4.2) = (4.2)^2 = 17.64$$
$$f(a) = a^2$$
$$f(2m) = (2m)^2 = 4m^2$$

In the latter four cases, we simply replaced $x$ by 2, 4.2, $a$, and $2m$, respectively.

Somewhat more complicated, but of critical importance later, are calculations like the following.

$$f(2 + h) = (2 + h)^2 = 4 + 4h + h^2$$
$$f(2 + h) - f(2) = 4 + 4h + h^2 - 4 = 4h + h^2$$
$$\frac{f(2 + h) - f(2)}{h} = \frac{4h + h^2}{h} = \frac{h(4 + h)}{h} = 4 + h$$

Because functional notation plays such a large role in calculus, we pause for three more examples.

### EXAMPLE A

If $f(x) = x^3 - 2x + 1$, find (a) $f(3)$, (b) $f(-2)$, (c) $f(\frac{1}{2})$.

**Solution** Perhaps it will help to think of the formula for $f$ in the following way

$$f(\quad) = (\quad)^3 - 2(\quad) + 1$$

Then putting the same number inside each pair of parentheses, we obtain

(a) $f(3) = (3)^3 - 2(3) + 1 = 27 - 6 + 1 = 22$
(b) $f(-2) = (-2)^3 - 2(-2) + 1 = -8 + 4 + 1 = -3$
(c) $f(\frac{1}{2}) = (\frac{1}{2})^3 - 2(\frac{1}{2}) + 1 = (\frac{1}{8}) - 1 + 1 = \frac{1}{8}$ ■

### EXAMPLE B

For $g(x) = x^2 - 4x$, find and simplify (a) $g(3)$, (b) $g(3 + h)$, (c) $[g(3 + h) - g(3)]/h$, (d) $[g(x + p) - g(x)]/p$.

**Solution**

(a) $g(3) = (3)^2 - 4(3) = 9 - 12 = -3$
(b) $g(3 + h) = (3 + h)^2 - 4(3 + h) = 9 + 6h + h^2 - 12 - 4h$
$\quad = -3 + 2h + h^2$

(c) $$\frac{g(3+h)-g(3)}{h} = \frac{-3+2h+h^2-(-3)}{h} = \frac{h(2+h)}{h} = 2+h$$

(d) $$\frac{g(x+p)-g(x)}{p} = \frac{(x+p)^2-4(x+p)-(x^2-4x)}{p} = \frac{x^2+2xp+p^2-4x-4p-x^2+4x}{p} = \frac{2xp+p^2-4p}{p} = \frac{p(2x+p-4)}{p} = 2x+p-4$$ ■

### EXAMPLE C

For $F(x) = 1/x$, find and simplify $[F(a+h)-F(a)]/h$.

**Solution**

$$\frac{F(a+h)-F(a)}{h} = \frac{\dfrac{1}{a+h}-\dfrac{1}{a}}{h} = \frac{\dfrac{a}{a(a+h)}-\dfrac{a+h}{a(a+h)}}{h} = \frac{a-(a+h)}{a(a+h)}\cdot\frac{1}{h} = \frac{-h}{a(a+h)h} = \frac{-1}{a(a+h)}$$ ■

**CAUTION**
The common error in Example C is to write

$$F(a+h) = \frac{1}{a}+h$$

which is wrong. Rather

$$F(a+h) = \frac{1}{a+h}$$

## Domain and Range

The **domain** for a function is the set of allowable values for the independent variable; the **range** is the set of values attained by the dependent variable. Thus for the "area of a square" function $y = f(x) = x^2$, we take the domain to be the set of positive numbers (a square can have any positive length). In this case, the range would also be the set of positive numbers. If no domain is specified either explicitly or implicitly (that lengths should be positive is an implicit restriction), then we take the domain to be the largest set of real numbers for which the formula for the function makes sense. Thus in the case of $F(x) = 1/x$ (Example C), we take the domain to be the set of all real numbers except 0 (0 is excluded to avoid division by zero). If $G(x) = \sqrt{x-1}$, we take the domain to be the set of all real numbers greater than or equal to 1. Numbers smaller than 1 are not allowed because we cannot take the square root of a negative number (imaginary numbers play no role in this book).

**RECALL**
If $a \geq 0$, then $\sqrt{a}$ denotes the nonnegative square root of $a$. For $G(x) = \sqrt{x-1}$,

$$G(5) = \sqrt{5-1} = 2$$

On the other hand,

$$G(-3) = \sqrt{-3-1} = \sqrt{-4}$$

is left undefined.

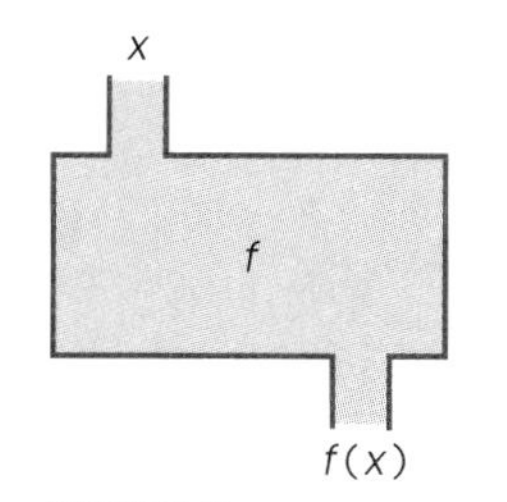


FIGURE 3

## Functions Viewed as Machines

A helpful way to think of a function $f$ is as a machine (Figure 3). The $f$ machine accepts an input $x$, works on it, and produces the output $f(x)$. The set of all input numbers is the domain of the function, while the set of output numbers is its range.

Your hand-held calculator nicely illustrates this view (Figure 4). Enter an input 3 and press the $x^2$ key, and the calculator displays 9 as output. Scientific models have built-in keys for many functions ($\sqrt{x}$, $1/x$, $\log x$, $\sin x$, and so on).

If you do not own a calculator, we urge you to purchase one. Be sure to get a scientific model and, if you can afford it, one that is programmable.

Almost every section of this book wil have some problems that are simplified by the use of a calculator. Such problems are marked with the symbol [C]. A few problems are best done on a programmable calculator or a personal computer. They bear the code [PC].

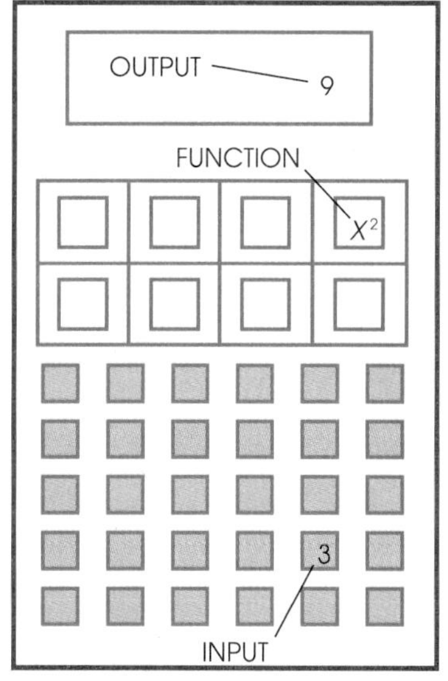


FIGURE 4

## Composite Functions

Simple machines can be hooked together in tandem to form more complicated machines; so can functions. Let the function $f$ take the input $x$ and produce the output $f(x)$. Then let $f(x)$ become the input for a second function $g$ which accordingly produces the output $g(f(x))$ as in Figure 5. The resulting coupled function $g(f)$ is called the **composite of $g$ with $f$.**

### EXAMPLE D

Let $f(x) = (x - 1)^2$ and $g(x) = 1/x$. Find (a) $g(f(3))$, (b) $g(f(\frac{1}{2}))$, (c) $g(f(1))$, (d) $g(f(x))$, (e) $g(g(x))$.

**Solution** Always begin by calculating the value of the function that first operates on the input; that is, begin by evaluating the inner function.

(a) $g(f(3)) = g(4) = \frac{1}{4}$
(b) $g(f(\frac{1}{2})) = g(\frac{1}{4}) = 4$
(c) $g(f(1)) = g(0) = \frac{1}{0}$, which is undefined
(d) $g(f(x)) = g[(x - 1)^2] = 1/(x - 1)^2$
(e) $g(g(x)) = g(1/x) = 1/(1/x) = x$ ■

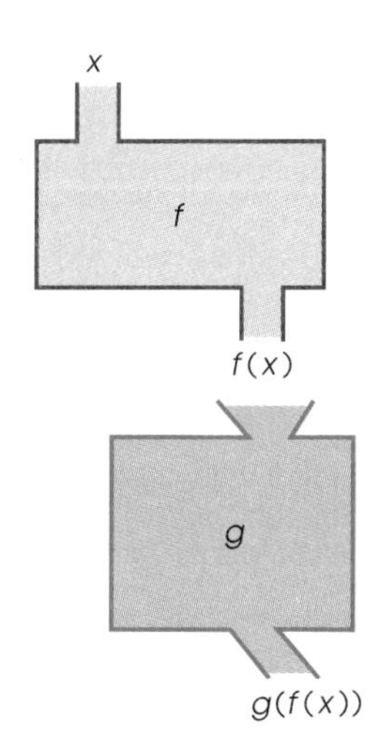


FIGURE 5

Now we are in for a mild surprise: $g(f)$ and $f(g)$ are in general not the same; composition of functions is not commutative.

### EXAMPLE E

Let $f(x) = (x - 1)^2$ and $g(x) = 3x + 1$. Find $g(f(x))$ and $f(g(x))$.

**Solution**

$$g(f(x)) = g[(x-1)^2] = 3(x-1)^2 + 1 = 3x^2 - 6x + 4$$

$$f(g(x)) = f(3x+1) = (3x+1-1)^2 = 9x^2 \qquad \blacksquare$$

In using your calculator, you will often, perhaps without thinking, calculate the composite of several functions. The next example actually involves the composite of $h(x) = 1/x$ with $g(x) = x^2$ with $f(x) = \log x$.

### EXAMPLE F

Use your calculator to evaluate $1/(\log 5)^2$.

**Solution** Press the following keys.

$$5 \; \boxed{\log} \; \boxed{x^2} \; \boxed{1/x}$$

Our calculator gives 2.0468354. Note that this result is really $h(g(f(5)))$. ■

## Problem Set 1.1

*In Problems 1–4, write an equation for the function that is described.*

1. The area $y$ of a rectangle that is twice as long as wide, in terms of the width $x$.
2. The distance $s$ in miles traveled by a car going 50 miles per hour, in terms of $t$, the number of hours spent traveling.
3. The volume $V$ of a box with a square base and height twice the base length, in terms of the base length $x$.
4. The total surface area $S$ of the box described in Problem 3, in terms of the base length $x$, assuming the box has a lid.
5. For $f(x) = 7x - 3$, evaluate each expression.
   (a) $f(2)$, (b) $f(-2)$, (c) $f(0)$, (d) $f(\frac{8}{7})$, (e) $f(\frac{3}{14})$
6. For $g(x) = 5x + 2$, evaluate each expression.
   (a) $g(1)$, (b) $g(-1)$, (c) $g(10)$, (d) $g(\frac{-2}{5})$, (e) $g(\frac{3}{10})$
7. For $F(x) = 2x^3 - 4x + 3$, evaluate each of the following.
   (a) $F(1)$, (b) $F(-2)$, (c) $F(5)$, (d) $F(\frac{1}{2})$, (e) $F(-\frac{3}{2})$
8. For $G(x) = x^4 - 2x^2 - 6$, evaluate each of the following.
   (a) $G(1)$, (b) $G(-2)$, (c) $G(3)$, (d) $G(-\frac{1}{2})$, (e) $G(\sqrt{3})$
9. For $f(x) = 3x - 2x^2$, find and simplify:
   (a) $f(2)$ (b) $f(2 + h)$
   (c) $[f(2 + h) - f(2)]/h$
   (d) $[f(x + d) - f(x)]/d$
10. For $g(x) = 4x - x^2$, find and simplify
    (a) $g(3)$ (b) $g(3 + h)$
    (c) $[g(3 + h) - g(3)]/h$
    (d) $[g(x + p) - g(x)]/p$
11. For $f(x) = (4/x) - 2$, find and simplify
    $$[f(a+h) - f(a)]/h$$
12. For $f(x) = 2/x^2$, find and simplify
    $$[f(a+h) - f(a)]/h$$
13. For $g(x) = 5 - 2x + 3x^2$, find and simplify
    $$[g(x+h) - g(x)]/h$$
14. For $g(x) = x^3$, find and simplify
    $$[g(c+h) - g(c)]/h$$
15. Let $f(x) = 2x - 4$ and $g(x) = 2/x^2$. Find:
    (a) $g(f(1))$ (b) $g(f(\frac{5}{2}))$ (c) $g(f(2))$
    (d) $g(f(x))$ (e) $g(g(x))$
16. Let $f(x) = x^2 - 2x - 3$ and $g(x) = 2/(x - 5)$. Find:
    (a) $g(f(0))$ (b) $g(f(2))$ (c) $g(f(-2))$
    (d) $g(f(x))$ (e) $g(g(x))$
17. Let $f(x) = x^2 + 1$ and $g(x) = \sqrt{x - 3}$. Find:
    (a) $g(f(4))$ (b) $g(f(\sqrt{2}))$
    (c) $g(f(\frac{3}{2}))$ (d) $g(f(x))$

18. Let $f(x) = \sqrt{x} + 1/\sqrt{x}$ and $g(x) = x^2$. Find:
(a) $g(f(1))$ (b) $g(f(4))$
(c) $g(f(2))$ (d) $g(f(x))$
19. Let $f(x) = 2x^2 + 1$ and $g(x) = 1 - x$. Find $f(g(x))$ and $g(f(x))$.
20. Let $f(x) = 3x + 1$ and $g(x) = x^2 - x$. Find $f(g(x))$ and $g(f(x))$.
21. For $f(x) = x^3 + 1$ and $g(x) = x^2$, find $g(f(x))$ and $f(g(x))$.
22. For $f(x) = (1/x) + 2$ and $g(x) = 1/(x - 2)$, find $g(f(x))$ and $f(g(x))$.
23. If $f(x) = 2x - 1$, $g(x) = x^2$, and $h(x) = 4/(x + 1)$, find $g(h(f(x)))$ and $f(h(g(x)))$.
24. If $f(x) = 1/x^2$, $g(x) = 4x^2$, and $h(x) = \sqrt{x}$, find $h(g(f(x)))$ and $f(g(h(x)))$.
C 25. Calculate (with your calculator in radian mode):
(a) $\ln(\sin\sqrt{0.5})$ (b) $\sqrt{\cos(\ln 3.2)}$
(c) $\ln\sqrt{\cos 1.2}$
C 26. Calculate:
(a) $\sqrt{1/\ln \pi}$ (b) $1/\ln\sqrt{\pi}$
(c) $1/(\ln \pi)^2$
27. If $h(g(f(x))) = \sqrt{\cos(\ln x)}$, give formulas for $f(x)$, $g(x)$, and $h(x)$.
28. If $h(g(f(x))) = 1/\ln\sqrt{x}$, give formulas for $f(x)$, $g(x)$, and $h(x)$.

---

29. Give the domain of each of the following functions.
(a) $f(x) = (x - 1)^2 + 3$ (b) $g(x) = 5/(x - 2)^2$
(c) $F(x) = 3/(x^2 - 9)$
(d) $G(x) = 3/[(x^2 + 1)^2 - 4]$
(e) $H(x) = \sqrt{1/(x - 1)}$
30. Let $f(x) = (x + 1)^2$, $g(x) = \sqrt{x}$, and $h(x) = 1/(x^2 + 1)$. Find and simplify:
(a) $f(-3)$ (b) $g(\frac{25}{4})$
(c) $h(\frac{5}{3})$ (d) $h(g(f(0)))$
(e) $f(g(h(\sqrt{5}/2)))$ (f) $h(g(f(x)))$
(g) $f(f(x))$
31. If $f(x) = x^4$, find and simplify

$$[f(x + h) - f(x)]/h$$

32. In a certain city, you can rent a car for \$18.00 per day and \$0.20 per mile. Write a formula for $f(x)$, the cost of renting a car for 10 days and driving $x$ miles.
33. Write a formula for the daily cost $f(x)$ of manufacturing $x$ pairs of speed skates if there are fixed costs (utilities, taxes, rent, and so on) of \$700 per day and variable costs (material and labor) of \$60 per pair of skates manufactured.
34. Susan sells major appliances. Her monthly salary is \$1200 plus \$20 for each appliance she sells. The store deducts 26% of her total salary for taxes and retirement and then \$10 that she has pledged to the United Way. Write a formula for $f(x)$, her take home pay, if she sells $x$ appliances in a month.
35. Annual union dues at a certain firm are $\frac{1}{2}$% of the first \$18,000 of salary and $1\frac{1}{2}$% of the portion of salary which exceeds \$18,000. Write a formula for the annual dues $f(x)$ of an employee whose annual salary is $x$ dollars assuming this salary exceeds \$18,000.
36. A tray is to be made from a square piece of sheet metal measuring 14 inches on a side by cutting equal squares out of the four corners and turning up the sides. If these squares are $x$ inches on a side, what is the formula for the volume $V(x)$ of the tray and what is the domain for this function?
37. A function $g$ takes an input $x$, doubles it, adds 3, squares the result, subtracts 4, and finally takes the square root. Write a formula for $g(x)$.
C 38. A function $F$ takes a positive input $x$, finds its square root, adds 3, raises the result to the third power, and finally subtracts 10.
(a) Evaluate $F(7)$ and $F(19.6)$.
(b) Write a formula for $F(x)$.
C 39. A study of a certain hospital showed that the percentage $p(t)$ of its patients discharged within $t$ days is given by $p(t) = 100\{1 - [10/(10 + t)]^3\}$.
(a) Calculate $p(0)$, $p(3)$, $p(20)$.
(b) Within how many days will 99% of its patients be discharged?
40. Let $f(x) = \dfrac{1}{1 - 1/(1 - x)}$. Find and simplify $\dfrac{f(1/x)}{f(x)}$.

## 1.2 GRAPHS OF FUNCTIONS

Two Frenchmen, René Descartes and Pierre de Fermat, are responsible for the main idea in this section. Although Descartes is better known and usually gets credit for it, Fermat introduced the concept first.

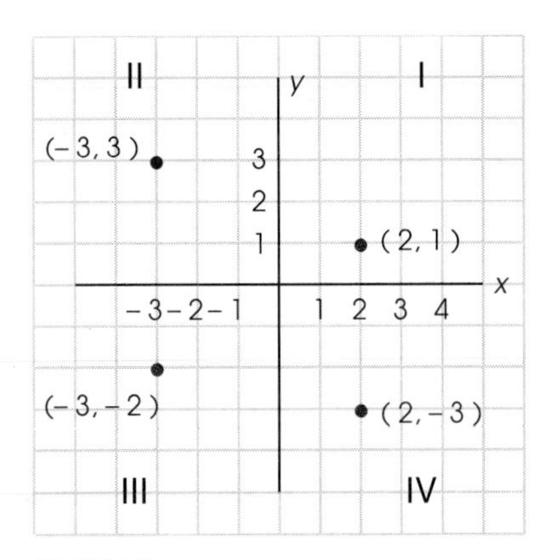


**FIGURE 6**

Place two real lines in a plane, one horizontal and the other vertical. Call the first the ***x*-axis** and the second the ***y*-axis;** call their intersection point the **origin.** Now any point in the plane is uniquely specified by its coordinates $(x, y)$. The first, called the ***x*-coordinate** or abscissa, measures the directed distance from the vertical axis. The second, called the ***y*-coordinate** or ordinate, measures the directed distance from the horizontal axis. A few examples of points and their coordinates are shown in Figure 6. Note that the two axes divide the plane into four quadrants, labeled I, II, III, and IV as shown.

## The Graph of a Function

The graph of the function $f$ is the set of all points whose coordinates $(x, y)$ satisfy the equation $y = f(x)$. But how does one draw a graph? We suggest a sequence of three steps.

1. Obtain the coordinates of a few points.
2. Plot these points in the plane.
3. Connect the points with a smooth curve in the order of increasing $x$-values.

$y = x^2 - 3$

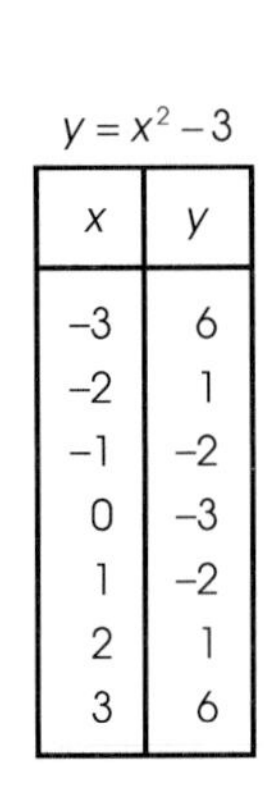

| x | y |
|---|---|
| −3 | 6 |
| −2 | 1 |
| −1 | −2 |
| 0 | −3 |
| 1 | −2 |
| 2 | 1 |
| 3 | 6 |

**Step 1**
Make a table of values

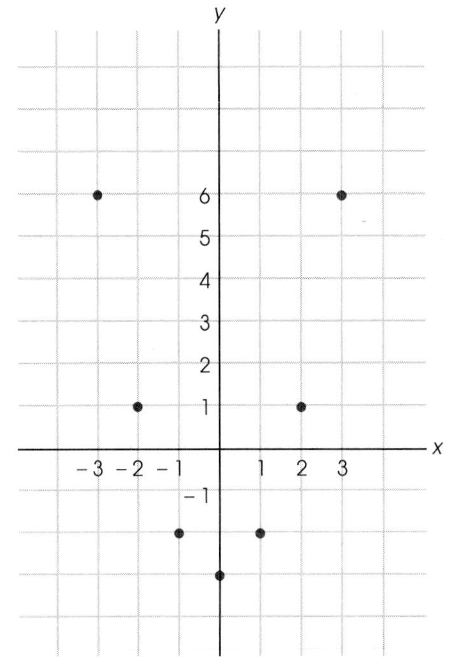


**Step 2**
Plot those points

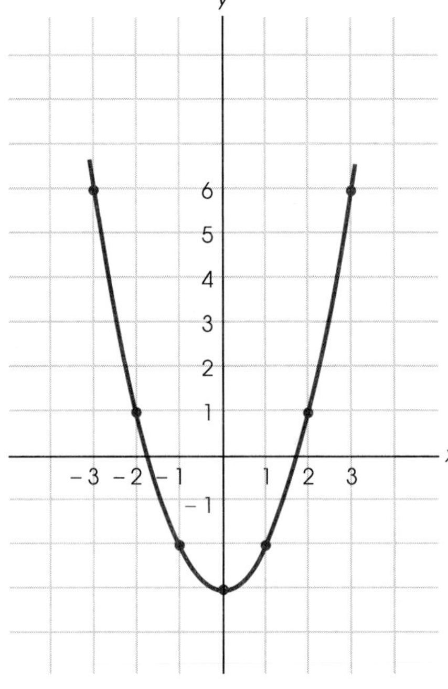


**Step 3**
Connect those points with a smooth curve

**FIGURE 7**

The best way to do Step 1 is to make a table of values. Assign values to $x$ and determine the corresponding values of $y = f(x)$; then list the pairs of values in tabular form. The whole three-step procedure is illustrated in Figure 7 for $y = f(x) = x^2 - 3$.

Of course, you must use some common sense in carrying out the three-step procedure. Be sure to plot enough points so that the special features of the graph are evident.

## A Partial Catalog of Functions

Our first class of functions is the class of linear functions, that is, functions of the form $f(x) = ax + b$. These functions will be the subject of the next section; here we give one example.

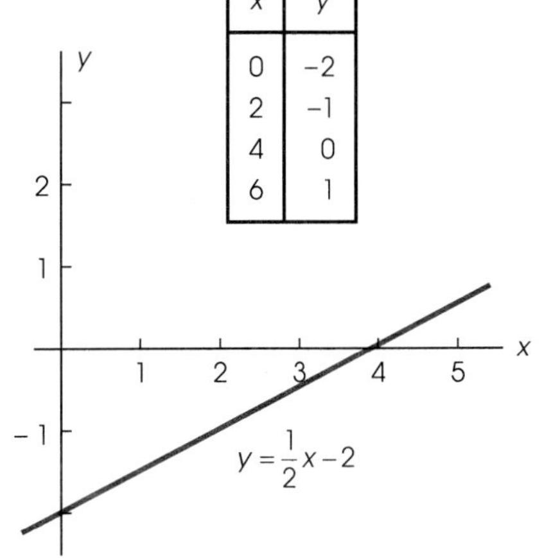

| x | y |
|---|---|
| 0 | −2 |
| 2 | −1 |
| 4 | 0 |
| 6 | 1 |



**FIGURE 8**

### EXAMPLE A

Sketch the graph of $y = f(x) = \frac{1}{2}x - 2$.

**Solution** A small table of values and the graph are shown in Figure 8. The graph is a straight line. ■

Next we mention the class of **quadratic functions,** that is, functions of the form $f(x) = ax^2 + bx + c$, $a \neq 0$. One such function was graphed in Figure 7; the graph was a cup-shaped curve called a **parabola.** The graph of a quadratic function is always a parabola, but it will open down if the leading coefficient is negative, as we note in our next example.

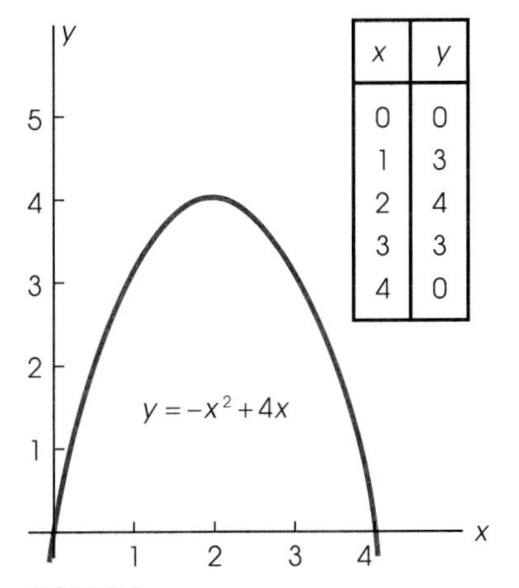

| x | y |
|---|---|
| 0 | 0 |
| 1 | 3 |
| 2 | 4 |
| 3 | 3 |
| 4 | 0 |



**FIGURE 9**

### EXAMPLE B

Sketch the graph of $y = f(x) = -x^2 + 4x$.

**Solution** A table of values and the graph are shown in Figure 9. ■

Linear and quadratic functions are special cases of what are called **polynomial functions,** functions whose formula consists of a sum of terms of the form $cx^m$, $c$ being a constant and $m$ a nonnegative integer. A typical example is

$$f(x) = 2x^4 - x^3 + 3x^2 + 19$$

In general, any function of the form

$$f(x) = a_n x^n + a_{n-1}x^{n-1} + \cdots + a_1 x + a_0, \qquad a_n \neq 0$$

where the $a$'s are real numbers and $n$ is a nonnegative integer, is a polynomial function of degree $n$.

The graph of a polynomial function is always a smooth curve with a number of hills and valleys—usually one less than the degree. For example, the graph of a typical third-degree polynomial has one hill and one valley for a total of two; the graph of a typical fourth-degree polynomial has a total of three hills and valleys.

### EXAMPLE C

Sketch the graphs of $y = f(x) = x^3 - 3x$ and $y = g(x) = x^4 - 2x^2 - 1$

**Solution** The required graphs are shown in Figures 10 and 11. ■

The quotient of two polynomial functions is called a **rational function.** An example is

$$f(x) = \frac{3x^4 - 2x^3 + 11x - 9}{3x^2 - 7x + 2}$$

Graphs of rational functions tend to be complicated; we postpone discussion of them until Chapter 3.

## Symmetry Properties

The graphs in Figures 10 and 11 illustrate two important kinds of symmetry. The graph of $y = f(x) = x^3 - 3x$ (Figure 10) is **symmetric with respect to the origin;** that is, for each point on the graph there is another point directly opposite on the other side of the origin and equally far away. More precisely, if $(x, y)$ is on the graph, so is $(-x, -y)$.

The graph of $y = g(x) = x^4 - 2x^2 - 1$ (Figure 11) is **symmetric with respect to the $y$-axis.** For each point on the graph, there is another directly

| x | y |
|---|---|
| −2 | −2 |
| −1 | 2 |
| 0 | 0 |
| 1 | −2 |
| 2 | 2 |

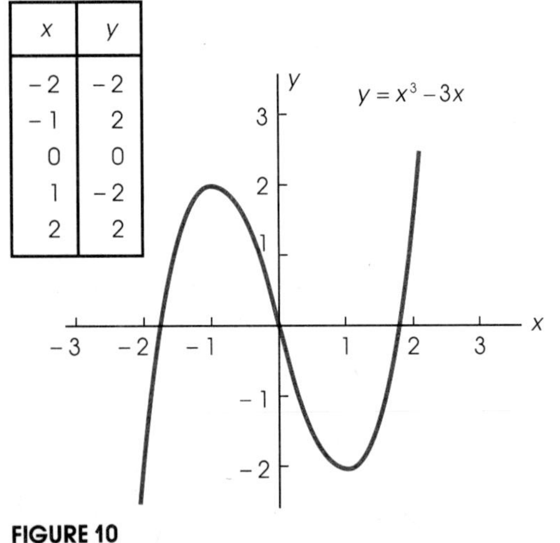


FIGURE 10

| x | y |
|---|---|
| −2 | 7 |
| −1 | −2 |
| 0 | −1 |
| 1 | −2 |
| 2 | 7 |

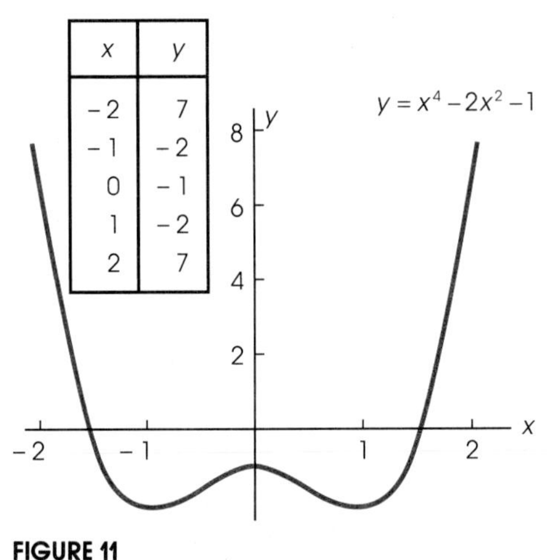


FIGURE 11

opposite on the other side of the $y$-axis and equally far away. Here the coordinate condition is that if $(x, y)$ is on the graph, so is $(-x, y)$.

We could have predicted the symmetry properties of $f$ and $g$ before we drew their graphs. If $f(-x) = -f(x)$, the graph will be symmetric with respect to the origin; if $f(-x) = f(x)$, the graph will be symmetric with respect to the $y$-axis. Note that for $f(x) = x^3 - 3x$,

$$f(-x) = (-x)^3 - 3(-x) = -x^3 + 3x = -(x^3 - 3x) = -f(x)$$

while for $g(x) = x^4 - 2x^2 - 1$,

$$g(-x) = (-x)^4 - 2(-x)^2 - 1 = x^4 - 2x^2 - 1 = g(x)$$

A function $f$ satisfying $f(-x) = -f(x)$ is called an **odd function,** probably because a polynomial with only odd powers is an odd function. If $f(-x) = f(x)$, then $f$ is called an **even function.** A polynomial with only even powers is an even function.

### EXAMPLE D

In each case, decide whether the given function is odd, even, or neither.

(a) $f(x) = 2/(x^2 + 1)$
(b) $g(x) = 3x^3 + 2x^2 + 4$
(c) $h(x) = (2x^6 + 1)/x$

### Solution

(a) $f(-x) = 2/[(-x)^2 + 1] = 2/(x^2 + 1) = f(x)$; $f$ is even.
(b) $g(-x) = 3(-x)^3 + 2(-x)^2 + 4 = -3x^3 + 2x^2 + 4$; $g$ is neither even nor odd.
(c) $h(-x) = [2(-x)^6 + 1]/(-x) = (2x^6 + 1)/(-x) = -(2x^6 + 1)/x = -h(x)$; therefore $h$ is an odd function.

Note that of these three functions, only $g$ is a polynomial. ■

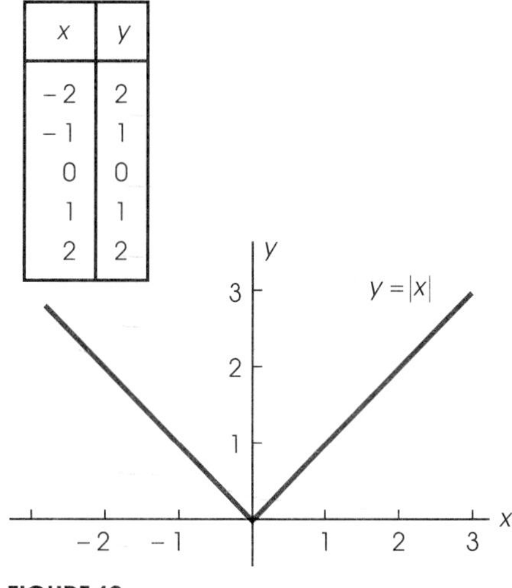

| x | y |
|---|---|
| -2 | 2 |
| -1 | 1 |
| 0 | 0 |
| 1 | 1 |
| 2 | 2 |



FIGURE 12

## Functions with Multipart Rules

A special function that will play a significant role in this book is the **absolute value function** $|\ldots|$, defined by

$$|x| = \begin{cases} x & \text{if } x \geq 0 \\ -x & \text{if } x < 0 \end{cases}$$

Its graph is shown in Figure 12. Note the sharp corner at the origin.

Occasionally, we will meet other functions defined by different rules on separate parts of their domains.

## EXAMPLE E

Sketch the graph of

$$f(x) = \begin{cases} -x + 1 & \text{if } x < 0 \\ -x^2 + 3x & \text{if } 0 \le x \le 3 \\ x - 2 & \text{if } x > 3 \end{cases}$$

**Solution** The three-part graph is shown in Figure 13. Note that at 0 and 3, it is $-x^2 + 3x$ that determines the value of the function. This is the reason for the open circles and the heavy dots at these points on the graph.

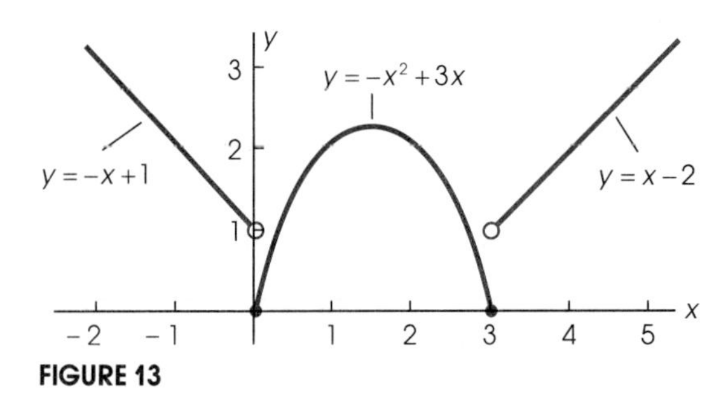


FIGURE 13

■

## EXAMPLE F

Dues for a certain professional organization are based on one's annual salary $S$ according to the following scheme: (a) \$30 if $S \le 20{,}000$, (b) \$30 plus 0.1% of salary in excess of 20,000 if $20{,}000 < S \le 50{,}000$, (c) \$60 plus 0.2% of salary in excess of 50,000 if $S > 50{,}000$. Write a multipart formula for the dues $D(S)$ and then graph the resulting function.

**Solution**

$$D(S) = \begin{cases} 30 & \text{if } S \le 20{,}000 \\ 30 + 0.001(S - 20{,}000) & \text{if } 20{,}000 < S \le 50{,}000 \\ 60 + 0.002(S - 50{,}000) & \text{if } S > 50{,}000 \end{cases}$$

$$= \begin{cases} 30 & \text{if } S \le 20{,}000 \\ 10 + 0.001S & \text{if } 20{,}000 < S \le 50{,}000 \\ -40 + 0.002S & \text{if } S > 50{,}000 \end{cases}$$

**FIGURE 14**

The graph, which consists of three line segments, is shown in Figure 14. ■

## A Final Note

We encourage you to study the chart that appears as Figure 15 at the end of the problem set. This chart supplements what we have said about graphing and illustrates how a graph changes under various transformations.

---

## Problem Set 1.2

1. Which of the following are polynomial functions?
   (a) $f(x) = \frac{3}{2}x^2 + 5x - 6$
   (b) $g(x) = 4x^3 - 5/x$
   (c) $h(x) = \pi x^4 - \sqrt{2}x + 3$
   (d) $F(x) = \sqrt{x^2 + 4}$
   (e) $G(t) = 2.4t^6 - 1.5t^2 + 9$
   (f) $H(s) = \frac{1}{5}s^5 - (2/\sqrt{3})s + 4$
2. Which of the following are polynomial functions?
   (a) $f(x) = 2x^3 - \frac{4}{5}x - 13$
   (b) $g(x) = 4x^2 + 2/x^2$
   (c) $h(x) = 3$
   (d) $F(x) = 2x + \sqrt{x} - 2$
   (e) $G(t) = (t^2 + 4)/t + 3t^2$
   (f) $H(s) = \sqrt{3}\, s^{10} - (15/\pi)s^2 + 10$
3. Identify each of the following functions as to whether it is even, odd, or neither even nor odd.
   (a) $f(x) = 5x^3 - 7x$
   (b) $g(x) = 5x^4 - 2x^2 + 11$
   (c) $h(x) = 3x^3 + 5x - 4$
   (d) $F(x) = x^6 - 2x^4 + 2x - 5$
   (e) $G(x) = 4/(x^2 + 5)$
   (f) $H(x) = (x^2 + 5)/(2x^3 + x)$
4. Identify each of the following functions as to whether it is even, odd, or neither.
   (a) $f(x) = x^5 + 14$
   (b) $g(x) = 3x^{11} - 2x^9 + 7x$
   (c) $h(x) = x^6 - 2x^2 + 14$
   (d) $F(x) = (x^3 - 2x)^2$
   (e) $G(x) = x^3/(x^4 - 2x^2)$
   (f) $H(x) = (x^4 + 3x^2 + 1)(x^3 - 3x)$

*In Problems 5–26, graph each function showing enough of the graph to bring out its essential features. Note either of the two kinds of symmetry discussed in the text.*

5. $f(x) = 2x - 5$
6. $f(x) = -3x + 2$
7. $f(x) = -\frac{3}{4}x + 2$
8. $f(x) = \frac{3}{2}x - 3$
9. $f(x) = -3$
10. $f(x) = 5$
11. $f(x) = -x^2 + 4$
12. $f(x) = -x^2 + 4x$
13. $f(x) = x^2 - 5x$
14. $f(x) = x^2 - 5$
15. $f(x) = x^2 + x - 4$
16. $f(x) = -x^2 + 3x + 2$
17. $f(x) = x^3 - 4x$
18. $f(x) = 9x - x^3$

19. $f(x) = x^4 - 4x^2 + 3$
20. $f(x) = x^4 - 5x^2 + 6$
21. $f(x) = x^3 - x^2$
22. $f(x) = x^3 - 2x^2 + 1$
23. $f(x) = x^5 - x$
24. $f(x) = 16x - x^5$
25. $f(x) = |x - 3|$
26. $f(x) = |2x + 4|$

*As in Example E, Problems 27–32 define functions by means of multipart rules. Sketch the graph of each function.*

27. $f(x) = \begin{cases} 2x & \text{if } x \le 1 \\ -3x & \text{if } x > 1 \end{cases}$

28. $f(x) = \begin{cases} x + 2 & \text{if } x \le 0 \\ 1 & \text{if } x > 0 \end{cases}$

29. $f(x) = \begin{cases} x^2 & \text{if } x < 1 \\ 2x & \text{if } x \ge 1 \end{cases}$

30. $f(x) = \begin{cases} 2x + 2 & \text{if } x < 0 \\ 1 - x^2 & \text{if } x \ge 0 \end{cases}$

31. $f(x) = \begin{cases} x & \text{if } x < -1 \\ x^2 - 2 & \text{if } -1 \le x \le 2 \\ -x^2 + 6 & \text{if } x > 2 \end{cases}$

32. $f(x) = \begin{cases} -x^2 & \text{if } x < 0 \\ -x^2 + 3x & \text{if } 0 \le x < 3 \\ x - 2 & \text{if } x \ge 3 \end{cases}$

33. In 1984, the United States domestic rates for first-class mail were 20 cents for the first ounce or fraction thereof, and 17 cents for each additional ounce or fraction. Let $C(x)$ be the cost of mailing $x$ ounces. Sketch the graph of $C(x)$ for $0 < x \le 6$.

34. In a certain city, the United Way offered the following guidelines for giving to its annual fund: (a) 1% for incomes under \$30,000, (b) \$300 plus 2% of income exceeding \$30,000 for incomes between \$30,000 and \$50,000, (c) \$700 plus 3% of income exceeding \$50,000 for incomes over \$50,000. Graph $G(I)$, the suggested gift corresponding to income $I$.

---

35. Let $f$ be an even function, $g$ an odd function. Decide whether $h$ is even or odd or neither in each of the following.
(a) $h(x) = f(x) \cdot g(x)$
(b) $h(x) = f(x)/g(x)$
(c) $h(x) = [g(x)]^2$
(d) $h(x) = [f(x)]^3$
(e) $h(x) = [f(x)]^2 + [g(x)]^2$
(f) $h(x) = f(x) + x^2 + 4$
(g) $h(x) = [f(x) + g(x)]^2$
(h) $h(x) = g(x) + g(-x)$

36. Graph each of the following.
(a) $f(x) = (x - 1)(x + 2)$
(b) $f(x) = 3(x - 1)^2$
(c) $f(x) = (x + 2)(x - 1)^2$
(d) $f(x) = x(x^2 - 4)$
(e) $f(x) = \begin{cases} -2 & \text{if } x \le -2 \\ 1 + x & \text{if } -2 < x \le 0 \\ 1 - x & \text{if } 0 < x \le 2 \\ -2 & \text{if } x > 2 \end{cases}$

37. Graph $y = x^2$, $y = x^2 + 2$, and $y = (x - 2)^2$ using the same axes. Then explain how the graphs of $y = f(x) + k$ and $y = f(x - k)$ are related to the graph of $y = f(x)$, assuming $k$ is a constant.

38. The RS Bus Company offered to take a church group to a lake resort for \$150 per person if 30 or fewer sign up. If more than 30 persons make the trip, each fare will be reduced by \$15. The capacity of the bus is 70 people. Write a two-part formula for $R(x)$, the total fare revenue if $x$ persons make the trip. Then calculate $R(35)$ and $R(70)$.

39. The UV Bus Company will charge the church group in Problem 38 a fare of \$160 each if 30 or fewer people sign up but will reduce the fare for everyone by \$2 for each person exceeding 30 who makes the trip. Write a two-part formula for $R(x)$, the total fare revenue if $x$ persons make the trip. Then calculate $R(35)$ and $R(70)$.

40. The accompanying table is taken from the 1983 Internal Revenue Service (IRS) tax form. It gives a multipart rule for a function representing tax $T$ in terms of taxable income $I$ for a single taxpayer. Sketch the graph of $T$ for $0 \le I \le 15{,}000$.

**Schedule X**
**Single Taxpayers**

Use this Schedule if you checked Filing Status Box 1 on Form 1040—

| *If the amount on Form 1040, line 37 is:* Over— | *But not over—* | *Enter on Form 1040, line 38* | *of the amount over—* |
|---|---|---|---|
| \$0 | \$2,300 | —0— | |
| 2,300 | 3,400 | 11% | \$2,300 |
| 3,400 | 4,400 | \$121 + 13% | 3,400 |
| 4,400 | 8,500 | 251 + 15% | 4,400 |
| 8,500 | 10,800 | 866 + 17% | 8,500 |
| 10,800 | 12,900 | 1,257 + 19% | 10,800 |
| 12,900 | 15,000 | 1,656 + 21% | 12,900 |
| 15,000 | 18,200 | 2,097 + 24% | 15,000 |
| 18,200 | 23,500 | 2,865 + 28% | 18,200 |
| 23,500 | 28,800 | 4,349 + 32% | 23,500 |
| 28,800 | 34,100 | 6,045 + 36% | 28,800 |
| 34,100 | 41,500 | 7,953 + 40% | 34,100 |
| 41,500 | 55,300 | 10,913 + 45% | 41,500 |
| 55,300 | .... | 17,123 + 50% | 55,300 |

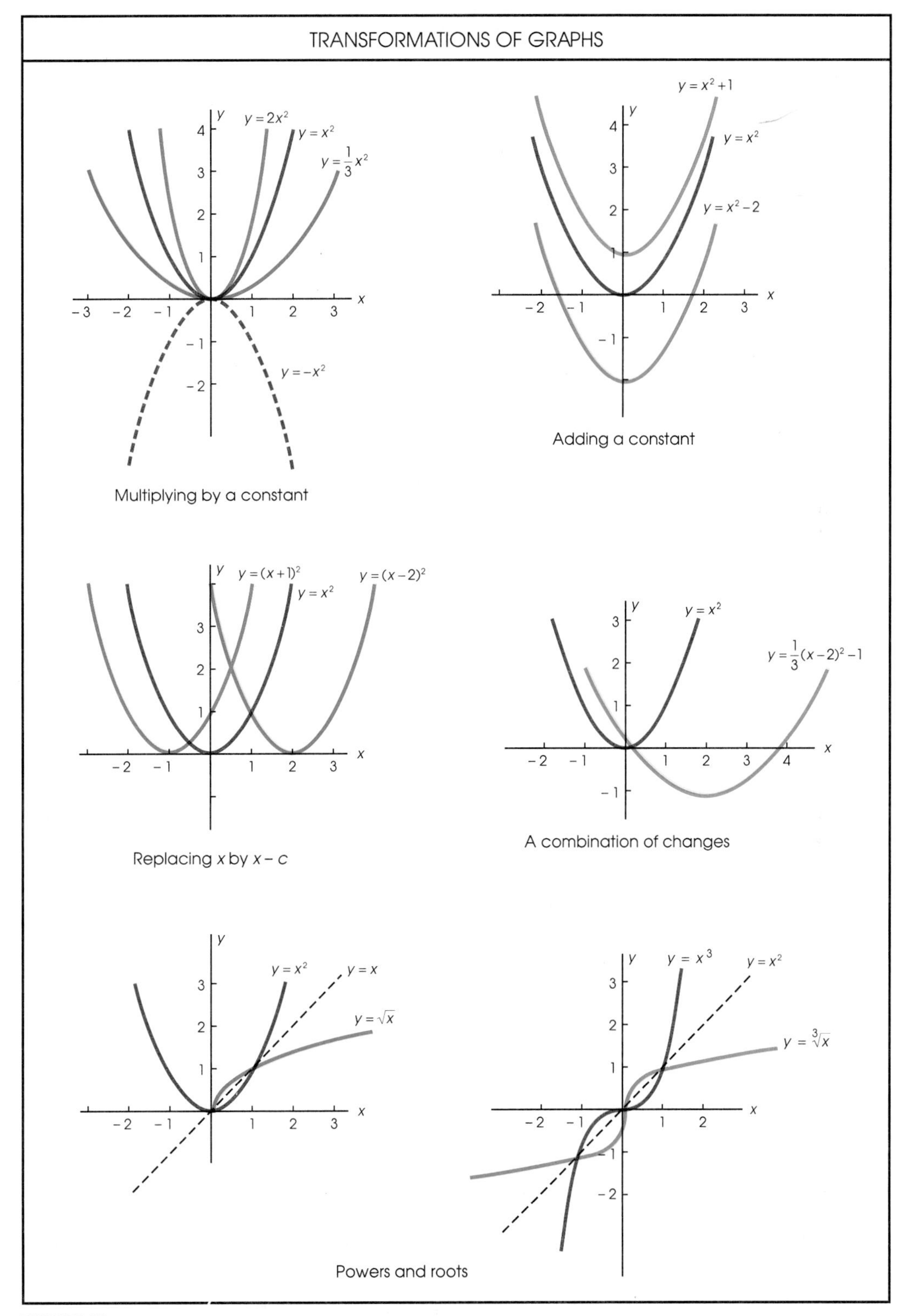
TRANSFORMATIONS OF GRAPHS
y = 2x²
y = x²
y = 1/3 x²
y = −x²
Multiplying by a constant
y = x² + 1
y = x²
y = x² − 2
Adding a constant
y = (x + 1)²
y = x²
y = (x − 2)²
Replacing x by x − c
y = x²
y = 1/3(x − 2)² − 1
A combination of changes
y = x²
y = x
y = √x
y = x³
y = x²
y = ∛x
Powers and roots

**FIGURE 15**

C 41. Use the information in Problem 40 to find the tax paid by a single person whose taxable income was (a) \$8751, (b) \$12,390, (c) \$34,981.

PC 42. In Utopia, there are no taxes on incomes of \$50,000 or less and only 6% taxes on that part of one's income which exceeds \$50,000. Write a program for computing the tax on any income in Utopia.

43. The 1986 Tax Reform Act sets federal taxes for single taxpayers (beginning in 1988) at 15% on a taxable income of \$17,850 or less and at 28% on the portion of a taxable income exceeding \$17,850. Work Problem 41 under this plan.

PC 44. (For computer buffs only.) Write a program for computing the tax on any income using the table in Problem 40.

## 1.3 LINEAR FUNCTIONS: CONSTANT SLOPE

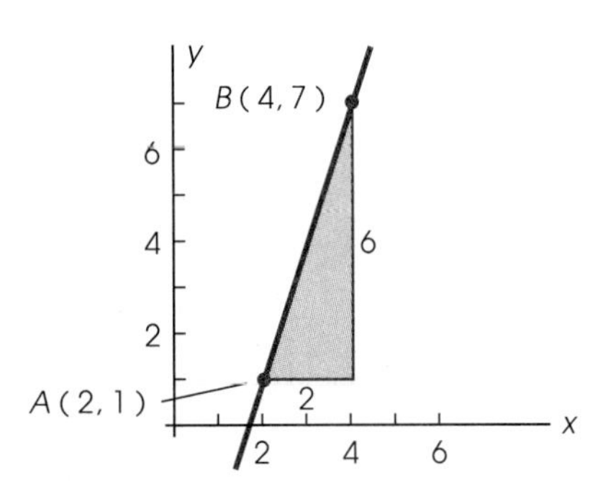


FIGURE 16

The simplest kind of function is a **linear function,** that is, a function of the form

$$f(x) = ax + b$$

where $a$ and $b$ are constants. Linear functions get their name from the fact that their graphs are lines. Thus, our study of linear functions begins with a complete discussion of lines.

We assume that you have a good intuitive feeling for the concept of line from looking along a taut string or the edge of a ruler. Let us agree on one thing: Two points determine a unique line. In Figure 16, there is exactly one line through points $A$ and $B$.

FIGURE 17

### The Slope of a Line

Consider the line in Figure 16. A move from $A$ to $B$ produces a vertical change of $7 - 1 = 6$ units and a corresponding horizontal change of $4 - 2 = 2$ units. The ratio $\frac{6}{2} = 3$ of these two numbers is called the slope of the line.

In general, if $y$ changes by an amount $\Delta y$ (read "delta $y$") while $x$ is changing by $\Delta x$ (Figure 17), we say that the slope of the line is $m = \Delta y / \Delta x$, that is,

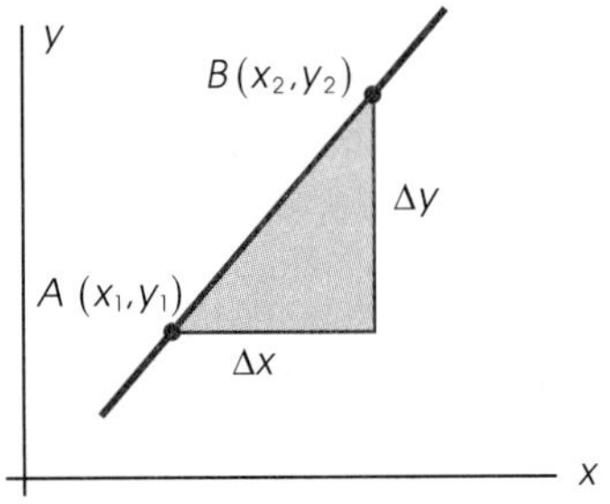

$$m = \frac{\text{change in } y}{\text{change in } x} = \frac{\Delta y}{\Delta x} = \frac{y_2 - y_1}{x_2 - x_1}$$

A careful reader will ask an immediate question. Since a line has many points, does the value we get for the slope depend on which pair of points we use for $A$ and $B$? The answer is no. The similar triangles in Figure 18 show us that

$$\frac{y_2' - y_1'}{x_2' - x_1'} = \frac{y_2 - y_1}{x_2 - x_1}$$

The points $A'$ and $B'$ give the same value for $m$ as do $A$ and $B$. It does not

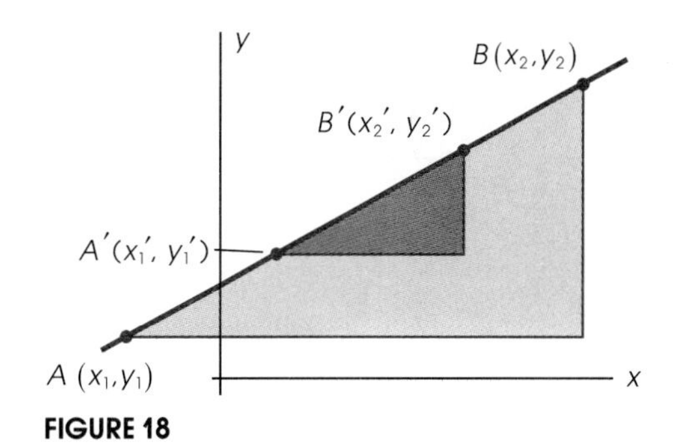


**FIGURE 18**

even matter whether $A$ is to the left or right of $B$, since

$$\frac{y_1 - y_2}{x_1 - x_2} = \frac{y_2 - y_1}{x_2 - x_1}$$

All that matters is that we subtract the coordinates in the same order in the numerator and the denominator.

The feature just noted, that $m = \Delta y/\Delta x$ is constant, distinguishes lines from all other curves. In fact, $m$ is a measure of the steepness of a line. A horizontal line has zero slope, and a line that rises to the right has positive slope. The larger this positive slope is, the more steeply the line rises. A line that falls to the right has negative slope. The concept of slope is meaningless for vertical lines, since it would involve division by zero. In Figure 19, we have computed the slope for a number of lines all going through the point (2, 1).

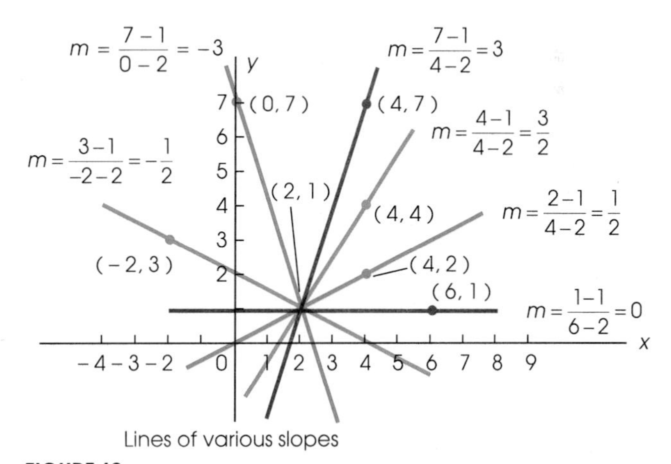


**FIGURE 19**

## EXAMPLE A

Find the slope of the line passing through (−2, 6) and (4, 3).

### Solution

$$m = (3 - 6)/[4 - (-2)] = -3/6 = -1/2$$

■

## The Point-Slope Form

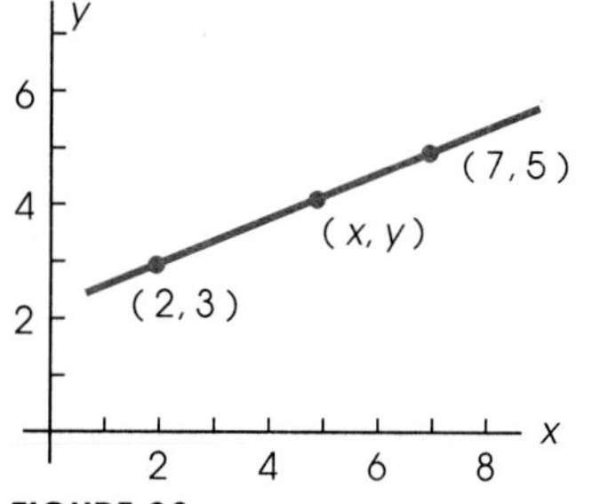
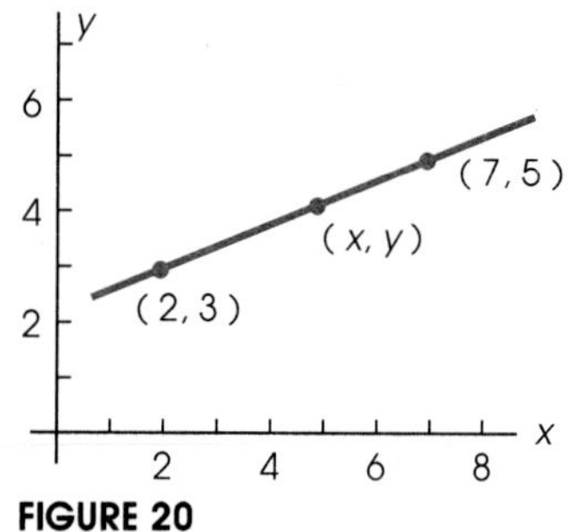

FIGURE 20

Our goal is to write an equation for the line in Figure 20. Note that this line

(i) passes through (2, 3).
(ii) has slope $(5 - 3)/(7 - 2) = \frac{2}{5}$.

Consider any other point $(x, y)$ on this line. When we use this point together with (2, 3) to measure slope, we must get $\frac{2}{5}$, that is,

$$\frac{y - 3}{x - 2} = \frac{2}{5}$$

or, after multiplying by $x - 2$,

$$y - 3 = \tfrac{2}{5}(x - 2)$$

This last equation is satisfied by all points on the line, even by (2, 3). Moreover, only points on the line can satisfy this equation.

What we have just done in an example can be done in general. The line passing through the fixed point $(x_0, y_0)$ with slope $m$ (Figure 21) has equation

FIGURE 21

$$\boxed{y - y_0 = m(x - x_0)}$$

We call this equation the **point-slope form** of the equation of a line.

### EXAMPLE B

Find the point-slope equation of the line through $(-2, 4)$ and $(4, 7)$. Show that you get equivalent equations using either $(-2, 4)$ or $(4, 7)$ as the fixed point $(x_0, y_0)$.

**Solution** First, we calculate the slope $m = (7 - 4)/(4 + 2) = \frac{1}{2}$. Then, using $(-2, 4)$ as the fixed point gives the equation

$$y - 4 = \tfrac{1}{2}(x + 2)$$

whereas (4, 7) yields

$$y - 7 = \tfrac{1}{2}(x - 4)$$

Solve either equation for $y$ and you will get $y = \frac{1}{2}x + 5$. ■

## The Slope-Intercept Form

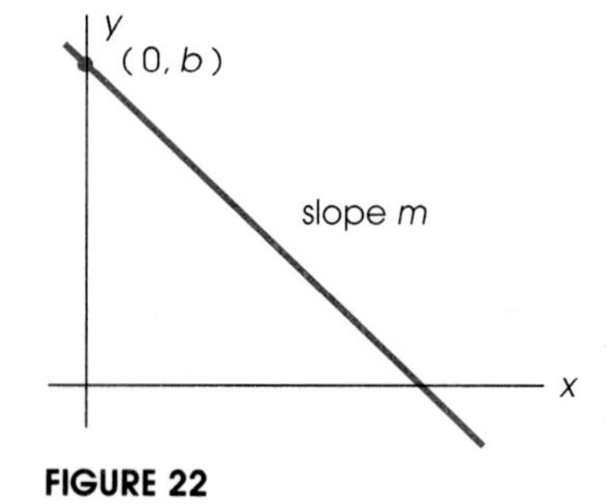

FIGURE 22

Suppose we are given the slope $m$ for a line and the $y$-intercept $b$, meaning that the line intersects the $y$-axis at $(0, b)$ as in Figure 22. Choosing $(0, b)$ as the fixed point of the point-slope form, we get

$$y - b = m(x - 0)$$

which we can write as

$$y = mx + b$$

The latter equation is called the **slope-intercept form.**

Of what use is this new form? Anytime we see an equation written this way, we immediately recognize it as an equation of a line and can read off its slope and $y$-intercept.

> **POINT OF INTEREST**
> It can be shown that two nonvertical lines are perpendicular if and only if their slopes are negative reciprocals, that is, if and only if their slopes $m_1$ and $m_2$ satisfy
> $$m_2 = -\frac{1}{m_1}$$

**EXAMPLE C**

Find an equation for the line through $(2, -3)$ that is parallel to the line with equation $3x - 4y = 7$.

**Solution** When $3x - 4y = 7$ is solved for $y$, we get successively

$$-4y = -3x + 7$$
$$y = \tfrac{3}{4}x - \tfrac{7}{4}$$

The latter is an equation of a line with slope $m = \frac{3}{4}$.

Now we call on a basic fact: *Two nonvertical lines are parallel if and only if they have the same slope*. It follows that the line we want has slope $\frac{3}{4}$. Since this line goes through $(2, -3)$, its equation in point-slope form is

$$y + 3 = \tfrac{3}{4}(x - 2)$$ ■

A derivation similar to that in Example C shows that any equation of the form $ax + by = c$ with $b \neq 0$ represents a nonvertical line. We will find plenty of uses for this fact.

**EXAMPLE D**

Find an equation for the line through the intersection of the lines $3x - 2y = 10$ and $2x + 6y = 14$ which has slope $m = -\frac{2}{3}$. Write your answer in the form $ax + by = c$.

**Solution** Our first task is to solve the two given equations simultaneously. Begin by multiplying the first equation by 3 and adding it to the second equation.

$$\begin{aligned} 9x - 6y &= 30 \\ 2x + 6y &= 14 \\ \hline 11x \qquad &= 44 \\ x &= 4 \end{aligned}$$

Now substitute $x = 4$ in either of the original equations and solve for $y$. You will get $y = 1$. The intersection point is $(4, 1)$.

The required line passes through $(4, 1)$ and has slope $-\frac{2}{3}$. In point-slope form, its equation is

$$y - 1 = -\tfrac{2}{3}(x - 4)$$

Finally, multiplying both sides by 3 and simplifying gives

$$3y - 3 = -2x + 8$$

or

$$2x + 3y = 11$$

FIGURE 23

The geometric interpretation of this example appears in Figure 23. ■

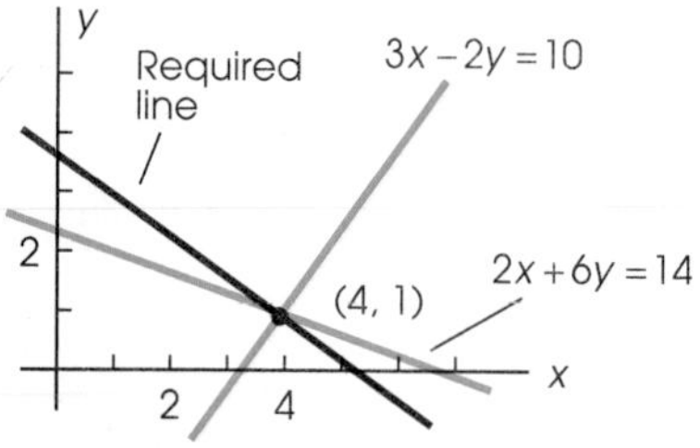

## Lines and Linear Functions

We claimed in our introduction that the graph of the linear function $f(x) = ax + b$ is a line. Our discussion shows that it is the line with slope $a$ and $y$-intercept $b$. For example, the graph of the function $f(x) = 4x + 3$ is the line with slope 4 and $y$-intercept 3. Conversely, each nonvertical line has a slope $m$ and a $y$-intercept $b$ and so is the graph of the linear function $f(x) = mx + b$.

**EXAMPLE E**

Find the formula for the linear function $f$ that satisfies $f(-1) = 4$ and $f(3) = 12$. Then calculate $f(2)$.

**Solution** The graph of $f$ (a line) goes through $(-1, 4)$ and $(3, 12)$ and so has slope $m = (12 - 4)/(3 + 1) = 2$. In point-slope form, the equation of this line is

$$y - 4 = 2(x + 1)$$

or solved for $y$,

$$y = 2x + 6$$

Thus, $f(x) = 2x + 6$ and $f(2) = 10$. ■

FIGURE 24

## Two Simple Examples of Slopes

In building a house, a carpenter may say he is putting up a roof with a 7:12 pitch (Figure 24). What he means is that the roof will rise 7 feet in a run of 12 feet; that is, the roof has slope $\frac{7}{12}$.

In the mountains, one often sees a road sign like that in Figure 25, indicating what is called the grade. A grade of 12% is simply a slope of $\frac{12}{100}$. Of course, if you are driving downhill, the slope is really $-\frac{12}{100}$.

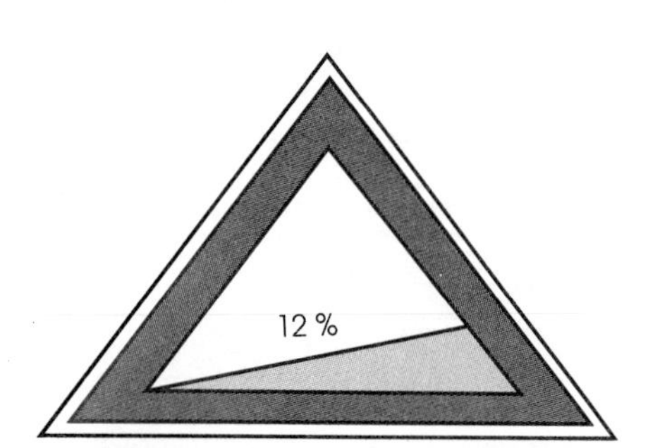


FIGURE 25

## Problem Set 1.3

*In Problems 1–6, find the slope of the line passing through the indicated points.*

1. (3, 1) and (5, 9)
2. (2, 3) and (10, 15)
3. (−4, 2) and (3, 0)
4. (2, −4) and (0, −6)
5. [C] (2.563, −9.108) and $(\sqrt{5}, -\sqrt{3})$
6. [C] $(\pi, \sqrt{3} + \sqrt{5})$ and $(1.642, \sqrt{2})$

*Find an equation for the indicated line.*

7. The line through (2, 1) with slope 3.
8. The line through (3, 5) with slope $\frac{3}{2}$.
9. The line through (3, 1) and (5, 9).
10. The line through (2, 3) and (10, 15).
11. The line through (1, −2) and (3, 7).
12. The line through (2, 3) and (−10, 15).

*In Problems 13–16 find an equation for the indicated line. Write your final answer in the form $ax + by = c$.*

13. The line with slope 4 and $y$-intercept −3.
14. The line with $y$-intercept 4 and slope $-\frac{2}{3}$.
15. The line through (2, −1) that is parallel to the line $5x - y = 12$.
16. The line through (5, 0) that is parallel to the line $2x + 3y = 13$.

*Here is an important fact.* ***Two nonvertical lines are perpendicular if and only if their slopes are negative reciprocals.*** *For example, the lines $y = 2x - 7$ and $y + 1 = \frac{-1}{2}(x - 5)$ are perpendicular because 2 and $-\frac{1}{2}$ are negative reciprocals. Use this information to obtain an equation for the indicated lines in Problems 17–20. Write your final answer in the form $ax + by = c$.*

17. The line through (2, −1) that is perpendicular to the line $y = -\frac{1}{2}x + 4$.
18. The line through (4, 1) that is perpendicular to the line $y = 3x - 5$.
19. The line through (2, 4) that is perpendicular to the line $x - 4y = 21$.
20. The line through (7, −3) that is perpendicular to the line $5x + 2y = -10$.

*Follow the ideas of Example D to find an equation for the line described in each of Problems 21–24. Write your final answer in the form $ax + by = c$.*

21. The line through the intersection of $x + 2y = -1$ and $3x + y = 7$ which has slope $\frac{3}{4}$.
22. The line through the intersection of $4x + 3y = 2$ and $x + 5y = -8$ which has slope $-\frac{2}{5}$.
23. The line through the intersection of $x + 2y = 5$ and $3x - 2y = -9$ which is parallel to the second line.
24. The line through the intersection of $2x + 4y = 0$ and $3x + 8y = -4$ which is perpendicular to the first line.

*In Problems 25–28, find a formula of the form $f(x) = mx + b$ for the linear function $f$ satisfying the given conditions. Then find $f(5)$. See Example E.*

25. $f(1) = 3, f(3) = 9$
26. $f(-2) = 5, f(8) = 25$
27. $f(-2) = 4, f(3) = 2$
28. $f(1) = -2, f(6) = 7$

---

29. Find the slope and $y$-intercept of each line.
    (a) $3y = -2x + 6$ (b) $4x + 3y = -36$
    (c) $y + 2 = \frac{3}{4}(x - 1)$
    (d) $3y + 2x = -7x + 9y + 16$
30. Find an equation for the line with $y$-intercept 2 and $x$-intercept 3.
31. Write (in the form $ax + by = c$) an equation of the line through (2, −5) which is:
    (a) parallel to the line $2y = -3x + 2$.
    (b) parallel to the line through (0, 4) and (3, −2).
    (c) parallel to the $x$-axis.
    (d) perpendicular to the line $3x - 5y = 6$.
32. Show that the equation of the line with $y$-intercept $b$ and $x$-intercept $a$ can be put in the form $x/a + y/b = 1$. Assume $a$ and $b$ are both nonzero.
33. A farm tractor costs $20,000 and will have a value of $2000 at the end of 10 years. Let $f(x)$ be the value of the tractor after $x$ years ($0 \leq x \leq 10$) and assume that $f$ is a linear function. Find a formula for $f(x)$.
34. Find the value of $k$ for which the line $4x + ky = 5$
    (a) passes through the point (1, −1).
    (b) is parallel to the line $y = 2x + 4$.
    (c) has equal $x$- and $y$-intercepts.
    (d) has $y$-intercept −10.

35. The Celsius and Fahrenheit temperature scales are related in a linear way; that is, $F = aC + b$.
    (a) Use the fact that $F = 32$ when $C = 0$ and $F = 212$ when $C = 100$ to determine $a$ and $b$.
    (b) Calculate $F$ when $C = 32$.
    (c) Calculate $C$ when $F = 81$.
36. Sketch the graph of the line $y = -2x + 4$ and the point (7, 5). Find:
    (a) an equation of the line through (7, 5) perpendicular to the given line.
    (b) the point of intersection of the given line and the line in part (a).
    (c) the distance between (7, 5) and the given line.

[C] 37. In pedaling up a mountain road with a grade of 12%, a bicyclist increased her altitude 1000 feet. How far did she pedal? *Hint:* Draw a right triangle and use the Pythagorean Theorem $a^2 + b^2 = c^2$.

[C] 38. A rectangular house 24 feet by 36 feet has a gable roof with a 7:12 pitch (to have a gable roof means the roof line on the shorter side of the house looks like a wide inverted vee). Assuming the roof has no overhanging eaves, calculate the area of the roof. *Hint:* You will need the Pythagorean Theorem (see Problem 37).

## 1.4 APPLICATIONS OF LINEAR FUNCTIONS

The concept of linearity finds numerous applications in the world of experience. A process is **linear** if a change $\Delta x$ in input produces a change $\Delta y$ in output that is a constant multiple of $\Delta x$; that is, $\Delta y = k\Delta x$. Examples abound. A change in temperature produces a proportional change in the height of mercury in a glass tube (the principle of the thermometer). A change in the depth of an underwater diver produces a proportional change in the pressure on his skin. A change in the number of refrigerators produced by a manufacturer normally produces a proportional change in the revenue received. Recently, a major automobile manufacturer advertised that its cars behave in a linear way. The claim was that an increase in force on the brakes or steering wheel produces a proportional response from the car which would therefore be very predictable.

Linear processes are the easiest of all processes to describe mathematically; this is because they are represented by linear functions. A number of such processes are discussed in this section. Later (Sections 1.5 and 1.6), we will take up the more difficult concept of nonlinear processes.

### Business Applications

Many readers of this book will take a job in sales; our first formal example should seem relevant to them.

#### EXAMPLE A (WORKING ON COMMISSION)

Alice Drinkwater has accepted a position at Compudyne selling personal computers. Her salary will be \$1500 per month plus a commission of 3% of her sales. Express her total salary $T(x)$ in terms of $x$, her monthly sales in dollars. Then calculate her salary based on sales of \$25,000 per month.

**Solution**

$$T(x) = 1500 + (0.03)x$$

$$T(25{,}000) = 1500 + (0.03)(25{,}000) = \$2250$$

Note that $T(x)$ does have the linear form $ax + b$. ■

The Internal Revenue Service allows several methods for figuring depreciation of equipment, one being *straight-line depreciation*.

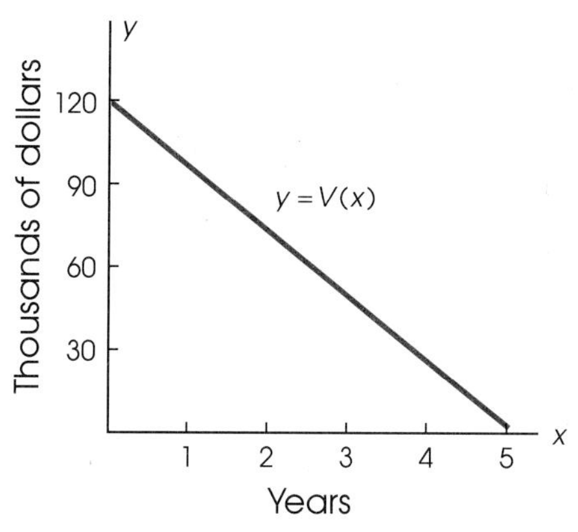


**FIGURE 26**

### EXAMPLE B (STRAIGHT-LINE DEPRECIATION)

A milling company bought a machine costing \$120,000, expecting it to depreciate at a constant yearly rate and to have a junk value of \$3000 after 5 years. What was its value at the end of $x$ years ($0 \le x \le 5$)? At the end of 3.5 years?

**Solution** The graph of $V$ (Figure 26) is the line through (0,120,000) and (5,3000) and so has slope

$$m = (3000 - 120{,}000)/(5 - 0) = -23{,}400$$

Thus

$$V(x) = -23{,}400x + 120{,}000$$

and

$$V(3.5) = (-23{,}400)(3.5) + 120{,}000 = \$38{,}100$$

Note that the slope $m = -23{,}400$ means that the machine loses \$23,400 in value each year. ■

### EXAMPLE C (COST OF PRODUCTION)

A manufacturer of vacuum cleaners has fixed monthly costs (warehouse rent, office utilities, insurance, and so on) of \$4500 and estimates that labor and material for each cleaner cost \$115. Find the cost $C(x)$ of producing $x$ vacuum cleaners per month.

**Solution** $C(x) = 4500 + 115x$. Note that when graphed, the line would have slope $m = 115$, exactly the cost of one vacuum cleaner. ■

### EXAMPLE D (BREAK-EVEN ANALYSIS)

The manufacturer in Example C expects to receive \$145 for each vacuum cleaner it sells. Assuming the manufacturer can sell all it makes, what is its monthly revenue $R(x)$ and how many units must it sell per month to just break even?

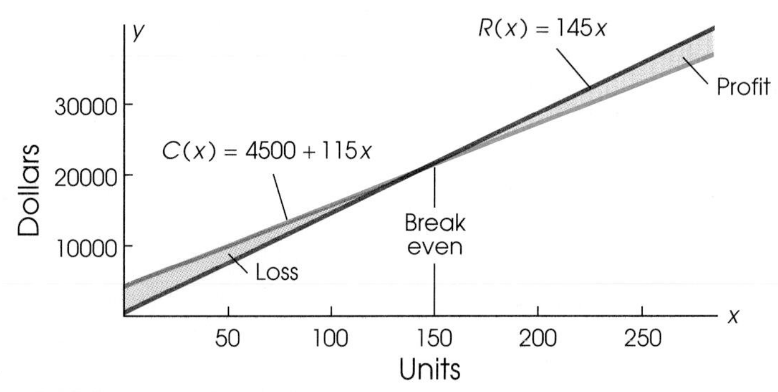


**FIGURE 27**

**Solution** $R(x) = 145x$. To exactly break even, $R(x) = C(x)$; that is,

$$145x = 4500 + 115x$$

$$30x = 4500$$

$$x = 150$$

Making and selling 150 units will allow the manufacturer to break even. Making and selling more than 150 units will yield a profit. The situation is shown graphically in Figure 27. ■

## Rates of Change

About half of calculus (the part called *differential calculus*) deals with rates of change. Given two related variables $x$ and $y$, we often want to know how fast one is changing with respect to the other. To ask how fast a car is going is to ask how the distance traveled $s$ is changing with respect to the elapsed time $t$. The price $p$ of pork is dependent on the number $h$ of hogs being raised on farms. An economist wants to know how $p$ changes with respect to $h$. These rates of change are constant precisely when the underlying functions are linear.

### EXAMPLE E

John Whitfield is 50 miles from Chicago and driving away from that city at 55 miles per hour. Express his distance $s$ from Chicago in terms of the elapsed time $t$ starting right now.

**Solution**

$$s = 50 + 55t$$

Be sure to note something. The speed (that is, the rate of change in $s$ with respect to $t$) 55 is exactly the slope of the line $s = 50 + 55t$. This is no accident. ■

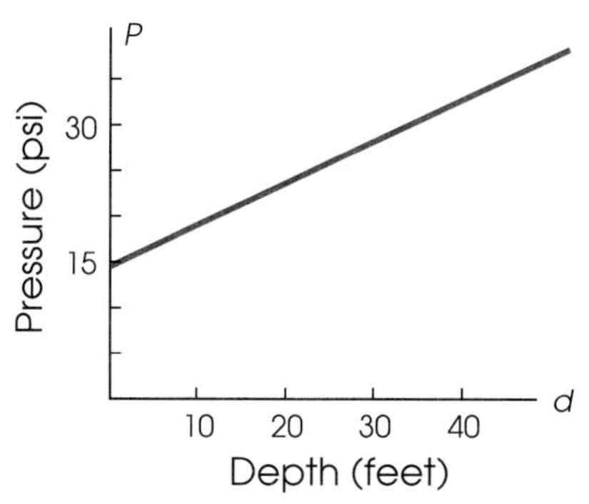


FIGURE 28

## EXAMPLE F

The pressure at the surface of a lake is 15 pounds per square inch, while at a depth of 33 feet it has doubled to 30 pounds per square inch (Figure 28). If pressure $p$ increases linearly with depth $d$ (it does), find the rate of change in $p$ with respect to $d$. Then write a formula for $p$ in terms of $d$.

**Solution**

$$\text{rate} = \frac{\Delta p}{\Delta d} = \frac{30 - 15}{33 - 0} = \frac{15}{33} \approx 0.4545 \text{ pounds per square inch per foot}$$

$$p = 0.4545d + 15$$

Note again that the rate 0.4545 is the same as the slope of the line. It tells us that the pressure increases 0.4545 pounds per square inch with each additional foot of depth. ■

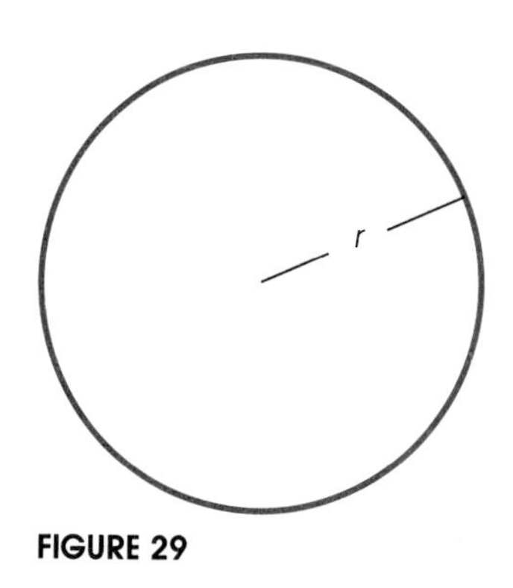


FIGURE 29

## EXAMPLE G

A circular ring (Figure 29) is heated, thus increasing the radius $r$ and circumference $C$ of the ring. How does $C$ change with respect to $r$?

**Solution** Since $C = 2\pi r$ (formula from geometry) is a linear relationship, the desired rate is simply the coefficient of $r$, namely, $2\pi \approx 6.2832$. If $r$ increases by 1 unit, the circumference will increase by a little more than 6 units. Note that this statement is true no matter what the size of the original ring (the ring on your finger or one of the rings of Saturn). ■

## Summary

We have said much about linear functions in the last two sections, but here is the essence of it all. A linear function is determined by an equation of the form $y = mx + b$. The graph of this equation is a straight line. The number $m$ (a constant) has two interpretations. It measures the slope (steepness) of the line, and it tells us how fast $y$ is changing with respect to $x$. That the geometrical notion of slope and the physical notion of rate of change are really the same idea is worth highlighting.

| slope = rate of change |
|---|

This fact, here observed for linear functions, is also true for nonlinear functions, the subject we take up in the next two sections.

## Problem Set 1.4

C 1. Anthropologists use the linear function $H(x) = 2.89x + 70.64$ to estimate the height (in centimeters) of a male whose humerus (bone from elbow to shoulder) is $x$ centimeters long. Among the old bones discovered at a desert site was a male humerus 43.52 centimeters long. How tall was this man?

C 2. Biologists use the linear function $E(R) = 1.235R + 3.805$ to estimate the energy yield (in kilocalories per liter of oxygen) where $R$ is the so-called respiratory quotient. Compute the energy yield for a person on a high-fat diet for which $R$ is 0.74.

3. Neil Thackwood sells appliances. He makes \$1200 per month plus a commission of 4% of his sales.
   (a) Express his total salary $S(x)$ in terms of $x$, his monthly sales in dollars.
   (b) Calculate his total salary for a month in which his sales amounted to \$36,000.

4. Marie Howser sells dresses at an exclusive women's shop. Her monthly salary is \$1100 plus \$30 for each dress she sells.
   (a) Express her monthly salary $f(x)$ in terms of $x$, the number of dresses she sells.
   (b) Calculate her salary for a month in which she sold 38 dresses.
   (c) If Marie hopes to earn \$2600 next month, how many dresses must she sell?

5. Airbus leases cars for \$150 per month plus 12¢ per mile.
   (a) Find the formula for $F(x)$, the cost of leasing a car for a month and driving it $x$ miles.
   (b) What is the total monthly cost for a person who leases a car and plans to drive 8000 miles each month?

6. A natural gas exploration company offered a Montana rancher \$10,000 for the right to drill on his ranch plus 9¢ for each thousand cubic feet of gas extracted. Find the formula for the amount $A(x)$ the rancher will receive expressed in terms of $x$, thousands of cubic feet of gas.

7. A road grader purchased today for \$75,000 will depreciate linearly to a scrap value of \$8000 after 10 years. See Example B.
   (a) Write a formula for its value $V(n)$ after $n$ years $(0 \leq n \leq 10)$.
   (b) Sketch the graph of $V(n)$ for $0 \leq n \leq 10$.
   (c) What is the slope of this graph and what is its significance?

8. A farmer purchases a combine today for \$120,000, knowing that it will depreciate (linearly) to a value of only \$10,000 at the end of 5 years.
   (a) Write a formula for the value $C(n)$ of the combine at the end of $n$ years $(0 \leq n \leq 5)$.
   (b) Calculate the value of the combine after 3.5 years.
   (c) When will the combine be worth \$80,000?

9. A manufacturer of color television sets has fixed monthly costs (utilities, insurance, and so on) of \$42,000, and the cost of labor and material for each set is \$250. Find the cost $C(x)$ of producing $x$ television sets per month. See Example C.

10. A manufacturer of freezers has fixed costs of \$28,000 per month, and the cost of labor and material is \$300 per freezer. Find the cost $C(x)$ of manufacturing $x$ freezers per month.

11. The manufacturer in Problem 9 sells each color television set at \$375. Assuming that every set that is made is sold, what is the monthly revenue $R(x)$ if $x$ sets are sold? How many sets must be sold per month in order to just break even? See Example D.

12. The manufacturer in Problem 10 sells the freezers at \$475. If every freezer that is made is sold, what is the monthly revenue from selling $x$ freezers? How many freezers must be sold per month for the company to just break even?

13. A sports company is planning to introduce a new line of cross-country skis. Set-up and overhead costs for the first year are estimated at \$20,000, and it will cost \$40 in labor and materials to make and market each pair of skis. The company expects to sell a pair of skis for \$100.
    (a) Determine $C(x)$, the cost of making and marketing $x$ pairs of skis during the first year.
    (b) Determine $R(x)$, the revenue from selling $x$ pairs of skis.
    (c) Determine $P(x) = R(x) - C(x)$, the profit realized from the production and sale of $x$ pairs of skis.
    (d) Calculate $P(1000)$, the profit from making and selling 1000 pairs of skis.

14. A consultant has warned the sports company in Problem 13 that it is unrealistic to expect to sell 1000 pairs of skis at \$100 per pair. She suggests that the demand $x$ for skis depends on the price $p$ (in dollars) according to the formula $x = 2000 - 15p$. Assume the consultant is right.
    (a) How many pairs of skis can be sold at $p = 100$?
    (b) What is the profit when $p = 100$?
    (c) Express the profit $P$ in terms of the price $p$.

15. Joan Appleseed is 480 miles from Los Angeles and is driving toward that city at 60 miles per hour. Express

her distance $s$ from Los Angeles in terms of $t$, the number of hours from now ($0 \leq t \leq 8$).

16. Jonathon Witherslack is 300 miles from home and driving away from it at 54 miles per hour. Express his distance $s$ from home $t$ hours from right now.

17. Sarah Newburg and John Plank are traveling east from Hays, Kansas. Sarah started at noon and is driving at a constant speed of 55 miles per hour. John started 2 hours later and is driving at a constant speed of 70 miles per hour.
    (a) Express the distance each has traveled as a function of $t$, the number of hours after 12:00 noon.
    (b) At what time does John pass Sarah?

18. The R-rating of fiberglass insulation varies linearly with its thickness. A 5-inch batt of insulation has an R-rating of 16. Roger Figham's house has an uninsulated ceiling with an R-rating of 4.
    (a) If he puts in $t$ inches of fiberglass insulation, what will the R-rating be?
    (b) To obtain an R-rating of 40, how thick will the insulation need to be?

[C] 19. A piece of steel railroad track was 64 feet long at 0°F and expanded to 64.114 feet at 50°F. Assume that the expansion of steel under heating is linear.
    (a) Express the length $L$ of the track in terms of the temperature $T$.
    (b) Determine $L$ when $T = 100$°F.
    (c) Find the rate of change in $L$ with respect to $T$.

20. The percentage of tomato seeds that germinate depends in a linear way on the temperature. At 12°C, 40% germinate, while at 15°C, 70% germinate.
    (a) Express the percentage $P$ that germinate in terms of the temperature $T$.
    (b) Find $P$ at $T = 17$°C.
    (c) Determine the rate of change in $P$ with respect to $T$.

21. The Fahrenheit and Celsuis temperature scales are related by the equation $F = \frac{9}{5}C + 32$.
    (a) Find the rate of change in $F$ with respect to $C$.
    (b) Solve for $C$ in terms of $F$ and find the rate of change in $C$ with respect to $F$.

22. An object moves in a straight line so that its velocity is $v = -88t + 100$ feet per second $t$ seconds from now. Find the rate of change in velocity with respect to time (that is, find the acceleration).

---

23. Records show that egg production in Halman County is growing linearly. In 1965 it was 1,200,000 cases, and in 1975 it was 1,680,000 cases.
    (a) Write a formula for $f(n)$, the number of cases produced $n$ years after 1965.
    (b) What is the rate of change in $f(n)$ with respect to $n$?

24. In a certain state 12% of the population was in college in 1935. By 1939 this number had risen to 17.4%. Assume that $P(t)$, the percentage of the population in college $t$ years after 1935, increases linearly with $t$.
    (a) Express $P(t)$ in terms of $t$.
    (b) Find the rate of change in $P(t)$ with respect to $t$.

25. The total cost of manufacturing $x$ snowsuits of a certain kind in one month was found to be a linear function of $x$. It cost \$30,000 to produce 500 snowsuits and \$82,500 to produce 2000 snowsuits.
    (a) Find the formula for $C(x)$, the total cost of manufacturing $x$ snowsuits in 1 month.
    (b) What are the fixed costs for the month?
    (c) Find the rate of change in the total monthly costs with respect to the number of snowsuits made.
    (d) What are the labor and material costs per snowsuit?

26. At 6:00 A.M. two motorists are 600 miles apart driving in opposite directions (on the same freeway). One motorist is driving at a constant speed of 62 miles per hour and the other at a constant speed of 43 miles per hour.
    (a) Find the formula for $s(t)$, the distance in miles between the two motorists $t$ hours after 6:00 A.M.
    (b) What is the rate of change in $s(t)$ with respect to $t$?

## 1.5 NONLINEAR FUNCTIONS: VARIABLE SLOPE

Most functions are nonlinear, and most physical processes, when examined closely, seem to be nonlinear too. For a function to be nonlinear means simply that its graph is a curve rather than a line. A typical example is shown in Figure 30. One look at this figure suggests a question. What do we mean by

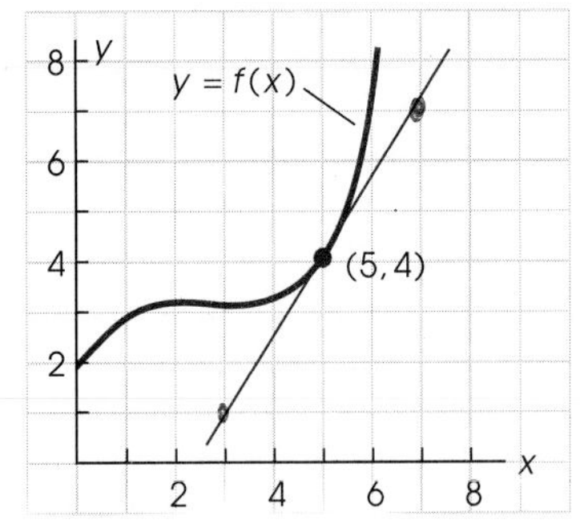


FIGURE 30

the slope of a curve? In contrast to a line, the steepness of a curve keeps changing. We can't talk about the slope of the whole curve; we will have to talk about its slope at a specific point. At such a point, it seems reasonable to define the slope of the curve to be the slope of the corresponding **tangent line.** For now, it is enough to say that the tangent line is the line that hugs the curve most closely at the point in question. We will be more precise about this concept shortly.

## EXAMPLE A

Find the slope of the curve in Figure 30 at the point (5, 4).

**Solution** The tangent line at (5, 4) is shown in Figure 30. This line appears to go through (3, 1) and (7, 7) and so has slope $m = (7 - 1)/(7 - 3) = \frac{6}{4}$. We conclude that the curve has slope $\frac{3}{2}$ at (5, 4). ■

## The Tangent Line

But what precisely is a tangent line? Euclid's notion of a tangent as a line touching a curve at just one point is all right for circles (Figure 31) but unsatisfactory for many other curves (Figure 32). The idea of a tangent to a curve at $P$ as the line which best approximates the curve near $P$ is better but still too vague for our purposes.

Let $P$ be a fixed point on a curve and let $Q$ be a nearby movable point on the curve. Consider the line, called a **secant line,** that goes through $P$ and $Q$. The tangent line at $P$ is the limiting position (if it exists) of the secant line as $Q$ moves toward $P$ along the curve (Figure 33).

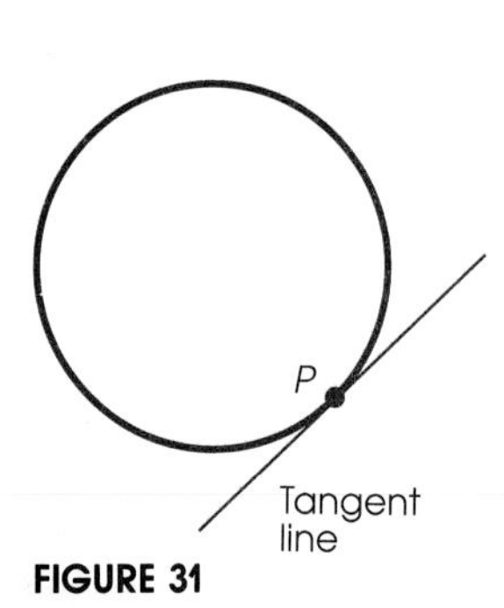


FIGURE 31

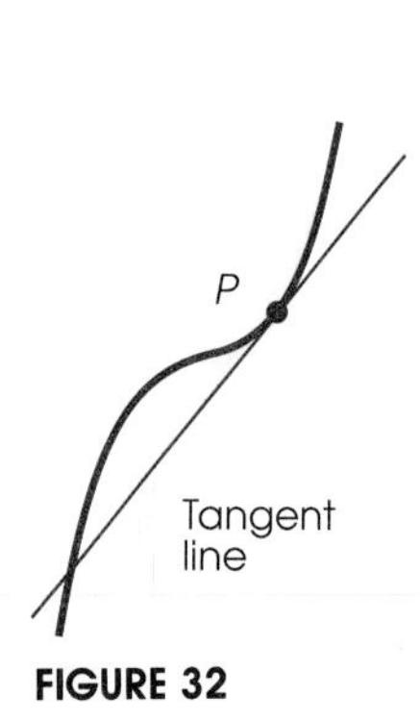


FIGURE 32

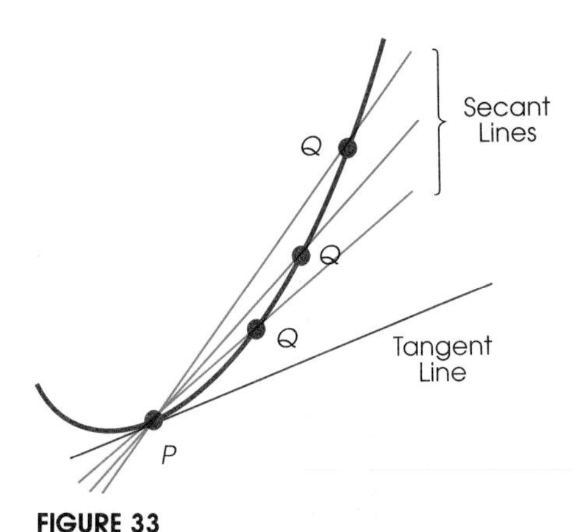


FIGURE 33

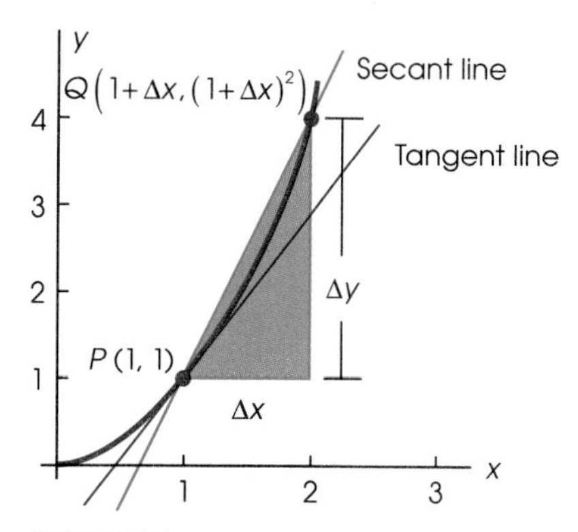


FIGURE 34

## The Slope of the Curve $y = x^2$

Consider as a first concrete example of a nonlinear curve, the graph of $y = x^2$ (Figure 34). Our goal is to find the slope of this curve at the point $P(1, 1)$. The nearby point $Q$ with $x$-coordinate $1 + \Delta x$ has $y$-coordinate $(1 + \Delta x)^2$. The secant line joining $P$ and $Q$ has slope

$$\frac{\Delta y}{\Delta x} = \frac{(1 + \Delta x)^2 - 1}{\Delta x} = \frac{1 + 2\Delta x + (\Delta x)^2 - 1}{\Delta x}$$

$$= \frac{\Delta x(2 + \Delta x)}{\Delta x} = 2 + \Delta x$$

Now let $\Delta x$ get smaller and smaller, thereby forcing $Q$ to approach $P$ along the curve and also forcing the secant line to rotate toward the tangent line. Correspondingly, the quantity $\Delta y/\Delta x = 2 + \Delta x$ will get nearer and nearer to 2. This limiting value of 2 is the slope of the curve at $P$.

What we have just done at the point $P(1, 1)$ can be done as well at any point $P(x, x^2)$ on the curve. The neighboring point $Q$ will have coordinates $[x + \Delta x, (x + \Delta x)^2]$, and the slope of the secant line $PQ$ will be

**CAUTION**
The symbol $\Delta x$ is used consistently to indicate a change in $x$. Thus, $\Delta x$ should be treated as a single symbol like $h$ or $t$. In particular, $\Delta x$ is not $\Delta$ times $x$. Thus, a term like $(\Delta x)^2$ simply indicates the square of a certain number. You should never write it as $\Delta^2 x^2$.

$$\frac{\Delta y}{\Delta x} = \frac{(x + \Delta x)^2 - x^2}{\Delta x} = \frac{2x\,\Delta x + (\Delta x)^2}{\Delta x} = \frac{\Delta x(2x + \Delta x)}{\Delta x} = 2x + \Delta x$$

As $\Delta x$ gets smaller and smaller, the quantity $\Delta y/\Delta x$ tends toward a limiting value of $2x$. In symbols, we write

$$\lim_{\Delta x \to 0} \frac{\Delta y}{\Delta x} = 2x$$

and say "the limit as $\Delta x$ approaches 0 of $\Delta y/\Delta x$ is $2x$."

We have achieved something very significant, our first true calculus result. We have shown that *the slope of the curve $y = x^2$ at $x$ has the value $2x$*. As we vary $x$, we vary the slope $m$. At $x = -1$, $m = 2(-1) = -2$; at $x = \frac{1}{2}$, $m = 2(\frac{1}{2}) = 1$; and at $x = 2$, $m = 2(2) = 4$ (Figure 35).

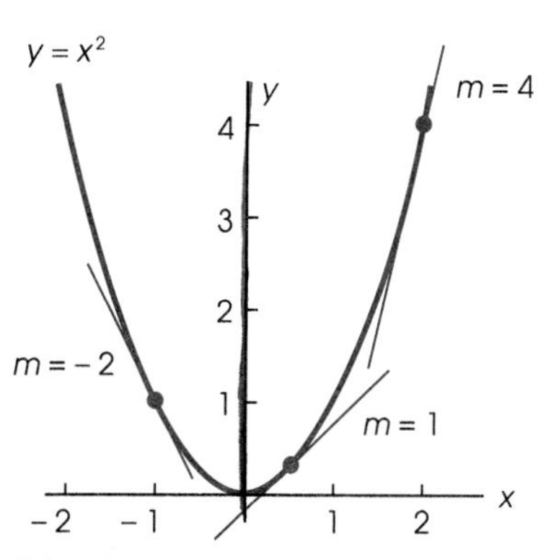


FIGURE 35

Find an equation for the tangent line to the curve $y = x^2$ at $x = -1$.

**Solution** The point on the curve corresponding to $x = -1$ is $(-1, 1)$, and the slope there (as we saw above) is $2(-1) = -2$. Thus, the equation of the tangent line (point-slope form) is

$$y - 1 = -2(x + 1)$$

■

## The Slope of the Curve $y = x^n$

Our goal for $y = x^n$ is to carry out the procedure that worked so well on $y = x^2$. Almost immediately we are faced with expanding $(x + \Delta x)^n$, something we should know how to do but may well have forgotten. It is the problem of raising a binomial (a two-term expression) to a power and goes by the name—the Binomial Theorem. Rather than state this theorem formally, we offer the following outline.

$$a + b = a + b$$

$$(a + b)^2 = a^2 + 2ab + b^2$$

$$(a + b)^3 = a^3 + 3a^2b + 3ab^2 + b^3$$

$$(a + b)^4 = a^4 + 4a^3b + 6a^2b^2 + 4ab^3 + b^4$$

$$\vdots$$

$$(a + b)^n = a^n + na^{n-1}b + \frac{n(n-1)}{2}a^{n-2}b^2 + \cdots + nab^{n-1} + b^n$$

Now for $y = x^n$, the slope of the secant line from $P(x, x^n)$ to $Q[x + \Delta x, (x + \Delta x)^n]$ is

$$\frac{\Delta y}{\Delta x} = \frac{(x + \Delta x)^n - x^n}{\Delta x}$$

$$= \frac{x^n + nx^{n-1}\,\Delta x + \frac{1}{2}n(n-1)x^{n-2}\,(\Delta x)^2 + \cdots + (\Delta x)^n - x^n}{\Delta x}$$

$$= \frac{\Delta x[nx^{n-1} + \frac{1}{2}n(n-1)x^{n-2}\,\Delta x + \cdots + (\Delta x)^{n-1}]}{\Delta x}$$

We conclude that

$$\lim_{\Delta x \to 0} \frac{\Delta y}{\Delta x} = nx^{n-1}$$

that is,

| slope of $y = x^n$ at $x$ is $nx^{n-1}$ |
|---|

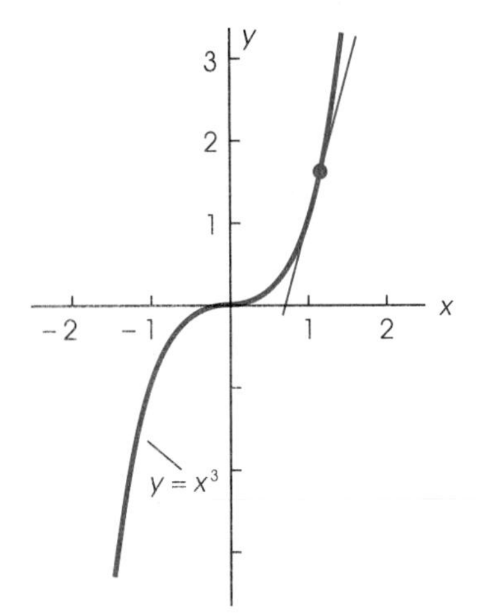


FIGURE 36

### EXAMPLE C

Find the slope of the curve $y = x^3$ at the point where $x = 1.2$ (Figure 36).

**Solution** The general formula for slope in this case is $3x^2$. At $x = 1.2$, the value of the slope is $3(1.2)^2 = 4.32$. ■

## Slope for Arbitrary Polynomials

We need a notation for the slope of a curve $y = f(x)$ at $x$. Following Leibniz, one of the founders of the calculus, we use the symbol $dy/dx$. Thus,

$$\frac{dy}{dx} = \lim_{\Delta x \to 0} \frac{\Delta y}{\Delta x}$$

For now, think of $d/dx$ as an operator. It operates on $y$ and gives the slope at $x$, namely, $dy/dx$. In particular, our slope formula may be written

$$\boxed{\frac{d(x^n)}{dx} = nx^{n-1}}$$

The operator $d/dx$ has two nice properties, properties that will be demonstrated in the next chapter. If $a$ and $b$ are constants, then

(i) $\dfrac{d(ax^n)}{dx} = a\dfrac{d(x^n)}{dx} = anx^{n-1}$

(ii) $\dfrac{d(ax^n + bx^m)}{dx} = \dfrac{d(ax^n)}{dx} + \dfrac{d(bx^m)}{dx} = anx^{n-1} + bmx^{m-1}$

These two results allow us to find the slope at $x$ for any polynomial. For example, if

$$y = 3x^4 + 2x^3 - 3x = 3x^4 + 2x^3 + (-3)x$$

then

$$\begin{aligned}\frac{dy}{dx} &= \frac{d(3x^4)}{dx} + \frac{d(2x^3)}{dx} + \frac{d(-3x)}{dx} \\ &= 3 \cdot 4x^3 + 2 \cdot 3x^2 + (-3)1 \\ &= 12x^3 + 6x^2 - 3\end{aligned}$$

Note one other fact. If $c$ is a constant, then the graph of $y = c$ is a horizontal line and so $dc/dx = 0$.

### EXAMPLE D

Find $dy/dx$ if $y = 2x^{10} - 11x^3 + 3x^2 - 5x + 11$.

**Solution**

$$\begin{aligned}\frac{dy}{dx} &= \frac{d(2x^{10})}{dx} + \frac{d(-11x^3)}{dx} + \frac{d(3x^2)}{dx} + \frac{d(-5x)}{dx} + \frac{d(11)}{dx} \\ &= 2 \cdot 10x^9 + (-11)3x^2 + 3 \cdot 2x + (-5)1 + 0 \\ &= 20x^9 - 33x^2 + 6x - 5\end{aligned}$$

■

## EXAMPLE E

Find an equation for the tangent line to the curve $y = 2x^4 - 3x^2 + 5x$ at the point (1, 4).

**Solution** The slope at $x$ is

$$\frac{dy}{dx} = 8x^3 - 6x + 5$$

and so the slope at $x = 1$ is $8 - 6 + 5 = 7$. The equation of the tangent line in point-slope form is

$$y - 4 = 7(x - 1)$$ ■

## EXAMPLE F

Find two points on the curve $y = x^3 - 2x$ where the slope is 1 (Figure 37).

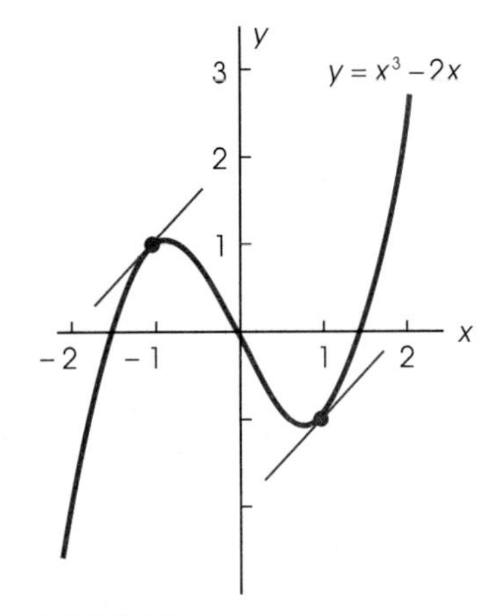

FIGURE 37

**Solution**

$$\frac{dy}{dx} = 3x^2 - 2$$

We want this slope to have the value 1, that is,

$$3x^2 - 2 = 1$$

$$3x^2 = 3$$

$$x^2 = 1$$

$$x = \pm 1$$

The $y$-coordinate corresponding to $x = 1$ is $y = -1$ and for $x = -1$ the $y$-coordinate is $y = 1$. We conclude that the required two points are $(1, -1)$ and $(-1, 1)$. ■

## Problem Set 1.5

*In Problems 1–8, you are to estimate the slope of the curve at the indicated point (see Example A). Note that you will first need to draw the tangent line in Problems 5–8.*

1.

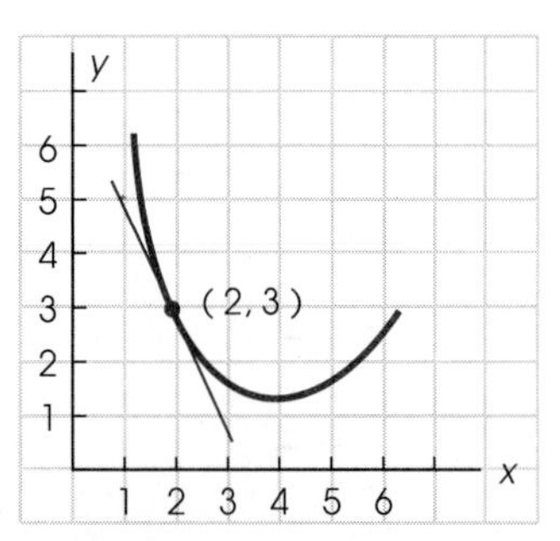


2.

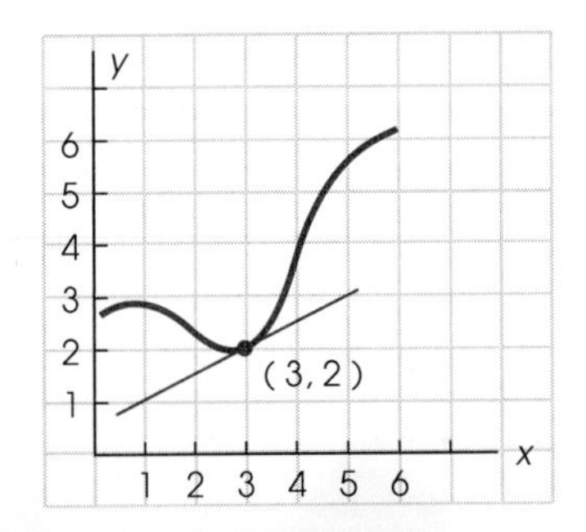

3.

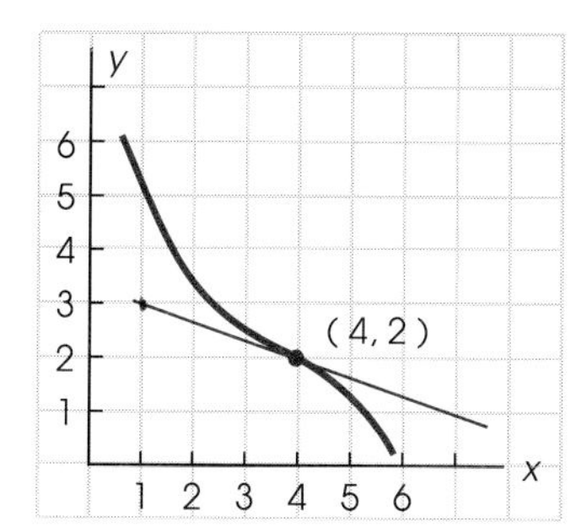


4.

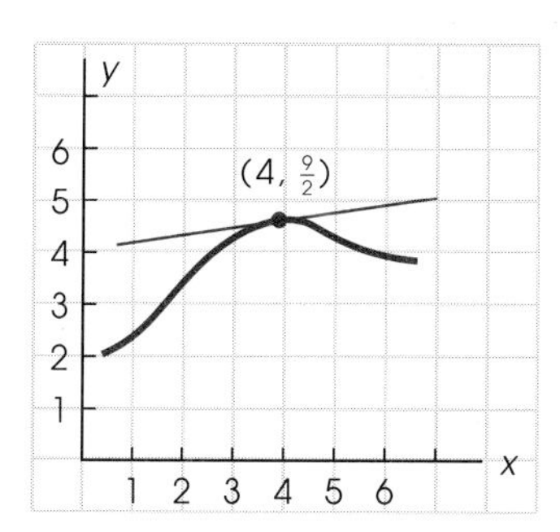


5.

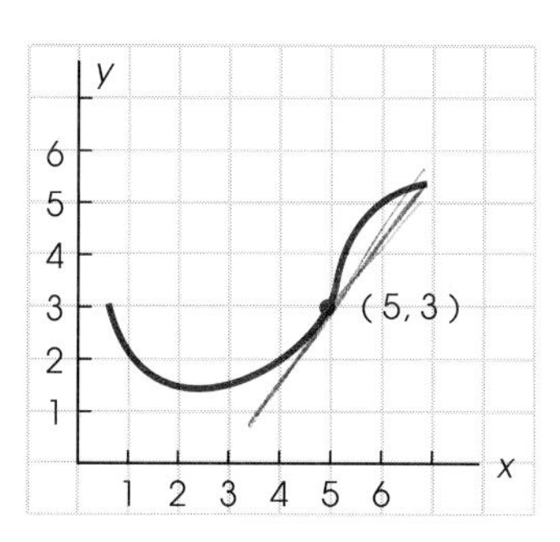


6.

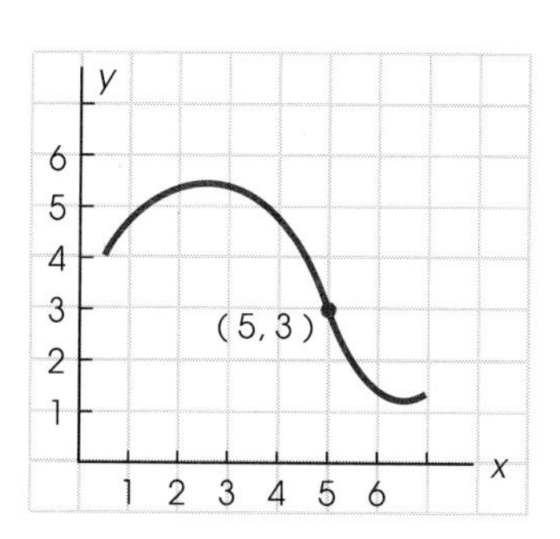


7.

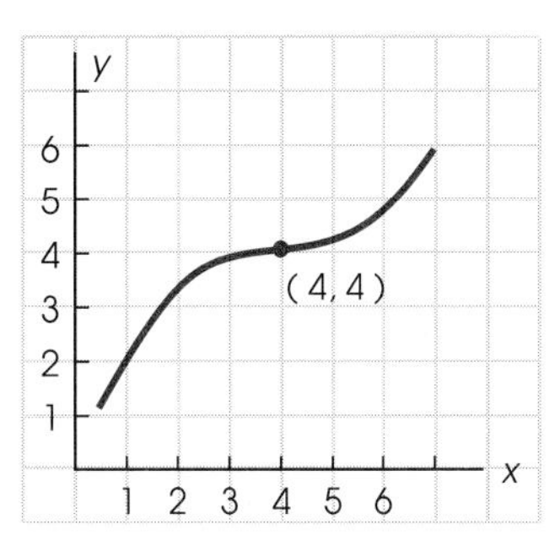


8.

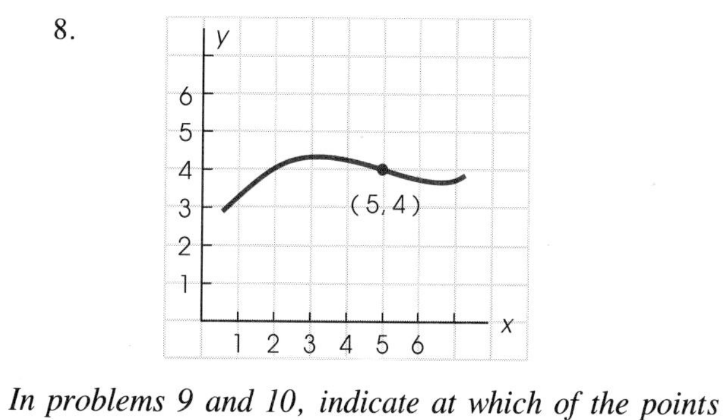


*In problems 9 and 10, indicate at which of the points A, B, C, D, and E, the curve has (a) positive slope, (b) negative slope, (c) zero slope.*

9.

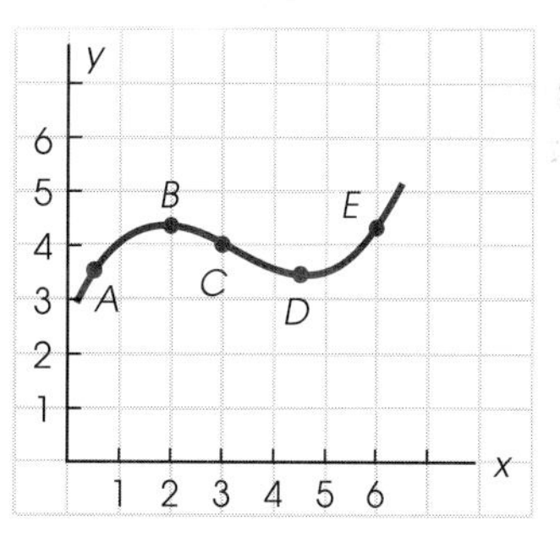


10.

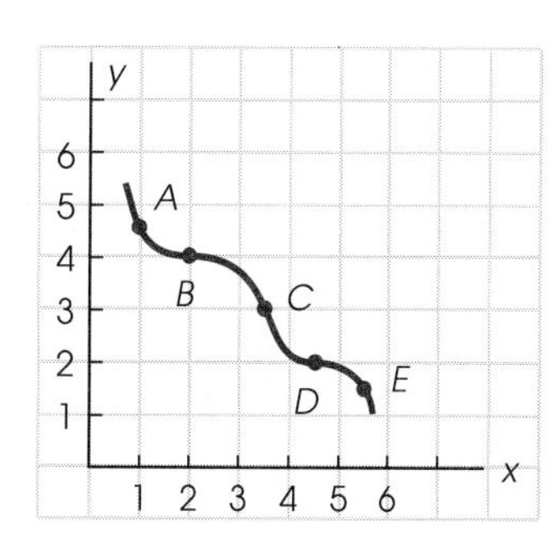

11. Sketch the graph of a function which has positive slope for $x < 2$, zero slope for $2 \le x \le 3$, and negative slope for $x > 3$.
12. Sketch the graph of a function which has negative slope for $x < -1$, positive slope for $-1 < x < 2$, and zero slope for $x \ge 2$.
13. Find an equation for the tangent line to the curve $y = x^2$ at the point where $x = 3$. See Example B.
14. Find an equation for the tangent line to the curve $y = x^2$ at the point where $x = -5$.

*In Problems 15–18, find the slope of the given curve at the indicated value of x (see Example C).*

15. $y = x^3$ at $x = 4$
16. $y = x^4$ at $x = -2$
17. $y = x^5$ at $x = -1$
18. $y = x^8$ at $x = 2$

*As illustrated in Example D, find dy/dx in Problems 19–24.*

19. $y = 5x^6 + 4x^3 - 12x$
20. $y = 3x^4 - 2x^3 + 5x - 11$
21. $y = 2x^4 - 3x^3 + 5x^2 + \pi$
22. $y = 2x^6 - 11x^2 + \pi^2$
23. $y = -5x^3 + \frac{1}{2}x^2 - \frac{3}{4}$
24. $y = \frac{3}{4}x^{12} - \frac{1}{3}x^6 + \frac{5}{7}$

*In Problems 25–28, find an equation for the tangent line to the given curve at the indicated point (see Example E).*

25. $y = 2x^5$ at $(-2, -64)$
26. $y = 4x - x^3$ at $(3, -15)$
27. $y = x^6 - 12x^3 + 5x - 8$ at the point where $x = 2$.
28. $y = 2x^5 + 11x^2 + 37$ at the point where $x = -2$.
29. Find two points on the curve $y = 2x^3 - 15x + 6$ where the slope is 9. See Example F.
30. Find two points on the curve $y = 2x^3 - 10x - 22$ where the slope is 44.

---

31. Consider the curve $y = 2x^3$. The points $P(x, 2x^3)$ and $Q[x + \Delta x, 2(x + \Delta x)^3]$ are both on this curve.
    (a) Find the slope $\Delta y/\Delta x$ of the secant line $PQ$ and simplify your answer.
    (b) Find the slope $dy/dx$ of the tangent line at $P$ by letting $\Delta x$ approach zero in your answer to part (a).
32. Follow the procedure in Problem 31 to find $dy/dx$ for the curve $y = 3x^2 - x$.
33. Find equations of the two tangent lines to the curve $y = x^3 - 9x + 10$ that have slopc 3.
34. Find an equation for the line perpendicular to the curve $y = 3x^2 - 7x + 2$ at the point $(0, 2)$. (You should be able to decide what the phrase "perpendicular to a curve" means).
35. At what values of $x$ is the slope of the curve $y = x^3 + 2x$ positive?
36. For what values of $x$ is the slope of the curve $y = 2x^2 + 4x - 5$ positive?
37. At what values of $x$ is the tangent line to the curve $y = 2x^2 - 9x + 7$ parallel to the line $y = 3x + 1$?
38. Sketch the two parabolas $y = x^2$ and $y = x^2 - 2x + 1$ and then show that they intersect at right angles.
39. [C] Find the slope of the curve $y = \pi x^3 - \sqrt{2}x^2 + \sqrt{3}$ at $x = 1.592$.

## 1.6 VARIABLE RATES OF CHANGE

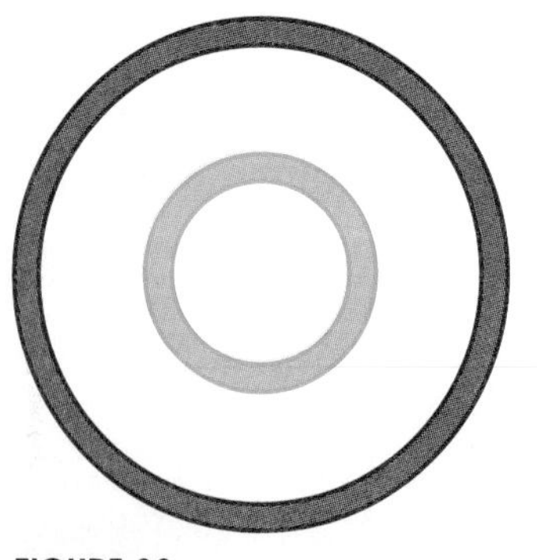
FIGURE 38

For linear functions, slope and rate of change turned out to be identical concepts, as we saw in Section 1.4. The same is true for nonlinear functions, as we shall try to make clear in this section. In contrast to the linear case where slope and rate of change were constant, slope and therefore rate of change are variable for nonlinear functions.

For a first concrete example, consider the (nonlinear) relationship between the area $A$ of a circle and its radius $r$, namely, $A = \pi r^2$. If we increase the radius of a small circle by a little, we increase the area by a certain amount. But if we increase the radius of a large circle by that same bit, we increase the area by a much larger amount (Figure 38). Quite clearly, the rate of change in the area with respect to the radius is variable; this rate increases as $r$ increases.

If $r$ increases from $r$ to $r + \Delta r$, then the area increases from $\pi r^2$ to $\pi(r + \Delta r)^2$. The change in area $\Delta A$ divided by the change in radius $\Delta r$ is called the **average rate of change** in $A$ with respect to $r$ on the given interval. Thus

$$\text{Average rate of change} = \frac{\Delta A}{\Delta r} = \frac{\pi(r + \Delta r)^2 - \pi r^2}{\Delta r}$$

The **rate of change** in area with respect to radius at $r$ is the limiting value of this average rate as $\Delta r$ shrinks to 0; that is,

$$\text{rate of change at } r = \lim_{\Delta r \to 0} \frac{\Delta A}{\Delta r}$$

But this latter expression is what we called slope at $r$ and denoted by $dA/dr$ in the previous section. We conclude that

$$\boxed{\frac{dA}{dr} = \text{slope at } r = \text{rate of change at } r}$$

Since we know how to calculate $dA/dr$, we can calculate rates of change with very little effort as we now illustrate.

### EXAMPLE A

Find for a circle the rate of change in area with respect to radius at $r = 4$ inches and at $r = 10$ inches.

**Solution**

$$A = \pi r^2$$

$$\frac{dA}{dr} = \pi \cdot 2r = 2\pi r$$

At $r = 4$, $dA/dr = 2\pi(4) \approx 25.13$ square inches per inch; at $r = 10$, $dA/dr = 2\pi(10) \approx 62.83$ square inches per inch.

Here is how to interpret these results. At $r = 4$, a small numerical increase in radius will produce a corresponding numerical increase in area of about 25 times that amount. However, at $r = 10$, a small numerical increase in radius produces almost 63 times that numerical increase in area. ■

## Speed and Velocity

College students don't need to be told about cars and speedometers; they have known about such things since they were children. But what does a speedometer measure?

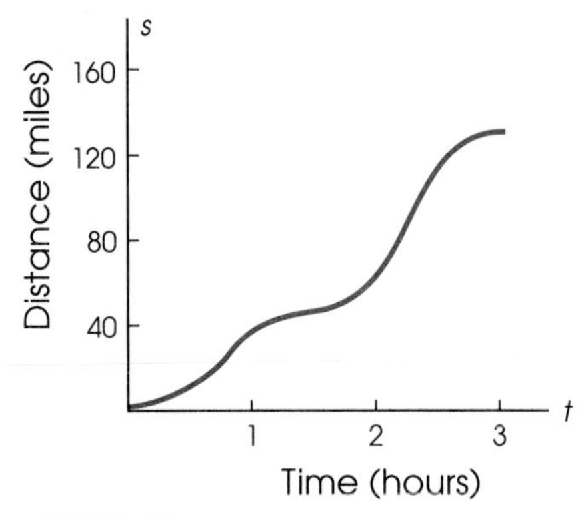


FIGURE 39

If I drove my car a distance of 135 miles in 3 hours, some would say that I drove at a speed of $\frac{135}{3} = 45$ miles per hour. That, in fact, would be my *average speed,* but this average speed has little to do with the speedometer reading. At the start of my trip, the speedometer registered 0; at times, it rose as high as 57; at the end, it fell back to 0. To put it simply, speed kept changing; it was a variable.

*Speed* (sometimes called instantaneous speed) is the rate of change in distance with respect to time. If I had charted $s$ (distance I traveled) in terms of $t$ (time elapsed), I would have obtained a graph something like Figure 39. My speed $ds/dt$ corresponds to the slope of the curve. Note that this slope is 0 at $t = 0$ and $t = 3$ (when I started and stopped). At $t = 2$, this slope measures about 55; I was going about 55 miles per hour at that instant.

What we have said so far assumes that distance $s$ is increasing with time $t$ (a car has a reverse gear but normally you don't use it on a highway). If the moving object can go backward as well as forward, then $ds/dt$ will be negative whenever distance is decreasing with time. In this situation, $ds/dt$ is given the name **velocity,** and **speed** is defined to be $|ds/dt|$. Thus, velocity can be positive, negative, or zero, but speed is always nonnegative.

## EXAMPLE B

Suppose an object moves along a line so that its directed distance $s$ (in feet) from a fixed point is given by

$$s = -t^3 + 6t^2$$

$t$ being measured in seconds. (a) Sketch the graph of $s$ for $0 \leq t \leq 6$. (b) Find the velocity $v = ds/dt$ at $t = 1$, 2, and 5.

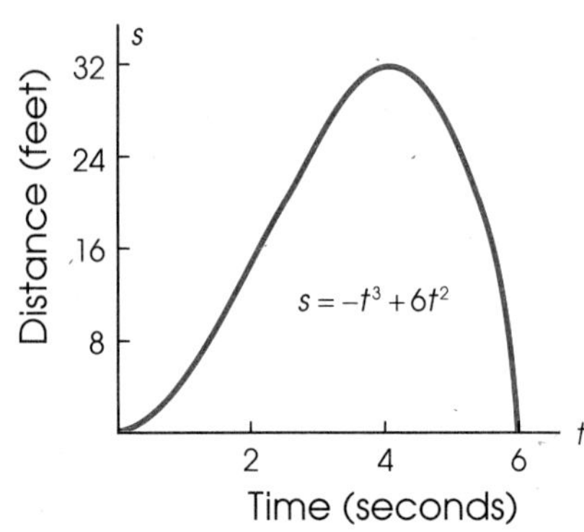


FIGURE 40

**Solution** (a) The graph is shown in Figure 40. (b) Since $v = ds/dt = -3t^2 + 12t$,

$$v(1) = -3 + 12 = 9 \text{ feet per second}$$

$$v(2) = -12 + 24 = 12 \text{ feet per second}$$

$$v(5) = -75 + 60 = -15 \text{ feet per second}$$

Note that the object is moving in reverse at $t = 5$. ■

## EXAMPLE C

A ball is thrown upward from an initial height of 100 feet and with an initial velocity of 64 feet per second. From physics, we learn that its height $s$ (in feet) after $t$ seconds is given by

$$s = 100 + 64t - 16t^2$$

(a) Find the ball's velocity at $t = 1$, 2, and 3.
(b) When did the ball reach maximum height?
(c) Determine the maximum height.

**Solution**

(a) Since $v = ds/dt = 64 - 32t$,

$$v(1) = 32 \text{ feet per second}$$
$$v(2) = 0 \text{ feet per second}$$
$$v(3) = -32 \text{ feet per second}$$

(b) The ball reaches maximum height when its velocity is 0, namely, at $t = 2$.

(c) $s(2) = 100 + 64(2) - 16(2)^2 = 164$ feet ■

## Other Time Rates

Since the independent variable is time, we refer to velocity as a *time rate of change*. Time rates of change come in many forms. A chemist wants to know how fast a reaction is occurring; a biologist is interested in how fast a drug is being absorbed; a sociologist studies the rate at which a rumor is spreading. Our next example might interest a city manager.

### EXAMPLE D

A small eastern city pumps water into its reservoir between 12:00 midnight and 6:00 A.M. (when electricity is cheapest and most people are asleep). During a recent day, a recording device charted the amount of water in the reservoir as shown in Figure 41. At what time were people using water most rapidly and what was the rate of usage at that time?

**Solution** The water level was falling most rapidly at about 8 A.M. At that time the slope of the curve is about $-50$, which means that water was being used at 50,000 gallons per hour. ■

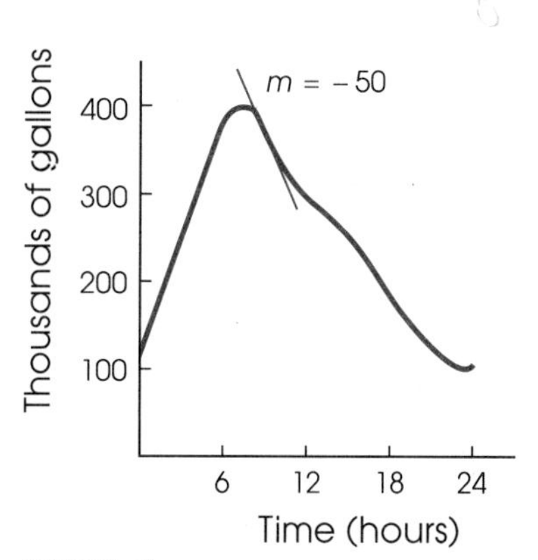


**FIGURE 41**

> **POINT OF INTEREST**
> "It is clear that Economics, if it is to be a science at all, must be a mathematical science. . . . As the complete theory of almost every other science involves the use of [differential calculus], so we cannot have a true theory of Economics without its aid."
>
> Economist W. Jevons

## Marginal Rates in Economics

Economists are especially interested in how certain variables (revenue, cost, profit) change with respect to the number of units produced. They call these rates **marginal** rates. If in manufacturing and selling $x$ units of its product, a company has total cost $C(x)$, total revenue $R(x)$, and total profit $P(x) = R(x) - C(x)$, then $dC/dx$, $dR/dx$, and $dP/dx$ are called marginal cost, marginal revenue, and marginal profit, respectively.

### EXAMPLE E

The ABC Company has fixed costs (office utilities, real estate taxes, and so on) of \$8000 each month and must pay \$48 in labor and materials for each tape recorder it builds. It expects to receive total revenue $R(x)$ from selling $x$ recorders per month, where

$$R(x) = 75x - 0.01x^2$$

(a) Find expressions for the marginal cost, marginal revenue, and marginal profit.
(b) Evaluate the marginal profit at $x = 100$, $x = 1000$, $x = 2000$ and interpret your results.

#### Solution

(a) The total cost is the sum of the fixed cost \$8000 and the variable cost \$$48x$; that is,

$$C(x) = 8000 + 48x$$

Since

$$R(x) = 75x - 0.01x^2$$

we see that

$$P(x) = R(x) - C(x) = 27x - 0.01x^2 - 8000$$

We conclude that the required rates are

$$\frac{dC}{dx} = 48$$

$$\frac{dR}{dx} = 75 - 0.02x$$

$$\frac{dP}{dx} = 27 - 0.02x$$

(b) At $x = 100$, $\dfrac{dP}{dx} = 27 - 0.02(100) = 25$

At $x = 1000$, $\dfrac{dP}{dx} = 27 - 0.02(1000) = 7$

At $x = 2000$, $\dfrac{dP}{dx} = 27 - 0.02(2000) = -13$

> **CAUTION**
> That marginal profit is only approximately equal to the increase in profit due to one unit increase in production should be apparent from the graph below.
>
> 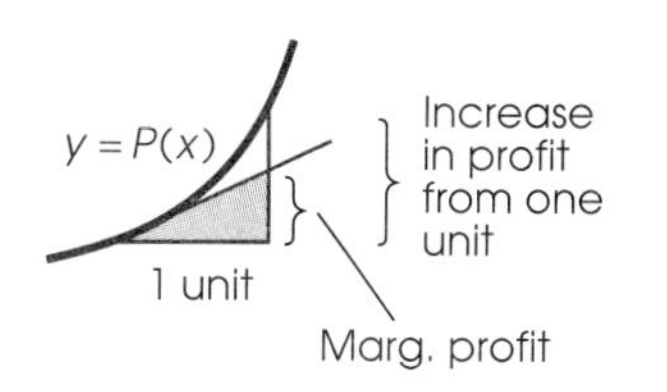
> 

These marginal profits have the following interpretation. Increasing production by one unit when $x = 100$ will yield (approximately) an additional profit of \$25. When $x = 1000$, this additional profit will be only \$7, and when $x = 2000$, there will be a reduction in profit of \$13. One might ask why very high production leads to lower profit. Mathematically, it is due to the term $-0.01x^2$ in the revenue function. This term reflects the reduction in price that usually comes from flooding the market. ■

**EXAMPLE F (MARGINAL TAX RATES)**

The graph in Figure 42 relates federal income tax to taxable income for a single taxpayer in 1983. Estimate the marginal tax rate for Susan Smith who had a taxable income of \$44,000 and interpret the result. (See Problem 40 in Section 1.2 for the basis of this graph.)

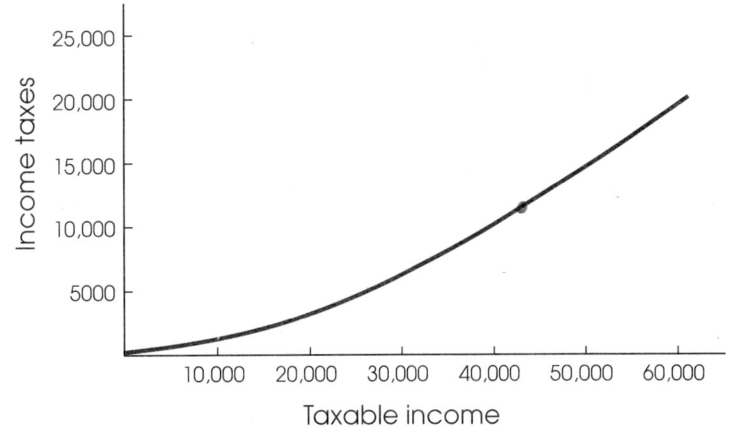


**FIGURE 42**

**Solution** The slope at 44,000 is about 0.45. Thus, Susan's marginal tax rate is 45%, which means that for an additional dollar of income she will pay an additional 45¢ in federal taxes. Note that the marginal tax rate increases with income. This is what we mean when we say that the United States had a progressive income tax in 1983. Some politicians argue for a flat income tax, meaning that the marginal tax rate should be constant. The 1986 Tax Reform Act has only two rates (15% and 28%) and thus federal income taxes are less progressive than formerly. ■

## Problem Set 1.6

1. The volume $V$ of a sphere of radius $r$ is given by $V = \frac{4}{3}\pi r^3$.
   (a) Find the rate of change in volume with respect to radius at $r = 2$ inches.
   (b) If the radius increased from 2 to 2.1 inches, about how much would the volume change?

2. The surface area $S$ of a sphere of radius $r$ is given by $S = 4\pi r^2$.
   (a) Find the rate of change in surface area with respect to radius at $r = 9$ centimeters.
   (b) If the radius increased from 9 to 9.2 centimeters, about how much would the surface area change?

3. A swimming pool is being filled in such a way that after $t$ minutes there are $100t^3 - 25t^2 + 50$ gallons of water in the tank. At what rate (in gallons per minute) is water entering the tank at the end of 2 minutes?
4. Between midnight and noon the temperature $T$ (in degrees Fahrenheit) rose according to the formula $T = \frac{1}{10}t^2 + 3t + 30$, where $t$ is the time in hours after midnight. At what rate was the temperature rising at 10.00 A.M.?

*In Problems 5 and 6, estimate the rate of change in y with respect to x at (a) x = 2, (b) x = 4.*

5.

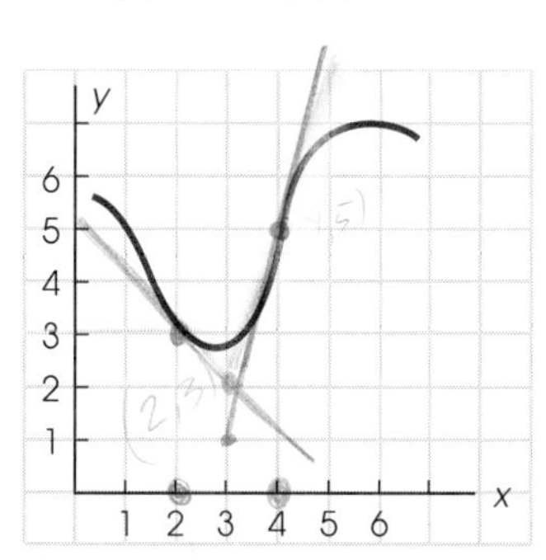


6.

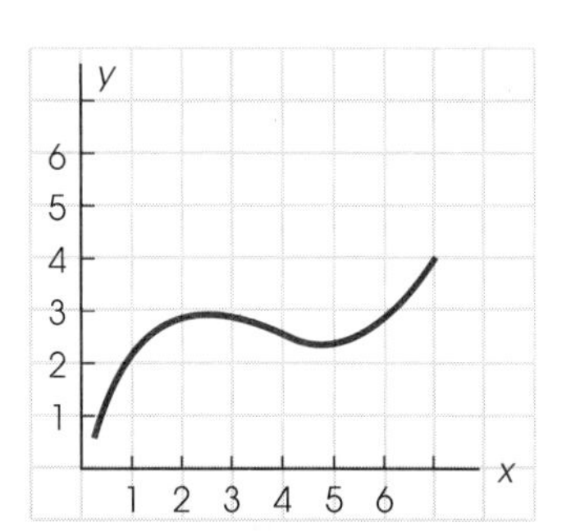


7. An object is moving along a line so that its directed distance $s$ (in feet) from a fixed point is given by

$$s = t^3 - 3t^2 + 2$$

where $t$ is measured in seconds.
(a) Sketch the graph of $s$ for $0 \leq t \leq 4$.
(b) Find the velocity $v = ds/dt$ when $t = 0, 1, 2, 4$.
(c) When is the velocity equal to zero?

8. Suppose that a particle moves along a line in such a way that its directed distance $s$ (in meters) from a fixed point after $t$ seconds is given by

$$s = 12t - t^3$$

(a) Sketch the graph of $s$ for $-3 \leq t \leq 3$.
(b) Find the velocity when $t = -3, -1, 0, 1, 3$.
(c) When is the velocity equal to zero?

C 9. Suppose an object moves along a line so that its directed distance $s$ (in feet) from a fixed point is given by $s = f(t)$, $t$ being measured in seconds. The **average velocity** of the object on the interval $t$ to $t + \Delta t$ is

$$v_{\text{ave}} = \frac{\Delta s}{\Delta t} = \frac{\text{change in } s}{\text{change in } t} = \frac{f(t + \Delta t) - f(t)}{\Delta t}$$

Thus, for $s = t^2$, the average velocity on the interval $t = 3$ to $t = 3.5$ is

$$v_{\text{ave}} = \frac{\Delta s}{\Delta t} = \frac{(3.5)^2 - 3^2}{3.5 - 3} = \frac{3.25}{0.5} = 6.5 \text{ feet per second}$$

Find:
(a) $v_{\text{ave}}$ on the interval 3 to 3.1.
(b) $v_{\text{ave}}$ on the interval 3 to 3.01.
(c) $v_{\text{ave}}$ on the interval 3 to 3.001.
(d) $v = ds/dt$ at $t = 3$. (Do you see the point of this problem?)

C 10. Follow the directions in Problem 9 for $s = 6t - t^2$.

11. When a ball is thrown upward from ground level at 128 feet per second, its height $s$ (in feet) after $t$ seconds is given by $s = 128t - 16t^2$. See Example C.
(a) Find the ball's velocity when $t = 2$ and $t = 6$.
(b) At what time does the ball reach its maximum height?
(c) What is this maximum height?

12. A ball is thrown downward at 96 feet per second from a helicopter which is 3000 feet above the ground. The height $s$ in feet of the ball after $t$ seconds is given by $s = -16t^2 - 96t + 3000$.
(a) Find the ball's velocity when $t = 2$ and $t = 4$.
(b) At what time is the ball falling at 224 feet per second?
(c) Find the height and the velocity of the ball when $t = 10$.

13. A ball thrown upward from the ground is at height $s$ (in feet) after $t$ seconds, where $s = 80t - 16t^2$. With what velocity does it hit the ground? (It hits the ground when $s = 0$.)

14. A ball is thrown upward from the rim of a building 96 feet high at 80 feet per second. Its height $s$ (in feet) after $t$ seconds is $s = -16t^2 + 80t + 96$. With what velocity does the ball hit the ground?

15. The accompanying graph (Figure 43) gives the population $P(t)$, in thousands, of a certain city at time $t$, the number of years after 1950. See Example D.

(a) Calculate the average rate of increase in population from 1960 to 1970.
(b) When was the rate of increase in population greatest?
(c) Estimate this maximum rate of increase.

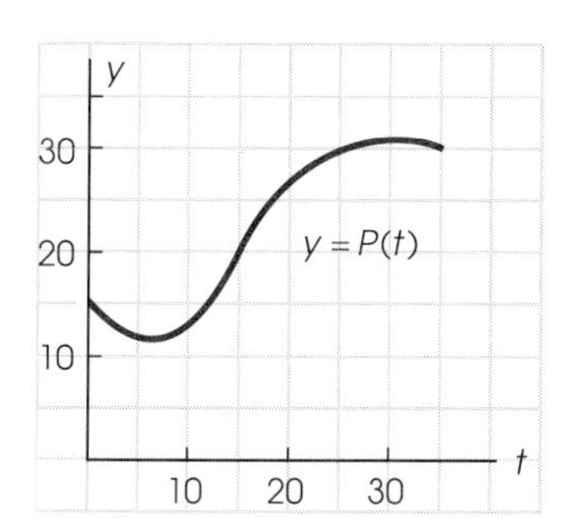


**FIGURE 43**

16. The graph in Figure 44 gives the number $f(t)$ of family farms in a certain Iowa county $t$ years after 1970.
(a) Calculate the average rate of change in the number of farms from 1970 to 1980.
(b) At what time was the number of farms decreasing most rapidly?
(c) At what rate was the number of farms changing at that time [the time in part (b)]?

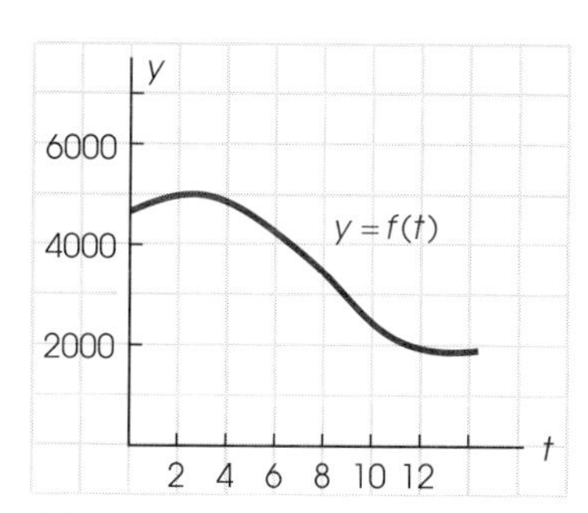


**FIGURE 44**

17. A small sports equipment manufacturer makes $x$ baseball bats per day. It estimates that its total profit (in cents) from the production and sale of these bats is $P(x) = 250x - x^2$. See Example E.
(a) What is the total profit per day at a production level of 80 bats per day?
(b) Find the marginal profit when the level of production is 80 bats per day.
(c) Approximately how much additional profit would be generated by increasing production to 81 bats per day?
(d) At what production level would it become foolish to consider increasing production?

18. A manufacturer of reclining chairs finds that the cost (in dollars) of manufacturing $x$ chairs per week is given by

$$C(x) = -0.05x^2 + 250x + 3000$$

(a) Calculate the marginal cost when the level of production is 500 chairs per week.
(b) Estimate the cost of making the 501st chair.

19. A manufacturer of toy wagons finds that the weekly cost and weekly revenue functions for manufacturing and selling $x$ wagons are given (both in dollars) by

$$C(x) = 400 + 3.5x - \tfrac{1}{300}x^2$$

$$R(x) = 7x - \tfrac{1}{200}x^2$$

(a) Find the marginal cost, marginal revenue, and marginal profit.
(b) What is the rate of change in cost when the level of production is 500 wagons per week?
(c) How fast does the revenue change when 500 wagons per week are sold?
(d) Calculate the rate of change in profit when $x = 500$ and when $x = 1000$.

20. A shoe manufacturer finds that the weekly cost and weekly revenue functions for manufacturing and selling $x$ pairs of shoes are given (both in dollars) by

$$C(x) = 1000 + 20x - 0.02x^2$$

$$R(x) = 50x - 0.10x^2$$

(a) Find the marginal cost, marginal revenue, and marginal profit.
(b) Estimate the cost of manufacturing the 201st pair of shoes.
(c) Find the rate of change in revenue when the level of production is 120 pairs of shoes.
(d) Find the marginal profit when 250 pairs of shoes are made (and sold) per week.
(e) At what level of production is it no longer profitable to increase production?

21. Suppose John MacDougal had a taxable income of $35,000 in 1983. According to the graph in Example F:
(a) How much income tax did he pay that year?
(b) What is the marginal tax (rate) at that income level?

22. Answer the same question as in Problem 21 for Lu Wong whose taxable income for 1983 was $48,000.

---

23. An object is moving along a line in such a way that its position (measured in centimeters) after $t$ seconds relative to a fixed point on the line is given by

$$s = \tfrac{1}{3}t^3 - 4t + 5$$

(a) Find the velocity when $t = 1$ and $t = 4$.
(b) At what instant is the velocity 96 centimeters per second?
(c) At what instant does the object change directions?
(d) Find the average velocity of the object from $t = 3$ to $t = 6$.

24. If a ball was thrown upward from the rim of a building 160 feet high with an initial speed of 48 feet per second, then its height above the ground after $t$ seconds is given by

$$s = -16t^2 + 48t + 160$$

(a) When did it reach maximum height?
(b) What was the maximum height?
(c) At what instant did the ball hit the ground?
(d) At what speed did the ball hit the ground?

25. Two particles moving along a line have the following positions relative to a fixed point on the line after $t$ seconds.

$$s_1 = 4t - 3t^2 \qquad s_2 = t^2 - 2t$$

(a) When do they have the same position?
(b) When do they have the same velocity?
(c) When do they have the same speed?

26. A manufacturer of blue jeans has learned that at $p$ dollars per pair he can sell $x$ pairs of blue jeans per month, where $x = 4000 - 125p$. The cost of manufacturing $x$ pairs of jeans per month (in dollars) is given by

$$C(x) = 25x + 1000$$

(a) Find the monthly revenue $R(x)$.
(b) Find the marginal cost.
(c) Find the marginal profit.
(d) Find the additional profit (or loss) that results from increasing production by one pair when the level of production is 200 pairs per month.
(e) At what level of production does it become inadvisable to increase production?

27. Suppose that the bacteria population in a certain medium $t$ hours from now is given by

$$N(t) = 25{,}000 + 1400t - 500t^2$$

(a) What is the present bacteria count?
(b) What is the growth rate of the population 30 minutes from now?
(c) At what time does the population start to decrease?

28. The radius of a spherical cancerous tumor is growing linearly at a rate of $\frac{1}{10}$ centimeter per week.
(a) If its radius is 3 centimeters now, find an expression for its radius $t$ weeks from now.
(b) Find a formula for the volume of the tumor $t$ weeks from now.
(c) At what rate will the volume of the tumor be increasing 5 weeks from now?

29. Here is the meat market manager's response to an angry customer: "I didn't say the price of meat was tapering off. I said the rate of increase was tapering off." Let $P = f(t)$ denote the price of meat at time $t$. What had the manager said about $P$ in the language of mathematics? What did the customer think he had said?

## CHAPTER 1 REVIEW PROBLEM SET

1. Let $f(x) = x^2 - 3x + 5$ and $g(x) = \sqrt{3x - 2}$.
(a) Find the domain of the function $g$.
(b) Calculate $f(g(2))$ and $g(f(1))$.
(c) Simplify $[f(x + h) - f(x)]/h$.

2. Let $f(x) = 4\sqrt{x - 2}$ and $g(x) = x^2 + 2$.
(a) Determine the domain of $f$.
(b) Calculate $g(f(3))$ and $f(g(2))$.
(c) Simplify $[g(x + h) - g(x)]/h$.

3. Identify each of the following functions as to whether it is even, odd, or neither.
(a) $f(x) = x^3 + 1$
(b) $g(x) = \sqrt{x^2 + 1/x}$
(c) $h(x) = |x|(2x^2 + 4)$

4. Identify each of the following functions as to whether it is even, odd, or neither.
(a) $f(x) = \sqrt{|x| + 1}$
(b) $g(x) = x/(x + 1)$
(c) $h(x) = \sqrt{x^3 + x}$

5. Sketch the graph of each of the following functions.
(a) $f(x) = -3x + 9$
(b) $g(x) = 9x - x^3$
(c) $h(x) = \begin{cases} 4 - x^2 & \text{if } x < 2 \\ x - 3 & \text{if } x \geq 2 \end{cases}$

6. Sketch the graph of each of the following functions.
(a) $f(x) = 4x - 12$
(b) $g(x) = x^3 - 4$

(c) $h(x) = \begin{cases} x^2 - 3 & \text{if } x \le 0 \\ 4 - x & \text{if } x > 0 \end{cases}$

7. Give formulas for $f(x)$, $g(x)$, and $h(x)$ if

$$h(g(f(x))) = \sqrt{(2x + 1)^2 + 3(2x + 1)}$$

8. Give formulas for $f(x)$, $g(x)$, and $h(x)$ if

$$h(g(f(x))) = 4/\sqrt{x^2 + 3x + 5}$$

9. Find the slope and $y$-intercept of each of the following.
   (a) $3x - 2y = 12$
   (b) $y - 2 = 3(x + 1)$
   (c) $\frac{x}{3} + \frac{y}{4} = 1$
10. Find the slope and $y$-intercept of each of the following lines.
   (a) $x + 2 = 4(y - 1)$
   (b) $6x - 5y = 15$
   (c) $2(x + y - 1) = 3(x - 2)$
11. Write in slope-intercept form the equation of the line that passes through $(-2, 1)$ and is parallel to the line $2x - 5y = 7$.
12. Write in slope-intercept form the equation of the line that passes through $(-2, 4)$ and is parallel to the line $3x - y = 8$.
13. Find the value of $k$ for which the line $kx - 3y = 10$
   (a) Passes through the point (2, 4).
   (b) Is parallel to the $x$-axis.
   (c) Is perpendicular to the line $y - 2 = 2(x - 1)$.
14. Find the value of $k$ for which the line $2x + ky = 15$
   (a) is parallel to the line $x - y = 4$;
   (b) passes through the point $(3/2, -4)$;
   (c) is perpendicular to the line $y - 1 = (-1/2)(x + 4)$.
15. Find an equation for the tangent line to the curve

$$y = 2x^5 - 10x^3 + 3x^2 - 2$$

at the point where $x = 1$.

16. Find an equation for the tangent line to the curve $y = -2x^3 + 3x^2 - 4x + 6$ at the point where $x = 2$.
17. Frank Wheeler leased a car for \$180 per month plus 15¢ per mile.
   (a) Write a formula for $F(x)$, the cost in dollars for a month in which Frank drove $x$ miles.
   (b) What will the car cost Frank for a month in which he drives 6000 miles?
18. When the Environmental Protection Agency found a certain company dumping sulfuric acid into the Missouri River, it fined the company \$320,000 plus \$2000 per day until the company complied with the agency's pollution regulations. Write a formula $F(x)$ for the total fine in terms of $x$, the number of days it took for compliance.
19. Phyllis, working at a department store, makes \$1800 per month plus a commission of 5% of her sales.
   (a) Express her total salary $S(x)$ in terms of $x$, her monthly sales in dollars.
   (b) What must her sales be in order for her to make \$2300 per month?
20. A medical insurance company pays \$1275 for a tonsillectomy and \$250 per day for a semiprivate room.
   (a) Write a formula $F(x)$ for the total amount the company will pay for a tonsillectomy that requires $x$ days of hospitalization.
   (b) Howard Hong, who carries this insurance, was billed \$3,200 after an operation and 4 days in the hospital. How much will he have to pay out of his own funds?
21. A bus, purchased today for \$90,000, will have a scrap value of \$2000 after 8 years. Let $f(t)$ be the value of the bus $t$ years from now $(0 \le t \le 8)$. Write a formula for $f(t)$ assuming that it is a linear function.
22. The amount in dollars $F(x)$ that the Great Plains Gas Company pays a property owner for extracting $x$ thousand cubic feet of natural gas from his or her land is a linear function of $x$. The company paid Helen \$8800 for extracting 5 million cubic feet of gas from her land; it paid John \$11,200 for extracting 20 million cubic feet of gas from his land.
   (a) Write a formula for $F(x)$.
   (b) How much does the company pay just for the *right* to extract natural gas from private property?
23. Figure 45 shows the graph of $y = f(x)$. For what values of $x$ is
   (a) $f(x) < 0$
   (b) $f(x) = 1$
   (c) the slope positive
   (d) the slope equal to 0
   (e) the slope equal to $-1$?

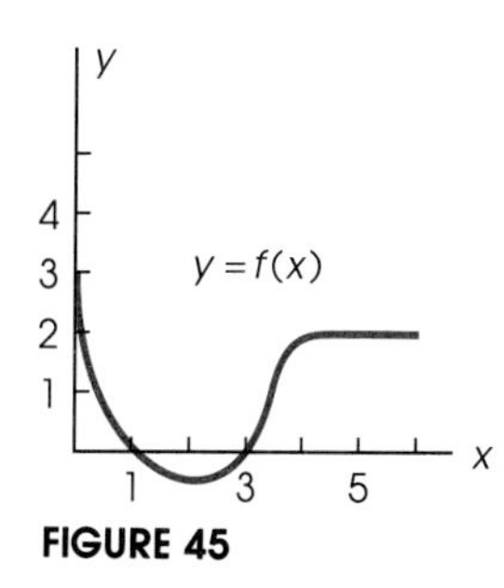


**FIGURE 45**

24. Sketch the graph of a function $y = f(x)$ satisfying all of the following conditions.

(a) $f(-1) = 2, f(3) = 4$, and $f(x) = 2$ for $x \geq 4$.
(b) The slope is 0 at $x = -1$ and at $x = 3$.
(c) The slope is negative for $x < -1$ and for $3 < x < 4$; the slope is positive for $-1 < x < 3$.

25. When a ball is thrown upward from ground level at 192 feet per second, its height $s$ in feet after $t$ seconds is given by $s = 192t - 16t^2$.
   (a) Determine the ball's velocity at $t = 3$ and $t = 10$.
   (b) At what time does the ball reach its maximum height?
   (c) What is the maximum height?
   (d) At what time does the ball hit the ground?
   (e) With what speed does it hit the ground?

26. When a ball was thrown upward from the top of a 512-foot building at 64 feet per second (Fig. 46), its height $s$ (in feet) above the ground after $t$ seconds was given by

$$s = -16t^2 + 64t + 512$$

   (a) Determine the ball's velocity at $t = 1$ and at $t = 5$.
   (b) At what time did the ball reach its greatest height?
   (c) When did the ball hit the ground?
   (d) With what speed did it hit the ground?

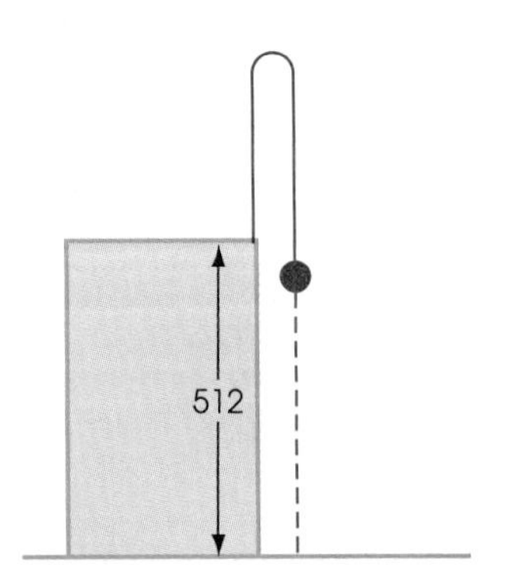


FIGURE 46

27. Figure 47 shows the percentage $f(t)$ of the total number of residents in a certain community that are 60 years old or older, $t$ years after January 1, 1975.
   (a) Calculate the average rate of increase in this percentage from January 1, 1977, to January 1, 1982.
   (b) Estimate the (instantaneous) rate of change in this percentage on January 1, 1982.

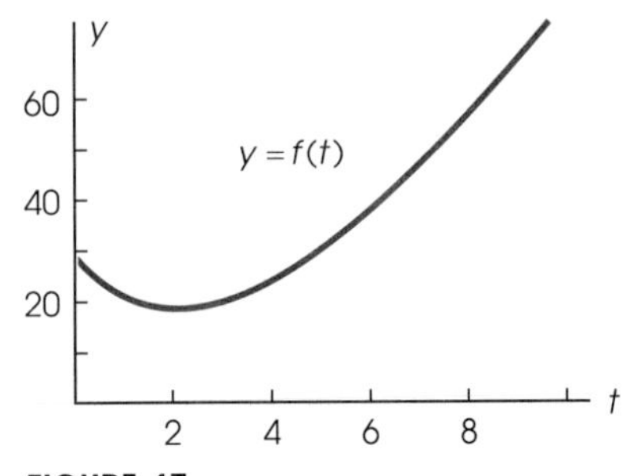


FIGURE 47

28. Mario made a 100-mile trip in his new car. Figure 48 shows $s$, the number of miles traveled, during the first $t$ hours.
   (a) What was his average velocity for the whole trip?
   (b) What was his average velocity during the first 30 minutes?
   (c) When was his velocity greatest?
   (d) Estimate his maximum velocity.
   (e) Estimate his velocity at $t = 1.5$ hours.

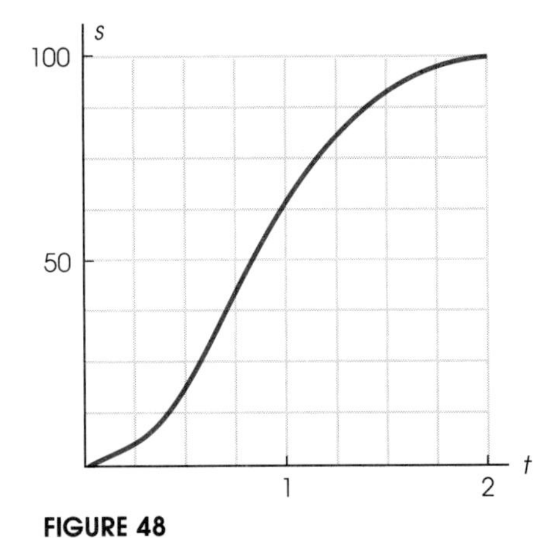


FIGURE 48

29. A manufacturer of tricycles estimates that the weekly cost and revenue functions for making and selling $x$ tricycles are given (in dollars) by

$$C(x) = 1200 + 43x - 0.008x^2$$

$$R(x) = 55x - 0.012x^2$$

   (a) Find the marginal cost, marginal revenue, and marginal profit when 1000 tricycles are made per week.

(b) Approximately what does it cost to make the 1001st tricycle?
(c) How many tricycles should be made and sold per week for maximum profit?

30. A manufacturer of ice skates estimates that the weekly cost and revenue (in dollars) for making and selling $x$ pairs of skates are given by

$$C(x) = 1500 + 46x - 0.01x^2$$

$$R(x) = 70x - 0.018x^2$$

(a) Find the marginal cost, marginal revenue, and marginal profit when 1200 pairs of skates are made per week.
(b) How many pairs should be made per week for maximum profit?

31. Figure 49 shows a square topped by a semicircle.
(a) Write formulas for the area $A(x)$ and perimeter $P(x)$ of the enclosed region in terms of $x$, the side length of the square.
(b) Find $dA/dx$ and $dP/dx$ when $x = 10$.

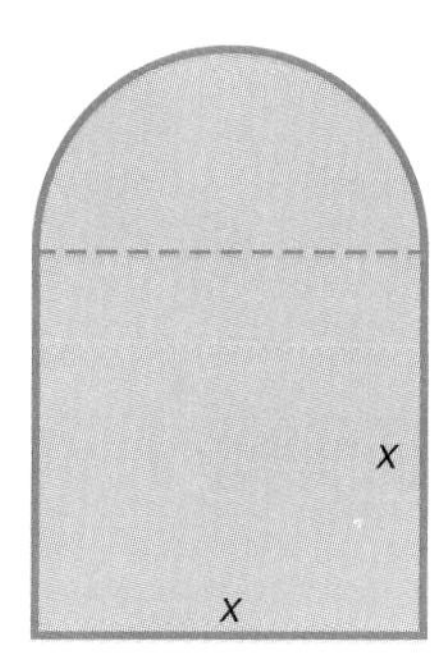


FIGURE 49

32. A rectangle twice as high as wide as topped by an equilateral triangle.
(a) Write a formula for the area $A(x)$ and the perimeter $P(x)$ of the enclosed region, given that it is $x$ units wide.
(b) Find $dA/dx$ and $dP/dx$ when $x = 12$.
(c) Use your answer to part (b) to estimate how much the area would increase if $x$ changed from 12 to 13.

FIGURE 50

*That game theory is today an important branch of applied mathematics is largely due to the influence of John von Neumann. He codified the principles that underlie parlor games such as chess and checkers and showed that these principles apply to competitive situations that arise in the world of business, politics, and the military.*

# 2

# The Derivative

John von Neumann
(1903–1957)

Hungary is small and land-locked but, for its size, it has produced in this century more first-rate mathematicians and physicists that any other country. Perhaps, the most brilliant of these was John von Neumann, born December 28, 1903, in Budapest. His work in pure mathematics was certainly of Nobel prize quality (unfortunately, there is no Nobel prize in mathematics), but that is not the reason for his place on our list. We honor rather his contributions to economics, physics, astronomy, biology, and computer science.

Because his genius was recognized early, he received private tutoring in mathematics and by age 17 had written his first paper with his teacher (Fekete). Before age 21, he had obtained two doctorates—one in mathematics, the other in chemistry.

After teaching at the Federal Institute of Technology in Zurich, the University of Berlin, and the University of Hamburg, he came to Princeton University in 1930. In 1933, he and Einstein were appointed two of the founding permanent members of the Institute for Advanced Study at Princeton, and there he continued working until his death.

His contributions to physics include *Mathematical Foundations of Quantum Mechanics* (1936). His pioneering work on game theory led in 1944 to the epoch-making treatise *Theory of Games and Economic Decisions* written jointly with the economist Oskar Morgenstern. Acting as a consultant to the U.S. Government, he worked at Los Alamos on the atomic bomb project, and this in turn pushed him into the development of the electronic computer. He is credited with the concept of the internally stored program. The analogy between the computer and the human brain inspired him to create a new field called automata theory.

Outside of science, his main interest was in history where he had an unbelievably detailed knowledge supported by a prodigous memory. When he was appointed to the U.S. Atomic Energy Commission in 1954, one of his colleagues is reported to have said, "Johnny will do all right, because he is probably the smartest man in the world."

## 2.1 THE LIMIT CONCEPT

Almost all that we do in calculus rests ultimately on the concept of limit. The word *limit* was used informally in Section 1.5. It is time to make the meaning of this word more precise. Perhaps an illustration from the business world can help us think about our task.

A few years ago, a Minneapolis bank announced with much fanfare that it would no longer compound interest on savings as most banks do—quarterly or monthly or daily. It would henceforth compound interest *continuously*, that is, instantaneously. What did the bank mean by this statement?

Suppose a bank's interest rate is 100% per year (this is unrealistically high, but it makes the calculations easier without spoiling the idea). Then in 1 year, \$1.00 grows to $1 + 1 =$ \$2.00 if interest is compounded annually, \$1.00 grows to $(1 + \frac{1}{12})^{12} = (1.08333)^{12} =$ \$2.61 if interest is compounded monthly, and \$1.00 grows to $(1 + \frac{1}{365})^{365} = (1.00274)^{365} =$ \$2.71 if interest is compounded daily. In general, if interest is compounded at the ends of periods of length $x$, \$1.00 will grow to

$$A(x) = (1 + x)^{1/x}$$

dollars at the end of 1 year.

In Figure 1, we have summarized a number of calculations made on a hand-held calculator. The question to be answered is, What is happening in column 3? That is, what number $L$ are the numbers in Column 3 approaching? Be sure to note that we can't get the answer by substituting $x = 0$ in the formula $A(x) = (1 + x)^{1/x}$. Why?

In Figure 2, we have displayed the same information in graphical form. The problem is to predict where the graph will hit the vertical axis based on its behavior to the right of this axis.

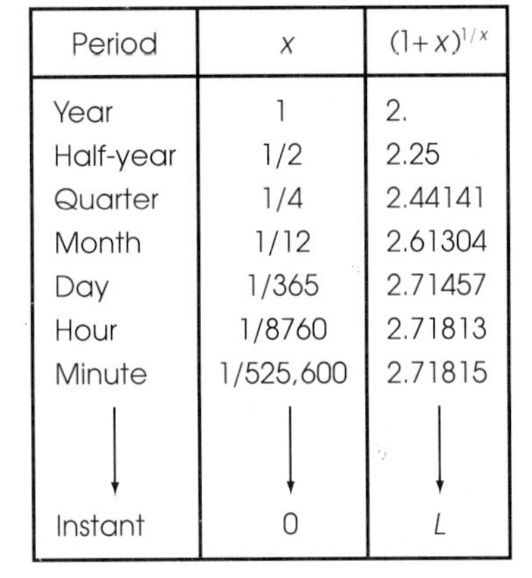

| Period | $x$ | $(1+x)^{1/x}$ |
|---|---|---|
| Year | 1 | 2. |
| Half-year | 1/2 | 2.25 |
| Quarter | 1/4 | 2.44141 |
| Month | 1/12 | 2.61304 |
| Day | 1/365 | 2.71457 |
| Hour | 1/8760 | 2.71813 |
| Minute | 1/525,600 | 2.71815 |
| ↓ | ↓ | ↓ |
| Instant | 0 | $L$ |

**FIGURE 1**

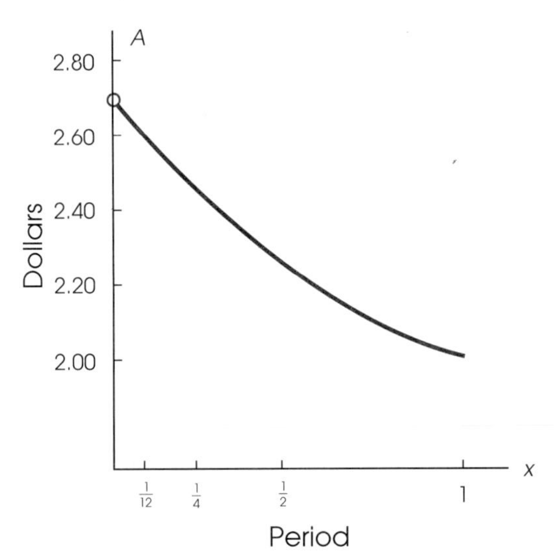


**FIGURE 2**

It turns out that the question we have posed has been studied in very great depth. The answer, an unending decimal traditionally denoted by the letter $e$, is known to thousands of decimal places. We write

$$\lim_{x\to 0} (1 + x)^{1/x} = e = 2.7182818\ldots$$

Of course, to the nearest penny, the answer to our question is \$2.72.

**POINT OF INTEREST**
Though the invention of calculus dates back to 1665–1666, its foundations lay in disarray until after 1800. A. Cauchy (1789–1857) set things right with a precise definition of limit. This Frenchman is credited with making calculus a coherent, logical subject.

## The General Problem

Let $c$ and $L$ be fixed numbers and $x$ a variable. Our aim is to give meaning to the statement

$$\lim_{x\to c} f(x) = L$$

which is read: *The limit as $x$ approaches $c$ of $f(x)$ is $L$*. Here is how we do it.

**DEFINITION (LIMIT)**
We say that $\lim_{x\to c} f(x) = L$ if $f(x)$ is arbitrarily close to $L$ whenever $x$ is sufficiently close to, but different from, $c$.

Note that we do not require anything to be true right at $x = c$. The function $f$ does not even need a value at $c$; it had none in the compound interest example above, and it will have none in Example D below. All that matters is the behavior of $f$ near $x = c$.

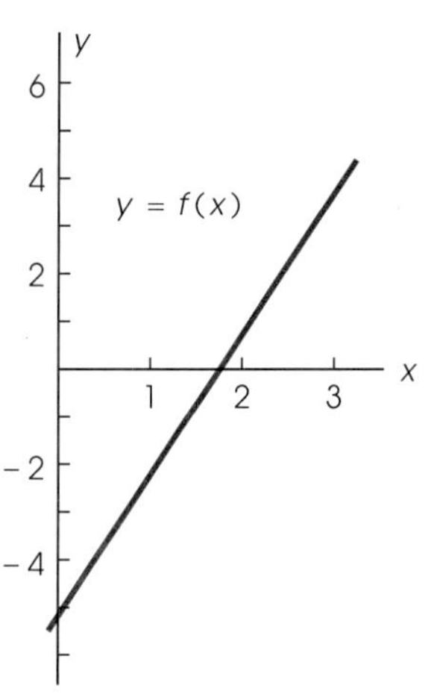

FIGURE 3

**EXAMPLE A**

Let $f(x) = 3x - 5$. Find $\lim_{x\to 2} f(x)$.

**Solution** The graph of $y = f(x)$ is shown in Figure 3. When $x$ is close 2, $3x - 5$ is close to $3 \cdot 2 - 5 = 1$. Thus,

$$\lim_{x\to 2} f(x) = \lim_{x\to 2} (3x - 5) = 1 \qquad ■$$

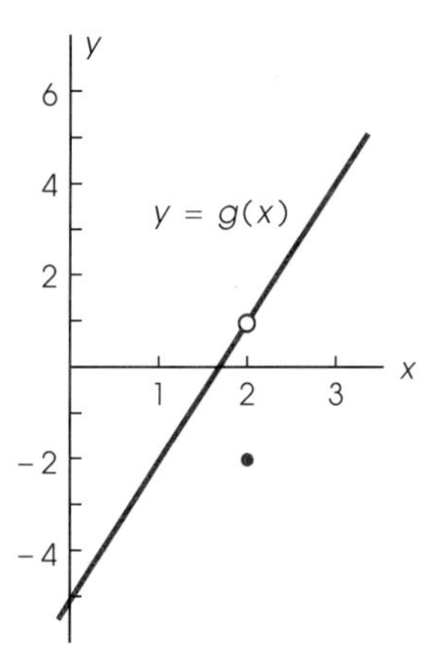

FIGURE 4

**EXAMPLE B**

Let

$$g(x) = \begin{cases} 3x - 5 & \text{if } x \neq 2 \\ -2 & \text{if } x = 2 \end{cases}$$

Find $\lim_{x\to 2} g(x)$.

**Solution** The function $g$ agrees with $f$ in Example A except at $x = 2$ (Figure 4). At the point $x = 2$, the behavior of $g$ is irrelevant. We conclude that

$$\lim_{x\to 2} g(x) = 1 \qquad ■$$

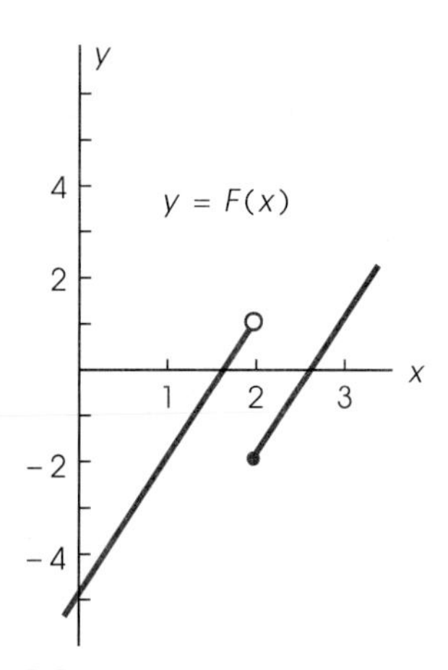


FIGURE 5

| $x$ | $\dfrac{3x^2-11x+10}{x-2}$ |
|---|---|
| 1 | -2 |
| 1.5 | -.5 |
| 1.9 | .7 |
| 1.99 | .97 |
| ↓ | ↓ |
| 2 | ? |
| ↑ | ↑ |
| 2.01 | 1.03 |
| 2.1 | 1.3 |
| 2.5 | 1.5 |
| 3 | 4 |

FIGURE 6

**RECALL**

Factoring a three-term expression is usually a matter of trial and error. We must make sure that the product of the factors we obtain gives the correct middle term. Here is the check for our example

$$\overbrace{(3x-\underbrace{5)(x}_{-5x}-2)}^{-6x} = 3x^2 + (\underset{-6x\,-\,5x}{\ }) + 10$$

$$= 3x^2 - 11x + 10$$

## EXAMPLE C

Let

$$F(x) = \begin{cases} 3x - 5 & \text{if } x < 2 \\ 3x - 8 & \text{if } x \ge 2 \end{cases}$$

Find $\lim_{x\to 2} F(x)$.

**Solution** Note the graph of $F$ in Figure 5. As $x$ approaches 2 from the left, $F(x)$ gets close to 1; but as $x$ approaches 2 from the right, $F(x)$ gets close to $-2$. Though we have not emphasized it before, we now point out that our definition of limit requires that we get the same answer from both sides. Since in this case we do not, we conclude that $F(x)$ doesn't have a limit as $x \to 2$. However, we do have a way of describing the behavior of $F$. We write

$$\lim_{x\to 2^-} F(x) = 1$$

and

$$\lim_{x\to 2^+} F(x) = -2$$

The symbol $x \to 2^-$ is read "$x$ approaches 2 from the left" while $x \to 2^+$ is read "$x$ approaches 2 from the right." ■

## EXAMPLE D

Find $\displaystyle\lim_{x\to 2} \frac{3x^2 - 11x + 10}{x - 2}$.

**Solution** There is no immediately obvious way to find this limit, since both numerator and denominator are tending toward 0 as $x \to 2$. We remind you that $0/0$ is not a number; this symbol is meaningless.

An idea that would occur to a good scientist is to experiment. Try evaluating the quotient for numbers nearer and nearer to 2 and see what happens. A hand-held calculator produced the results shown in Figure 6. The evidence there suggests, but does not prove, that the limit is 1. In fact, it could be 1.01 or 0.998 for all that Figure 6 tells us.

We need a method more reliable than punching calculator keys; this method is provided by algebra—the algebra of factoring. Note that

$$3x^2 - 11x + 10 = (3x - 5)(x - 2)$$

Thus,

$$\lim_{x\to 2} \frac{3x^2 - 11x + 10}{x - 2} = \lim_{x\to 2} \frac{(3x - 5)(x - 2)}{x - 2}$$

$$= \lim_{x\to 2} (3x - 5) = 1 \qquad ■$$

Perhaps you think we were just lucky in the example above. Not so. Whenever taking the limit as $x \to c$ of the quotient of two polynomials leads to the $0/0$ form, you can be sure that both numerator and denominator have the common factor $x - c$. Here is another example.

### EXAMPLE E

Find $\lim_{x \to 3} \dfrac{x^2 - x - 6}{x^2 - 9}$.

### Solution

$$\lim_{x \to 3} \frac{x^2 - x - 6}{x^2 - 9} = \lim_{x \to 3} \frac{(x - 3)(x + 2)}{(x - 3)(x + 3)}$$
$$= \lim_{x \to 3} \frac{x + 2}{x + 3}$$

Now the numerator tends toward 5 and the denominator toward 6. It seems clear that the quotient should approach $\frac{5}{6}$. That it does depends on one of the fundamental properties of limits, a matter we take up next. ■

## Properties of Limits

We need to know how sums, products, and quotients behave under the limit operations. For example, if $f(x) \to L$ and $g(x) \to M$ as $x \to c$, it seems reasonable that $f(x) + g(x) \to L + M$. This and the other properties of limits stated below can be proved and are proved in more advanced courses.

**MAIN LIMIT THEOREM**

Let $n$ be a positive integer, $k$ be a constant, and $f$ and $g$ be functions which have limits at $c$. Then

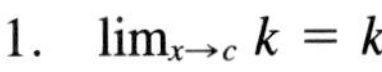

1. $\lim_{x \to c} k = k$
2. $\lim_{x \to c} x = c$
3. $\lim_{x \to c} kf(x) = k \lim_{x \to c} f(x)$
4. $\lim_{x \to c} [f(x) \pm g(x)] = \lim_{x \to c} f(x) \pm \lim_{x \to c} g(x)$
5. $\lim_{x \to c} [f(x) \cdot g(x)] = \lim_{x \to c} f(x) \cdot \lim_{x \to c} g(x)$
6. $\lim_{x \to c} \dfrac{f(x)}{g(x)} = \dfrac{\lim_{x \to c} f(x)}{\lim_{x \to c} g(x)}$ *provided* $\lim_{x \to c} g(x) \neq 0$
7. $\lim_{x \to c} [f(x)]^n = [\lim_{x \to c} f(x)]^n$
8. $\lim_{x \to c} \sqrt[n]{f(x)} = \sqrt[n]{\lim_{x \to c} f(x)}$ *provided* $\lim_{x \to c} f(x) > 0$ when $n$ is even

Note that Property 4 states two results, one using the plus sign, the other using the minus sign.

We give two further examples to illustrate the use of the Main Limit Theorem. The numbers in circles refer to the corresponding parts of the theorem.

## EXAMPLE F

Find $\lim_{x\to -2} (3x^4 - 2x)$, justifying each step by appealing to one of the properties of the Main Limit Theorem.

**Solution**

$$\lim_{x\to -2} (3x^4 - 2x) \overset{④}{=} \lim_{x\to -2} 3x^4 - \lim_{x\to -2} 2x$$

$$\overset{③}{=} 3 \lim_{x\to -2} x^4 - 2 \lim_{x\to -2} x$$

$$\overset{⑦}{=} 3\left(\lim_{x\to -2} x\right)^4 - 2 \lim_{x\to -2} x$$

$$\overset{②}{=} 3(-2)^4 - 2(-2) = 52$$

■

## EXAMPLE G

If $\lim_{x\to 3} f(x) = 2$ and $\lim_{x\to 3} g(x) = -4$, find

$$\lim_{x\to 3} \frac{2f(x) + 5}{3g(x)}$$

**Solution**

$$\lim_{x\to 3} \frac{2f(x) + 5}{3g(x)} \overset{⑥}{=} \frac{\lim_{x\to 3} [2f(x) + 5]}{\lim_{x\to 3} [3g(x)]}$$

$$\overset{④,③}{=} \frac{\lim_{x\to 3} 2f(x) + \lim_{x\to 3} 5}{3 \lim_{x\to 3} g(x)}$$

$$\overset{(3,1)}{=} \frac{2 \lim_{x\to 3} f(x) + 5}{3(-4)} = \frac{2\cdot 2 + 5}{-12} = -\frac{3}{4}$$

■

## Problem Set 2.1

*In Problems 1 and 2, find each limit.*

1. (a) $\lim_{x\to 3} (2x + 5)$ (b) $\lim_{x\to -4} (-3x + 10)$
   (c) $\lim_{x\to 2} (3x^2 - 2x + 1)$ (d) $\lim_{x\to -3} (x^2 - x + 5)$

2. (a) $\lim_{x\to -2} (-x + 9)$ (b) $\lim_{x\to 3} (7x - 8)$
   (c) $\lim_{x\to -1} (x^2 - 3x + 9)$ (d) $\lim_{x\to 4} (2x^2 - 2x + 1)$

*Determine whether $\lim_{x \to 3} f(x)$ exists in Problems 3–8 and, if so, give its value.*

3.

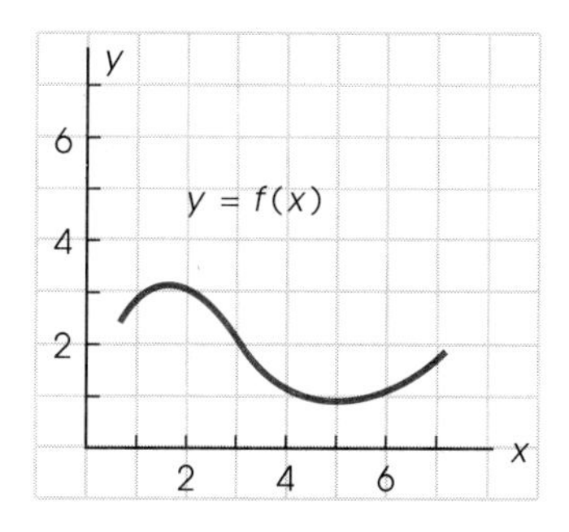


4.

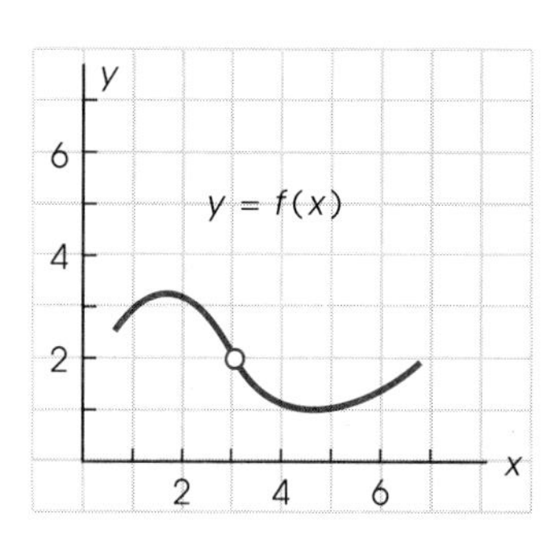


5.

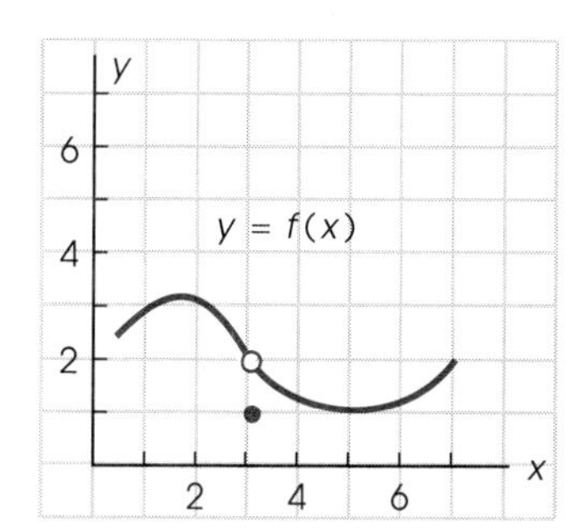


6.

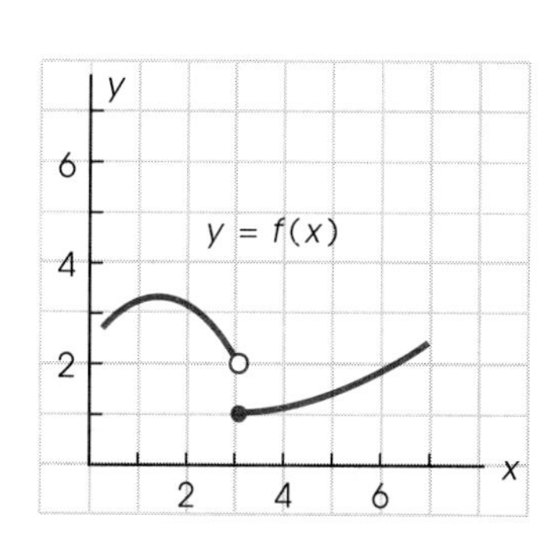


7.

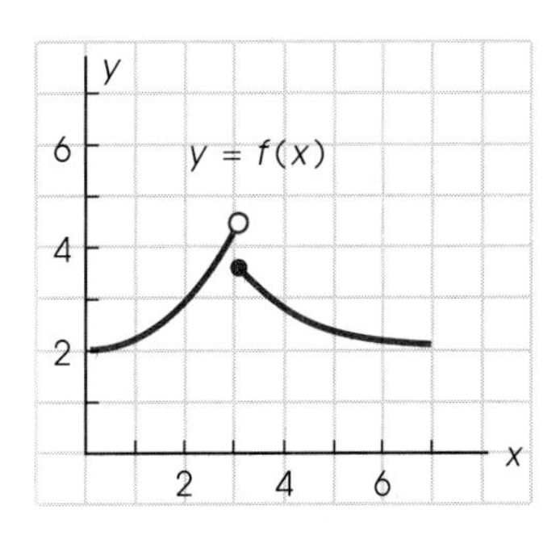


8.

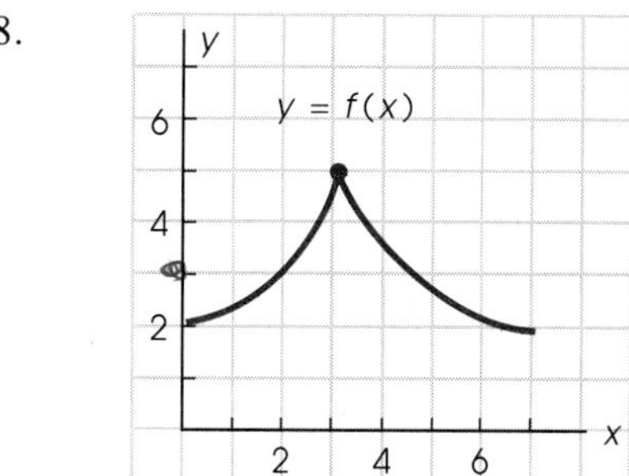


9. Let $f(x) = \begin{cases} 4x - 3 & \text{if } x \neq 3 \\ 5 & \text{if } x = 3 \end{cases}$

(a) Sketch the graph of $f$.
(b) Find $\lim_{x \to 3} f(x)$.

10. Let $g(x) = \begin{cases} -2x + 5 & \text{if } x \neq 4 \\ 2 & \text{if } x = 4 \end{cases}$

(a) Sketch the graph of $g$.
(b) Find $\lim_{x \to 4} g(x)$.

11. Sketch the graph of $F(x) = \begin{cases} x + 2 & \text{if } x \leq 1 \\ 2x & \text{if } x > 1 \end{cases}$.

Then find (if possible) each of the following.

(a) $F(1)$
(b) $\lim_{x \to 1} F(x)$
(c) $\lim_{x \to 2} F(x)$
(d) $\lim_{x \to 1^+} F(x)$
(e) $\lim_{x \to 1^-} F(x)$
(f) $\lim_{x \to -1} F(x)$

12. Sketch the graph of $G(x) = \begin{cases} 3x - 1 & \text{if } x < 2 \\ 2x - 2 & \text{if } x \geq 2 \end{cases}$.

Then find (if possible) each of the following.

(a) $G(2)$ (b) $\lim_{x\to 1} G(x)$

(c) $\lim_{x\to 3} G(x)$ (d) $\lim_{x\to 2} G(x)$

(e) $\lim_{x\to 2^-} G(x)$ (f) $\lim_{x\to 2^+} G(x)$

*In Problems 13–28, evaluate each limit (see Example D and E).*

13. $\lim_{x\to 4} \dfrac{x^2 - 4x}{x - 4}$
14. $\lim_{x\to -3} \dfrac{-3x^2 - 9x}{x + 3}$
15. $\lim_{x\to -3} \dfrac{x^2 + x - 6}{x + 3}$
16. $\lim_{x\to 6} \dfrac{x^2 - 4x - 12}{x - 6}$
17. $\lim_{x\to 3} \dfrac{3x^2 - 7x - 6}{x - 3}$
18. $\lim_{x\to 5} \dfrac{2x^2 - 7x - 15}{x - 5}$
19. $\lim_{x\to -2} \dfrac{x^2 - 4}{x^2 + 2x}$
20. $\lim_{x\to 3} \dfrac{3x^2 - 9x}{x^2 - 9}$
21. $\lim_{x\to 5} \dfrac{x^2 - 5x}{x^2 - 2x - 15}$
22. $\lim_{x\to -3} \dfrac{x^2 - 9}{x^2 + 2x - 3}$
23. $\lim_{x\to 4} \dfrac{x^2 + x - 12}{x^2 - 16}$
24. $\lim_{x\to 0} \dfrac{x^3 - x^2}{x^2 + 2x}$
25. $\lim_{x\to 0} \dfrac{x^4 - x^2}{3x^3 + 7x^2}$
26. $\lim_{x\to 3} \dfrac{x^2 - 6x + 9}{2x^2 - 5x - 3}$
27. $\lim_{x\to 2} \dfrac{(x - 1)(x^2 + 2x + 4)}{x^2 + 5x - 6}$
28. $\lim_{x\to 1} \dfrac{(x - 1)(x^2 + 2x + 4)}{x^2 + 5x - 6}$
29. Find $\lim_{x\to 2} (5x^3 - 4x^2 + 3x + 5)$ justifying each step by appealing to one of the properties in the Main Limit Theorem. See Example F.
30. Follow the instructions in Problem 29 for $\lim_{x\to 3} (2x^4 - 7x^2 + 9x - 1)$.

*Use the Main Limit Theorem to find the limits in Problems 31–42.*

31. $\lim_{x\to -2} (x^3 + 7)(x^2 + x - 1)$
32. $\lim_{x\to 3} (2x^3 - 50)(2x^2 - 3x + 1)$
33. $\lim_{x\to 5} (x^2 + 2x + 5)/(x + 3)$
34. $\lim_{x\to -4} (2x^2 + 7x + 4)/(3x + 4)$
35. $\lim_{x\to 3} (x^3 - 8x - 1)^6$
36. $\lim_{x\to 6} (2x^2 - 10x - 9)^4$
37. $\lim_{x\to 3} \sqrt[3]{x^4 - 10x^2 + 1}$
38. $\lim_{x\to -4} \sqrt[3]{x^3 + 2x^2 + 5}$
39. $\lim_{x\to 2} (\sqrt{x^2 + 5} + 2\sqrt[3]{2x^3 + 11})$
40. $\lim_{x\to 3} (2x - 5)^4(x - \sqrt{x^2 - 5})$
41. $\lim_{x\to 1} \dfrac{(x - 2)^3(x + 3)(x + 1)}{(x + 5)(x^2 + 2x - 2)}$
42. $\lim_{x\to 2} \dfrac{(x^2 + x + 1)(2x - 1)(x + 3)}{(3x - 5)^4(x^2 + 1)}$

*Find the limits in Problems 43 and 44, justifying each step by appealing to one of the properties in the Main Limit Theorem (see Example G).*

43. If $\lim_{x\to 2} f(x) = -3$ and $\lim_{x\to 2} g(x) = 4$, find
$$\lim_{x\to 2} \frac{3f(x) + 2x - 1}{2g(x)}$$
44. If $\lim_{x\to 5} f(x) = 6$ and $\lim_{x\to 5} g(x) = -2$, find
$$\lim_{x\to 5} \frac{f(x) \cdot g(x) + 30}{f(x) - 2x + 3}$$
45. If $\lim_{x\to 2} f(x) = 5$ and $\lim_{x\to 2} g(x) = -4$ find each of the following limits.
(a) $\lim_{x\to 2} [f(x) - x - 1]^4$
(b) $\lim_{x\to 2} \left(\dfrac{f(x) - 2x}{g(x) + 6}\right)^4$
(c) $\lim_{x\to 2} \sqrt[3]{7 - 5g(x)}$
(d) $\lim_{x\to 2} [x^3 + x^2 + x + 1 - 2f(x)]^3$
46. If $\lim_{x\to 3} f(x) = -4$ and $\lim_{x\to 3} g(x) = 3$, find each of the following limits.
(a) $\lim_{x\to 3} [g(x) - 2x + 1]^5$
(b) $\lim_{x\to 3} \left[\dfrac{f(x) + x^2}{3g(x) + 1}\right]^3$
(c) $\lim_{x\to 3} \sqrt{x^2 + 11 - 4f(x)}$
(d) $\lim_{x\to 3} [4g(x) - x^2 - x + 2]^6$

---

*In each of Problems 47–58 find the required limit or state that it doesn't exist.*

47. $\lim_{x\to 3} (2x - 4 + \sqrt{3x - 5})$
48. $\lim_{x\to -4} (2x + 10)\sqrt{5 - 5x}$
49. $\lim_{x\to 2} \dfrac{2x - 5}{(3x^2 - 4x + 3)^2}$
50. $\lim_{x\to 5} \dfrac{(x - 5)^2}{x^2 + 4}$
51. $\lim_{x\to 2} \dfrac{2x^3 - 16}{3x^2 - 12}$

52. $\lim_{x\to -1} \frac{4x^3 + 4}{x^2 - 4x - 5}$

53. $\lim_{x\to 3} \frac{2x - 5}{x^2 - 9}$

54. $\lim_{x\to 4} \frac{3x + 1}{2x^2 - 32}$

55. $\lim_{x\to 0} \frac{|x|}{x}$

56. $\lim_{x\to 2^+} \frac{|x - 2|}{x - 2}$

57. $\lim_{x\to 1} F(x)$, where $F(x) = \begin{cases} 2x + 1 & \text{if } x < 1 \\ 3x & \text{if } x \geq 1 \end{cases}$

58. $\lim_{x\to 2} G(x)$, where $G(x) = \begin{cases} x^2 & \text{if } x \leq 2 \\ x + 4 & \text{if } x > 2 \end{cases}$

[C] 59. By using successively $x = 0.1$, $0.01$, and $0.001$, estimate the following limits.
(a) $\lim_{x\to 0^+} x^x$
(b) $\lim_{x\to 0^+} \{[\ln(2 + x) - \ln 2]/x\}$

[C] 60. Follow the instructions in Problem 59 for
(a) $\lim_{x\to 0^+} [\ln(x + 1)/(x^2 + 2x)]$
(b) $\lim_{x\to 0} [(\sqrt{2 + x} - \sqrt{2})/x]$

61. Suppose $\lim_{x\to a} f(x) = 4$. Calculate each of the following or state that it doesn't exist.
(a) $\lim_{x\to a} [f(x) - 4]$
(b) $\lim_{x\to a} [f(x)/(x - a)]$
(c) $\lim_{x\to a} [\sqrt{f(x)} + f(x)]^2$
(d) $\lim_{x\to a} \{[f(x) - 4]/[f^2(x) - 16]\}$

## 2.2 THE DERIVATIVE

We have seen (Sections 1.5 and 1.6) that the symbol $dy/dx$ can represent many different physical quantities (slope of a curve, rate of change, velocity, marginal profit, and so on). Good sense suggests that we study the underlying concept independently of these specialized vocabularies and diverse applications. Mathematicians have chosen the neutral name *derivative*.

**DEFINITION (DERIVATIVE)**

Let $y = f(x)$. Then $dy/dx$, called the **derivative** of $y$ with respect to $x$, is given by

$$\frac{dy}{dx} = \lim_{\Delta x\to 0} \frac{f(x + \Delta x) - f(x)}{\Delta x} = \lim_{h\to 0} \frac{f(x + h) - f(x)}{h}$$

provided this limit exists.

We have used the definition above to introduce another bit of mathematical sleight of hand. Until now we have consistently used the symbol $\Delta x$ to represent a small change in $x$. This double symbol will become increasingly cumbersome as our calculations get more complicated, and we shall often replace it by the single letter $h$. There's nothing mysterious here. It is analagous to calling Abraham Lincoln simply Abe.

While the symbol $dy/dx$ is the one favored by most scientists, mathematicians are more likely to use another symbol that emphasizes the functional character of the derivative, namely, $f'(x)$. Thus, $dy/dx$ and $f'(x)$ are symbols for the same thing. We will use both interchangeably.

We begin with a simple example that shows how the definition above is used to find derivatives.

**EXAMPLE A**

Use the definition of derivative to find $f'(x)$ if $f(x) = x^2 - 3x$.

**Solution** You know from the previous chapter that the answer has to be $2x - 3$. Here is the required demonstration.

$$\begin{aligned}
f'(x) &= \lim_{h\to 0} \frac{f(x+h) - f(x)}{h} \\
&= \lim_{h\to 0} \frac{(x+h)^2 - 3(x+h) - (x^2 - 3x)}{h} \\
&= \lim_{h\to 0} \frac{x^2 + 2xh + h^2 - 3x - 3h - x^2 + 3x}{h} \\
&= \lim_{h\to 0} \frac{2xh + h^2 - 3h}{h} \\
&= \lim_{h\to 0} \frac{h(2x + h - 3)}{h} \\
&= \lim_{h\to 0} (2x + h - 3) \\
&= 2x - 3
\end{aligned}$$

■

## Two Key Examples

We use two examples to lead up to something very important.

**EXAMPLE B**

Find $f'(x)$ if $f(x) = 1/x$.

**Solution**

$$f'(x) = \lim_{h\to 0} \frac{f(x+h) - f(x)}{h} = \lim_{h\to 0} \frac{\dfrac{1}{x+h} - \dfrac{1}{x}}{h}$$

$$= \lim_{h\to 0} \frac{\dfrac{x - (x+h)}{(x+h)x}}{h} = \lim_{h\to 0} \frac{-\cancel{h}}{(x+h)x} \cdot \frac{1}{\cancel{h}}$$

$$= \lim_{h\to 0} \frac{-1}{(x+h)x}$$

All we have done so far is some rather straightforward algebra. Now we are ready to take the limit. Keep in mind that $x$ is kept fixed; it is $h$ that tends to 0. Using the properties of limits from Section 2.1, we obtain

$$f'(x) = \frac{\lim\limits_{h\to 0}(-1)}{\lim\limits_{h\to 0}(x+h)\cdot\lim\limits_{h\to 0}x} = \frac{-1}{x\cdot x} = \frac{-1}{x^2}$$ ■

**EXAMPLE C**

Find $g'(x)$ if $g(x) = \sqrt{x}$.

**Solution**

$$g'(x) = \lim_{h\to 0}\frac{g(x+h)-g(x)}{h} = \lim_{h\to 0}\frac{\sqrt{x+h}-\sqrt{x}}{h}$$

Here we face a problem in that this limit takes the nonsensical form 0/0, since both numerator and denominator tend to 0. This is typical; it also happened in Example A, though there we finessed the problem with some simple algebra that allowed us to cancel $h$ between the numerator and denominator. Our goal is the same here, but the required algebra is fancier. We rationalize the numerator—a result that is achieved when we multiply numerator and denominator by $\sqrt{x+h}+\sqrt{x}$. See what happens.

> RECALL
>
> $(a-b)(a+b) = a^2 - b^2$
>
> $(\sqrt{x+h}-\sqrt{x})(\sqrt{x+h}+\sqrt{x})$
>
> $= x + h - x$

$$g'(x) = \lim_{h\to 0}\frac{\sqrt{x+h}-\sqrt{x}}{h}\cdot\frac{\sqrt{x+h}+\sqrt{x}}{\sqrt{x+h}+\sqrt{x}}$$

$$= \lim_{h\to 0}\frac{x+h-x}{h(\sqrt{x+h}+\sqrt{x})} = \lim_{h\to 0}\frac{\not{h}}{\not{h}(\sqrt{x+h}+\sqrt{x})}$$

$$= \lim_{h\to 0}\frac{1}{\sqrt{x+h}+\sqrt{x}} = \frac{1}{\sqrt{x}+\sqrt{x}} = \frac{1}{2\sqrt{x}}$$ ■

If we rewrite the results just obtained using exponential notation (a subject we will review at the end of this section), we can observe something significant.

> RECALL
>
> $\frac{1}{x} = x^{-1}$
>
> $\sqrt{x} = x^{1/2}$

$$(i): \quad \frac{d(x^{-1})}{dx} = -1\cdot x^{-2}$$

$$(ii): \quad \frac{d(x^{1/2})}{dx} = \frac{1}{2}\cdot x^{-1/2}$$

Both $(i)$ and $(ii)$ have the form

$$\frac{d(x^n)}{dx} = nx^{n-1}$$

a result familiar from Section 1.5 but demonstrated there only for $n$ a positive integer. We are onto something truly marvelous.

## The Power Rule Revisited

We are going to use the symbol $r$ for any real number. It might be $-1$ or $\frac{1}{2}$ or even $\pi$ or $e$.

> **POWER RULE**
> Let $y = f(x) = x^r$, where $r$ is any real number. Then $f'(x) = rx^{r-1}$; that is,
>
> $$\frac{d(x^r)}{dx} = rx^{r-1}$$

We emphasize again that $r$ can be any real number. Thus,

$$\frac{d(x^\pi)}{dx} = \pi x^{\pi-1}$$

and

$$\frac{d(x^{2.0134})}{dx} = 2.0134x^{1.0134}$$

Unfortunately we can't prove the Power Rule in the generality stated (see Purcell/Varberg: *Calculus with Analytic Geometry*, 5th Ed., Section 7.4 for this proof); we haven't developed enough tools. However, we will give more circumstantial evidence in Problems 29 and 30.

**EXAMPLE D**

If $f(x) = x^2\sqrt{x}$, find $f'(x)$.

**Solution**

$$f(x) = x^2x^{1/2} = x^{2+1/2} = x^{5/2}$$

and so

$$f'(x) = \tfrac{5}{2}x^{3/2}$$ ■

Note two special cases of the Power Rule that we will use constantly. Since $x^0 = 1$ and $x^1 = x$,

> $$\frac{d(1)}{dx} = 0 \quad \text{and} \quad \frac{d(x)}{dx} = 1$$

Next we ask about constant multiples of powers. How do we find their derivatives? For example, what is the derivative of $\pi x^{5.4}$? The Constant Multiple Rule gives us the answer.

**CONSTANT MULTIPLE RULE**

Suppose that $f(x)$ has a derivative and that $k$ is a constant. Then

$$\frac{d[k\,f(x)]}{dx} = k\frac{d[\,f(x)]}{dx}$$

*Proof* Let $F(x) = k\,f(x)$. Then

$$\begin{aligned} F'(x) &= \lim_{h\to 0}\frac{F(x+h)-F(x)}{h} = \lim_{h\to 0}\frac{k\,f(x+h)-k\,f(x)}{h} \\ &= \lim_{h\to 0} k\frac{f(x+h)-f(x)}{h} = k\lim_{h\to 0}\frac{f(x+h)-f(x)}{h} \\ &= k\,f'(x) \end{aligned}$$

The key step was moving $k$ past the limit sign; this is legal because of part 3 of the Main Limit Theorem. ■

**EXAMPLE E**

Find the derivatives of each of the following. (a) $k$, (b) $3x^{4/3}$, (c) $\pi x^{5.4}$.

**Solution**

(a) $\dfrac{d(k)}{dx} = \dfrac{d(k\cdot 1)}{dx} = k\dfrac{d(1)}{dx} = k\cdot 0 = 0$

(b) $\dfrac{d(3x^{4/3})}{dx} = 3\dfrac{d(x^{4/3})}{dx} = 3\cdot\dfrac{4}{3}x^{1/3} = 4x^{1/3}$

(c) $\dfrac{d(\pi x^{5.4})}{dx} = \pi\dfrac{d(x^{5.4})}{dx} = \pi(5.4)x^{4.4} = 5.4\pi x^{4.4}$ ■

## A Review of Exponents

We assume the reader is familiar with the basic definitions for exponents. In particular, if $m$ and $n$ are positive integers,

$$x^m = \underbrace{x\cdot x\cdot\ \ldots\ \cdot x}_{m\text{ factors}} \qquad x^{-m} = \frac{1}{x^m}$$

$$x^0 = 1 \qquad x^{m/n} = (\sqrt[n]{x})^m$$

For example,

$$5^{-2} = \frac{1}{5^2} = \frac{1}{25} \qquad 27^{2/3} = (\sqrt[3]{27})^2 = 3^2 = 9$$

The behavior of exponents is governed by five fundamental laws.

**LAWS OF EXPONENTS**

Let $r$ and $s$ be any real numbers. Then

1. $x^r x^s = x^{r+s}$
2. $\dfrac{x^r}{x^s} = x^{r-s}$
3. $(x^r)^s = x^{rs}$
4. $(xy)^r = x^r y^r$
5. $\left(\dfrac{x}{y}\right)^r = \dfrac{x^r}{y^r}$

**EXAMPLE F**

Simplify

(a) $(27x^6)^{1/3}$

(b) $\dfrac{\sqrt[3]{x}\,x^2}{x^{3/4}}$

(c) $\dfrac{x^{2/3}(8x)^{-2/3}}{x^{-2}}$

(d) $(x^{1/2} + x^{3/2})^2$

**Solution**

(a) $(27x^6)^{1/3} = (27)^{1/3}(x^6)^{1/3} = 3x^2$

(b) $\dfrac{\sqrt[3]{x}\,x^2}{x^{3/4}} = \dfrac{x^{1/3}x^2}{x^{3/4}} = x^{1/3+2-3/4} = x^{4/12+24/12-9/12} = x^{19/12}$

(c) $\dfrac{x^{2/3}(8x)^{-2/3}}{x^{-2}} = \dfrac{x^{2/3}8^{-2/3}x^{-2/3}}{x^{-2}} = \dfrac{x^0 8^{-2/3}}{x^{-2}}$

$= \dfrac{x^2}{8^{2/3}} = \dfrac{x^2}{(\sqrt[3]{8})^2} = \dfrac{x^2}{4}$

(d) $(x^{1/2} + x^{3/2})^2 = (x^{1/2})^2 + 2x^{1/2}x^{3/2} + (x^{3/2})^2$

$= x + 2x^2 + x^3$ ■

## Problem Set 2.2

*In Problems 1–10, find $f'(x)$ by calculating $\lim_{h\to 0} [f(x+h) - f(x)]/h$.*

1. $f(x) = x^2 + 2x$
2. $f(x) = 3x^2 + 1$
3. $f(x) = x^3$
4. $f(x) = x^2 - x + 5$
5. $f(x) = 4/x$
6. $f(x) = 1/(x+1)$
7. $f(x) = x/(x+1)$
8. $f(x) = (2x+1)/x$
9. $f(x) = \sqrt{x+2}$
10. $f(x) = \sqrt{x-1}$

*In each of Problems 11–14, first write $f(x)$ as a single power of $x$ and then find $f'(x)$ using the Power Rule (see Example D).*

11. $f(x) = x\sqrt[3]{x}$
12. $x^2/\sqrt{x}$
13. $f(x) = x^3/x^{1.1}$
14. $f(x) = x^3x^{1.5}x^{-2}$

*Use the Power Rule and the Constant Multiple Rule to find $f'(x)$ in Problems 15–22 (see Example E).*

15. $f(x) = 6\sqrt[3]{x}$
16. $f(x) = 6\sqrt{x}$
17. $f(x) = -2x^{5/4}$
18. $f(x) = \sqrt{2}x^{\sqrt{2}}$
19. $f(x) = 3x^{1.09}$
20. $f(x) = (2/\pi)x^{2\pi}$
21. $f(x) = -11x^{\sqrt{3}}$
22. $f(x) = 4x^{2.11}$
23. Simplify each of the following expressions.
    (a) $(16x^8)^{1/4}$ (b) $x^2\sqrt{x^3}$
    (c) $x^2/\sqrt{x^3}$ (d) $\sqrt[4]{x}\,x^3/x^{3/2}$
    (e) $x^{3/4}(16x)^{-3/4}/x^{-3}$ (f) $(x^{3/2} - x^2)^2$
24. Simplify each of the following expressions.
    (a) $(x^4/9)^{3/2}$ (b) $x^2\sqrt[4]{x}$
    (c) $x^3/\sqrt{x}$ (d) $\sqrt{x}\,x^2/x^{2/3}$
    (e) $x^{3/2}(9x)^{-3/2}/x^{-2}$ (f) $(x^{1/3} + x^{5/3})^2$
25. First simplify; then find $dy/dx$.
    (a) $y = (9x^8)^{1/2}$ (b) $y = x^3\sqrt{4x^3}$
    (c) $y = 2x^4/\sqrt{x^5}$ (d) $y = \sqrt[4]{x}\,x^5x^{-5/4}$
26. First simplify; then find $dy/dx$.
    (a) $y = (x^6/16)^{3/2}$ (b) $y = x^5\sqrt[4]{16x}$
    (c) $y = x^3/2\sqrt{x}$ (d) $y = \sqrt{9x}\,x^2/x^{3/2}$

---

27. Find $dy/dx$ for each of the following using the Constant Multiple Rule and the Power Rule.
    (a) $y = 14$ (b) $y = 4x^7$
    (c) $y = 10x^{3.1}$ (d) $y = 12x^{7/6}$
    (e) $y = -3x^{2\pi+1}$ (f) $y = \pi^{3.1}$
28. Find $f'(8)$ for each of the following functions.
    (a) $f(x) = 2x^{4/3}$ (b) $f(x) = 3/\sqrt[3]{x}$
    (c) $f(x) = 2^{4/3}$ (d) $f(x) = x^{2/3}x^{5/3}x^{-3}$
29. Use the definition of the derivative $f'(x) = \lim_{h\to 0} [f(x+h) - f(x)]/h$ to find the derivative of $f(x) = x^{-2} = 1/x^2$.
30. Use the definition of the derivative to find $f'(x)$ if $f(x) = x^{1/3} = \sqrt[3]{x}$. *Hint:* Study Example C and recall that $(a - b)(a^2 + ab + b^2) = a^3 - b^3$.
31. The volume of a sphere is $\pi/6$ times the cube of the diameter. Find the rate of change in volume with respect to the diameter when the diameter is 10 centimeters.
32. Suppose that the size of a certain bacteria culture at time $t$ (in minutes) is given by $N(t) = 3t^2\sqrt{t}$ (in milligrams). Find the rate of growth at $t = 1$ and at $t = 9$.
33. According to the Inverse Square Law for electric charges, the force of repulsion between charges of sizes $q_1$ and $q_2$ statcoulombs is $F = q_1q_2/r^2$ dynes when two charges are $r$ centimeters apart. If $q_1 = 5$ and $q_2 = 80$, find the rate of change in force with respect to distance where $r = 10$.
34. It is estimated that a normal person can memorize $N = 30\sqrt{t}$ lines of a certain poem in $t$ hours of concentrated study $(0 \le t \le 9)$. What is the rate of learning when $t = 4$?
35. At what point on the graph of $y = 3x^{4/3}$ is the slope equal to 12?
36. Find the equation of the tangent line to $y = 4x^{3/2}$ at the point where $x = \frac{1}{4}$.

[C] 37. For common functions, $[f(4.001) - f(4)]/0.001$ and $f'(4)$ should be nearly equal. Why?
Calculate both of these quantities for
(a) $f(x) = \sqrt{x}$ (b) $f(x) = 2/\sqrt{x}$

[C] 38. While common sense suggests that the electoral college method of electing presidents gives more power per voter to small states, an analysis of the swing power of the large states has led some political scientists to the opposite conclusion. They propose the "$\frac{3}{2}$ rule" according to which the effective voting strength of a state in a presidential election is proportional to its population to the $\frac{3}{2}$ power. Based on this model, the amount $A$ a candidate spends in a state should vary with the population $P$ according to the formula

$$A = kP^{3/2} \qquad (k \text{ a constant})$$

(a) Alfonso Hornblower plans to spend \$1 million in his home state with a population of 4.5 million. Use this information to determine $k$.
(b) How much should Hornblower plan to spend in New York (17.5 million people)?
(c) Find the rate of change in spending with respect to population when $P = 4.5$ million.

## 2.3 RULES FOR DIFFERENTIATION

Our title introduces a new word—differentiation. The process of finding a derivative is called **differentiation.** If a function has a derivative, it is said to be **differentiable.** And the part of the calculus that deals with derivatives is called **differential calculus.**

We have already given two rules for differentiation—the Power Rule and the Constant Multiple Rule. This section develops several other rules that will greatly simplify our work.

## The Sum Rule

How can we find the derivative of $6x^4 + 12x^{1/2} - 2x^{-3}$? The Sum Rule will help. Some would say it should be called the Sum and Difference Rule, but that name is too long.

**SUM RULE**

Let $u = u(x)$ and $v = v(x)$ be differentiable functions and let $F(x) = u(x) \pm v(x)$. Then $F'(x) = u'(x) \pm v'(x)$; that is,

$$\frac{d(u \pm y)}{dx} = \frac{du}{dx} \pm \frac{dv}{dx}$$

*Proof* We give the proof for a sum; the case for a difference is completely similar.

$$\begin{aligned}
F'(x) &= \lim_{h\to 0} \frac{F(x+h) - F(x)}{h} \\
&= \lim_{h\to 0} \frac{u(x+h) + v(x+h) - [u(x) + v(x)]}{h} \\
&= \lim_{h\to 0} \left[\frac{u(x+h) - u(x)}{h} + \frac{v(x+h) - v(x)}{h}\right] \\
&= \lim_{h\to 0} \frac{u(x+h) - u(x)}{h} + \lim_{h\to 0} \frac{v(x+h) - v(x)}{h} \\
&= u'(x) + v'(x)
\end{aligned}$$

The key step was the next to last one; it is a consequence of part 4 of the Main Limit Theorem. ■

**EXAMPLE A**

If $F(x) = 6x^4 + 12x^{1/2} - 2x^{-3}$, find $F'(x)$.

**Solution**

$$F'(x) = \frac{d(6x^4)}{dx} + \frac{d(12x^{1/2})}{dx} - \frac{d(2x^{-3})}{dx} \qquad \text{(Sum Rule)}$$

$$= 6\frac{d(x^4)}{dx} + 12\frac{d(x^{1/2})}{dx} - 2\frac{d(x^{-3})}{dx} \qquad \text{(Constant Multiple Rule)}$$

$$= 6(4x^3) + 12(\tfrac{1}{2}x^{-1/2}) - 2(-3x^{-4}) \qquad \text{(Power Rule)}$$

$$= 24x^3 + 6x^{-1/2} + 6x^{-4}$$

■

If you see how to write the answer in a problem like Example A without any intermediate steps—fine. We don't mean to create extra work for you.

**EXAMPLE B**

Find an equation of the tangent line to the graph of $y = 2\sqrt{x} + 5$ at $x = 4$.

**Solution** Let $f(x) = 2\sqrt{x} + 5 = 2x^{1/2} + 5$. Then

$$f'(x) = 2(\tfrac{1}{2})x^{-1/2} + 0 = 1/\sqrt{x}$$

and so

$$f'(4) = 1/\sqrt{4} = \tfrac{1}{2}$$

At $x = 4$, $y = 2\sqrt{4} + 5 = 9$. Thus the desired equation (in point-slope form) is

$$y - 9 = \tfrac{1}{2}(x - 4)$$

■

**CAUTION**

A common error in Example B is to write

$$y - 9 = \frac{1}{\sqrt{x}}(x - 4)$$

This is *not* the equation of a line. The slope at $x = 4$ is

$$m = f'(4) = \tfrac{1}{2}$$

The equation of the tangent line at $x = 4$ is

$$y - 9 = \tfrac{1}{2}(x - 4)$$

One of the advantages of the $f'(x)$ notation appears in the solution above. We can write $f'(4)$ for the value of the derivative at $x = 4$. In $dy/dx$ notation, we would have to write something like

$$\left.\frac{dy}{dx}\right|_{x=4} \quad \text{or} \quad \left(\frac{dy}{dx}\right)_{x=4}$$

**CAUTION**

$$\frac{d(uv)}{dx} \neq \frac{du}{dx}\frac{dv}{dx}$$

If $u = x^3$ and $v = x^4$,

$$\frac{du}{dx} = 3x^2 \quad \text{and} \quad \frac{dv}{dx} = 4x^3$$

but

$$\frac{d(uv)}{dx} = \frac{d(x^7)}{dx} = 7x^6$$

$$7x^6 \neq 3x^2 \cdot 4x^3$$

## The Product Rule

Now we are in for a mild surprise. The derivative of a product is *not* equal to the product of the derivatives (see CAUTION). How then are we to find the derivative of something like

$$F(x) = (x^2 + 4)(x^3 + 3x + 2)$$

With the tools so far developed, we would have to first multiply the two factors together and then differentiate. Thus,

$$F(x) = x^5 + 7x^3 + 2x^2 + 12x + 8$$

and

$$F'(x) = 5x^4 + 21x^2 + 4x + 12$$

At least we could do it, but how would you like to try

$$F(x) = (3x^{4/3} + 9x^{10/9} + x^3)(x^{-3} + 7x^{-1} + 5x)$$

Discouraging isn't it? The Product Rule bypasses the difficulty.

**PRODUCT RULE**

Let $u = u(x)$ and $v = v(x)$ be differentiable and let $F(x) = u(x)v(x)$. Then $F'(x) = u(x)v'(x) + v(x)u'(x)$; that is,

$$\frac{d(uv)}{dx} = u\frac{dv}{dx} + v\frac{du}{dx}$$

Learn this rule in words. *The derivative of a product is the first times the derivative of the second plus the second times the derivative of the first.*

*Proof*

$$F'(x) = \lim_{h\to 0} \frac{F(x+h) - F(x)}{h}$$

$$= \lim_{h\to 0} \frac{u(x+h)v(x+h) - u(x)v(x)}{h}$$

$$= \lim_{h\to 0} \frac{u(x+h)v(x+h) - u(x+h)v(x) + u(x+h)v(x) - u(x)v(x)}{h}$$

$$= \lim_{h\to 0} \left[ u(x+h)\frac{v(x+h) - v(x)}{h} + v(x)\frac{u(x+h) - u(x)}{h} \right]$$

$$= \lim_{h\to 0} u(x+h) \lim_{h\to 0} \frac{v(x+h) - v(x)}{h} + \lim_{h\to 0} v(x) \lim_{h\to 0} \frac{u(x+h) - u(x)}{h}$$

$$= u(x)v'(x) + v(x)u'(x)$$ ■

**SHARP CURVE**

Subtracting $u(x+h)v(x)$ and then adding it right back in is a legal maneuver. It is this simple trick that makes the proof of the Product Rule work.

**EXAMPLE C**

Find the derivative of

$$F(x) = (x^2 + 4)(x^3 + 3x + 2)$$

**Solution** We use the Product Rule and then simplify.

$$\begin{aligned} F'(x) &= (x^2 + 4)\frac{d}{dx}(x^3 + 3x + 2) + (x^3 + 3x + 2)\frac{d}{dx}(x^2 + 4) \\ &= (x^2 + 4)(3x^2 + 3) + (x^3 + 3x + 2)(2x) \\ &= 3x^4 + 15x^2 + 12 + 2x^4 + 6x^2 + 4x \\ &= 5x^4 + 21x^2 + 4x + 12 \end{aligned}$$

Of course, our answer agrees with what we got earlier. ■

### EXAMPLE D

If $F(x) = (x^2 - 2x + 3)(x^3 + 3\sqrt{x} - 10)$, find $F'(1)$.

**Solution**

$$F'(x) = (x^2 - 2x + 3)(3x^2 + 1.5x^{-1/2}) + (x^3 + 3\sqrt{x} - 10)(2x - 2)$$

There is no need to simplify this; just substitute $x = 1$.

$$\begin{aligned} F'(1) &= (1 - 2 + 3)(3 + 1.5) + (1 + 3 - 10)(2 - 2) \\ &= (2)(4.5) + (-6)(0) = 9 \end{aligned}$$

■

## The Quotient Rule

Suppose we need the derivative of

$$F(x) = \frac{x^2 + 3x - 10}{x^2 + 9}$$

Then the next rule is very helpful.

**QUOTIENT RULE**

Let $u = u(x)$ and $v = v(x)$ be differentiable and let $F(x) = u(x)/v(x)$. Then

$$F'(x) = \frac{v(x)u'(x) - u(x)v'(x)}{[v(x)]^2}$$

or, equivalently,

$$\frac{d(u/v)}{dx} = \frac{v\dfrac{du}{dx} - u\dfrac{dv}{dx}}{v^2}$$

Learn this rule in words too. *The derivative of a quotient is equal to the denominator times the derivative of the numerator minus the numerator times the derivative of the denominator, all divided by the square of the denominator.*

> **CAUTION**
> Be sure to notice in Example E that the minus sign applies to the whole second term
> $$2x^3 + 6x^2 - 20x$$

### EXAMPLE E

Find $F'(x)$ if $F(x) = (x^2 + 3x - 10)/(x^2 + 9)$.

**Solution**

$$F'(x) = \frac{(x^2 + 9)(2x + 3) - (x^2 + 3x - 10)(2x)}{(x^2 + 9)^2}$$

$$= \frac{2x^3 + 3x^2 + 18x + 27 - (2x^3 + 6x^2 - 20x)}{(x^2 + 9)^2}$$

$$= \frac{-3x^2 + 38x + 27}{(x^2 + 9)^2}$$ ■

### EXAMPLE F

If $F(x) = \dfrac{x^2 + 4}{x^{1/2}}$ find $F'(4)$ in two different ways.

**Solution**

Method 1. We use the Quotient Rule.

$$F'(x) = \frac{x^{1/2} \cdot 2x - (x^2 + 4) \cdot \frac{1}{2}x^{-1/2}}{x}$$

$$F'(4) = \frac{2 \cdot 8 - (16 + 4) \cdot \frac{1}{2} \cdot \frac{1}{2}}{4} = \frac{16 - 5}{4} = \frac{11}{4}$$

Method 2. We first simplify.

$$F(x) = \frac{x^2 + 4}{x^{1/2}} = \frac{x^2}{x^{1/2}} + \frac{4}{x^{1/2}} = x^{3/2} + 4x^{-1/2}$$

$$F'(x) = \tfrac{3}{2}x^{1/2} - 2x^{-3/2}$$

$$F'(4) = \frac{3}{2} \cdot 2 - 2 \cdot \frac{1}{4^{3/2}} = 3 - \frac{2}{8} = \frac{11}{4}$$ ■

### EXAMPLE G

Let $n$ be a positive integer. Show using the Quotient Rule that

$$\frac{d}{dx}(x^{-n}) = -nx^{-n-1}$$

**Solution**

$$\frac{d(x^{-n})}{dx} = \frac{d\left(\frac{1}{x^n}\right)}{dx} = \frac{x^n \cdot 0 - 1 \cdot nx^{n-1}}{x^{2n}}$$

$$= \frac{-nx^{n-1}}{x^{2n}} = -nx^{n-1-2n} = -nx^{-n-1}$$

This example provides more evidence for the Power Rule in Section 2.2. ■

## Proof of the Quotient Rule

We do the proof in two stages. First let $G(x) = 1/v(x)$. Then

$$G'(x) = \lim_{h \to 0} \frac{\frac{1}{v(x+h)} - \frac{1}{v(x)}}{h} = \lim_{h \to 0} \frac{\frac{v(x) - v(x+h)}{v(x+h)v(x)}}{h}$$

$$= \lim_{h \to 0} \left[ -\frac{v(x+h) - v(x)}{h} \cdot \frac{1}{v(x+h)v(x)} \right]$$

$$= -v'(x) \cdot \frac{1}{[v(x)]^2} = -\frac{v'(x)}{[v(x)]^2}$$

Next let $F(x) = \frac{1}{v(x)} \cdot u(x)$ and apply the Product Rule together with what we just learned.

$$F'(x) = \frac{1}{v(x)} \cdot u'(x) + u(x) \cdot \frac{-v'(x)}{[v(x)]^2}$$

$$= \frac{v(x)u'(x)}{[v(x)]^2} - \frac{u(x)v'(x)}{[v(x)]^2}$$

$$= \frac{v(x)u'(x) - u(x)v'(x)}{[v(x)]^2}$$

## Problem Set 2.3

*In Problems 1–8, find $f'(x)$ as in Example A. In many cases you should rewrite $f(x)$ before trying to find its derivative. For example, in Problem 7, write $f(x) = \sqrt{x}(x^3 - 2\sqrt{x} + 5) = \sqrt{x}x^3 - 2x + 5\sqrt{x} = x^{7/2} - 2x + 5x^{1/2}$.*

1. $f(x) = 2x^3 - 3x^{2/3} + 5x^{-2}$
2. $f(x) = -6x^{5/3} + 2x^{-3} + 2x^{3/2}$
3. $f(x) = 2x^3 + 5\sqrt{x} - 4/x^5$
4. $f(x) = 3x^4 - 3\sqrt[4]{x} - 5/x^3$
5. $f(x) = 3\sqrt[3]{x} - 4/\sqrt{x}$

6. $f(x) = x\sqrt[3]{x} - 2/\sqrt[3]{x}$
7. $f(x) = \sqrt{x}(x^3 - 2\sqrt{x} + 5)$
8. $f(x) = \sqrt[3]{x}(2x + 4 - 5/\sqrt[3]{x})$

*In Problems 9–12, find an equation for the tangent line to the given curve at the point with the indicated x-coordinate (see Example B).*

9. $y = 2x^3 - 4x^2 - 3/x;\ x = -1$
10. $y = 3x^2 - 4x + 2\sqrt{x};\ x = 4$
11. $y = 8\sqrt[4]{x} - 15 + 16/x;\ x = 16$
12. $y = 3x^{2/3} - 8/\sqrt[3]{x};\ x = 8$
13. Let $F(x) = (x^2 - 9)(x^2 + 4)$. Find $F'(x)$ in two ways.
    (a) Use the Product Rule and then simplify.
    (b) First multiply out and then differentiate.
14. Let $G(t) = (t + 3)(t^2 - 3t + 1)$. Find $G'(t)$ in two ways as in Problem 13.

*Use the Product Rule in Problems 15–18 to find $F'(x)$. Simplify your answer if you can.*

15. $F(x) = (x^2 + 5)(2x^3 - 3x + 9)$
16. $F(x) = (3x^2 + 5x + 1)(2x^2 - 3)$
17. $F(x) = (\sqrt{x} + 4)(x^{3/2} + 2\sqrt{x} + 2)$
18. $F(x) = (3 + 5x^{-1})(7 + 4x^2)$

*For the given function, find the indicated derivative (see Example D).*

19. $F(x) = (x^2 + x - 4)(x + \sqrt[3]{x});\ F'(1)$
20. $G(x) = (3x - 1)(2\sqrt[4]{x} + 3\sqrt{x});\ G'(1)$
21. $f(t) = (t - 16/t)(3\sqrt{t} - t\sqrt{t});\ f'(4)$
22. $g(t) = (\sqrt[3]{t} + 1)(t^2 - 8t);\ g'(8)$
23. Let $F(x) = (2x^2 + x - 3)/x$. Find $F'(x)$ in two ways.
    (a) Use the Quotient Rule and then simplify.
    (b) First divide and then differentiate.
24. Follow the directions in Problem 23 for $F(x) = (3x - 2)/\sqrt{x}$.

*In Problems 25–28, use the Quotient Rule to find $f'(x)$.*

25. $f(x) = \dfrac{2x^2 + 5}{2x - 1}$
26. $f(x) = \dfrac{4x^2}{x^3 - 2x + 1}$
27. $f(x) = \dfrac{4\sqrt{x}}{3 + \sqrt{x}}$
28. $f(x) = \dfrac{x^2 - 4x + 3}{x^2 + 2x}$

---

*In Problems 29–34, find $dy/dx$ by using the rules we have developed.*

29. $y = 2x + 3 - 4\sqrt{x} + 5/x^3$
30. $y = (3x^2 + x - 7)(5 - 4x^3)$
31. $y = (x^{-3} + 4x)(2 + 3x^{-1})$
32. $y = (3x - 8)(x^2 + 7)$
33. $y = 2x^2 + 3x - x/(x - 4)$
34. $y = (x^2 - 3x + 1)(x^2 - 4) + x^3 - 3x^2 + 4x - 9$
35. Find an equation for the tangent line to the curve $y = 2x^2 - 3 + \sqrt[3]{x}$ at the point where $x = 1$.
36. Let $H(x) = [G(x)]^2 = G(x) \cdot G(x)$. Use the Product Rule to show that $H'(x) = 2G(x) \cdot G'(x)$.
37. Use the result from Problem 36 to find $H'(x)$ for each of the following.
    (a) $H(x) = (3x + 1)^2$
    (b) $H(x) = (x^2 - x + 4)^2$
    (c) $H(x) = (4\sqrt{x} + 5)^2$
    (d) $H(x) = (4x + 3/x^2)^2$
38. If the density of algae in a water tank $t$ hours from now is expected to be $2\sqrt{t}/(\sqrt{t} + 10)$ algae per cubic centimeter, find the rate of change in density with respect to time 9 hours from now.
39. If an object moves along a line so that its position $s$ in centimeters is given by $s = (2t^2 - 3\sqrt{t})/(t + 1)$ after $t$ seconds, find its velocity after 4 seconds.
40. The number of units of oxygen per gallon of water in a lake $t$ days after a sewage plant began dumping untreated sewage was found to be $N = 100\{2 - [t/(t + 8)]\}$. Find the rate of change in units of oxygen with respect to time at (a) $t = 2$, (b) $t = 92$.
41. A manufacturer of riding mowers expects to sell each mower for \$720. The cost of making and selling $x$ mowers per week is

$$C(x) = 0.02x^2 + 500x + 3000$$

    (a) Find the profit $P(x)$ from producing and selling $x$ mowers per week.
    (b) Find the marginal profit and evaluate it when $x = 1000$.
    (c) For what values of $x$ is the marginal profit zero? What is the economic significance of this value of $x$?

C 42. Economists often talk about average cost, that is, cost per unit of output. If the total cost of making $x$ units of something is $C(x)$, then the average cost is $C(x)/x$. Suppose that a manufacturer of jackets finds that the total cost in dollars of making $x$ jackets per week is

$$C(x) = 2000 + 12x - 8\sqrt{x} + 0.05x^{3/2}$$

If this manufacturer plans to make 500 jackets per week, what is
(a) The total cost?
(b) The average cost?
(c) The marginal cost?
(d) The marginal average cost?

# 2.4 THE GENERAL POWER RULE

Imagine trying to find the derivative of

$$F(x) = (x^2 + 4x - 9)^{50}$$

We would first multiply 50 factors of $x^2 + 4x - 9$ together, thus obtaining a polynomial with 101 terms. Then we would differentiate the polynominal. If we were very accurate and exceedingly lucky, we would get the right answer after about 5 hours of work. There is a better way.

## The Derivative of $u^n$

**POINT OF INTEREST**
Inferring a general law from a few special cases is, of course, not a proof. Called *induction,* it is commonly used in science but should be sharply distinguished from *deduction,* the method of proof in mathematics.

Let $u = u(x)$ be a differentiable function. Our goal is to find a simple formula for $d(u^n)/dx$; our tool is the Product Rule. We begin with simple cases, looking for a pattern.

$$\frac{d(u^2)}{dx} = \frac{d(u \cdot u)}{dx} = uu' + uu' = 2uu'$$

$$\frac{d(u^3)}{dx} = \frac{d(u \cdot u^2)}{dx} = u \cdot 2uu' + u^2 \cdot u' = 3u^2u'$$

$$\frac{d(u^4)}{dx} = \frac{d(u \cdot u^3)}{dx} = u \cdot 3u^2u' + u^3 \cdot u' = 4u^3u'$$

The pattern seems clear and suggests that

$$\frac{d(u^n)}{dx} = nu^{n-1}\frac{du}{dx}$$

In fact, this formula is correct when $n$ is any real number, but our proof will have to wait until Section 5.3.

**GENERAL POWER RULE**
Let $F(x) = [u(x)]^r$, where $u = u(x)$ is a differentiable function and $r$ is a real number. Then

$$F'(x) = r[u(x)]^{r-1}u'(x)$$

Equivalently,

$$\frac{d(u^r)}{dx} = ru^{r-1}\frac{du}{dx}$$

Note that if $u(x) = x$, the General Power Rule becomes

$$\frac{d(x^r)}{dx} = rx^{r-1}\frac{dx}{dx} = rx^{r-1}(1) = rx^{r-1}$$

which is the ordinary power rule. The force of the General Power Rule is that we can easily find the derivative of a power of any function including the one in the introduction to this section.

### EXAMPLE A

Find the derivative of

$$F(x) = (x^2 + 4x - 9)^{50}$$

**Solution** Let $u = x^2 + 4x - 9$. Then

$$F'(x) = \frac{d(u^{50})}{dx} = 50u^{49}\frac{du}{dx} = 50(x^2 + 4x - 9)^{49}(2x + 4)$$ ■

### EXAMPLE B

Find $F'(x)$ if $F(x) = \sqrt{3x + 7}$.

**Solution** Let $u = 3x + 7$. Then

$$F'(x) = \frac{d(u^{1/2})}{dx} = \frac{1}{2}u^{-1/2}\frac{du}{dx} = \frac{1}{2\sqrt{u}}\frac{du}{dx}$$

$$= \frac{1}{2\sqrt{3x + 7}}(3) = \frac{3}{2\sqrt{3x + 7}}$$ ■

### EXAMPLE C

If $y = \sqrt[3]{(2x^3 + 9x)^4}$, find $dy/dx$.

**Solution** Write $y = (2x^3 + 9x)^{4/3}$. Then

$$\frac{dy}{dx} = \frac{4}{3}(2x^3 + 9x)^{1/3}(6x^2 + 9)$$

If you can make the mental substitution $u = 2x^3 + 9x$ (as we did) without actually writing it down, that is fine. ■

> **CAUTION**
> A very common error in Example C is to forget the factor
> $$6x^2 + 9$$

### EXAMPLE D

If $y = 2/(2x + 3)^4$, find $dy/dx$.

**Solution** We could use the Quotient Rule, but it is probably better to first write $y = 2(2x + 3)^{-4}$. Then from the Constant Multiple Rule followed by the General Power Rule, we obtain

$$\frac{dy}{dx} = 2 \cdot \frac{d(2x+3)^{-4}}{dx} = 2 \cdot (-4)(2x+3)^{-5}(2)$$

$$= -16(2x+3)^{-5} = \frac{-16}{(2x+3)^5}$$ ■

## Combining the Rules

We must be able to use the Sum Rule, Product Rule, Quotient Rule, and General Power Rule in all kinds of combinations. Often the key will be choosing the right rule to apply first.

### EXAMPLE E

Find the derivative of $y = \left(\frac{2x+3}{3x-7}\right)^4$.

**Solution** At least mentally, let $u = (2x+3)/(3x-7)$. Then apply the General Power Rule followed by the Quotient Rule.

$$\qquad \mathbf{4\,u^3} \qquad\qquad\qquad \mathbf{\frac{du}{dx}}$$

$$\frac{dy}{dx} = 4\left(\frac{2x+3}{3x-7}\right)^3 \cdot \frac{(3x-7)2 - (2x+3)3}{(3x-7)^2}$$

$$= 4\left(\frac{2x+3}{3x-7}\right)^3 \frac{-23}{(3x-7)^2}$$

$$= -92\frac{(2x+3)^3}{(3x-7)^5}$$ ■

### EXAMPLE F

Find the derivative of $y = 4x^3(5x^2+1)^{3/2}$.

**Solution** We begin with the Product Rule and then use the Power Rule.

$$\frac{dy}{dx} = 4x^3\frac{d}{dx}(5x^2+1)^{3/2} + (5x^2+1)^{3/2}\frac{d}{dx}(4x^3)$$

$$= 4x^3 \cdot \frac{3}{2}(5x^2+1)^{1/2}(10x) + (5x^2+1)^{3/2}(12x^2)$$

$$= 60x^4(5x^2+1)^{1/2} + 12x^2(5x^2+1)^{3/2}$$

It would not be wrong to leave the answer as shown. On the other hand, some old-fashioned algebra will improve the form of the result. Note that

**SHARP CURVE**
Here we have used

$$au + av = a(u + v)$$

but in a sophisticated form.

$12x^2(5x^2 + 1)^{1/2}$ is a common factor of both terms. Thus,

$$\frac{dy}{dx} = 12x^2(5x^2 + 1)^{1/2}(5x^2 + 5x^2 + 1)$$

$$= 12x^2(5x^2 + 1)^{1/2}(10x^2 + 1) \quad \blacksquare$$

### EXAMPLE G

If $y = \dfrac{(x^2 + 1)^4}{(x + 5)^3}$, find $\dfrac{dy}{dx}$.

**Solution** We begin with the Quotient Rule.

$$\frac{dy}{dx} = \frac{(x + 5)^3 \dfrac{d}{dx}(x^2 + 1)^4 - (x^2 + 1)^4 \dfrac{d}{dx}(x + 5)^3}{(x + 5)^6}$$

$$= \frac{(x + 5)^3 \cdot 4(x^2 + 1)^3 \cdot 2x - (x^2 + 1)^4 \cdot 3(x + 5)^2}{(x + 5)^6}$$

It is tempting to leave the answer in this form. However, we can greatly simplify it if we factor out $(x + 5)^2(x^2 + 1)^3$ from both terms of the numerator. We get

$$\frac{dy}{dx} = \frac{(x + 5)^2(x^2 + 1)^3[8x(x + 5) - 3(x^2 + 1)]}{(x + 5)^6}$$

$$= \frac{(x + 5)^2(x^2 + 1)^3(8x^2 + 40x - 3x^2 - 3)}{(x + 5)^6}$$

$$= \frac{(x^2 + 1)^3(5x^2 + 40x - 3)}{(x + 5)^4} \quad \blacksquare$$

## Applications

Don't forget that the derivative allows us to calculate slopes and rates of change.

### EXAMPLE H

Find the slope of the curve $y = x^2\sqrt{x^2 + 3x + 6}$ at the point where $x = 2$.

**Solution** Let $F(x) = x^2(x^2 + 3x + 6)^{1/2}$. Using first the Product Rule, then the General Power Rule, we obtain

$$F'(x) = x^2(\tfrac{1}{2})(x^2 + 3x + 6)^{-1/2}(2x + 3) + (x^2 + 3x + 6)^{1/2}(2x)$$

There is no need to simplify this before we evaluate at $x = 2$.

$$F'(2) = 4(\tfrac{1}{2})(16)^{-1/2}(7) + (16)^{1/2}(4)$$
$$= 4(\tfrac{1}{2})(\tfrac{1}{4})(7) + (4)(4) = \tfrac{39}{2}$$ ■

### EXAMPLE I

Comtrol makes fancy programmable calculators at a cost of \$150.00 each and is presently making 100 units per day. The president knows that if she increases production, she will have to lower the price. A market study she commissioned estimated that Comtrol can sell $x$ units per day at a price of $p = 400/\sqrt{1 + 0.02x}$ dollars. Determine the marginal profit, evaluate it at $x = 100$, and interpret the result.

**Solution** Let $C(x)$, $R(x)$, and $P(x)$ denote the daily cost, daily revenue, and daily profit in producing $x$ units. Then

$$C(x) = 150x$$

$$R(x) = px = \frac{400x}{\sqrt{1 + 0.02x}}$$

$$P(x) = \frac{400x}{\sqrt{1 + 0.02x}} - 150x$$

Thus, the marginal profit $P'(x)$ is

$$P'(x) = \frac{(1 + 0.02x)^{1/2}(400) - (400x)(\frac{1}{2})(1 + 0.02x)^{-1/2}(0.02)}{1 + 0.02x} - 150$$

and

$$P'(100) = \frac{(3)^{1/2}(400) - (400)(3)^{-1/2}}{3} - 150$$

$$\approx 153.96 - 150 = \$3.96$$

This means that increasing production by one unit will only increase the profit by about \$3.96.

Note that at $x = 100$, Comtrol will receive $p = 400/\sqrt{3} \approx \$230.94$ for each unit at a cost of only \$150.00 each, which means a profit of \$80.94 each. How can you reconcile this figure with the \$3.96 obtained above? ■

## Problem Set 2.4

*In Problems 1–16, find $f'(x)$.*

1. $f(x) = (3x - 4)^{12}$
2. $f(x) = (7x - 8)^9$
3. $f(x) = (3x^2 + x + 1)^8$
4. $f(x) = (2x^2 + 5x - 6)^{14}$
5. $f(x) = (x + 2/x^2)^5$
6. $f(x) = (x^{2/3} - 2x)^6$
7. $f(x) = (x^2 - 2\sqrt{x})^{10}$
8. $f(x) = (2\sqrt{x} - 5/x)^4$

9. $f(x) = \sqrt{2 + 3x - x^2}$
10. $f(x) = \sqrt[3]{9x - 1}$
11. $f(x) = (2x^3 - 7x + 6)^{2/3}$
12. $f(x) = (x^4 - 2x^3 + 2)^{3/4}$
13. $f(x) = \sqrt[4]{(2x^3 + x)^3}$
14. $f(x) = (\sqrt{3x} - x^5)^5$
15. $f(x) = 4/(3x - 1)^5$
16. $f(x) = 2/\sqrt{2x + 5}$
17. $f(x) = -2/\sqrt[3]{(2x - 1)^2}$
18. $f(x) = 4/(\sqrt{2x} + 3)^3$
19. Find the slope of the curve $y = (t^2 + 3t - 1)^{-5/2}$ at $t = 2$.
20. Find an equation for the tangent line to the curve $y = (t^3 + t + 1)^{-3/4}$ at the point (0, 1).

*In Problems 21–32, find $dy/dx$. You will need to use various combinations of the rules we have learned (see Examples E, F, and G).*

21. $y = \left(\frac{2x - 1}{3x + 2}\right)^3$ 22. $y = \left(\frac{4x + 1}{3x - 2}\right)^4$
23. $y = \left(\frac{x^2 + 4}{2x - 5}\right)^5$ 24. $y = \left(\frac{3x - 1}{x^2 + x - 4}\right)^3$
25. $y = (8x + 5)(2x - 1)^5$ 26. $y = (3x - 1)^8(4x + 3)$
27. $y = 3x^2(2x^2 + 3)^{5/3}$ 28. $y = 5x^3(x^2 + 4)^{5/2}$
29. $y = (x^2 + 1)^3(x^2 + 4)^2$
30. $y = (2x^3 + 1)^4(x^3 + x)^3$
31. $y = \frac{x^2}{(3x - 1)^4}$ 32. $y = \frac{(4x + 1)^4}{(5 + 2x)^6}$
33. Calculate $F'(1)$ if $F(x) = (2x - 1)^8(6\sqrt[3]{x} + 1)$
34. Calculate $g'(0)$ if $g(x) = (8x + 4)^{3/2}(2x - 1)^5$
35. Find an equation for the tangent line to the curve $y = (\sqrt{x} - 2)^4/(2x + 4)$ at $x = 1$.
36. Find an equation for the tangent line to the curve $y = (x^2 + 3)(x^2 + 2x - 14)^{-2/5}$ at $x = 3$.

[C] 37. An appliance manufacturer makes small refrigerators at a cost of \$120 each. A study shows that $x$ refrigerators can be sold per day if the price $p$ is set at $p = 600/\sqrt{1 + 0.04x}$ dollars.
(a) Determine the marginal daily profit when $x = 100$.
(b) Interpret your answer to part (a).
(c) Calculate the average profit per unit when $x = 100$.

[C] 38. An object moves along a line so that its position $s$ in feet at time $t$ seconds is given by

$$s = 16\sqrt{t^4 - 0.08t^2 + 9}$$

(a) Determine the position of the object at $t = 10$.
(b) Find the average velocity on the interval $t = 10$ to $t = 10.5$.
(c) Find the velocity at $t = 10$.

---

*In problems 39–42, find $F'(4)$.*

39. $F(x) = (4x - 10 - 8/x)^{3/2}$
40. $F(x) = \sqrt{x\sqrt{x} + 1} + \sqrt{x}$
41. $F(x) = [(\sqrt{x} + 3)(2x - 5)]^2$
42. $F(x) = (2x - 7)^5/\sqrt{2x + 1}$
43. At what value of $x$ does the curve $y = (x^4 - 32x + 57)^{1/2}$ have a horizontal tangent?
44. Find an equation for the tangent line to the curve $y = (2x^3 - 4x^2 + 5x - 8)^5$ at the point having $x = 2$.
45. If \$5000 is invested today at an interest rate of $r$ compounded twice a year, it will grow to an amount $A$ at the end of 4 years, where

$$A = 5000\left(1 + \frac{r}{2}\right)^8$$

(a) Find $dA/dr$, the rate at which $A$ changes with respect to $r$.
[C] (b) Calculate $A$ and $dA/dr$ when $r = 0.12$.
46. The population $N$ of a certain city $t$ years from now is projected to follow the formula

$$N = 24 - \frac{4}{t + 3} + \frac{36}{(t + 3)^2}$$

where $N$ is measured in thousands.
(a) Calculate the population 7 years from now.
(b) Will the population be growing or declining at that time?
47. Based on a study of potato yields, a biologist has obtained the formula

$$y = \frac{1{,}000{,}000}{8.5r + 40{,}000}$$

where $y$ is the yield per plant in kilograms and $r$ is the density (number of plants per hectare).
(a) Find the formula for $Y$, the yield per hectare.
(b) Show that $dy/dr$ is always negative but that $dY/dr$ is always positive.
(c) Interpret the results from part (b). Do they make sense?
48. An object $P$ is placed directly between a strong light $S_1$ and a weak light $S_2$, which are 20 feet apart (see Figure 7). When $P$ is $x$ feet from $S_1$, the intensity of

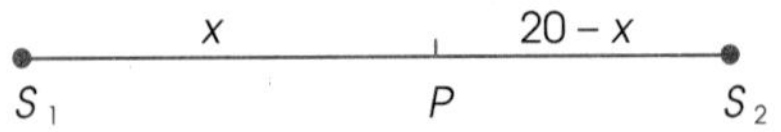

x 20 − x
$S_1$ P $S_2$

**FIGURE 7**

illumination at $P$ is given by

$$I = \frac{4000}{x^2} + \frac{1000}{(20 - x)^2}$$

(a) Calculate $I$ when $x = 10$ and $x = 15$.
(b) Calculate $dI/dx$ when $x = 10$ and $x = 15$.
(c) If $P$ is moved a bit toward $S_2$ from $x = 15$, will $I$ increase or decrease?

## 2.5 HIGHER-ORDER DERIVATIVES

No object from everyday life is more familiar to Americans than the automobile. With this in mind, we use a car to introduce the next idea. Let us suppose that Sylvia drives from her home to the office along a straight road—a distance of exactly 6 miles. On one particular day, a device was attached to her car which recorded the distance $s$ from her home at each time $t$. The results, shown as a graph, appear in Figure 8.

Let's try to interpret this graph. One obvious fact is that it took Sylvia 13 minutes to get to work. Of more interest are the linear segments between $t = 2$ and $t = 8$ and again between $t = 9$ and $t = 11$. On these intervals, the slope has a constant value of $\frac{1}{2}$, which means that the car was going at a constant speed of $\frac{1}{2}$ mile per minute (that is, 30 miles per hour).

Of still more interest are the curved parts of the graph. On the interval $0 \leq t \leq 2$, the slope (and hence the speed) was increasing. We say the car was accelerating, due, of course, to the fact that Sylvia was pressing down on the accelerator until she reached the desired speed of 30 miles per hour. At $t = 8$, she again began to accelerate, perhaps to pass a car, and then slowed down (decelerated) to 30 miles per hour. Finally at $t = 11$ she started slowing down to reach the desired speed of zero at her destination.

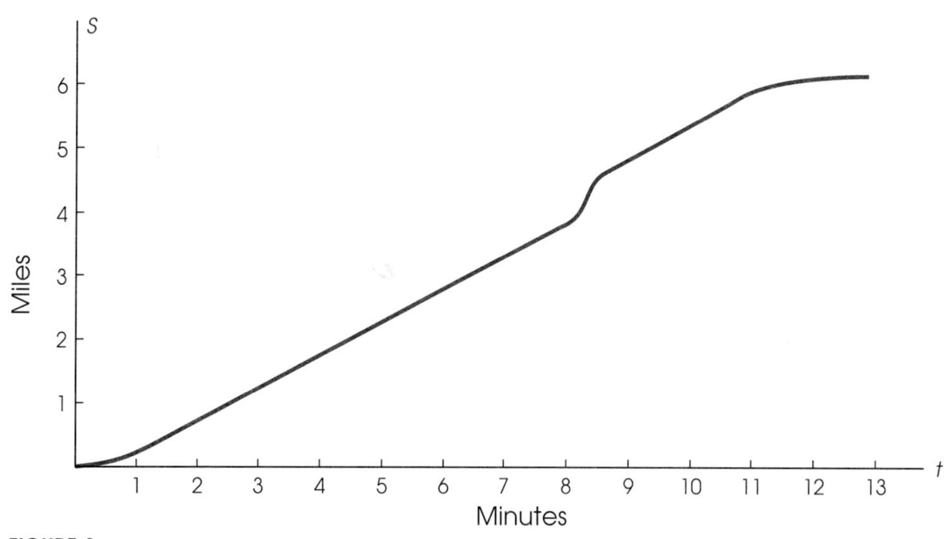


FIGURE 8

What then is acceleration? It is a measure of how speed is changing. More specifically, **acceleration** $a$ is the rate of change in velocity $v$ with respect to time $t$, that is,

$$a = \frac{dv}{dt} = \frac{d}{dt}\left(\frac{ds}{dt}\right)$$

When $a$ is positive, velocity is increasing; when $a$ is negative, velocity is decreasing. And the larger $a$ is in magnitude, the more rapidly velocity is changing.

## Higher Derivatives

Because time will play such an important role in this section, we use $t$ rather than $x$ as the independent variable. Suppose that $f(t) = t^5 + 4t^3 + 1$, so that $f'(t) = 5t^4 + 12t^2$. Notice that $f'$ is itself a function. Its derivative, called the **second derivative of $f$** and denoted by $f''$, is given by $f''(t) = 20t^3 + 24t$. Now we see that the process can be repeated again and again. Here are the results.

$$\begin{aligned}
f(t) &= t^5 + 4t^3 + 1 \\
f'(t) &= 5t^4 + 12t^2 \\
f''(t) &= 20t^3 + 24t \\
f'''(t) &= 60t^2 + 24 \\
f^{(4)}(t) &= 120t \\
f^{(5)}(t) &= 120 \\
f^{(6)}(t) &= 0 \\
f^{(7)}(t) &= 0
\end{aligned}$$

Pay attention to the notation. More than three primes becomes clumsy, so we use a numeral enclosed in parentheses to indicate derivatives of higher order than 3.

Recall that if $y = f(t)$, we have two alternative symbols for the derivative, namely, $f'(t)$ and $dy/dt$. What corresponds to $f''(t)$? We could write

$$f''(t) = \frac{d}{dt}\left(\frac{dy}{dt}\right)$$

and we sometimes do. However, Gottfried Leibniz, one of the founders of the calculus, decreed that we might also write $d^2y/dt^2$. We have the following correspondence.

**POINT OF INTEREST**

Gottfried Wilhelm Leibniz (1646–1716) was a universal genius—an expert in law, religion, politics, history, and philosophy as well as mathematics. He gave us the notation $dy/dx$ for the derivative and $\int$ for the integral.

| Name | Prime Notation | Leibniz Notation |
|---|---|---|
| (First) derivative | $f'(t)$ | $\frac{dy}{dt}$ |
| Second derivative | $f''(t)$ | $\frac{d^2y}{dt^2}$ |
| Third derivative | $f'''(t)$ | $\frac{d^3y}{dt^3}$ |
| Fourth derivative | $f^{(4)}(t)$ | $\frac{d^4y}{dt^4}$ |
| Fifth derivative | $f^{(5)}(t)$ | $\frac{d^5y}{dt^5}$ |
| $n$th derivative | $f^{(n)}(t)$ | $\frac{d^ny}{dt^n}$ |

### EXAMPLE A

If $y = 3t^2 - 2t^{1/2}$, find $d^3y/dt^3$.

**Solution**

$$\frac{dy}{dt} = 6t - t^{-1/2}$$

$$\frac{d^2y}{dt^2} = 6 + \frac{1}{2}t^{-3/2}$$

$$\frac{d^3y}{dt^3} = -\frac{3}{4}t^{-5/2}$$

■

### EXAMPLE B

If $f(t) = 1/\sqrt{2t + 3}$, find $f'''(3)$.

**Solution**

$$f(t) = (2t + 3)^{-1/2}$$

$$f'(t) = -\tfrac{1}{2}(2t + 3)^{-3/2}(2) = -(2t + 3)^{-3/2}$$

$$f''(t) = \tfrac{3}{2}(2t + 3)^{-5/2}(2) = 3(2t + 3)^{-5/2}$$

$$f'''(t) = 3(-\tfrac{5}{2})(2t + 3)^{-7/2}(2) = -15(2t + 3)^{-7/2}$$

$$f'''(3) = -15(9)^{-7/2} = -15/(\sqrt{9})^7 = -\tfrac{5}{729}$$

■

### EXAMPLE C

If $y = (x^3 + 3)^4$, find $d^2y/dx^2$.

**Solution**

$$\frac{dy}{dx} = 4(x^3 + 3)^3 \cdot 3x^2 = 12x^2 \cdot (x^3 + 3)^3$$

$$\begin{aligned}\frac{d^2y}{dx^2} &= 12x^2 \cdot 3(x^3 + 3)^2 \cdot 3x^2 + (x^3 + 3)^3 \cdot 24x \\ &= 12x(x^3 + 3)^2[9x^3 + 2(x^3 + 3)] \\ &= 12x(x^3 + 3)^2(11x^3 + 6)\end{aligned}$$

■

## Applications of the First and Second Derivatives

Differential calculus was developed to describe and predict motion. The first derivative measures velocity (rate of change in distance with respect to time); the second derivative measures acceleration (rate of change in velocity with respect to time). It is appropriate that we begin with a motion problem—the problem of a falling body.

If an object is thrown straight up from an initial height $s_0$ feet with an initial velocity $v_0$ feet per second, then its height $s$ above the ground in feet after $t$ seconds is

$$s = -16t^2 + v_0t + s_0$$

This is what the physicists tell us, though they are quick to say that the formula is accurate only near the surface of the earth (not out in space) and for objects that are dense enough so air resistance is minimal (no feathers please).

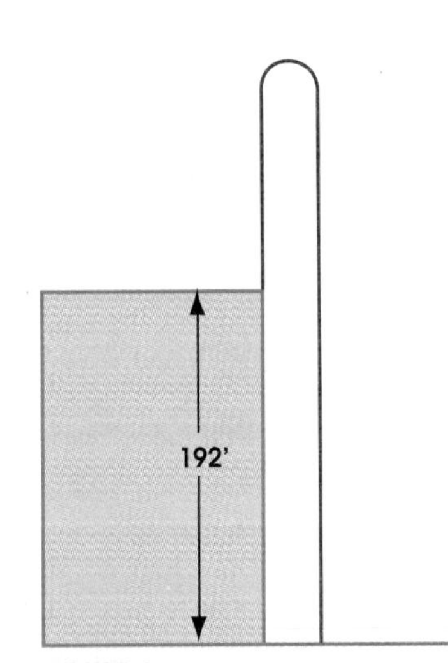


FIGURE 9

### EXAMPLE D

Suppose that a baseball was thrown upward from the roof of a building 192 feet high with an initial velocity of 64 feet per second (Figure 9).

(a) What was its velocity at $t = 1$? At $t = 5$?
(b) When did it reach maximum height?
(c) What was the maximum height?
(d) When did the ball hit the ground?
(e) With what velocity did it hit the ground?
(f) What was the ball's acceleration at $t = 1$? At $t = 5$?

**Solution** Here $s_0 = 192$ and $v_0 = 64$, so

$$s = -16t^2 + 64t + 192$$

$$v = \frac{ds}{dt} = -32t + 64$$

$$a = \frac{dv}{dt} = -32$$

(a) At $t = 1$, $v = -32 \cdot 1 + 64 = 32$ feet per second.
At $t = 5$, $v = -32 \cdot 5 + 64 = -96$ feet per second.
The negative sign means that the ball is falling at $t = 5$.

(b) The ball reached maximum height when the velocity was zero, that is, when $-32t + 64 = 0$. This occurred when $t = 2$.

(c) At $t = 2$,

$$s = -16(2)^2 + 64(2) + 192 = 256 \text{ feet}$$

(d) The ball hit the ground when $s = 0$.

$$-16t^2 + 64t + 192 = 0$$
$$t^2 - 4t - 12 = 0$$
$$(t - 6)(t + 2) = 0$$
$$t = 6, -2$$

Clearly, the solution $t = 6$ is the one we want.

(e) At $t = 6$, $v = -32(6) + 64 = -128$ feet per second.

(f) The acceleration is constant: $-32$ feet per second per second. This means that the velocity decreases by 32 feet per second every second. ■

To introduce our next example we paraphrase an article from the business page of a large metropolitan newspaper (*St. Paul Dispatch,* Aug. 27, 1984).

> A worldwide debt explosion is threatening the United States and other wealthy countries . . . . In the United States the ratio of government debt to national income remained unchanged at around 28 percent up to 1981 but then it began to increase sharply, reaching almost 36 percent in 1983 . . . . The IMF released a table showing that the speed of debt increase was greater in the United States than in Japan . . . .

## EXAMPLE E

Interpret the facts in the article above in terms of derivatives. Sketch a graph reflecting these facts.

**Solution** Let $R$ be the ratio of government debt to national income in the United States at time $t$.

(a) $R = 0.28$ $\quad (t < 1981)$
(b) $R = 0.36$ $\quad (t = 1983)$
(c) $\dfrac{dR}{dt} = 0$ $\quad (t < 1981)$
(d) $\dfrac{dR}{dt} > 0$ $\quad (1981 \le t \le 1983)$
(e) $\dfrac{d^2R}{dt^2} > 0$ $\quad (1981 \le t \le 1983)$
(f) $\left(\dfrac{d^2R}{dt^2}\right)_{\text{US}} > \left(\dfrac{d^2R}{dt^2}\right)_{\text{Japan}}$ $\quad (1981 \le t \le 1983)$

A graph that fits facts (a) through (e) is shown in Figure 10. ■

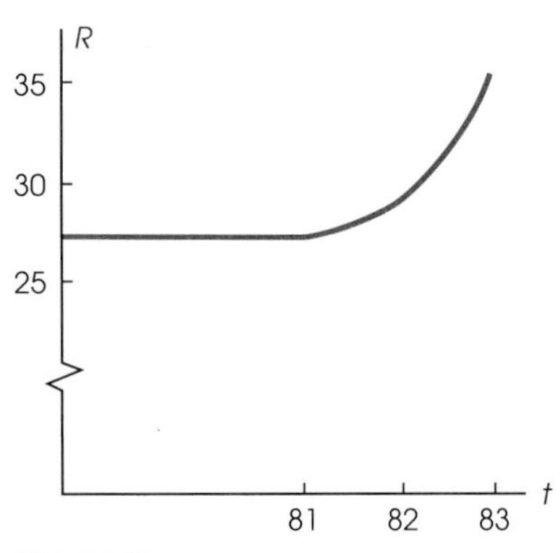


FIGURE 10

## EXAMPLE F

*GTS News* reported in the spring of 1984 that unemployment in the United States was continuing to decrease but at a slower and slower rate. Interpret this statement in the language of derivatives.

**Solution** Let $U$ denote the number of people unemployed at time $t$ in the United States. Then in the spring of 1984, $dU/dt < 0$ and $d^2U/dt^2 > 0$. A graph of $U$ against $t$ would look like Figure 11. ■

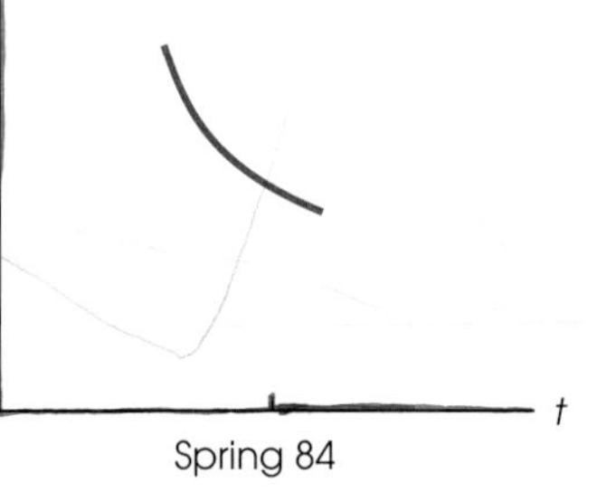


FIGURE 11

Occasionally, you will hear bankers and business people say, "We've got to watch that second derivative." Do you see why?

## Problem Set 2.5

*In Problems 1–6, find $d^3y/dx^3$.*

1. $y = x^5 + 3x^3$
2. $y = 2x^3 - x^2 + 2$
3. $y = 4/x$
4. $y = 6\sqrt{x}$
5. $y = 2x^{3/2} - x^4$
6. $y = \frac{3}{20}x^5 + 3/x^2$
7. For $f(t) = \sqrt{3t + 7}$, find $f''(3)$.
8. For $g(t) = \sqrt[3]{2t + 6}$, find $g''(1)$.
9. For $h(s) = 3/\sqrt{s + 4}$, find $h'''(0)$.
10. For $F(u) = \sqrt{2u + 3}$, find $F'''(3)$.
11. Find $d^2y/dx^2$ for $y = (x^2 - 3x + 4)^3$.
12. Find $d^2y/dx^2$ for $y = (x^3 - x + 2)^2$.
13. Find $f''(4)$ if $f(t) = t/(t + 4)$.
14. Find $g''(0)$ if $g(u) = u^2/(u^2 + 4)$.
15. Suppose that a rock was thrown upward from the roof of a building 256 feet high with an initial velocity of 96 feet per second. See Example D.
    (a) Write a formula for its height $s$ (in feet) above the ground after $t$ seconds.
    (b) What was its height after 2 seconds?
    (c) What was its velocity after 2 seconds?
    (d) When did it reach its maximum height?
    (e) When did the rock hit the ground?
    (f) With what velocity did it hit the ground?
    (g) What was the rock's acceleration at $t = 2$? At $t = 6$?
16. Answer the same questions as in Problem 15 for the case where a rock is thrown downward from the roof of the same building with an initial speed of 64 feet per second.
17. A particle is moving in a straight line according to the equation $s = 2t^3 - 3t^2 - 12t + 8$ ($s$ in centimeters, $t$ in seconds).
    (a) Find the position, the velocity, and the acceleration of the particle when $t = 3$.
    (b) Find the velocity at the instant when the acceleration is zero.
    (c) Find the acceleration at the instant when $t > 0$ and the velocity is zero.
    (d) On what interval of time is the velocity negative?
18. Suppose that a particle is moving in a straight line so that its position $s$ (in meters) at time $t$ (in minutes) is given by
$$s = 3t - 12\sqrt{t} + 2$$
    (a) Find the position, the velocity, and the acceleration of the particle when $t = 1$.
    (b) Where is the particle when the velocity is zero?
    (c) What value does the velocity approach as $t$ gets larger and larger?
    (d) What value does the acceleration approach as $t$ gets larger and larger?
19. The daily revenue $R$ of Tony's Ice Cream Shop climbed during the period 1/1/72 to 12/31/77 and then began to fall. The rate of change in daily revenue increased from 1/1/72 to 6/30/74 but then began a decline.
    (a) When was $dR/dt > 0$?
    (b) When was $d^2R/dt^2 < 0$?
    (c) When was $dR/dt = 0$?
    (d) When was $d^2R/dt^2 = 0$?
    (e) Sketch a graph which reflects all the above facts.
20. The number $N(t)$ of bacteria in a culture was carefully monitored over a period of 10 hours. During all this time $N(t)$ increased—at a constant rate during the first 2 hours, then at a slower and slower rate after a drug was administered at $t = 2$, and finally at an increasing rate when the effect of the drug began to wear off 4 hours after it was administered.
    (a) When was $N'(t) > 0$?
    (b) When was $N''(t) = 0$?
    (c) When was $N''(t) < 0$?
    (d) Sketch a graph that reflects all the above facts.
21. Interpret each of the following statements in the language of derivatives.
    (a) Inflation $I$ held steady during all of last year but is expected to increase more and more rapidly during the years ahead.
    (b) At the present time the price $P$ of oil is declining rapidly, but this trend is expected to slow and then reverse directions in about 2 years.
    (c) John's temperature $T$ is still rising, but the penicillin seems to be taking effect.
    (d) Nancy has the accelerator of her sports car pressed to the floor, but after another 2 minutes that won't do any good.
22. Let $T(n)$ be the time it takes a person to learn $n$ French words. One experiment indicated that one learns words more and more rapidly until a vocabulary of 200 words is reached. Then the rate of learning remains constant until 400 words are reached, after which the rate of learning slows.
    (a) For what values of $n$ is $T'(n) > 0$?
    (b) For what values of $n$ is $T''(n) > 0$?
    (c) For what values of $n$ is $T''(n) = 0$?

---

*In Problems 23–26, find $f'''(1)$.*

23. $f(x) = 6/\sqrt{2x - 1}$
24. $f(x) = 12x^{-2/3} + \sqrt{x}$
25. $f(x) = (x^2 + 1)^3$
26. $f(x) = x\sqrt{3 + x}$
27. Find the rate of change in the slope of $y = (4 + x)^3$ at the point $(1, 125)$.

28. At what point on the curve $y = 12\sqrt{x}$ is $d^2y/dx^2 = -\frac{1}{9}$?
29. A ball thrown upward from the top of a building 336 feet high hits the ground after 7 seconds.
    (a) What was the initial velocity? *Hint*: $s = -16t^2 + v_0t + 336$.
    (b) With what speed did the ball strike the ground?
30. In a small town of 400 people, the number $N(t)$ of people who had heard a certain rumor after $t$ days is indicated by the graph in Figure 12.
    (a) When is the rate at which the rumor is spreading positive?
    (b) When is the rate of spread decreasing?
    (c) When is the rumor spreading most rapidly?

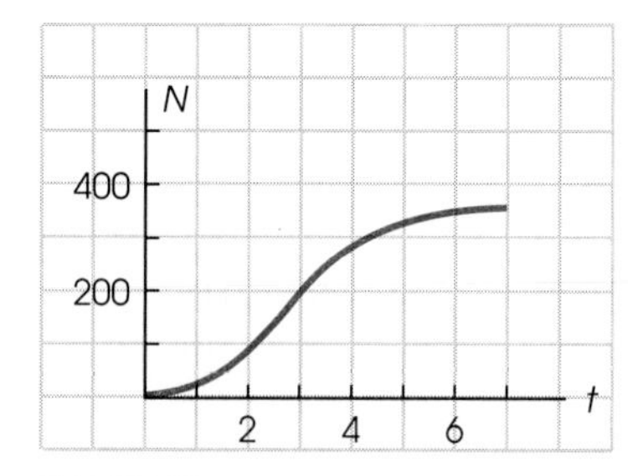


**FIGURE 12**

## CHAPTER 2 REVIEW PROBLEM SET

1. Sketch the graph of $f(x) = \begin{cases} x + 2 & \text{if } x < 2 \\ x^2 - 4 & \text{if } x \geq 2 \end{cases}$
   Then find (if possible) each of the following.
   (a) $f(2)$ (b) $f(f(2))$ (c) $\lim_{x\to 4} f(x)$
   (d) $\lim_{x\to 2^-} f(x)$ (e) $\lim_{x\to 2^+} f(x)$ (f) $\lim_{x\to 2} f(x)$
2. Sketch the graph of $f(x) = \begin{cases} (x-2)^2 & \text{if } x < 2 \\ 4 - x & \text{if } x \geq 2 \end{cases}$
   Find (if possible) each of the following.
   (a) $f(2)$ (b) $f(f(3))$ (c) $\lim_{x\to 5} f(x)$
   (d) $\lim_{x\to 2^-} f(x)$ (e) $\lim_{x\to 2^+} f(x)$ (f) $\lim_{x\to 2} f(x)$

*Find each limit in Problems 3–30.*

3. $\lim_{x\to -2} (3x + 8)$
4. $\lim_{x\to 3} (3x - 1)$
5. $\lim_{x\to 5} (x^2 - 4x + 5)$
6. $\lim_{x\to -3} [(x-2)^2 + \sqrt{3}]$
7. $\lim_{x\to 4} (x\sqrt{x} - 5)$
8. $\lim_{x\to 2} x^2\sqrt{x+7}$
9. $\lim_{x\to 5} \dfrac{2x - 10}{x^2 - 5x}$
10. $\lim_{x\to -3} \dfrac{3x + 9}{x^2 + 3x}$
11. $\lim_{x\to 3} \dfrac{x^2 - 4x + 3}{x^2 - 3x}$
12. $\lim_{x\to 5} \dfrac{x^2 - 3x - 10}{x^2 - 25}$
13. $\lim_{x\to -4} \dfrac{(x+4)^2}{x^2 + 2x - 8}$
14. $\lim_{x\to 4} \dfrac{(x-4)^3}{x^2 - x - 12}$
15. $\lim_{x\to -2} \dfrac{x^2 - x - 6}{x^2 + 7x + 10}$
16. $\lim_{x\to -6} \dfrac{x^2 + 5x - 6}{x^2 - x - 42}$
17. $\lim_{x\to 2} \dfrac{x^3 - 8}{x^2 + 4x - 12}$
18. $\lim_{x\to 5} \dfrac{x^3 - 125}{x^2 - 2x - 15}$
19. $\lim_{x\to 3} \dfrac{x^3 - 8}{x^2 + 4x - 12}$
20. $\lim_{x\to -5} \dfrac{x^3 - 125}{x^2 - 2x - 15}$
21. $\lim_{x\to 3} \dfrac{2^x}{x}$
22. $\lim_{x\to -1} \dfrac{2x + 3}{3^x}$
23. $\lim_{x\to 2} (3x^2 - 11)\sqrt{x^2 + 5}$
24. $\lim_{x\to 2} (2x^3 - 2x - 7)\sqrt{5x^2 + 7}$
25. $\lim_{x\to -4} \dfrac{3x + 2}{x^2 - 10}$
26. $\lim_{x\to 10} \dfrac{3x^2 - 2x - 5}{2x^2 + 5x + 25}$
27. $\lim_{x\to 3} \dfrac{(2x-5)^6(x-3)}{(x^2 - 9)(x^2 + x - 11)}$
28. $\lim_{x\to 1} \dfrac{(x^2 + x - 2)(4x - 3)^{12}}{(x^2 - 1)\sqrt{x^2 + 10x - 2}}$
29. $\lim_{x\to 10} [\sqrt{4x + 9} - 2\sqrt[3]{x^2 - 36}]$
30. $\lim_{x\to 5} \left[\dfrac{10}{\sqrt{4x + 5}} - 3\sqrt[3]{x^2 - 8x + 7}\right]$

*In Problems 31–34, find $f'(x)$ by using the definition of the derivative, that is, by calculating*

$$\lim_{h\to 0} \frac{f(x+h) - f(x)}{h}$$

31. $f(x) = x^2 - 3x - 2$
32. $f(x) = 2x^2 + 3x + 1$
33. $f(x) = 2/\sqrt{x}$
34. $f(x) = \sqrt{x^2 + 1}$

*In Problems 35–58, find $f'(x)$ using the rules for differentiation.*

35. $f(x) = 3\sqrt[3]{x} = 3x^{1/3}$
36. $f(x) = 8\sqrt[4]{x}$
37. $f(x) = 6x^2\sqrt[3]{x}$
38. $f(x) = 6x\sqrt[3]{x^2}$
39. $f(x) = x^{5/4}(16x)^{1/4}$
40. $f(x) = x^{3/2}\sqrt{25x^5}$
41. $f(x) = \sqrt{x\sqrt{x}}$
42. $f(x) = \sqrt[3]{x^4\sqrt{x}}$

43. $f(x) = (2x - 1)^6$
44. $f(x) = (3x + 1)^3$
45. $f(x) = \sqrt{x^2 - 2x + 5}$
46. $f(x) = \sqrt{x^2 + 8x - 10}$
47. $f(x) = \left(2x - \dfrac{4}{\sqrt{x}}\right)^3$
48. $f(x) = (2x + 4/x)^4$
49. $f(x) = (x^3 - 3x)^{4/3}$
50. $f(x) = (x^5 - 5x^2)^{7/5}$
51. $f(x) = \left(\dfrac{2x^2 + 1}{x^3 - 2x + 4}\right)^3$
52. $f(x) = \left(\dfrac{3x - 2}{x^3 + 3}\right)^5$
53. $f(x) = (2x^2 + 5)\sqrt{2x + 5}$
54. $f(x) = (2x^2 + x + 1)\sqrt{4x + 1}$
55. $f(x) = (2x^2 + 5)\sqrt{2x^2 + 5}$
56. $f(x) = (2x^2 + x + 1)\sqrt{2x^2 + x + 1}$
57. $f(x) = [(x^2 + 2x)^4 + 3]^5$
58. $f(x) = [(2x + 1)^5 - 2]^4$
59. If $g(t) = 64\sqrt{t + 3}$, find $g'''(1)$.
60. If $h(s) = 54(2s + 1)^{1/3}$, find $h'''(-1)$.
61. Find an equation for the tangent line to the graph of $y = 5\sqrt[5]{x} - 6/x^2$ at the point with $x = 1$.
62. Find an equation of the tangent line to the graph of

$$y = 4/(2x - 1)^2 + 3(x^2 + 6x + 2)^{1/2}$$

at the point where $x = 1$.
63. At what point on the graph of $y = 4\sqrt{x}$ is the tangent line parallel to the line $x - 2y = 5$?
64. At what point of the graph of $y = 2x^{3/2}$ is the tangent line parallel to $6x - y = 5$?
65. If \$10,000 is invested today at an interest rate of $x$ percent compounded 12 times a year, it will grow to an amount $A$ by the end of 10 years, where

$$A = 10{,}000\left(1 + \frac{x}{1200}\right)^{120}$$

(a) Find the rate at which $A$ changes with respect to $x$.
C (b) Evaluate $A$ and $dA/dx$ when $x = 9$.
66. The temperature in degrees Fahrenheit during the period from $t = 0$ to $t = 6$ (hours) was given by

$$F(t) = 72 - 15t + 6/(t + 2)^2$$

How fast was the temperature falling at $t = 2$?
67. Suppose that a particle is moving along a coordinate line in such a way that its position $s$ relative to the origin after $t$ seconds is given by $s = 12t - 4t^{3/2} + 6$ (in feet).
(a) Find the position, velocity, and acceleration of the particle at $t = 1$.
(b) Find the acceleration at the instant when the velocity is zero.
(c) During what time interval is the velocity positive?
68. A particle is moving along a cordinate line so that its position $s$ (in feet) relative to the origin after $t$ seconds satisfies $s = 4t + 16t^{1/2} - 8t^{3/2} + 24$.
(a) Find its position, velocity, and acceleration at $t = 4$.
(b) Find its acceleration at the instant when its velocity is 0.
69. A flu epidemic hit a certain city. The number $y$ of people who came down with the disease within $t$ days of the outbreak is indicated in Figure 13.
(a) When is the rate of spread of the disease positive?
(b) When is the rate of spread increasing?
(c) When is the disease spreading most rapidly?

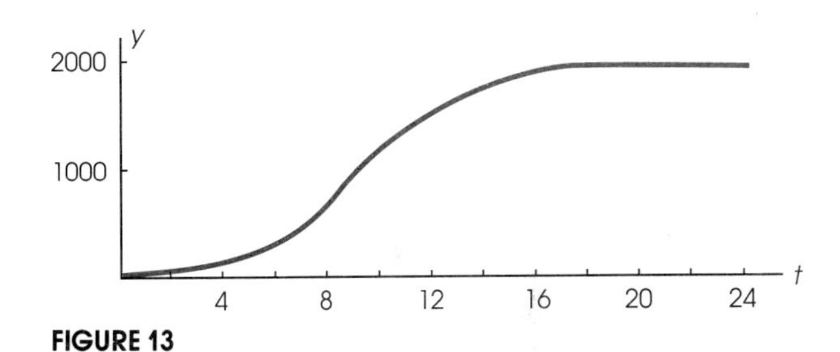


**FIGURE 13**

70. Suppose that a rumor spread through a city of 20,000 so that after 8 days almost everyone had heard it. During the first 4 days, the rate at which the rumor spread was increasing; after that the rate decreased. Sketch a graph of $y = N(t)$, where $N(t)$ represents the number of people who had heard the rumor after $t$ days, given that $N(4) = 12{,}000$.

*Do computers think? Is the human brain just a giant computer—a very complex jumble of electronic connections, switches, and diodes? These questions continue to fascinate scientists and they fascinated Norbert Wiener. A new branch of applied mathematics called artificial intelligence considers these questions.*

# 3

# Applications of the Derivative to Graphing



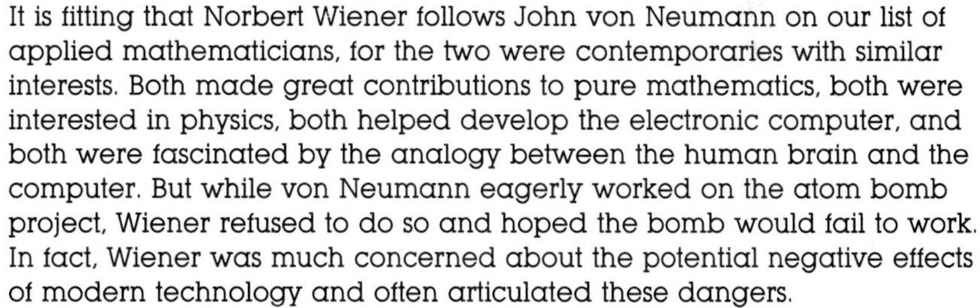

It is fitting that Norbert Wiener follows John von Neumann on our list of applied mathematicians, for the two were contemporaries with similar interests. Both made great contributions to pure mathematics, both were interested in physics, both helped develop the electronic computer, and both were fascinated by the analogy between the human brain and the computer. But while von Neumann eagerly worked on the atom bomb project, Wiener refused to do so and hoped the bomb would fail to work. In fact, Wiener was much concerned about the potential negative effects of modern technology and often articulated these dangers.

Norbert Wiener
(1894–1964)

Born in Missouri, Norbert was the son of Harvard language professor, Leo Wiener. Leo quickly recognized the talents of his precocious son and drove him mercilously. Thus, Norbert was reading and writing by age 3, had finished college at 14, and had obtained his Ph.D. from Harvard at 18. It was a regimen that left scars on the young man from which he never recovered. In spite of world recognition, he suffered from feelings of insecurity and social inferiority.

After postdoctoral study in England and Germany and some years of trying to find a suitable job, he was hired by the newly established Massachusetts Institute of Technology where he spent the rest of his life. He helped propel this institution to world renown, partly by his own contributions and also by his influence on many budding young scientists.

Wiener's early work dealt with the mathematics of Brownian motion and harmonic analysis and turned to the theory of information and communication. He can be called the founder of control theory, to which he gave the name *cybernetics.* With Arturo Rosenblueth, he made contributions to neurophysiology. His books include *Ex-prodigy, I Am a Mathematician, Cybernetics,* and *The Human Use of Human Beings.*

The authors of the present volume are two of Wiener's mathematical grandchildren. Both worked under Robert Cameron who had studied with Wiener. Both wrote their dissertations on the "Wiener integral."

## 3.1 CONTINUITY AND DIFFERENTIABILITY

The old maxim—one picture is worth a thousand words—holds also in calculus. The aim of a picture is to show the qualitative features of an object and thereby to reveal relationships that may be only dimly perceived in a mass of words. In calculus, the most important pictures are graphs of functions. This topic attracted our attention very early, especially in Section 1.2. In this chapter, we can probe much deeper because we have developed a powerful tool, the derivative. This tool acts as a microscope allowing us to examine the fine structure of a graph—to accurately locate its peaks, its valleys, and its turning points. All of this will become clear as we move through the chapter.

Our starting point is very basic, having to do with the fundamental nature of a curve. The subject is continuity.

### Continuity

The words *continuity* and *continuous* are in our everyday vocabulary. We say that a manufacturing plant was in continuous production for 10 years, meaning that it operated throughout that whole period with no interruptions. We remark about the continuity of a certain author's writings, meaning that the ideas expressed follow a consistent path; if they change over time, they do so in a smooth, regular way.

Mathematicians use the words continuous and continuity in a very similar fashion. A function is continuous on a certain interval if its graph has no jumps or wild fluctuations. Put another way, a function is continuous if its graph can be drawn without lifting the pencil from the paper (or the chalk from the chalk board). These informal descriptions help us, but they are not adequate for what follows. A precise definition rests on the fundamental notion of limit.

**DEFINITION**

The function $f$ is **continuous** at the point $c$ if

$$\lim_{x \to c} f(x) = f(c)$$

and $f$ is continuous on a set if it is continuous at each point of the set.

Perhaps we can understand this definition best by asking what it means for a function to be discontinuous at a point. The function $f$ is discontinuous at $c$ if one or more of the following are true.

1. $\lim_{x\to c} f(x)$ doesn't exist (Figure 1).
2. $f(c)$ is not defined (Figure 2).
3. $\lim_{x\to c} f(x)$ and $f(c)$ both exist but are unequal (Figure 3).

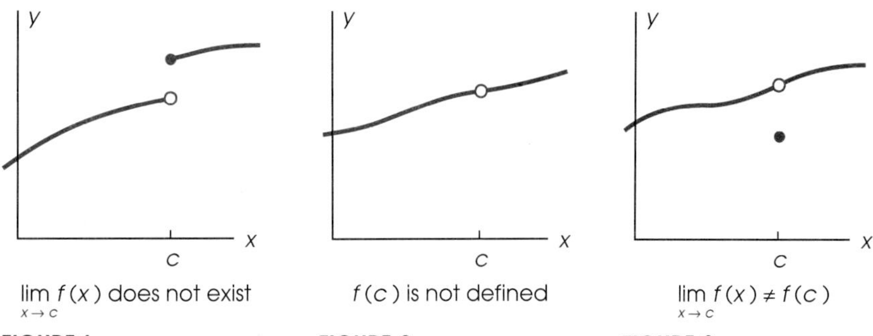


$\lim_{x\to c} f(x)$ does not exist

**FIGURE 1**

$f(c)$ is not defined

**FIGURE 2**

$\lim_{x\to c} f(x) \neq f(c)$

**FIGURE 3**

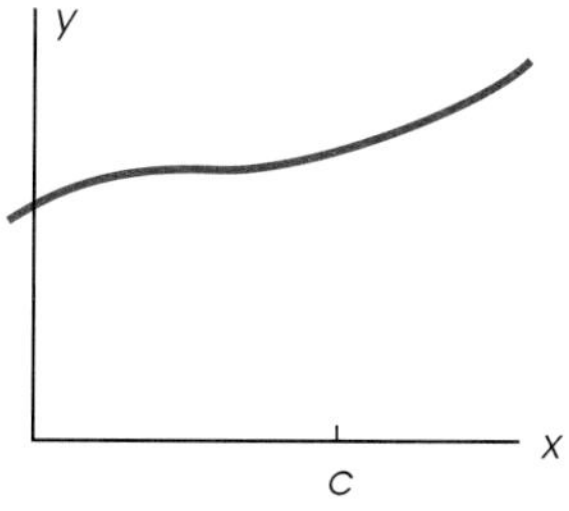

Figure 4 exhibits a function which is continuous at $c$. Note that the condition $\lim_{x\to c} f(x) = f(c)$ holds for this function.

$\lim_{x\to c} f(x) = f(c)$

**FIGURE 4**

## EXAMPLE A

Show that $f(x) = (x^2 - 4)/(x - 2)$ is continuous at $x = 1$ but is discontinuous at $x = 2$.

**Solution** By using the Main Limit Theorem (Section 2.1), we obtain

$$\lim_{x\to 1} f(x) = \frac{\lim_{x\to 1} (x^2 - 4)}{\lim_{x\to 1} (x - 2)} = \frac{-3}{-1} = 3$$

Since $f(1)$ also equals 3, the required condition for continuity holds.

At $x = 2$, the function $f$ is undefined, so $f$ is discontinuous there. Note the graph of $f$ shown in Figure 5. ■

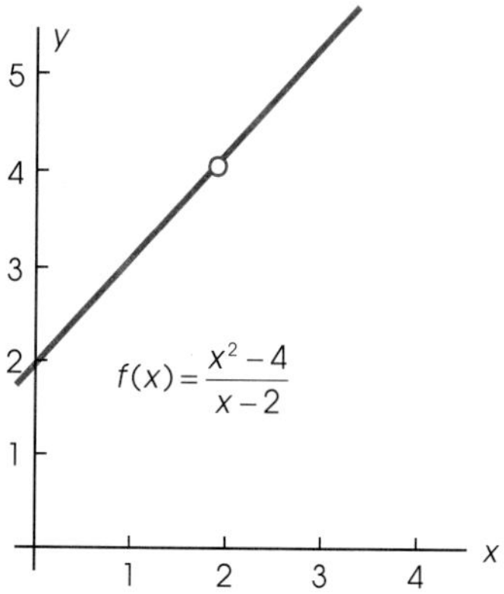

The function $f$ in Example A is a rational function—a quotient of two polynomials. It is easy to show (using the Main Limit Theorem) that a polynomial is continuous everywhere and that a rational function is continuous except where its denominator is zero.

Sometimes a discontinuity can be removed by the simple device of defining (or redefining) the function at the point of discontinuity. This is the case with the function $f$ in Example A. If we define $f(2)$ to be 4, the function $f$ thereby becomes continuous at 2 (see Figure 5).

$f(x) = \dfrac{x^2 - 4}{x - 2}$

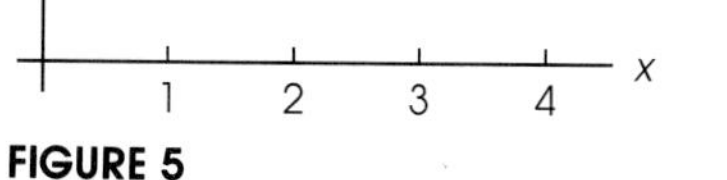

**FIGURE 5**

## EXAMPLE B

Show that the discontinuity of

$$g(x) = \frac{x^2 - 4x - 5}{x + 1}$$

at $x = -1$ can be removed by defining $g(-1)$ appropriately.

**Solution**

$$\lim_{x \to -1} g(x) = \lim_{x \to -1} \frac{(x-5)(x+1)}{x+1} = \lim_{x \to -1} (x-5) = -6$$

Thus if we define $g(-1) = -6$, the function $g$ becomes continuous at $-1$. ■

## EXAMPLE C

The function $F(x) = (x^2 - x - 6)/(x^2 - 9)$ has discontinuities at $-3$ and 3. One of these discontinuities can be removed; the other cannot. Determine which can be removed and define $F$ there to make $F$ continuous.

**Solution**

$$\lim_{x \to -3} F(x) = \lim_{x \to -3} \frac{(x-3)(x+2)}{(x-3)(x+3)} = \lim_{x \to -3} \frac{x+2}{x+3}$$

Clearly this limit does not exist, since the numerator is approaching $-1$ and the denominator is approaching 0. Thus, the discontinuity at $-3$ cannot be removed.

In contrast,

$$\lim_{x \to 3} F(x) = \lim_{x \to 3} \frac{(x-3)(x+2)}{(x-3)(x+3)} = \lim_{x \to 3} \frac{x+2}{x+3} = \frac{5}{6}$$

Defining $F(3)$ to be $\frac{5}{6}$ will remove the discontinuity at 3. Figure 6 displays the graph of $F$ and makes clear the nature of the two discontinuities. ■

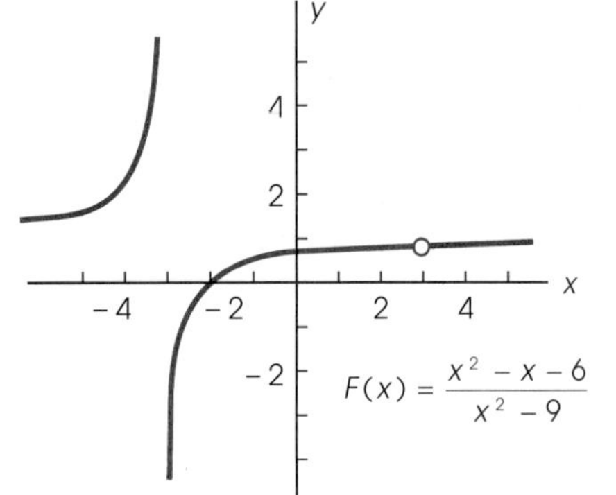


FIGURE 6

## EXAMPLE D

Discuss the continuity of

$$f(x) = \begin{cases} x - 1 & \text{if } x \le 2 \\ -x^2 + 5 & \text{if } x > 2 \end{cases}$$

**Solution** The two parts of $f$ are polynomials and therefore continuous everywhere. The only question is whether they have been pasted together correctly at 2 so that $\lim_{x \to 2} f(x) = f(2)$. The graph (Figure 7) suggests that the answer is yes, but it is also easy to give an algebraic argument. Note first that $f(2) = 2 - 1 = 1$. Also,

$$\lim_{x \to 2^-} f(x) = \lim_{x \to 2^-} (x - 1) = 1$$

and

$$\lim_{x \to 2^+} f(x) = \lim_{x \to 2^+} (-x^2 + 5) = 1$$

so $\lim_{x \to 2} f(x) = 1 = f(2)$. We conclude that $f$ is continuous at 2 and therefore everywhere. ■

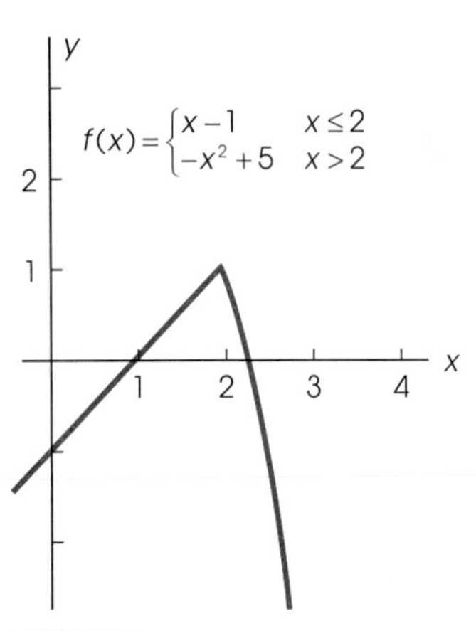


FIGURE 7

## Differentiability

For a function to be differentiable means that it has a derivative, and this in turn means that the graph of the function has a slope. How can a function fail to have a slope, say, at the point $c$? There are several rather obvious ways.

1. The function fails to even be continuous at $c$ (Figure 8).
2. The graph of the function has a corner at $c$ and therefore a different slope from the right and from the left (Figure 9).
3. The graph of $f$ has a vertical tangent at $c$ (Figure 10).

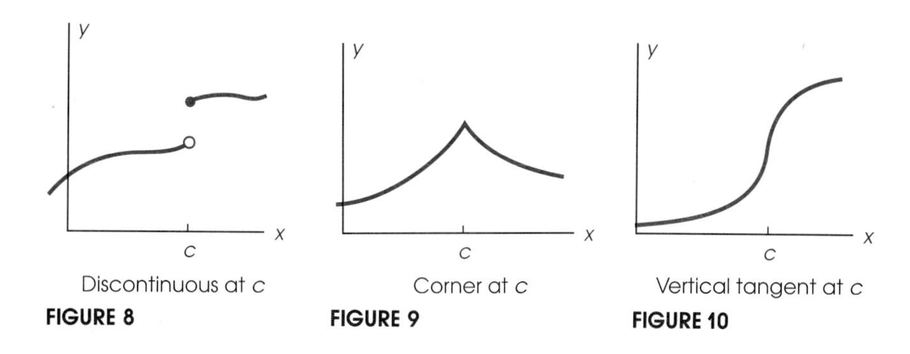


**FIGURE 8** **FIGURE 9** **FIGURE 10**

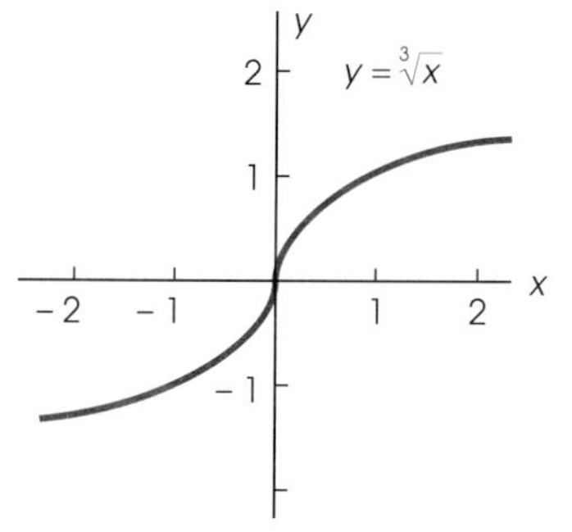


**FIGURE 11**

### EXAMPLE E

At what points is $f(x) = \sqrt[3]{x} = x^{1/3}$ differentiable?

**Solution** $f'(x) = \frac{1}{3}x^{-2/3} = 1/3(\sqrt[3]{x})^2$, a formula valid for $x \neq 0$. Thus, $f$ is differentiable for all $x \neq 0$. At $x = 0$, the formula for $f'(x)$ suggests that there may be a vertical tangent, and this is confirmed when we draw a graph (Figure 11). Note that $f$ is continuous everywhere, even at $x = 0$. ■

### EXAMPLE F

At what points is $f(x) = |x|$ differentiable?

**Solution** The graph of $f$ (Figure 12) makes the answer clear immediately. In fact, $f'(x) = -1$ if $x < 0$ and $f'(x) = 1$ if $x > 0$. At $x = 0$, there is no tangent line, hence no slope, and therefore no derivative. ■

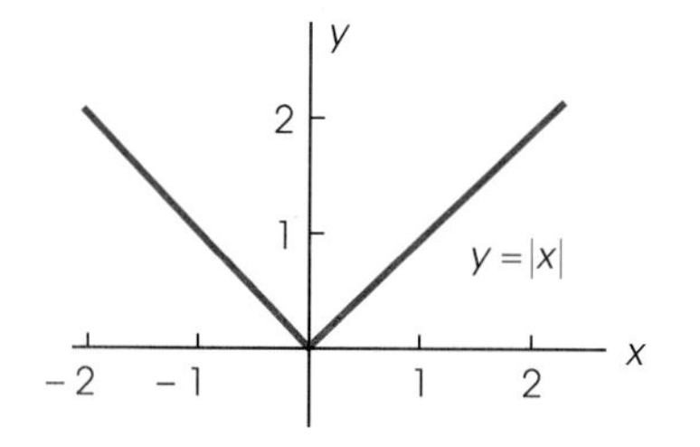


**FIGURE 12**

A moment's thought should convince you that a polynomial [for example, $P(x) = 3x^5 - 7x^4 + 11x - 16$] is differentiable everywhere and that a rational function [for example, $f(x) = (3x^3 - 2x + 11)/(x^2 - 9)$] is differentiable everywhere except where its denominator is zero. The reason is straightforward; we have formulas for their derivatives.

Refer again to Examples E and F. Note that in both cases, the given function was continuous at 0 but not differentiable there. Continuity does not

**CAUTION**
A statement and its converse are not equivalent. All men are mortals but not all mortals are men. Differentiability implies continuity, but continuity does *not* imply differentiability.

imply differentiability. The converse, however, is true, as we now make clear.

**THEOREM**
If $f$ is differentiable at $c$, then $f$ is continuous at $c$.

*Proof* For $h \neq 0$

$$f(c+h) = \frac{f(c+h) - f(c)}{h} \cdot h + f(c)$$

Thus by the Main Limit Theorem,

$$\lim_{h \to 0} f(c+h) = \lim_{h \to 0} \frac{f(c+h) - f(c)}{h} \cdot \lim_{h \to 0} h + \lim_{h \to 0} f(c)$$
$$= f'(c) \cdot 0 + f(c)$$

that is,

$$\lim_{h \to 0} f(c+h) = f(c)$$

This is equivalent to saying

$$\lim_{x \to c} f(x) = f(c)$$

the condition for continuity. ■

We summarize by noting two implications, but we emphasize that neither implication is reversible.

$$f \text{ is differentiable at } c$$
$$\Downarrow$$
$$f \text{ is continuous at } c$$
$$\Downarrow$$
$$\lim f(x) \text{ exists at } c$$

**EXAMPLE G**

Decide by studying Figure 13 which of the three conditions above holds at $c_1$, $c_2$, $c_3$, and $c_4$.

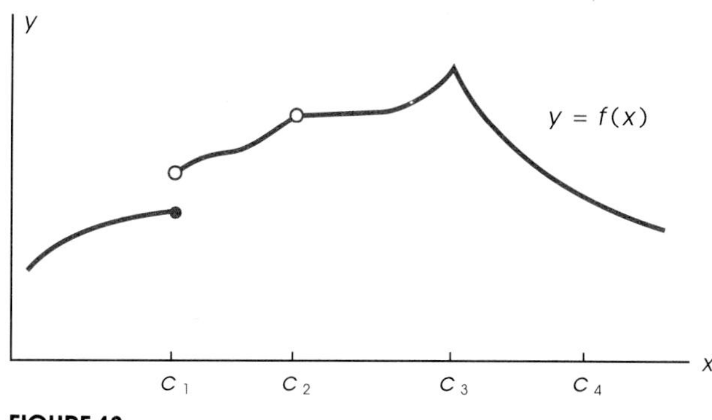


FIGURE 13

**Solution**

$f$ is differentiable at $c_4$.
$f$ is continuous at $c_3$ and $c_4$.
$\lim f(x)$ exists at $c_2$, $c_3$, and $c_4$.
None of the three conditions holds at $c_1$. ■

## Problem Set 3.1

*In Problems 1–8, indicate whether or not the given physical process is continuous for all $t > 0$.*

1. The depth of water in a bathtub $t$ hours after it was installed in a house.
2. The number of dollars in your checking account $t$ hours from now.
3. The time showing on your digital wristwatch $t$ hours after you bought it.
4. Your height $t$ hours after you were born.
5. The number of hours you will spend studying mathematics during the next $t$ hours.
6. The number of people inside the city limits of Chicago $t$ hours from now.
7. The Celsius temperature of your body $t$ hours from now as shown by a mercury thermometer.
8. The Celsius temperature that will be reported $t$ hours from now on the electronic sign outside the local bank.

*In Problems 9–12, show that the function $f$ is continuous at $x = a$ but discontinuous at $x = b$ (see Example A).*

9. $f(x) = \dfrac{x^2 - 4x}{x - 4}$; $a = 1$, $b = 4$
10. $f(x) = \dfrac{x^2 - x - 2}{x - 2}$; $a = 1$, $b = 2$
11. $f(x) = \dfrac{x^2 - 4}{3x + 6}$; $a = 4$, $b = -2$
12. $f(x) = \dfrac{2x^2 - 10x}{3x - 15}$; $a = 10$, $b = 5$

*Each function in Problems 13–16 has a removable discontinuity at $x = a$. How should $f(a)$ be defined to remove this discontinuity? See Example B.*

13. $f(x) = \dfrac{x^2 - 2x - 8}{3x - 12}$; $a = 4$
14. $f(x) = \dfrac{x^2 - 3x - 10}{2x + 4}$; $a = -2$
15. $f(x) = \dfrac{x^2 - 1}{2 - 2x}$; $a = 1$
16. $f(x) = \dfrac{x^2 - x - 6}{6 - 2x}$; $a = 3$

*Each function in Problems 17–22 has discontinuities at two values of x. Find these values and, if the discontinuity is removable, indicate how f should be defined at that value to remove the discontinuity. See Example C.*

17. $f(x) = \dfrac{x^2 - 4x - 12}{x^2 - 5x - 6}$
18. $f(x) = \dfrac{x^2 - x - 6}{x^2 - 2x - 3}$
19. $f(x) = \dfrac{x + 4}{x^2 - 5x + 4}$
20. $f(x) = \dfrac{x}{x^2 - x - 30}$
21. $f(x) = \dfrac{3x^2 - 12}{x^2 - 4}$
22. $f(x) = \dfrac{x^3 - x}{x^2 - 1}$

*In Problems 23–32, determine all values of x where the given function is discontinuous. For the functions defined by multipart rules, see Example D.*

23. $f(x) = x^5 - 5x + 2$
24. $f(x) = 4/(x^2 + 1)$
25. $f(x) = \sqrt{x^2 + 1}$
26. $f(x) = \sqrt{1 - x^2}$
27. $f(x) = 1/\sqrt{x}$
28. $f(x) = 1/\sqrt{x - 3}$
29. $f(x) = \begin{cases} 2x & \text{if } x \le 3 \\ x^2 - 3 & \text{if } x > 3 \end{cases}$
30. $f(x) = \begin{cases} \frac{1}{4}x^2 & \text{if } x \le 2 \\ 3 - x & \text{if } x > 2 \end{cases}$
31. $f(x) = \begin{cases} x^2 + 1 & \text{if } x \le 1 \\ 5x - \frac{2}{x} & \text{if } x > 1 \end{cases}$
32. $f(x) = \begin{cases} \frac{1}{3}x^3 & \text{if } x < 2 \\ 2x + 3 & \text{if } x \ge 2 \end{cases}$

*In Problems 33–44, indicate the values of x (if any) where f fails to be differentiable (see Problems E and F).*

33. $f(x) = x^{4/3}$
34. $f(x) = 2x^{1.2} + 5x$
35. $f(x) = 1/(x + 1)$
36. $f(x) = 1/(x^2 + 1)$
37. $f(x) = \sqrt[3]{x}(x + 1)$
38. $f(x) = \sqrt[5]{x}$
39. $f(x) = \sqrt[5]{2x + 1}$
40. $f(x) = \sqrt[3]{2x - 1}$
41. $f(x) = |x - 2|$
42. $f(x) = |x + 3|$
43. $f(x) = \begin{cases} x & \text{if } x \le 2 \\ 2 & \text{if } x > 2 \end{cases}$
44. $f(x) = \begin{cases} x^2 & \text{if } x \le 1 \\ x & \text{if } x > 1 \end{cases}$

*Examine the graphs in Problems 45 and 46. As in Example G, decide at which of the points p, q, r, s, and t:*

*(a) f has a value.*
*(b)* $\lim f(x)$ *exists.*
*(c) f is continuous.*
*(d) f is differentiable.*

45.

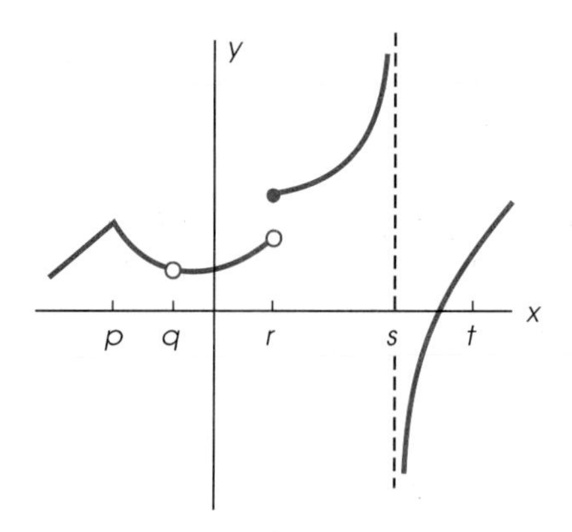


46.

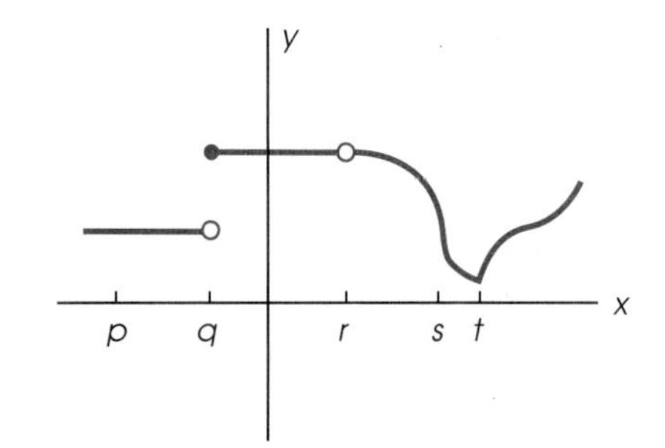


*In Problems 47–54, decide whether or not the given function is continuous at x = 3. If it is continuous at x = 3, decide whether or not it is differentiable there.*

47. $f(x) = (x - 3)^2 + \sqrt[3]{x - 3}$
48. $f(x) = |x - 3|$
49. $f(x) = |x - 3|/(x - 3)$
50. $f(x) = (2x - 6)/(x - 3)$
51. $f(x) = \dfrac{x^2 - 3x}{2x - 6}$
52. $f(x) = \begin{cases} \dfrac{x^2 - 3x}{2x - 6} & \text{if } x \ne 3 \\ 1.5 & \text{if } x = 3 \end{cases}$
53. $f(x) = \begin{cases} x^2 - 3x & \text{if } x \le 3 \\ x - 3 & \text{if } x > 3 \end{cases}$
54. $f(x) = \begin{cases} x + 3 & \text{if } x < 3 \\ x^2 - 3 & \text{if } x \ge 3 \end{cases}$
55. Let $[x]$ denote the *greatest integer in x*. For example, $[3.4] = 3$, $[3] = 3$, and $[-1.2] = -2$. Sketch the graph of $f(x) = [x]$. At what values of $x$ is this function discontinuous?
56. Let $\langle x \rangle$ denote the *distance from x to the nearest integer*. For example, $\langle 3.4 \rangle = 0.4$, $\langle 3.7 \rangle = 0.3$, and

$\langle 3 \rangle = 0$. Sketch the graph of $f(x) = \langle x \rangle$. At what values of $x$ is this function discontinuous? Not differentiable?

57. Sketch the graph of a function $f$ defined on the interval $0 \le x \le 6$ and satisfying all of the following conditions.
    (a) $f(0) = f(2) = f(4) = f(6) = 2$.
    (b) $f$ is continuous everywhere except at $x = 2$.
    (c) $f$ is differentiable on $0 < x < 6$ except at $x = 2$ and $x = 4$.

## 3.2 MONOTONICITY AND CONCAVITY

Several kinds of intervals will arise in this section, and we introduce special notation to describe them. The double inequality $a < x < b$ determines the **open interval** consisting of all numbers between $a$ and $b$, not including $a$ and $b$. This interval is denoted by $(a, b)$. In contrast, the inequality $a \le x \le b$ determines the corresponding **closed interval,** which does include $a$ and $b$. It is denoted by $[a, b]$. The two situations are illustrated in Figure 14.

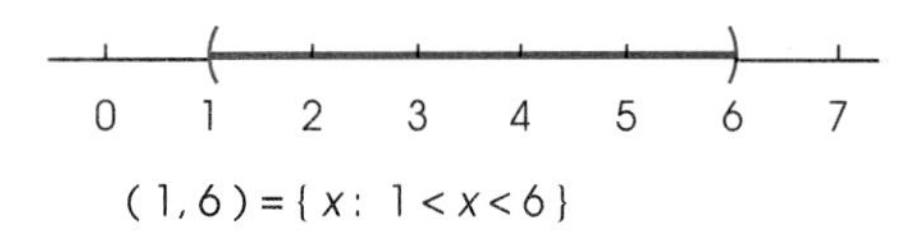


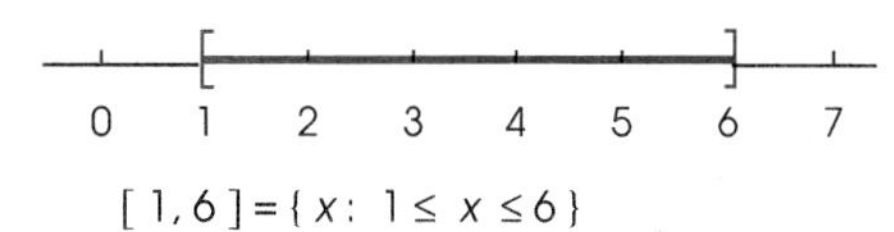


**FIGURE 14**

Nine types of intervals can occur; they are shown in the accompanying table.

| Inequality | Interval Notation | Graph |
|---|---|---|
| $a < x < b$ | $(a, b)$ | a b |
| $a \le x \le b$ | $[a, b]$ | a b |
| $a \le x < b$ | $[a, b)$ | a b |
| $a < x \le b$ | $(a, b]$ | a b |
| $-\infty < x \le b$ | $(-\infty, b]$ | b |
| $-\infty < x < b$ | $(-\infty, b)$ | b |
| $a \le x < \infty$ | $[a, \infty)$ | a |
| $a < x < \infty$ | $(a, \infty)$ | a |
| $-\infty < x < \infty$ | $(-\infty, \infty)$ | |

## Solving Inequalities

Solving inequalities is much like solving equations. Here are the allowable operations—operations that do not change the solution set.

1. We may add the same number to both sides of an inequality.
2. We may multiply both sides of an inequality by a positive number.
3. We may multiply both sides of an inequality by a negative number provided we reverse the direction of the inequality sign.

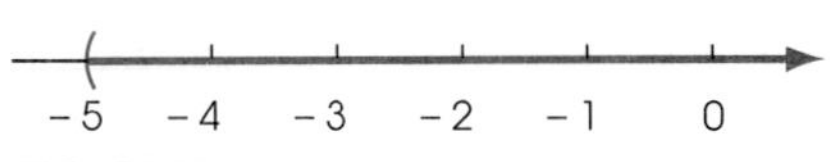

FIGURE 15

> **RECALL**
> If we cannot factor
> $$ax^2 + bx + c$$
> we can still find the split points by means of the **Quadratic Formula**
> $$x = \frac{-b \pm \sqrt{b^2 - 4ac}}{2a}$$

### EXAMPLE A

Solve the inequality $3x - 8 < 5x + 2$ and write the solution set in interval notation.

**Solution**

$$3x - 8 < 5x + 2$$
$$3x < 5x + 10 \qquad \text{(adding 8)}$$
$$-2x < 10 \qquad \text{(adding } -5x\text{)}$$
$$x > 10/-2 \qquad \text{(multiplying by } -\tfrac{1}{2}\text{)}$$
$$x > -5$$

The solution set is $(-5, \infty)$ (see Figure 15). ■

### EXAMPLE B

Solve the quadratic inequality $x^2 - 2x \leq 8$ and write the solution set in interval notation.

**Solution** As with quadratic equations, we bring all nonzero terms to one side and then factor (if possible).

$$x^2 - 2x \leq 8$$
$$x^2 - 2x - 8 \leq 0$$
$$(x - 4)(x + 2) \leq 0$$

The two numbers $-2$ and 4 where the left side is 0 are called **split points.** They split the real line into the three intervals $(-\infty, -2)$, $(-2, 4)$, and $(4, \infty)$. On each of them, $f(x) = (x - 4)(x + 2)$ has a constant sign, either positive or negative. We can determine which by picking any point on the interval as a **test point**; $f$ will be positive or negative on the interval according as $f$ at the test point is positive or negative. The whole process is displayed in Figure 16; it yields the solution set $[-2, 4]$. The endpoints $-2$ and 4 are included because the inequality $\leq$ includes equality. ■

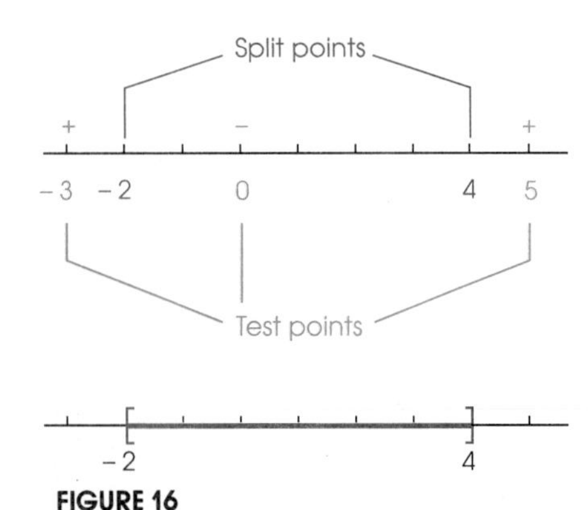

FIGURE 16

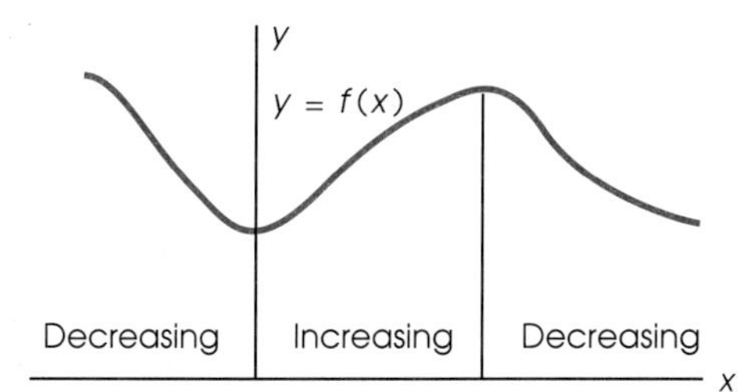


**FIGURE 17**

## Increasing or Decreasing

We say that $f$ is **increasing** on an interval $I$ if the graph of $f$ rises to the right on $I$. Similarly $f$ is **decreasing** on $I$ if the graph of $f$ falls to the right on $I$ (see Figure 17). We use the adjective **monotonic** to describe a function which is either increasing or decreasing on an interval.

If the slope of a curve is positive, then the corresponding function is increasing; if the slope is negative, the function is decreasing. When stated in terms of derivatives, this gives a very important theorem.

> **MONOTONICITY THEOREM**
>
> Let $f$ be differentiable on the open interval $(a, b)$
>
> (i) If $f'(x) > 0$ for all $x$ in $(a, b)$, then $f$ is increasing on $(a, b)$.
>
> (ii) If $f'(x) < 0$ for all $x$ in $(a, b)$, then $f$ is decreasing on $(a, b)$.

### EXAMPLE C

Find the intervals on which $f(x) = 2x^2 - 3x + 11$ is increasing or decreasing.

**Solution** Since $f'(x) = 4x - 3$, our problem reduces to finding out where $4x - 3 < 0$ and where $4x - 3 > 0$. The first inequality is solved as follows.

$$4x - 3 < 0$$

$$4x < 3$$

$$x < \tfrac{3}{4}$$

From this we conclude that $f$ is decreasing on $(-\infty, \frac{3}{4})$. It is easy to see in a similar manner that $f$ is increasing on $(\frac{3}{4}, \infty)$. Actually we could have anticipated a result like this, since we know that the graph of $f$ is a parabola that turns up (Figure 18). ■

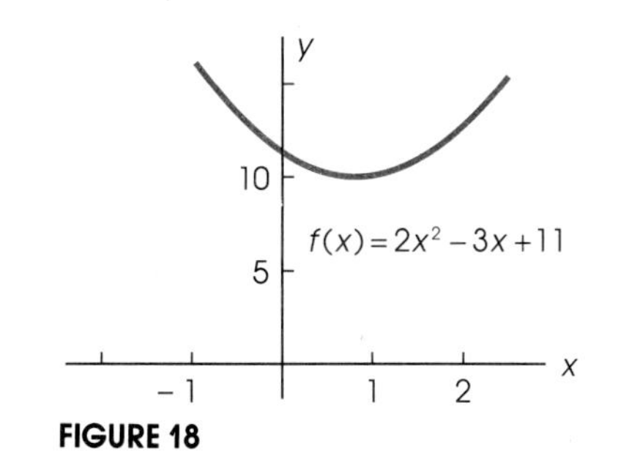


**FIGURE 18**

### EXAMPLE D

Find the intervals on which $g(x) = 2x^3 - 3x^2 - 12x$ is increasing or decreasing. Then sketch the graph of $f$.

**Solution** Since $g'(x) = 6x^2 - 6x - 12 = 6(x - 2)(x + 1)$, our problem is to find where $6(x - 2)(x + 1)$ is positive and where it is negative. The split points $-1$ and $2$ split the real line into the three intervals $(-\infty, -1)$, $(-1, 2)$, and $(2, \infty)$. When tested (Figure 19), we find that $g'$ is positive on the first and third of these intervals and negative on the second one. We conclude that $g$ is increasing on $(-\infty, -1)$ and $(2, \infty)$, and decreasing on $(-1, 2)$. With this information plus the values of $g$ at the split points,

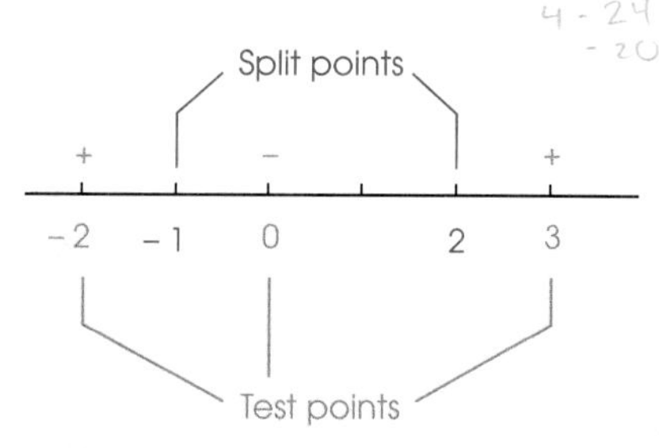


**FIGURE 19**

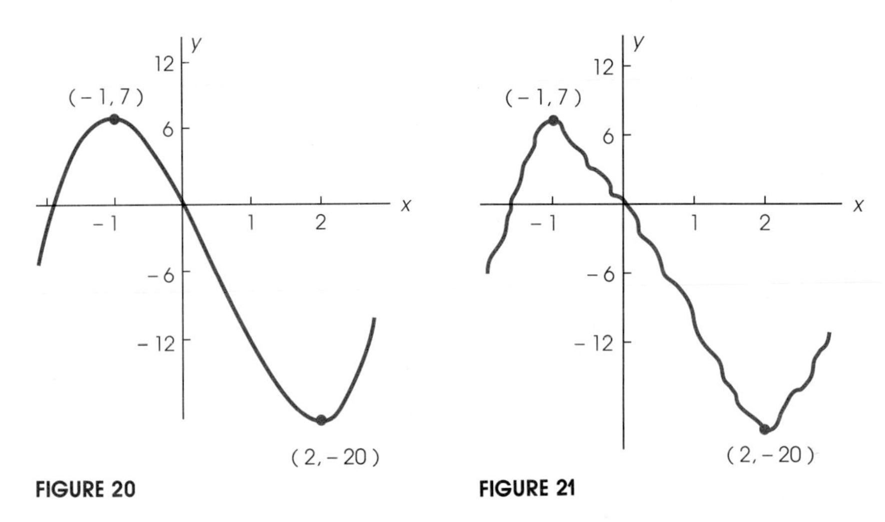


FIGURE 20 FIGURE 21

namely, $g(-1) = 7$ and $g(2) = -20$, we can quickly sketch a graph. But does it look like Figure 20 or Figure 21? Both increase and decrease on the right intervals and have their peak and pit in the right spots.

Since both we and nature have a predisposition toward simplicity, 99% of us would bet money on Figure 20. But to make certain, we need another concept called concavity. ■

## Concave Up or Concave Down

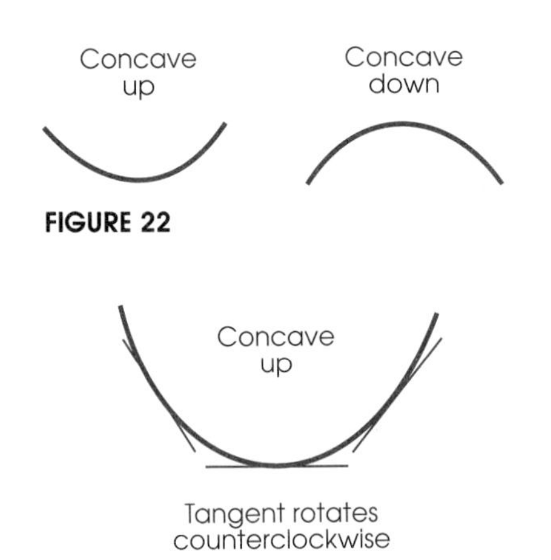


FIGURE 22

FIGURE 23

We are all familiar with one manifestation of concavity. At some time, most of us have worn glasses, at least sunglasses. The inner surface of a typical glass lens is concave. For a graph to be concave means that it has this shape, but it may be **concave up** or **concave down** (Figure 22).

Of course, our aim is to relate concavity to the calculus, in particular, to the derivative. Let us agree first of all that what we really mean by saying that the graph of a function $f$ is concave up is that the tangent line is rotating counterclockwise as it moves from left to right (Figure 23). This in turn means that the slope is increasing; that is, $f'$ is an increasing function. But we can guarantee that $f'$ is increasing by requiring that its derivative, namely, $f''$, be positive. We formulate the result as a theorem.

**CONCAVITY THEOREM**

Let $f$ be twice differentiable on the open interval $(a, b)$.

(i) If $f''(x) > 0$ for all $x$ in $(a, b)$, then the graph of $f$ is concave up on $(a, b)$.

(ii) If $f''(x) < 0$ for all $x$ in $(a, b)$, then the graph of $f$ is concave down on $(a, b)$.

**POINT OF INTEREST**

We can summarize what we have learned by drawing four pictures.

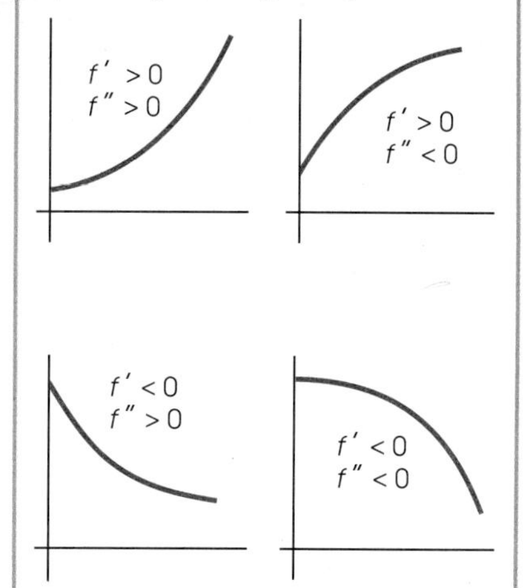


Consider again $f(x) = 2x^2 - 3x + 11$ in Example C. Note that $f'(x) = 4x - 3$ and $f''(x) = 4$. Since the second derivative is always positive, the graph is concave up on the whole real line, confirming that the graph we drew in Figure 18 is correct.

Of more interest is $g(x) = 2x^3 - 3x^2 - 12x$ in Example D. Since $g'(x) = 6x^2 - 6x - 12$, we obtain $g''(x) = 12x - 6 = 6(2x - 1)$. It is easy to see that $g''(x) < 0$ on $(-\infty, \frac{1}{2})$ and $g''(x) > 0$ on $(\frac{1}{2}, \infty)$. We conclude that the graph of $g$ is concave down on the first interval, and concave up on the second interval. Now we can assert with confidence that the graph of $g$ is much like Figure 20; it cannot wriggle as does the graph in Figure 21.

## EXAMPLE E

Sketch the graph of a function $f$ which satisfies all of the following conditions.

(i) It is continuous everywhere.

(ii) $(-2, 2)$ and $(3, 7)$ are points on the graph.

(iii) $f'(-2) = 0$ and $f'(3) = 0$.

(iv) $f''(x) > 0$ on $(-\infty, 1)$ and $f''(x) < 0$ on $(1, \infty)$.

**Solution** We draw the graph in two stages. First, we plot the two points $(-2, 2)$ and $(3, 7)$ and show that the slope is 0 at these two points (shown in black in Figure 24). Second, condition (iv) means the graph must be concave up on $(-\infty, 1)$, concave down on $(1, \infty)$. This allows us to draw the colored curve.

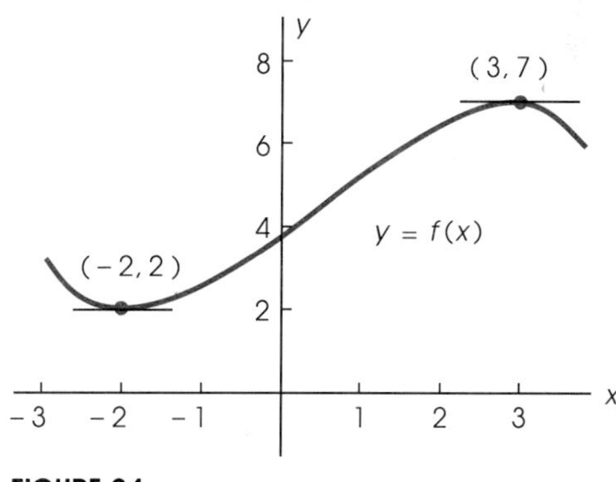


FIGURE 24

■

## EXAMPLE F

Sketch the graph of a function $f$ which satisfies all of the following conditions

(i) It is continuous except at $x = 0$.

(ii) $\lim_{x\to 0^-} f(x) = 1$ and $\lim_{x\to 0^+} f(x) = 2$
(iii) $f'(x) = -1$ on $(-\infty, 0)$ and $f'(x) > 0$ on $(0, \infty)$
(iv) $f''(x) < 0$ on $(0, \infty)$

**Solution** The graph is shown in Figure 25. Clearly (i) and (ii) hold. Condition (iii) means that the graph is a line of slope $-1$ on $(-\infty, 0)$ and that the graph rises on $(0, \infty)$. Finally (iv) forces the graph to be concave down on $(0, \infty)$.

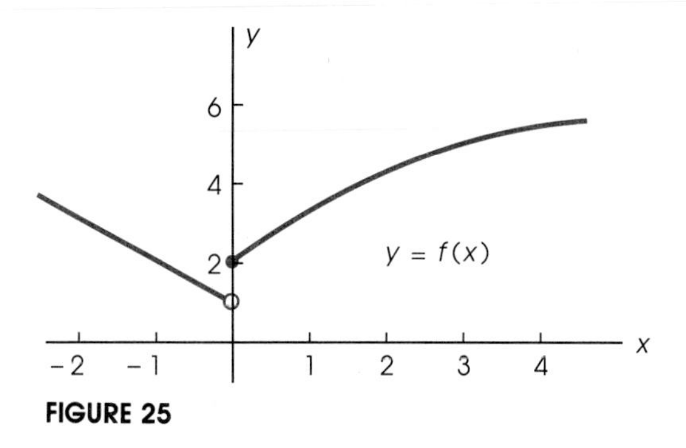


FIGURE 25

■

## Problem Set 3.2

*In Problems 1–6, show the indicated set on the real line. Recall that the symbol ∪ denotes union. Thus, A ∪ B is the set of points in A or B.*

1. $[-3, 6)$
2. $(-2, 3]$
3. $[-2, 0) \cup [1, 4]$
4. $(-\infty, 1] \cup [2, 3]$
5. $\{x: -2 < x \leq 5\}$
6. $\{x: 1 \leq x < 4\}$

*In Problems 7–12, solve the given inequality and write the solution set in interval notation (see Example A). In Problems 11 and 12, we suggest that you begin by clearing fractions. Thus, in Problem 11, begin by multiplying both sides by 6.*

7. $4x - 11 < 7x + 7$
8. $6x + 13 \geq x - 17$
9. $2(x - 3) \geq 5x + 1$
10. $4 - 3x \leq 5(x + 3)$
11. $\frac{1}{2}x - 2 < -\frac{2}{3}x + 5$
12. $\frac{3}{4}x + 3 > \frac{2}{5}x + 10$

*Solve the quadratic inequalities in Problems 13–20 as illustrated in Example B. Then write the solution set in interval notation and show this set on the number line.*

13. $(x + 7)(x - 3) \leq 0$
14. $(x - 6)(x + 4) < 0$
15. $x^2 + 6x - 16 < 0$
16. $x^2 + 2x - 8 \leq 0$
17. $x^2 - x \geq 6$
18. $x^2 - 5x > 6$
19. $2x^2 + 7x - 4 < 0$
20. $3x^2 + 13x - 10 \leq 0$

*In Problems 21–28, find the interval(s) on which f(x) is increasing and the interval(s) on which f(x) is decreasing (see Examples C and D).*

21. $f(x) = 3x^2 - 6x + 11$
22. $f(x) = 2x^2 + 8x - 2$
23. $f(x) = -4x^2 + 16x - 9$
24. $f(x) = -x^2 + 8x - 7$
25. $f(x) = x^3 - 12x$
26. $f(x) = x^3 - 27x$
27. $f(x) = x^3 - 3x^2 - 12x$
28. $f(x) = 2x^3 + 3x^2 - 36x$

*In Problems 29–34, determine the interval(s) on which the graph of f(x) is concave up and the interval(s) on which it is concave down (see the discussion preceding Example E).*

29. $f(x) = x^3 - 12x$
30. $f(x) = x^3 - 27x$

31. $f(x) = x^3 - 3x^2 - 12x$
32. $f(x) = 2x^3 + 3x^2 - 36x$
33. $f(x) = x^4 - 24x^2 + 12x - 24$
34. $f(x) = x^4 - 6x^3 - 12x + 24$
35. Sketch the graph of $f(x) = x^3 - 12x$. Be sure to make use of what you learned in Problems 25 and 29.
36. Sketch the graph of $f(x) = x^3 - 27x$ (see Problems 26 and 30).
37. Sketch the graph of $f(x) = x^3 - 3x^2 - 12x$ (see Problems 27 and 31).
38. Sketch the graph of $f(x) = 2x^3 + 3x^2 - 36x$ (see Problems 28 and 32).

*In each of Problems 39–42, we have shown the graph of a function $y = f(x)$. Indicate where (a) $f'(x) = 0$, (b) $f'(x) < 0$, (c) $f''(x) > 0$.*

39.

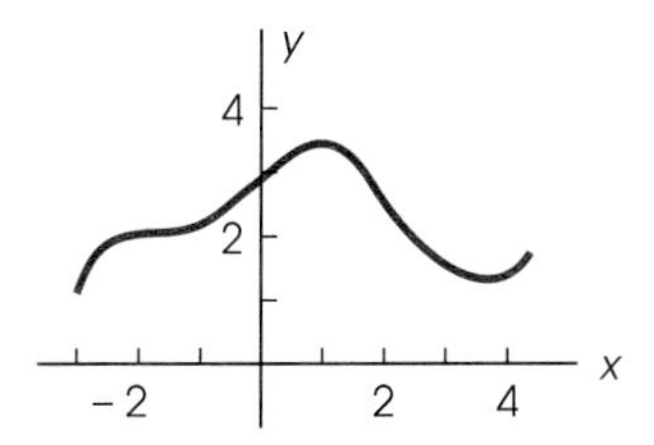


40.

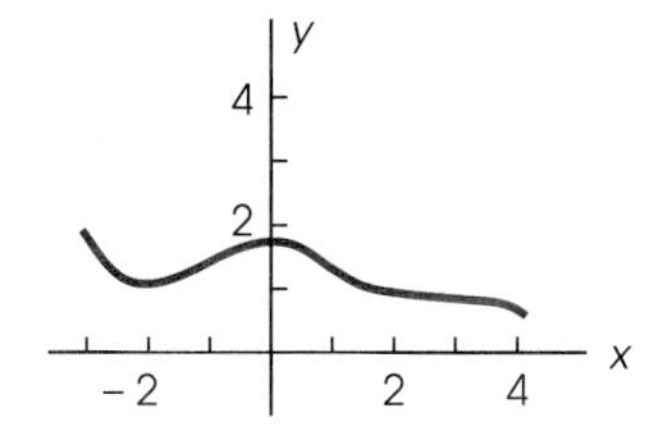


41.

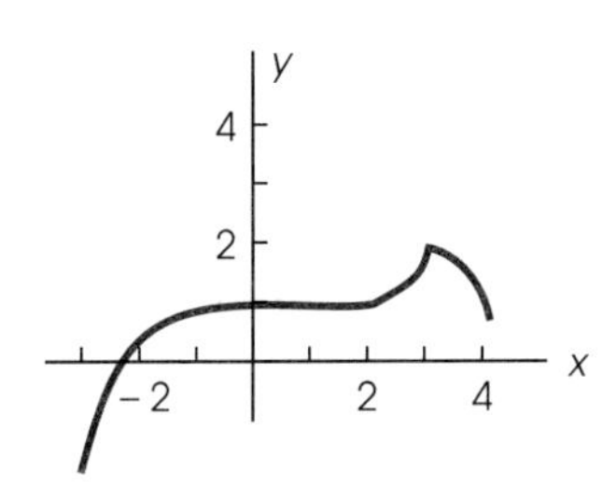


42.

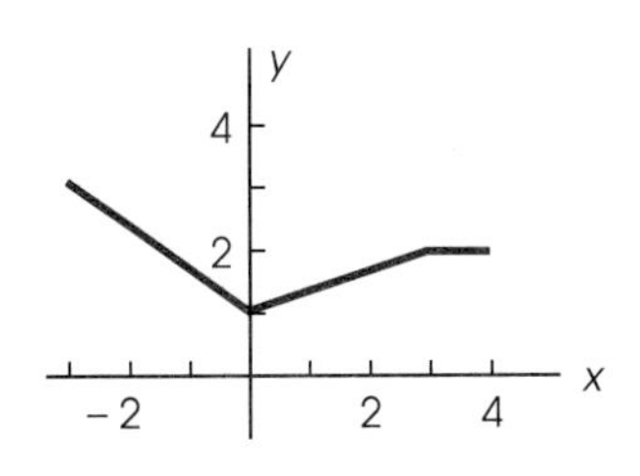


*In Problems 43–50, sketch the graph of a function that satisfies all of the given conditions (see Examples D and E).*

43. (i) $f$ is continuous everywhere.
(ii) $f(-2) = 4$ and $f(4) = -2$.
(iii) $f'(-2) = f'(4) = 0$.
(iv) $f''(x) < 0$ on $(-\infty, 1)$ and $f''(x) > 0$ on $(1, \infty)$.
44. (i) $f$ is continuous everywhere.
(ii) $f(0) = 2$ and $f(6) = 7$.
(iii) $f'(0) = f'(6) = 0$.
(iv) $f''(x) > 0$ on $(-\infty, 3)$ and $f''(x) < 0$ on $(3, \infty)$.
45. (i) $f$ is continuous everywhere.
(ii) $f(-1) = 2$, $f(3) = 5$, and $f(7) = 1$.
(iii) $f'(-1) = f'(3) = f'(7) = 0$.
(iv) $f'(x) > 0$ on $(-1, 3) \cup (7, \infty)$.
(v) $f''(x) > 0$ on $(-\infty, 1) \cup (5, \infty)$ and $f''(x) < 0$ on $(1, 5)$.
46. (i) $f$ is continuous everywhere.
(ii) $f(-5) = 5$, $f(0) = 3$, and $f(5) = 10$.
(iii) $f'(-5) = f'(0) = f'(5) = 0$.
(iv) $f'(x) < 0$ on $(-5, 0) \cup (5, \infty)$.
(v) $f''(x) > 0$ on $(-2, 3)$ and $f''(x) < 0$ on $(-\infty, -2) \cup (3, \infty)$.
47. (i) $f$ is continuous everywhere except at $x = 6$.
(ii) $\lim_{x\to 6^-} f(x) = 6$ and $\lim_{x\to 6^+} f(x) = 4$.
(iii) $f'(x) = 0$ on $(-\infty, 2)$.
(iv) $f''(x) < 0$ on $(2, 6)$ and $f''(x) > 0$ on $(6, \infty)$.
48. (i) $f$ is continuous everywhere except at $x = 2$.
(ii) $\lim_{x\to 2^-} f(x) = 2$ and $\lim_{x\to 2^+} f(x) = 0$.
(iii) $f'(x) = 1$ on $(2, \infty)$.
(iv) $f''(x) > 0$ on $(-\infty, -2)$ and $f''(x) < 0$ on $(-2, 2)$.
49. (i) $f$ is continuous everywhere.
(ii) $f'(x) = 1$ on $(-2, 1)$ and $f'(x) = -1$ on $(1, 3)$.
(iii) $f''(x) > 0$ on $(-\infty, -2)$ and $f''(x) < 0$ on $(3, \infty)$.
50. (i) $f$ is continuous everywhere.
(ii) $f'(x) = 0$ on $(2, 5)$ and $f'(x) < 0$ on $(5, \infty)$.
(iii) $f''(x) > 0$ on $(-\infty, 2)$ and $f''(x) < 0$ on $(5, \infty)$.

---

*Solve the inequalities in Problems 51–58, writing your answers in interval notation.*

51. $3(x - \frac{1}{2}) < 5(x + \frac{2}{3})$
52. $\frac{2}{3}(x + 2) > \frac{3}{4}(2x - 1)$
53. $x^2 - 10x \geq -16$
54. $x^2 - 7x \leq 18$
55. $(x + 3)(x - 2)^2(x - 4) < 0$
56. $(x + 1)^3(x - 1)(x - 2) > 0$

57. $(x + 3)/(x - 2) > 0$
58. $(x + 1)^2/(x - 3) < 0$

*In each of Problems 59–62, interpret the given statement in terms of derivatives and sketch a possible graph.*

59. From 1970 to 1990, unemployment $U = f(t)$ in a certain country continued to rise, but at a slower and slower rate.
60. The number $N = g(t)$ of cases of AIDS in the United States grew more and more rapidly from 1980 to 1990.
61. When given an antibiotic, John's temperature $T = f(t)$ fell rapidly at first but then more and more slowly and it finally leveled off at 99°F after 10 hours.
62. The number $I = g(t)$ of items produced per hour after $t$ weeks by a new production worker tends to grow rapidly at first but then more and more slowly, leveling off after about 15 weeks.
63. In the presence of an unlimited food supply, the population $y = g(t)$ of ants in a certain area at time $t$ will tend to grow in such a way that $dy/dt = ky$, where $k$ is a positive constant. Sketch a possible graph for $y$. *Hint:* Note that $dy/dt$ is always positive, so $y$ is increasing with $t$. What happens to $dy/dt$ as $t$ increases?
64. The amount $A = f(t)$ of the radioactive element radium in a sample at time $t$ satisfies the equation $dA/dt = -kA$, where $k$ is a positive constant. Sketch a possible graph for $A$.
65. In a city with population $P$, the number of people $N = f(t)$ who have heard a certain rumor $t$ days after it was started tends to satisfy the equation $dN/dt = kN(P - N)$, where $k$ is a positive constant. Sketch a possible graph for $N$.

## 3.3 EXTREME POINTS AND INFLECTION POINTS

We begin with the question: What are the most important points on a graph? A moment's reflection may suggest the correct answer: the points where the graph changes character. Chief among these are the *extreme points,* where the graph shifts from rising to falling (or vice versa), and the *inflection points,* where the graph changes its concavity from up to down (or vice versa). It is these points that play the major role in many applications of calculus to problems of the real world (see Chapter 4). Our task for this section is to locate precisely all extreme points and inflection points of a graph. Naturally, our tools are the first and second derivatives.

### Theory of Extrema

Extreme values are of two types—maximum values and minimum values. These words are so important that we define them formally.

> DEFINITION
> Let $I$, the domain of $f$, contain the point $c$. We say that
>
> (i) $f(c)$ is the **maximum value** of $f$ on $I$ if $f(c) \geq f(x)$ for all $x$ in $I$.
> (ii) $f(c)$ is the **minimum value** of $f$ on $I$ if $f(c) \leq f(x)$ for all $x$ in $I$.

Since the maximum value is the largest value of $f$ on the whole domain, it is sometimes called the **global maximum value.** This contrasts with a **local**

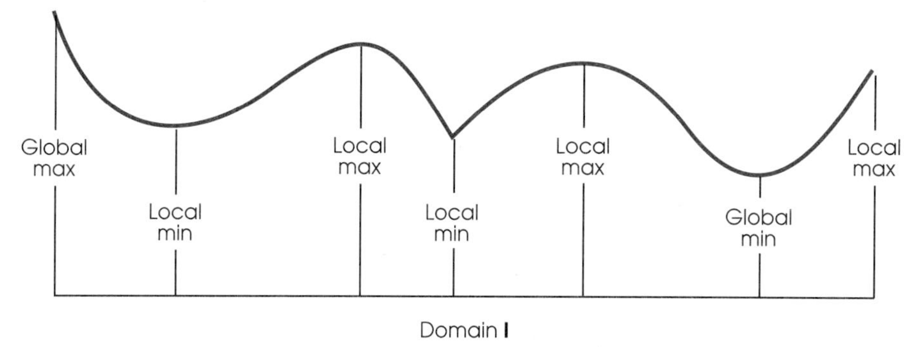


**FIGURE 26**

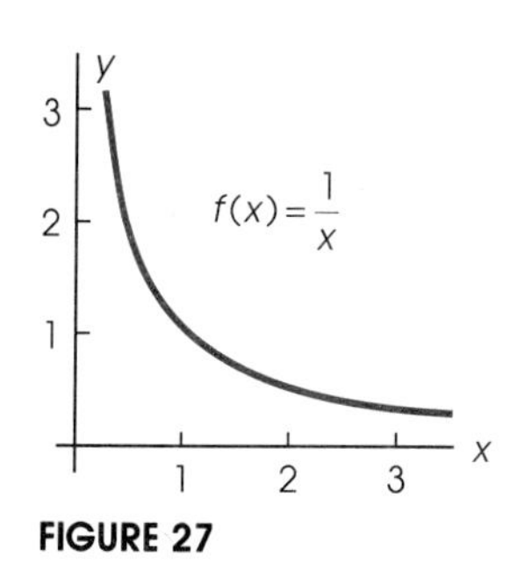


**FIGURE 27**

**maximum** value; for it we only require that $f(c)$ be the largest value of $f$ in a neighborhood (local region) about $c$. All of this is illustrated in Figure 26.

Not every function has a maximum value. Consider, for example, $f(x) = 1/x$ on the interval $(0, \infty)$ which clearly has no maximum value (Figure 27). Or consider the same function on the interval $(1, 2)$; it attains no maximum value even though it is bounded. Note that our definition requires that the function assume the maximum value. The function $f(x) = 1/x$ gets arbitrarily close to the value 1 on $(1, 2)$, but it never takes on this value. Actually, closed intervals are the natural domains to use in extrema theory because of the following theorem.

> **POINT OF INTEREST**
> For $f$ to be continuous on a closed interval $[a, b]$, it is sufficient that $f$ be continuous at each point of $(a, b)$ and that
>
> $$\lim_{x \to a^+} f(x) = f(a)$$
>
> and
>
> $$\lim_{x \to b^-} f(x) = f(b)$$

> **MAX-MIN EXISTENCE THEOREM**
> If $f$ is continuous on a closed interval $[a, b]$, then $f$ attains both a maximum value and a minimum value there.

Knowing that a function has a maximum value and a minimum value is helpful, but what we really want to know is how to locate the points where these values occur. A careful look at Figure 26 suggests the following result.

> **MAX-MIN LOCATION THEOREM**
> Let $f$ be defined on an interval $I$ containing the point $c$. If $f(c)$ is a maximum or minimum value (global or local), then $c$ is one of the following:
>
> (i) an endpoint of $I$,
> (ii) a point where the derivative is zero $[f'(c) = 0]$,
> (iii) a point where the derivative doesn't exist.

Putting the two theorems together, we can offer a simple procedure for finding the maximum and minimum values for a continuous function $f$ on a closed interval $[a, b]$. Simply find all the **critical points** [points of the domain that are of type (i), (ii), or (iii) above]; then evaluate $f$ at each of these points. The largest of these values is the maximum value, and the smallest is the minimum value.

## EXAMPLE A

Find the maximum and minimum values of

$$f(x) = x^3 - 3x^2 + 1$$

on the closed interval $[-\frac{1}{2}, 4]$.

**Solution** The critical points are the endpoints $-\frac{1}{2}$ and 4 and the points where the derivative is zero. To locate the latter we find $f'(x)$, set it equal to zero, and solve.

$$f'(x) = 3x^2 - 6x$$

$$3x(x - 2) = 0$$

$$x = 0, 2$$

The values at the four critical points are $f(-\frac{1}{2}) = \frac{1}{8}$, $f(0) = 1$, $f(2) = -3$, and $f(4) = 17$. Thus the minimum value is $-3$ and the maximum value is 17. Figure 28 makes the situation completely clear.

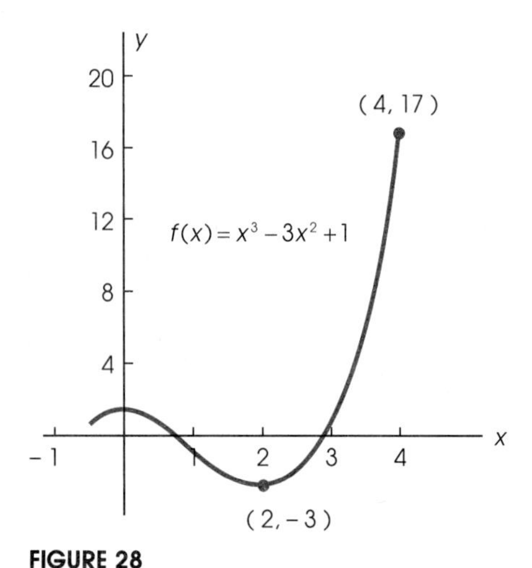


FIGURE 28

■

## EXAMPLE B

Find the maximum and minimum values of $f(x) = x^{2/3}$ on $[-1, 8]$.

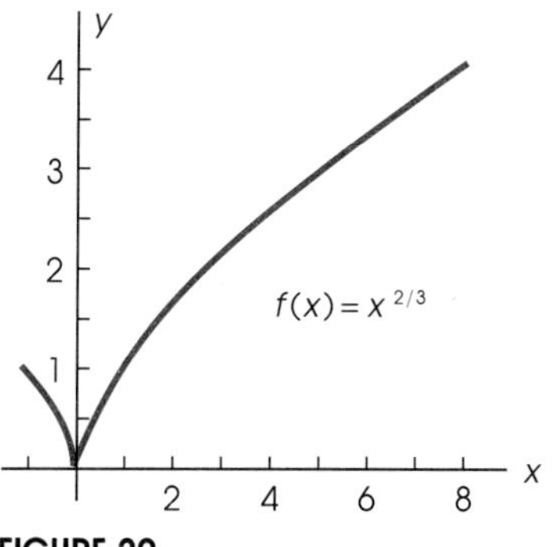


**FIGURE 29**

**Solution** $f'(x) = \frac{2}{3}x^{-1/3} = 2/(3\sqrt[3]{x})$ which is never 0. However, 0 is a critical point since $f'(0)$ fails to exist. In addition there are the two endpoints $-1$ and 8. Since $f(-1) = 1$, $f(0) = 0$, and $f(8) = 4$, the minimum value is 0 and the maximum value is 4. We don't need to go further but, to develop confidence in our methods, we show the graph in Figure 29. ■

### EXAMPLE C

Find the maximum and minimum values of

$$f(x) = \frac{x}{x^2 + 2}$$

on $[-1, 4]$.

**Solution**

$$f'(x) = \frac{(x^2 + 2)1 - x(2x)}{(x^2 + 2)^2} = \frac{-x^2 + 2}{(x^2 + 2)^2}$$

When we set this expression equal to 0 and solve, we get the two solutions $\pm\sqrt{2}$, but only $\sqrt{2}$ is in the interval $[-1, 4]$. The critical points are therefore $-1$, $\sqrt{2}$, and 4. Since $f(-1) = -\frac{1}{3} \approx -0.3333$, $f(\sqrt{2}) = \sqrt{2}/4 \approx 0.3536$, and $f(4) = \frac{4}{18} \approx 0.2222$, we conclude that the maximum value is $\sqrt{2}/4$ and the minimum value $-\frac{1}{3}$. ■

## Tests for Local Extrema

While it is global maxima and minima that are of most interest, we often wish to locate and evaluate local maxima and minima too. Fortunately, the location theorem applies to local extrema as well as global extrema; local extrema, too, can occur only at critical points. But once we have found the critical points, we would like a simple way of deciding whether they yield local maxima, local minima, or neither. The three graphs in Figure 30 suggest an important test.

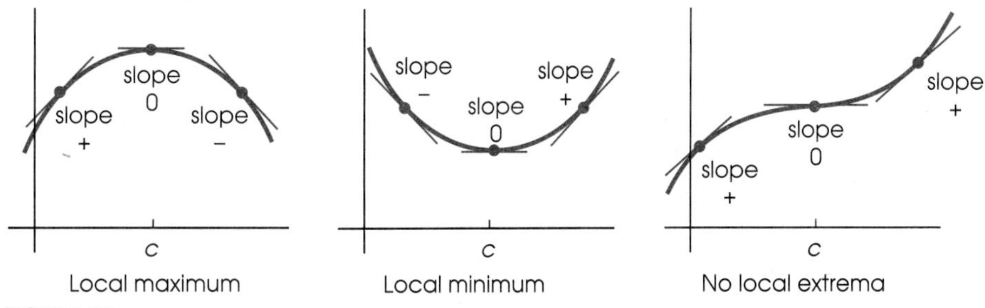


**FIGURE 30**

**FIRST DERIVATIVE TEST FOR LOCAL EXTREMA**

Let $f$ be continuous on an interval containing the critical point $c$ in its interior.

(i) If $f'(x) > 0$ to the left of $c$ and $f'(x) < 0$ to the right of $c$, then $f(c)$ is a local maximum value.
(ii) If $f'(x) < 0$ to the left of $c$ and $f'(x) > 0$ to the right of $c$, then $f(c)$ is a local minimum value.
(iii) If $f'(x)$ has the same sign on both sides of $c$, then $f(c)$ is neither a local maximum nor a local minimum value.

## EXAMPLE D

Find all local extrema of

$$f(x) = \tfrac{1}{3}x^3 - \tfrac{1}{2}x^2 - 2x + 1$$

on $(-\infty, \infty)$ and identify them correctly.

### Solution

$$f'(x) = x^2 - x - 2 = (x - 2)(x + 1)$$

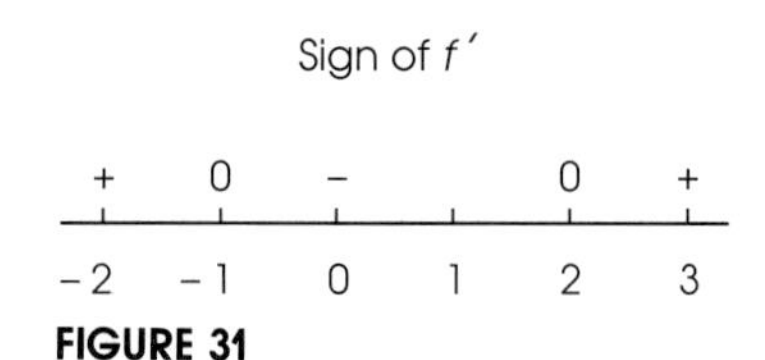

FIGURE 31

The critical points $-1$ and 2 split the real line into the three intervals $(-\infty, -1)$, $(-1, 2)$, and $(2, \infty)$. On testing these intervals with the test points $-2$, 0, and 3, we find $f'(-2) > 0$, $f'(0) < 0$, and $f'(3) > 0$ (Figure 31). We conclude that $f(-1) \approx 2.17$ is a local maximum value and $f(2) \approx -2.33$ is a local minimum value. ■

Look at Figure 30 again and note that the graph of $f$ is concave down at the local maximum and concave up at the local minimum. This suggests another test.

**SECOND DERIVATIVE TEST FOR LOCAL EXTREMA**

Let $f'$ and $f''$ exist on an interval containing $c$ in its interior and suppose $f'(c) = 0$.

(i) If $f''(c) < 0$, then $f(c)$ is a local maximum value.
(ii) If $f''(c) > 0$, then $f(c)$ is a local minimum value.
(iii) If $f''(c) = 0$, the test fails.

Refer again to Example D for which

$$f(x) = \tfrac{1}{3}x^3 - \tfrac{1}{2}x^2 - 2x + 1$$

$$f'(x) = x^2 - x - 2 = (x - 2)(x + 1)$$

$$f''(x) = 2x - 1$$

At the critical points $-1$ and 2, $f''(-1) = -3 < 0$ and $f''(2) = 3 > 0$. The Second Derivative Test says that $f(-1)$ is a local maximum value and $f(2)$ is a local minimum value.

### EXAMPLE E

Identify the critical points of

$$f(x) = \tfrac{1}{4}x^4 - x^3 - 5x^2 + 1$$

and use the Second Derivative Test to determine which yield local maximum values and which yield local minimum values.

**Solution**

$$f'(x) = x^3 - 3x^2 - 10x = x(x - 5)(x + 2)$$

$$f''(x) = 3x^2 - 6x - 10$$

The critical points are $-2$, 0, and 5. Since $f''(-2) = 14 > 0$, $f''(0) = -10 < 0$, and $f''(5) = 35 > 0$, we conclude that

$$f(-2) = -7 \text{ is a local minimum value.}$$

$$f(0) = 1 \text{ is a local maximum value.}$$

$$f(5) = -92.75 \text{ is a local minimum value.}$$

■

## Inflection Points

A point on the graph of a continuous function $f$ where the concavity changes from up to down, or vice versa, is called an **inflection point.** Candidates for inflection points are points where

(i) $f''(x) = 0$ (Figure 32), or

(ii) $f''(x)$ does not exist (Figure 33).

We can establish that a candidate is legal by checking that $f''$ changes sign there.

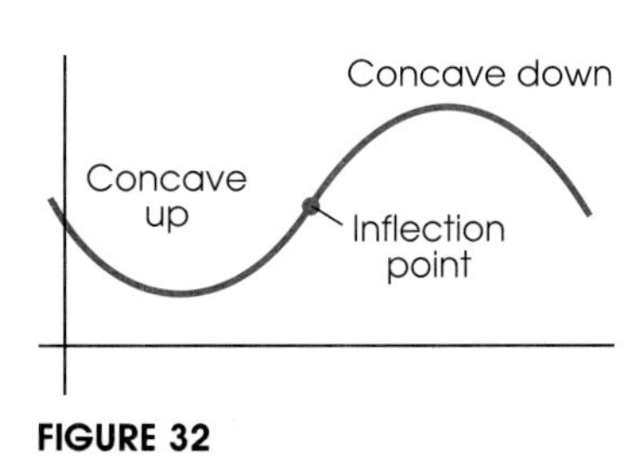


FIGURE 32

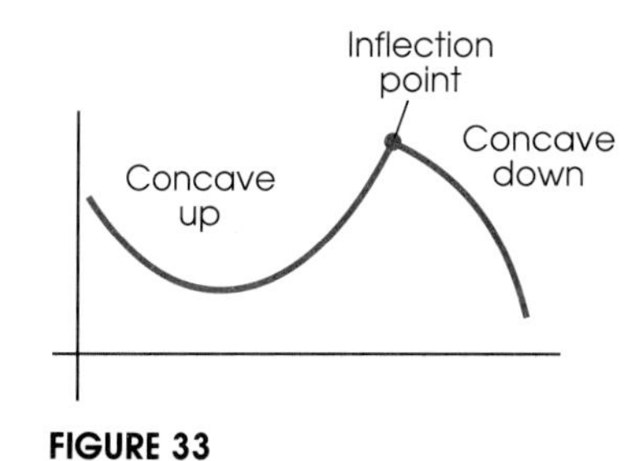


FIGURE 33

## EXAMPLE F

Find the inflection points on the graph of

$$f(x) = \tfrac{1}{12}x^4 + \tfrac{1}{3}x^3 + 2x + 3$$

### Solution

$$f'(x) = \tfrac{1}{3}x^3 + x^2 + 2$$

$$f''(x) = x^2 + 2x = x(x + 2)$$

The only candidates for inflection points are at $x = 0$ and $x = -2$ where $f''(x) = 0$. A quick check shows that the sign of $f''$ does change at these points (Figure 34). The inflection points with both coordinates given are $(-2, -2.3)$ and $(0, 3)$. ■

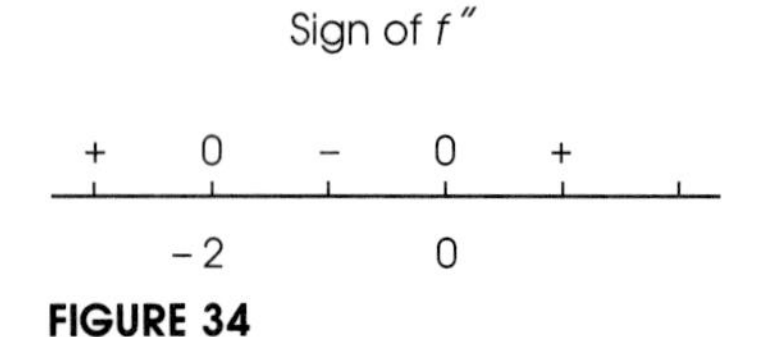


FIGURE 34

Our final example wraps together most of the ideas of this section.

## EXAMPLE G

Sketch the graph of

$$f(x) = 3x^4 - 4x^3 + 2$$

showing all extreme points and inflection points.

### Solution

$$f'(x) = 12x^3 - 12x^2 = 12x^2(x - 1)$$

$$f''(x) = 36x^2 - 24x = 12x(3x - 2)$$

The critical points (candidates for extrema) are $x = 0$ and $x = 1$. However $f'$ does not change sign at 0, so there is no extremum there. At $x = 1$, there is a sign change from negative to positive, and so by the First Derivative Test there is a local minimum at $x = 1$.

The candidates for inflection points are at $x = 0$ and $x = \frac{2}{3}$, and since $f''$ does change sign at both places, we declare their legitimacy.

The graph is sketched in Figure 35 with both coordinates of the key points noted.

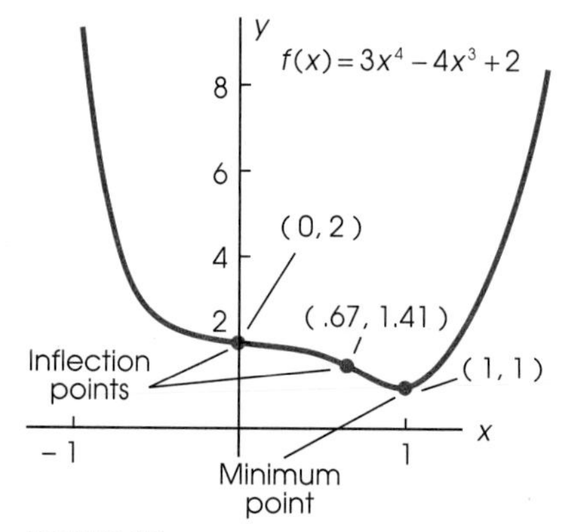


FIGURE 35 ■

## Problem Set 3.3

*Problems 1 and 2 show the graph of a function $f$ on the interval $[p, u]$ with critical values $p$, $q$, $r$, $s$, $t$, and $u$. Indicate which of these critical values will, when $f$ is evaluated there, give the global maximum value and the global minimum value. Also indicate which of the others will give a local maximum value and a local minimum value.*

1.

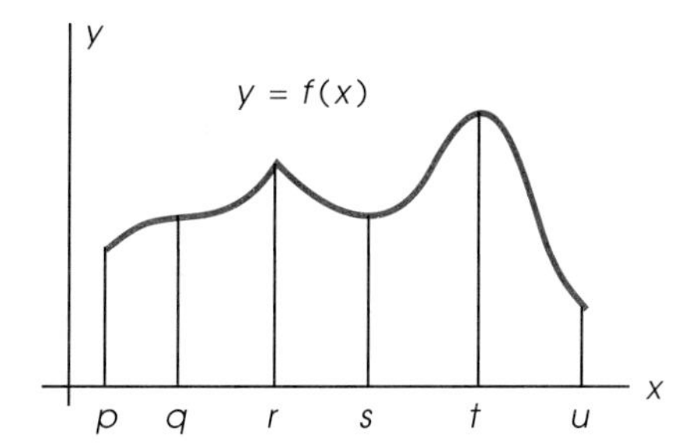


2.

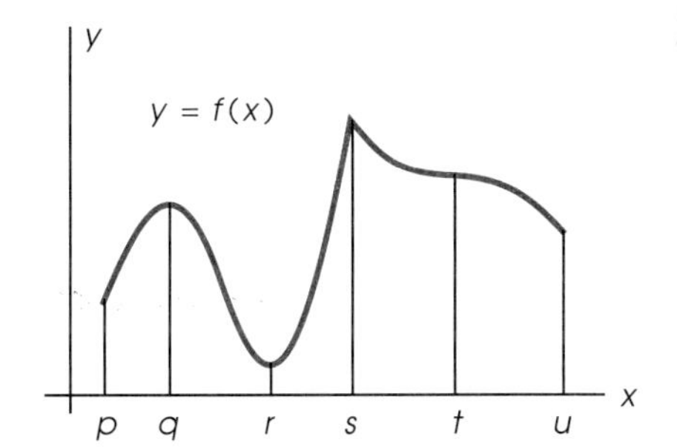


*Problems 3–8 show the graph of a function $f$ on the interval $[0, \infty)$. Indicate whether $f$ assumes a (global) maximum value and/or a (global) minimum value and, if so, give these value(s).*

3.

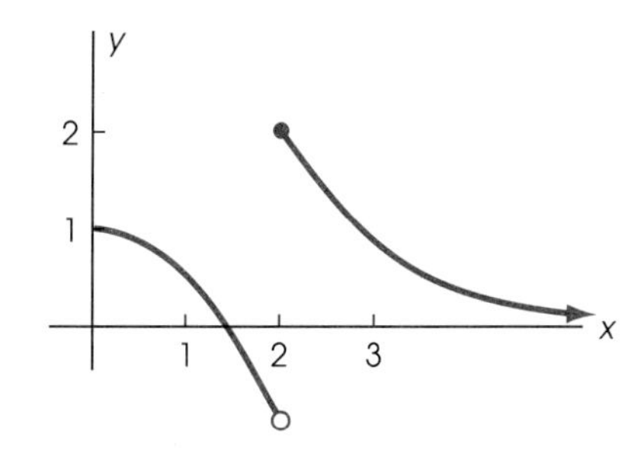


4.

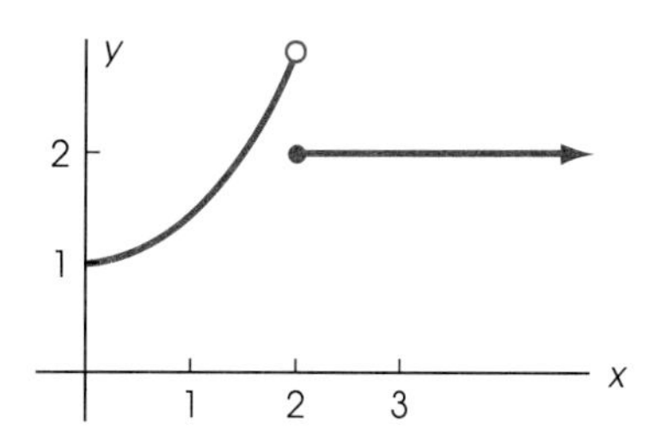


5.

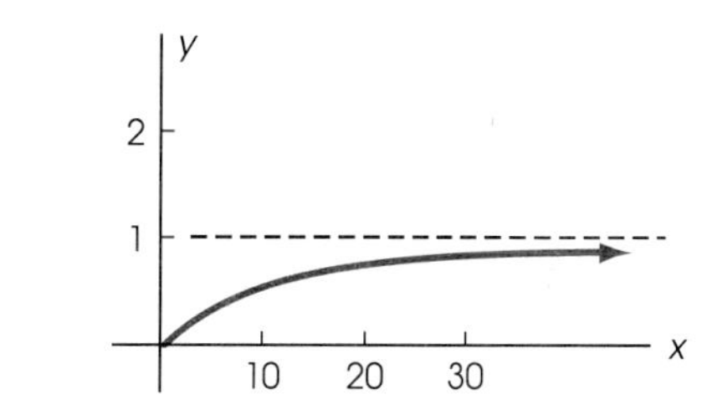


6.

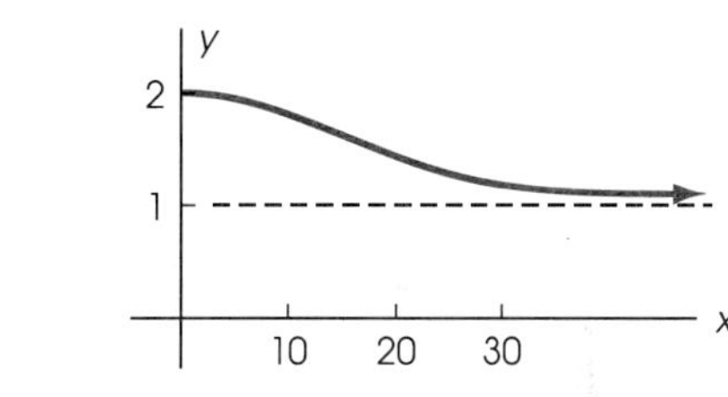


7.

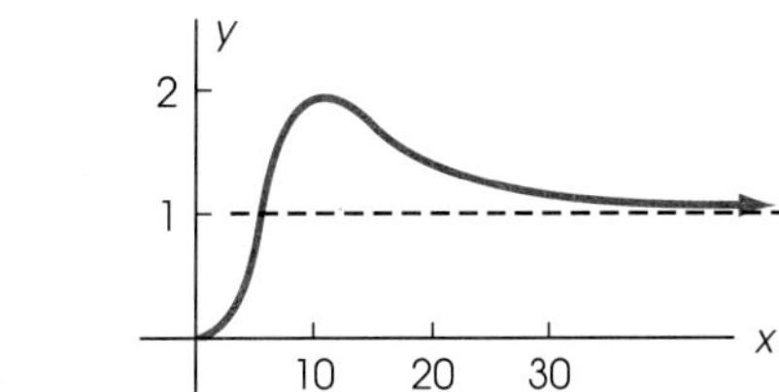


8.

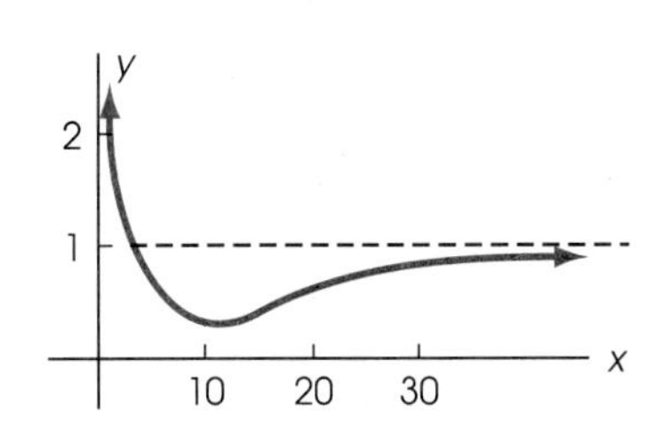

*In Problems 9–26, find the maximum and minimum values of f(x) on the indicated closed interval. See Examples A, B, and C.*

9. $f(x) = 2x^2 - 8x$; $[0, 3]$
10. $f(x) = 3x^2 + 6x$; $[-3, 2]$
11. $f(x) = x^2 - 6x + 3$; $[0, 5]$
12. $f(x) = 2x^2 - 8x + 3$; $[-1, 3]$
13. $f(x) = x^3 - 3x + 1$; $[-2, 3]$
14. $f(x) = x^3 - 12x + 3$; $[-3, 4]$
15. $f(x) = 2x^3 - 3x^2 - 12x + 2$; $[-3, 3]$
16. $f(x) = x^3 + 3x^2 - 24x + 10$; $[-5, 4]$
17. $f(x) = 2x^{1/3}$; $[-8, 1]$
18. $f(x) = x^{2/5}$; $[-1, 32]$
19. $f(x) = 3(x - 2)^{1/3}$; $[-6, 3]$
20. $f(x) = 2(x + 3)^{2/3}$; $[-4, 5]$
21. $f(x) = 3(x - 2)^{1/3} - x$; $[-6, 3]$
22. $f(x) = 2(x + 3)^{2/3} - \frac{4}{3}x$; $[-4, 5]$
23. $f(x) = 3x/(x^2 + 1)$; $[-2, 3]$
24. $f(x) = 2x/(x^2 + 4)$; $[-2, 3]$
25. $f(x) = x^2/(x^2 + 1)$; $[-2, 3]$
26. $f(x) = 2x^2/(x^2 + 4)$; $[-4, 2]$

*In Problems 27–34, find all critical points and use the First Derivative Test to determine which yield local maximum values and which yield local minimum values (see Example D).*

27. $f(x) = 2x^3 - 3x^2 + 4$
28. $f(x) = x^3 - 3x + 3$
29. $f(x) = -\frac{1}{4}x^3 + 3x - 2$
30. $f(x) = \frac{2}{27}x^3 - 2x + 1$
31. $f(x) = 2x^3 - 15x^2 + 24x + 10$
32. $f(x) = 2x^3 - 15x^2 + 36x - 20$
33. $f(x) = x + 4/x$
34. $f(x) = x^2 - 2/x$

*In Problems 35–40, find all critical points and use the Second Derivative Test to decide which give local maximum values and which give local minimum values (see Example E).*

35. $f(x) = x^3 - 6x^2 + 9x - 3$
36. $f(x) = x^3 + 3x^2 + 3x - 3$
37. $f(x) = x^4 - 8x^2 + 12$
38. $f(x) = 3x^4 + 8x^3 - 90x^2 + 4$
39. $f(x) = x - 4/x^2$
40. $f(x) = x^2 + 16/x$

*In Problems 41–48, find all inflection points as in Example F.*

41. $f(x) = x^3 - 12x$
42. $f(x) = x^3 - 27x$
43. $f(x) = x^3 - 6x^2 + 9x - 3$
44. $f(x) = x^3 + 3x^2 + 3x - 3$
45. $f(x) = x^4 - 4x^3 - 18x^2 + 10$
46. $f(x) = x^4 + 8x^3 + 24x - 20$
47. $f(x) = 3x^2 - x^{-2}$
48. $f(x) = 4(x^2 - x^{-1})$

*Sketch the graph of f in each of Problems 49–52, showing the coordinates of all local extreme points and inflection points.*

49. $f(x) = x^3 - 6x^2 + 9x - 3$ (Use the results from Problems 35 and 43.)
50. $f(x) = x^3 + 3x^2 + 3x - 3$ (Use the results from Problems 36 and 44.)
51. $f(x) = x^4 - 4x^3 + 10$
52. $f(x) = x^4 - 2x^3 - 5$

---

*In Problems 53–56, find the maximum and minimum values of f on the indicated closed interval.*

53. $f(x) = 2x + 8/x$; $[1, 4]$
54. $f(x) = x^2 + 4/\sqrt{x}$; $[\frac{1}{4}, 4]$
55. $f(x) = \begin{cases} x + 6 & \text{if } -1 \le x \le 0 \\ x^2 - 4x + 6 & \text{if } 0 < x \le 3 \end{cases}$
56. $f(x) = x^{4/3} - 4x^{1/3} + 2$; $[\frac{1}{8}, 8]$

*In Problems 57–60, sketch the graph of f, showing the coordinates of all local extreme points and all inflection points.*

57. $f(x) = x^3 - 3x^2 + 8$
58. $f(x) = x^5 - \frac{20}{3}x^3 - 10$
59. $f(x) = \begin{cases} -x^2 + 4 & \text{if } x \le 1 \\ x^2 - 4x + 6 & \text{if } x > 1 \end{cases}$
60. $f(x) = x + 9/x$
61. Sketch the graph of a continuous function $f$ with domain $[0, 4)$ having its global minimum at $x = 0$, an inflection point at $x = 1$, and a global maximum at $x = 2$.

62. Sketch the graph of a continuous function $f$ with domain $[0, 4)$ which has a global maximum at $x = 0$, a local minimum at $x = 1$, a local maximum at $x = 3$, an inflection point at $x = 2$, and no global minimum.

*Suppose that $f$ has the graph shown in Problems 63 and 64. Sketch a possible graph for $f'$.*

63.

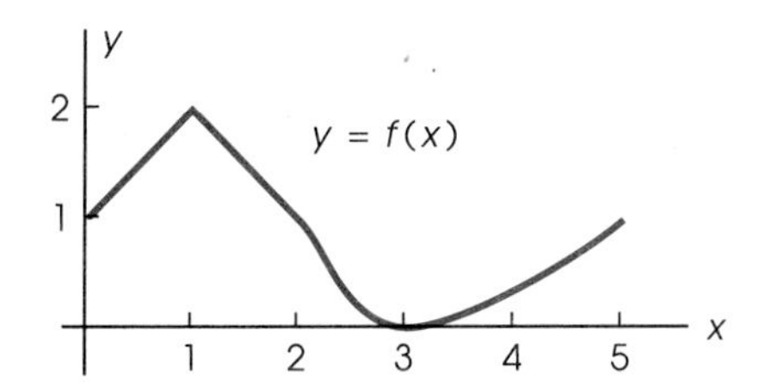


64.

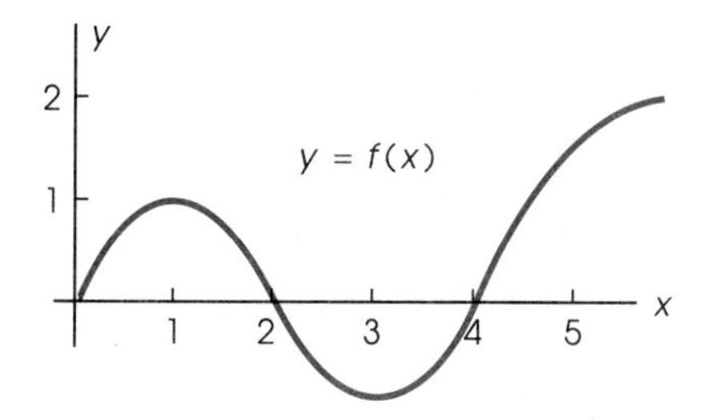


*Suppose that $f(0) = 0$ and that $f'$ has the graph shown in Problems 65 and 66. Sketch a possible graph of $f$.*

65.

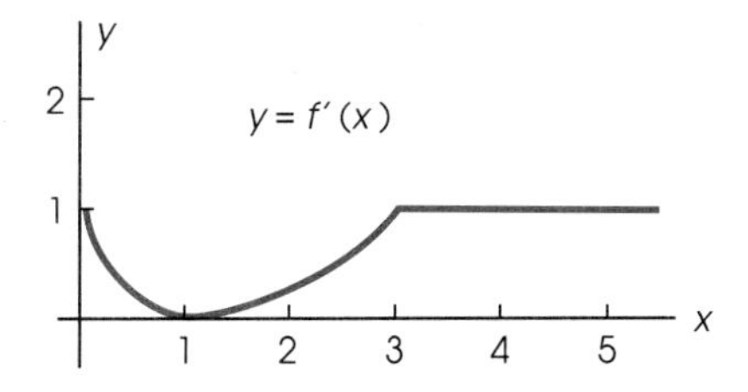


66.

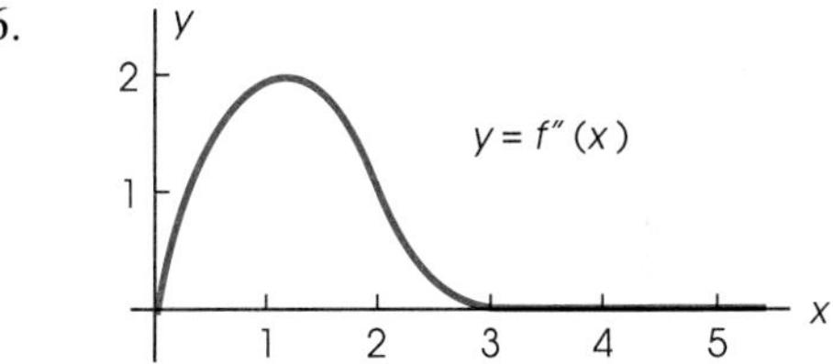


67. The concentration $f(t)$ of a certain chemical in the blood of a patient $t$ hours after injection was given by $f(t) = 4t^2/(64 + t^4)$. Sketch the graph of $f$ and determine when the concentration was greatest.
68. The cost $C$ in dollars of producing $x$ items of a certain product is given by the formula $C = 200 + 5x + 0.1x^2$. Find the value of $x$ that will minimize the *average cost* of the product.

## 3.4 LIMITS INVOLVING INFINITY AND ASYMPTOTES

Our notation for infinite intervals (for example, $(4, \infty)$ and $(-\infty, 6]$) has alerted you to one use for the symbols $\infty$ and $-\infty$. They provide a shorthand way of indicating that an interval is unbounded to the right or to the left. Please note that we have not referred to $\infty$ and $-\infty$ as numbers. For example, we have never tried to add $\infty$ to a number or to divide $\infty$ by a number.

In this section, we develop other uses for the symbols $\infty$ and $-\infty$. We write

$$\lim_{x\to\infty} f(x) = L$$

and call it a **limit at infinity.** In writing $x \to \infty$, we simply mean that $x$ gets larger and larger without bound. And in writing the limit statement above, we mean to say that $f(x)$ gets arbitrarily close to $L$ as $x$ gets larger and larger.

Similarly, in writing

$$\lim_{x \to a} f(x) = \infty$$

called an **infinite limit,** we are saying that as $x$ approaches $a$, $f(x)$ gets larger and larger without bound. All of this will become clearer as we work a number of examples.

## Limits at Infinity

It seems rather obvious that

$$\lim_{x \to \infty} \frac{1}{\sqrt{x}} = 0$$

After all, the numerator is fixed while the denominator grows without bound as $x \to \infty$. The table in Figure 36 confirms our reasoning. The same kind of thinking leads to the conclusion that if $p > 0$, then

$$\lim_{x \to \infty} \frac{1}{x^p} = 0$$

| $x$ | $1/\sqrt{x}$ |
|---|---|
| 200 | .07091 |
| 2000 | .02236 |
| 20000 | .00707 |
| 200000 | .00224 |
| 2000000 | .00071 |
| 20000000 | .00022 |

**FIGURE 36**

Here is another example with a similar theme.

### EXAMPLE A

Find $\lim_{x \to \infty} (3x^2 + 2)/(2x^3 - 1)$.

**Solution** We first divide numerator and denominator by $x^3$, the highest power appearing in the denominator. Then we use the properties of limits stated in the Main Limit Theorem.

$$\lim_{x\to\infty}\frac{3x^2+2}{2x^3-1}=\lim_{x\to\infty}\frac{\dfrac{3x^2+2}{x^3}}{\dfrac{2x^3-1}{x^3}}=\lim_{x\to\infty}\frac{\dfrac{3}{x}+\dfrac{2}{x^3}}{2-\dfrac{1}{x^3}}$$

$$=\frac{\lim\limits_{x\to\infty}\dfrac{3}{x}+\lim\limits_{x\to\infty}\dfrac{2}{x^3}}{\lim\limits_{x\to\infty}2-\lim\limits_{x\to\infty}\dfrac{1}{x^3}}$$

$$=\frac{0+0}{2-0}=0$$ ■

The same procedure works in a problem where the limit is nonzero.

## EXAMPLE B

Find $\lim_{x\to\infty}(2x^2-x+5)/(3x^2+2)$.

**Solution** This time we divide numerator and denominator by $x^2$, which amounts to dividing every term by $x^2$.

$$\lim_{x\to\infty}\frac{2x^2-x+5}{3x^2+2}=\lim_{x\to\infty}\frac{2-\dfrac{1}{x}+\dfrac{5}{x^2}}{3+\dfrac{2}{x^2}}$$

$$=\frac{2-0+0}{3+0}=\frac{2}{3}$$ ■

## EXAMPLE C

Find $\lim\limits_{x\to-\infty}\dfrac{\sqrt{4x^4+1}}{x^2+\sqrt[3]{x}}$.

**Solution**

$$\lim_{x\to-\infty}\frac{\sqrt{4x^4+1}}{x^2+\sqrt[3]{x}}=\lim_{x\to-\infty}\frac{\dfrac{\sqrt{4x^4+1}}{x^2}}{\dfrac{x^2+x^{1/3}}{x^2}}$$

$$=\lim_{x\to-\infty}\frac{\dfrac{\sqrt{4x^4+1}}{\sqrt{x^4}}}{1+\dfrac{1}{x^{5/3}}}=\lim_{x\to-\infty}\frac{\sqrt{\dfrac{4x^4+1}{x^4}}}{1+\dfrac{1}{x^{5/3}}}$$

$$=\lim_{x\to-\infty}\frac{\sqrt{4+\dfrac{1}{x^4}}}{1+\dfrac{1}{x^{5/3}}}=\frac{\sqrt{4+0}}{1+0}=2$$

■

If $\lim_{x\to\infty} f(x) = L$ or if $\lim_{x\to-\infty} f(x) = L$, we call the line $y = L$ a **horizontal asymptote** for the graph of $f$. A horizontal asymptote provides a guideline for drawing the graph of $f$, as we demonstrate in our next example.

## EXAMPLE D

Find the horizontal asymptotes (if any) for the graph of $f(x) = 2x^2/(1 + x^2)$ and show them as dotted lines. Then sketch the graph of $f$.

**Solution**

$$\lim_{x\to\infty}\frac{2x^2}{1+x^2}=\lim_{x\to\infty}\frac{2}{\dfrac{1}{x^2}+1}=\frac{2}{0+1}=2$$

Similarly,

$$\lim_{x\to-\infty}\frac{2x^2}{1+x^2}=2$$

Thus the line $y = 2$ is a horizontal asymptote for the graph (in both directions). With this information and a brief table of values, we can sketch the graph (Figure 37). Note how the graph snuggles up to the line $y = 2$ as $x \to \pm\infty$.

| $x$ | $\dfrac{2x^2}{1+x^2}$ |
|---|---|
| 0 | 0 |
| $\pm 1$ | 1 |
| $\pm 3$ | 1.8 |
| $\pm 10$ | 1.98 |

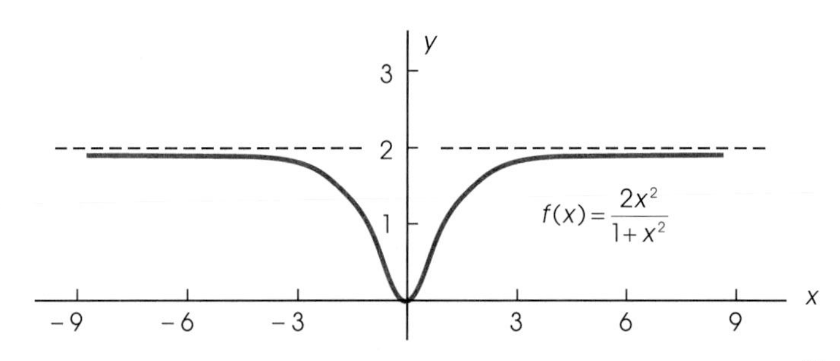


FIGURE 37

■

## Infinite Limits

Consider the graph of $f(x) = 1/(x - 1)$ which is shown in Figure 38. It

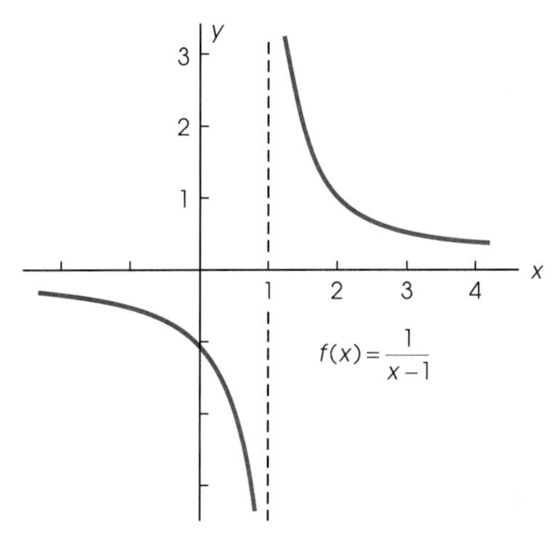


FIGURE 38

makes no sense to ask for $\lim_{x\to 1} 1/(x - 1)$, but it is reasonable to write

$$\lim_{x\to 1^-} \frac{1}{x-1} = -\infty \qquad \lim_{x\to 1^+} \frac{1}{x-1} = \infty$$

We say that the line $x = 1$ is a **vertical asymptote,** since the graph gets nearer and nearer to this line as $x \to 1^-$ and as $x \to 1^+$.

On the other hand, we write

$$\lim_{x\to 1} \frac{1}{(x-1)^2} = \infty$$

since whether $x \to 1^-$ or $x \to 1^+$, $g(x) = 1/(x - 1)^2$ gets larger and larger in the positive direction. Again it is correct to say that the line $x = 1$ is a vertical asymptote (Figure 39). Note in both Figures 38 and 39 that the line $y = 0$ is a horizontal asymptote.

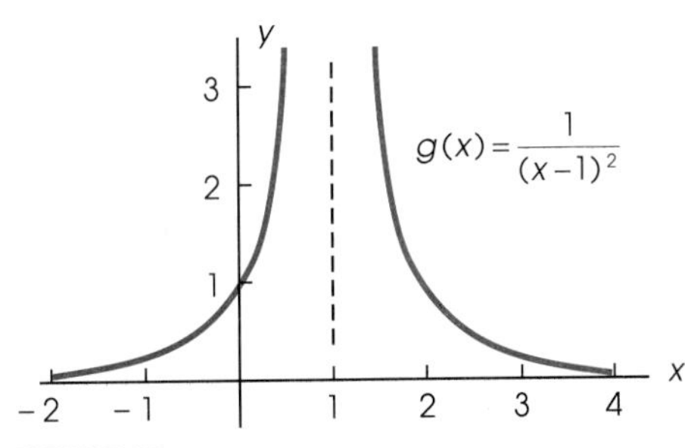


FIGURE 39

## EXAMPLE E

Find the horizontal and vertical asymptotes of the graph of

$$f(x) = \frac{x^2}{x^2 - x - 6} = \frac{x^2}{(x - 3)(x + 2)}$$

Then sketch the graph.

**Solution** Since

$$\lim_{x\to\infty} \frac{x^2}{x^2 - x - 6} = 1 = \lim_{x\to-\infty} \frac{x^2}{x^2 - x - 6}$$

it follows that $y = 1$ is a horizontal asymptote (in both directions).

We expect vertical asymptotes where the denominator is zero, that is, at $x = -2$ and $x = 3$. After noting the sign of the denominator near $-2$ and 3 (Figure 40) and the fact that the numerator is positive, we deduce

$$\lim_{x\to-2^-} f(x) = \infty \qquad \lim_{x\to-2^+} f(x) = -\infty$$
$$\lim_{x\to3^-} f(x) = -\infty \qquad \lim_{x\to3^+} f(x) = \infty$$

Sign of $(x-3)(x+2)$

+    −    +

−2    3

**FIGURE 40**

All of this information together with a brief table of values allows us to sketch the graph shown in Figure 41.

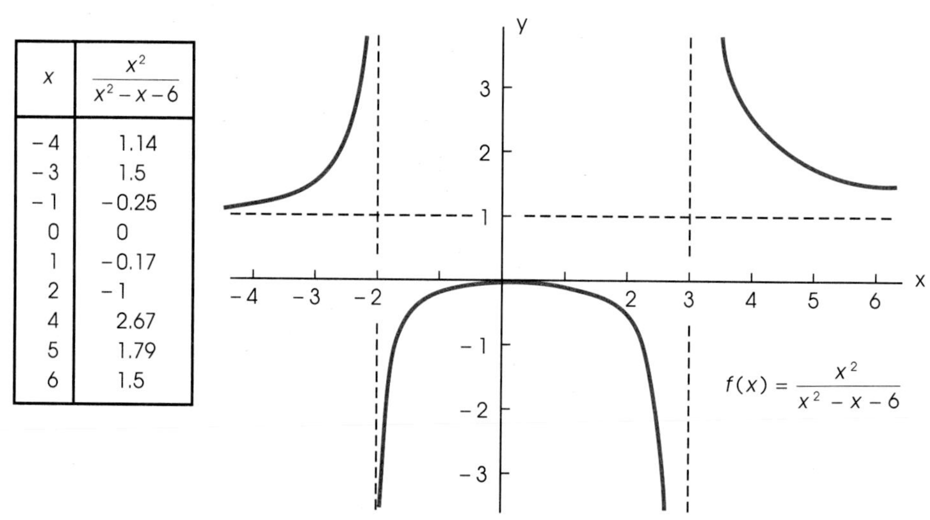

| $x$ | $\frac{x^2}{x^2-x-6}$ |
|---|---|
| −4 | 1.14 |
| −3 | 1.5 |
| −1 | −0.25 |
| 0 | 0 |
| 1 | −0.17 |
| 2 | −1 |
| 4 | 2.67 |
| 5 | 1.79 |
| 6 | 1.5 |



**FIGURE 41**

## Putting It All Together

In our study of graphs, we have developed a number of important concepts. Now it is time to summarize. We do this by offering a suggested graphing procedure.

1. Note any discontinuities.
2. Identify and draw any horizontal or vertical asymptotes.
3. Find $f'(x)$ and $f''(x)$. Use them to locate extreme points and inflection points and to determine intervals of monotonicity and concavity.
4. Make a table of values which includes the extreme points and inflection points.
5. Plot the points in Step 4 and sketch the graph.

**EXAMPLE F**

Sketch the graph of $f(x) = x^2/(x - 1)$.

**Solution**

1. There is a discontinuity at $x = 1$.
2. The line $x = 1$ is a vertical asymptote; there are no horizontal asymptotes.
3. 

$$f'(x) = \frac{(x-1)2x - x^2}{(x-1)^2} = \frac{x^2 - 2x}{(x-1)^2} = \frac{x(x-2)}{(x-1)^2}$$

$$f''(x) = \frac{(x-1)^2(2x-2) - (x^2 - 2x)2(x-1)}{(x-1)^4} = \frac{2}{(x-1)^3}$$

The critical points are 0 and 2, and on checking the sign of $f'$ (Figure 42) we conclude that there is a local maximum at $x = 0$ and a local minimum at $x = 2$. The sign of $f''$ is also shown in Figure 42 demonstrating that the graph of $f$ is concave down on $(-\infty, 1)$ and concave up on $(1, \infty)$.

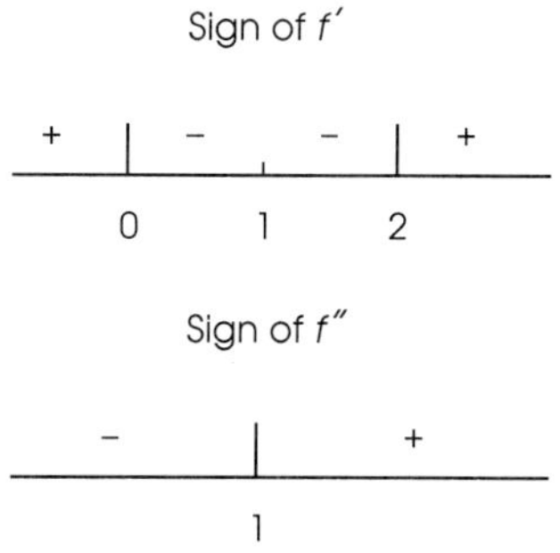

**FIGURE 42**

4. See Figure 43.
5. See Figure 43.

| $x$ | $\frac{x^2}{x-1}$ |
|---|---|
| −2 | −1.33 |
| −1 | −0.5 |
| 0 | 0 |
| 0.5 | −0.5 |
| 1.5 | 4.5 |
| 2 | 4 |
| 3 | 4.5 |
| 4 | 5.33 |

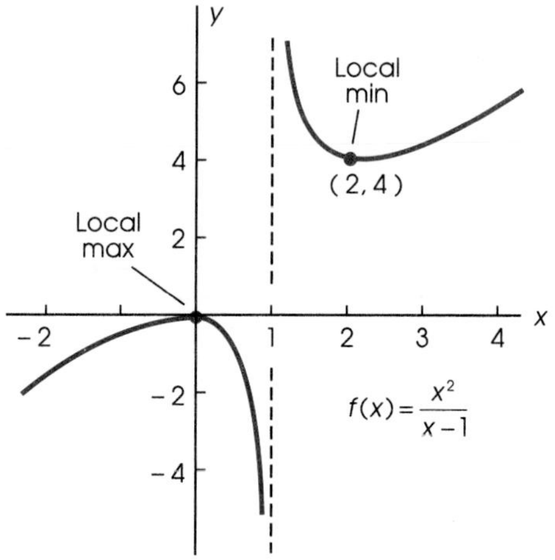

**FIGURE 43**

■

## Problem Set 3.4

*Recall that the statement $\lim_{x\to\infty} f(x) = L$ means that $f(x)$ gets closer and closer to $L$ as $x$ gets larger and larger. Using similar language, give meaning to each of the statements in Problems 1–6.*

1. $\lim_{x\to-\infty} f(x) = 10$
2. $\lim_{x\to-3} f(x) = -\infty$
3. $\lim_{x\to\infty} f(x) = \infty$
4. $\lim_{x\to-4} f(x) = \infty$
5. $\lim_{x\to2^+} f(x) = -\infty$
6. $\lim_{x\to2^-} f(x) = \infty$

*Find each of the limits in Problems 7–20.*

7. $\lim_{x\to\infty} \frac{1}{\sqrt{x}}$
8. $\lim_{x\to\infty} \frac{2}{x^{4/3}}$
9. $\lim_{x\to0} \frac{1}{x^2}$
10. $\lim_{x\to0} \frac{1}{x}$
11. $\lim_{x\to0^+} \frac{1}{\sqrt[3]{x}}$
12. $\lim_{x\to0^-} \frac{1}{\sqrt[3]{x}}$
13. $\lim_{x\to4^-} \frac{1}{x-4}$
14. $\lim_{x\to4^+} \frac{1}{x-4}$
15. $\lim_{x\to3} \frac{1}{(x-3)^4}$
16. $\lim_{x\to-2} \frac{1}{(x+2)^2}$
17. $\lim_{x\to\infty} \left(4 + \frac{3}{x}\right)$
18. $\lim_{x\to-\infty} \left(2 - \frac{2}{x^2}\right)$
19. $\lim_{x\to\infty} \dfrac{5 + \frac{3}{x}}{4 - \frac{2}{x^2}}$
20. $\lim_{x\to\infty} \dfrac{\frac{2}{x}}{1 + \frac{3}{x^2}}$

*As in Examples A, B and C, find each of the limits in Problems 21–38.*

21. $\lim_{x\to\infty} \frac{2x+5}{3x^2-2}$
22. $\lim_{x\to\infty} \frac{3x-2}{x^2-2x}$
23. $\lim_{x\to-\infty} \frac{3-x}{5-x^2}$
24. $\lim_{x\to-\infty} \frac{5+3x}{2x^2-4}$

25. $\lim_{x\to\infty} \frac{2 - 4x}{6 + 3x}$

26. $\lim_{x\to\infty} \frac{3x + 6}{8 - x}$

27. $\lim_{x\to\infty} \frac{2x^2 - x + 11}{x^2 + 70}$

28. $\lim_{x\to\infty} \frac{3x^2 + 25}{-x^2 + 5x}$

29. $\lim_{x\to-\infty} \frac{2x^3 - 2x^2 + x + 7}{4x^3 - 10x + 2}$

30. $\lim_{x\to-\infty} \frac{x^3 + 11x - 4}{5x^3 + 4x^2 + 11}$

31. $\lim_{x\to\infty} \frac{x^4 + 10}{2x^6 - 4x + 1}$

32. $\lim_{x\to\infty} \frac{2x^3 + x - 10}{x^4 + 3x^3 + 2x}$

33. $\lim_{x\to\infty} \frac{5\sqrt{x} - 2}{\sqrt{x} + 1}$

34. $\lim_{x\to\infty} \frac{2x^{3/2} + 1}{x^{3/2} + 5}$

35. $\lim_{x\to-\infty} \frac{1 - \sqrt[3]{x}}{\sqrt[3]{x} + 2}$

36. $\lim_{x\to-\infty} \frac{2\sqrt[3]{x}}{4 + \sqrt[3]{x}}$

37. $\lim_{x\to-\infty} \frac{\sqrt{x^6 + 2x^2 + 7}}{2x^3 - x + 4}$

38. $\lim_{x\to\infty} \frac{\sqrt{9x^4 + 3x + 7}}{(2x + 3)^2}$

*In Problems 39–50, find the horizontal and vertical asymptotes (if any) for the graph of each function (see Examples D and E).*

39. $f(x) = -\frac{3}{x^2}$

40. $f(x) = \frac{3x^2}{x^2 + 4}$

41. $f(x) = \frac{3 - x}{x + 2}$

42. $f(x) = \frac{2x + 7}{4x - 1}$

43. $f(x) = \frac{2 + 3x + 2x^2}{-x^2 + 5x}$

44. $f(x) = \frac{x^2 - 5x + 4}{2x^2 - 18}$

45. $f(x) = \frac{x^2 - 4}{x^2 - 2x - 3}$

46. $f(x) = \frac{4 - x^2}{2x^2}$

47. $f(x) = \frac{2x^3 + 7x - 3}{x^2 + x + 2}$

48. $f(x) = \frac{3x^2}{2x^2 - 2x + 1}$

49. $f(x) = \frac{x^2 + 2x - 3}{x^2 - 1}$

50. $f(x) = \frac{x^2 - x - 6}{x^2 - 2x - 3}$

*In Problems 51–60, sketch the graph of f using the five-step procedure in Example F.*

51. $f(x) = \frac{x}{x - 2}$

52. $f(x) = \frac{x - 3}{x - 2}$

53. $f(x) = \frac{x}{x^2 - 4}$

54. $f(x) = \frac{x + 1}{x^2 - 9}$

55. $f(x) = \frac{3x^2}{x^2 + 2}$

56. $f(x) = \frac{x^2 - 4}{x^2 + 4}$

57. $f(x) = \frac{x^2 + 1}{x}$

58. $f(x) = \frac{x^2 + 9}{2x}$

59. $f(x) = x - \frac{4}{x}$

60. $f(x) = x^2 - \frac{1}{x}$

---

61. Find the horizontal and vertical asymptotes (if any) for the graph of $f(x) = (2x^2 - 3)/[x(x^2 - 9)]$.

62. Find the horizontal and vertical asymptotes (if any) for the graph of $f(x) = 3x/(\sqrt{x} - 2)^2$.

63. Give the formula $f(x) = \ldots$ for a function that has a horizontal asymptote at $y = 4$ and vertical asymptotes at $x = -2$ and $x = 3$.

64. Give the formula $f(x) = \ldots$ for a function that has a horizontal asymptote at $y = -2$ and vertical asymptotes at $x = 0$, $x = 1$, and $x = 2$.

65. Find each of the following limits.

(a) $\lim_{x\to\infty} \frac{\sqrt{x} + 2}{x}$ (b) $\lim_{x\to\infty} \frac{2x - 3}{\sqrt{x^2 + 1}}$

66. Graph the function $f(x) = \frac{2x^2 - x - 1}{x^2 - 3x - 4}$

67. A manufacturer estimates that the total cost of producing $x$ television sets per month is $C(x) = 42{,}000 + 250x$ dollars. Find the limiting value of the average cost per set, $C(x)/x$, as $x$ gets larger and larger.

68. A manufacturer of freezers has fixed monthly costs of \$28,000 and estimates the cost of labor and materials for each freezer to be \$300. Assuming that it can make $x$ such freezers each month and sell them at \$475 each, find the limiting value of the average profit per freezer as $x$ gets larger and larger.

69. Experience shows that an assembly line worker can perform a certain task in $12[1 + 1/(4\sqrt{n})]$ minutes after doing this task $n$ times. How long will this task take in the long run?

70. A right triangle has one leg of length 10, the other leg of length $x$, and a hypotenuse of length $h$. Find (a) the limit of the ratio $x/h$ as $x$ gets larger and larger, (b) the limit of the ratio of the perimeter of the triangle to its area as $x$ gets larger and larger.

71. For an equilateral triangle of side length $x$, find the limit of the ratio of its perimeter to its area as $x$ tends to 0.

72. Show that the graph of $f(x) = (2x^2 + x + 1)/x$ has the line $y = 2x + 1$ as a slant asyinptote.

## CHAPTER 3 REVIEW PROBLEM SET

1. Let $f(x) = (x^3 - 8)/(x^2 - 4)$.
   (a) Show that $f$ is continuous at $x = 1$.
   (b) Is $f$ continuous at $x = -2$?
   (c) How should $f(2)$ be defined to remove the discontinuity of $f$ at $x = 2$?
2. Let $f(x) = (x^2 - x - 6)/(x^2 + x - 2)$.
   (a) Show that $f$ is continuous at $x = 2$.
   (b) Is $f$ continuous at $x = 1$?
   (c) How should $f(-2)$ be defined to remove the discontinuity at $x = -2$?
3. Solve each of the following inequalities, writing the solution set in interval notation.
   (a) $7x - 6 \leq 12x + 1$
   (b) $2x^2 + 5x < 3$
4. Solve the following inequalities, writing the solution in interval notation.
   (a) $3(x - 2) \geq 2x + 14$
   (b) $3x^2 - 10x < 8$

*In Problems 5–12, indicate whether or not the given function is continuous at $x = 1$. If it is continuous at $x = 1$, indicate whether or not it is differentiable there.*

5. $f(x) = \dfrac{x^2 - 3x + 2}{x^2 + 4}$
6. $f(x) = \dfrac{x^2 + 4x - 5}{x^2 - x - 2}$
7. $f(x) = \dfrac{x^2 - 3x + 2}{x^2 + 2x - 3}$
8. $f(x) = \dfrac{x^2 + 3x - 4}{2x^2 + 3x - 5}$
9. $f(x) = |3x - 3|$
10. $f(x) = |6 - 3x|$
11. $f(x) = \begin{cases} x^3 & \text{if } x \leq 1 \\ 2x^2 - x & \text{if } x > 1 \end{cases}$
12. $f(x) = \begin{cases} 2x^2 - x + 2 & \text{if } x \leq 1 \\ 1.5x^2 + 1.5 & \text{if } x > 1 \end{cases}$
13. Find the global maximum and minimum values of
$$f(x) = \sqrt[3]{x} - \tfrac{1}{3}x + 10$$
on the interval $[0, 27]$.
14. Find the global maximum and minimum values of $f(x) = \sqrt[5]{x} - \frac{1}{5}x + 8$ on the interval $[-1, 32]$.
15. Find all critical points for $f(x) = x^3 + 3x^2 - 24x + 7$ and determine which yield local maximum values and which yield local minimum values.
16. Find all critical points for $f(x) = 4x^3 + 15x^2 - 18x - 200$ and determine which ones yield local maximum values and which ones yield local minimum values.
17. Find both coordinates for all inflection points of the graph of $f(x) = x^4 - 24x^2 + 36x - 10$.
18. Find both coordinates for all inflection points of the graph of $f(x) = x^4 - 12x^3 + 48x^2 + 4x - 200$.
19. Let $f(x) = x^4 - 4x^3 - 3$.
   (a) Where is $f$ increasing? Decreasing?
   (b) Where is the graph of $f$ concave up? Concave down?
   (c) Sketch the graph of $f$.
20. Let $f(x) = 3x^4 - 4x^3 + 8$.
   (a) Where is $f$ increasing? Decreasing?
   (b) Where is the graph of $f$ concave up? Concave down?
   (c) Sketch the graph of $f$.

*Find each of the limits in Problems 21–28.*

21. $\displaystyle\lim_{x\to 2^+} \frac{4}{x^2 - 4}$
22. $\displaystyle\lim_{x\to 1^+} \frac{2}{x^3 - 1}$
23. $\displaystyle\lim_{x\to 0^-} \frac{2}{\sqrt[3]{x}}$
24. $\displaystyle\lim_{x\to 2^-} \frac{5}{\sqrt[3]{x - 2}}$
25. $\displaystyle\lim_{x\to \infty} \frac{3x - 2}{x^2 + 2}$
26. $\displaystyle\lim_{x\to \infty} \frac{2x^3 + x - 1}{5x^2 + 4}$
27. $\displaystyle\lim_{x\to \infty} \frac{4 - \sqrt{x}}{\sqrt{x} + 2}$
28. $\displaystyle\lim_{x\to \infty} \frac{3 + 2x\sqrt{x}}{-x\sqrt{x} + 2}$
29. Consider $f(x) = x + (25/x)$ on $(0, \infty)$.
   (a) Where is $f$ increasing? Decreasing?
   (b) Where is the graph of $f$ concave up? Concave down?
   (c) Find any local maximum or local minimum values.
   (d) Sketch the graph of $f$, showing all asymptotes.

30. Consider $f(x) = x - 4/x^2$ on $(-\infty, 0) \cup (0, \infty)$.
    (a) Where is $f$ increasing? Decreasing?
    (b) Where is the graph of $f$ concave up? Concave down?
    (c) Determine any local maximum or local minimum values.
    (d) Sketch the graph of $f$, showing all asymptotes.
31. Sketch the graph of $f(x) = 2x^2/(x^2 - 9)$, showing all asymptotes and clearly labeling any local extreme points.
32. Sketch the graph of $f(x) = 3x^2/(x^2 - 1)$, showing all asymptotes and labeling any local extreme points.
33. Give a formula for $f(x)$ if $f$ is a function whose graph has a horizontal asymptote at $y = 2$ and vertical asymptotes at $x = -1$, $x = 0$, and $x = 2$.
34. Give a formula for $f(x)$ if $f$ is a function whose graph has a horizontal asymptote at $y = -2$ and vertical asymptotes at $x = -2$, $x = 0$, and $x = 3$.
35. Let $f(x) = \sqrt[3]{x}/(x + 2)$.
    (a) What is the domain of $f$?
    (b) Where is $f$ discontinuous?
    (c) Find $\lim_{x\to\infty} f(x)$.
    (d) Where is $f$ not differentiable?
    (e) Evaluate $f'(8)$.
36. Let $f(x) = \sqrt[3]{x - 2}/(x - 4)$.
    (a) What is the domain of $f$?
    (b) Where is $f$ discontinuous?
    (c) Find $\lim_{x\to\infty} f(x)$.
    (d) Where does $f$ fail to be differentiable?
    (e) Find $f'(3)$.
37. Evaluate

$$\lim_{h\to 0} \frac{\sqrt[3]{8 + h} - \sqrt[3]{8}}{h}$$

by recognizing this to be the value of a certain derivative.

38. Evaluate

$$\lim_{h\to 0} \frac{(16 + h)^{3/4} - 2(16 + h)^2 - [16^{3/4} - 2(16)^2]}{h}$$

by recognizing this to be the value of a certain derivative.

*The chart below shows economic growth in the United States above and below the historical trend of $2\frac{1}{2}\%$ per year (represented by the zero line) and is typical of the many charts that appear in your economics textbook. Economists analyze such charts and propose mathematical models to explain them.*

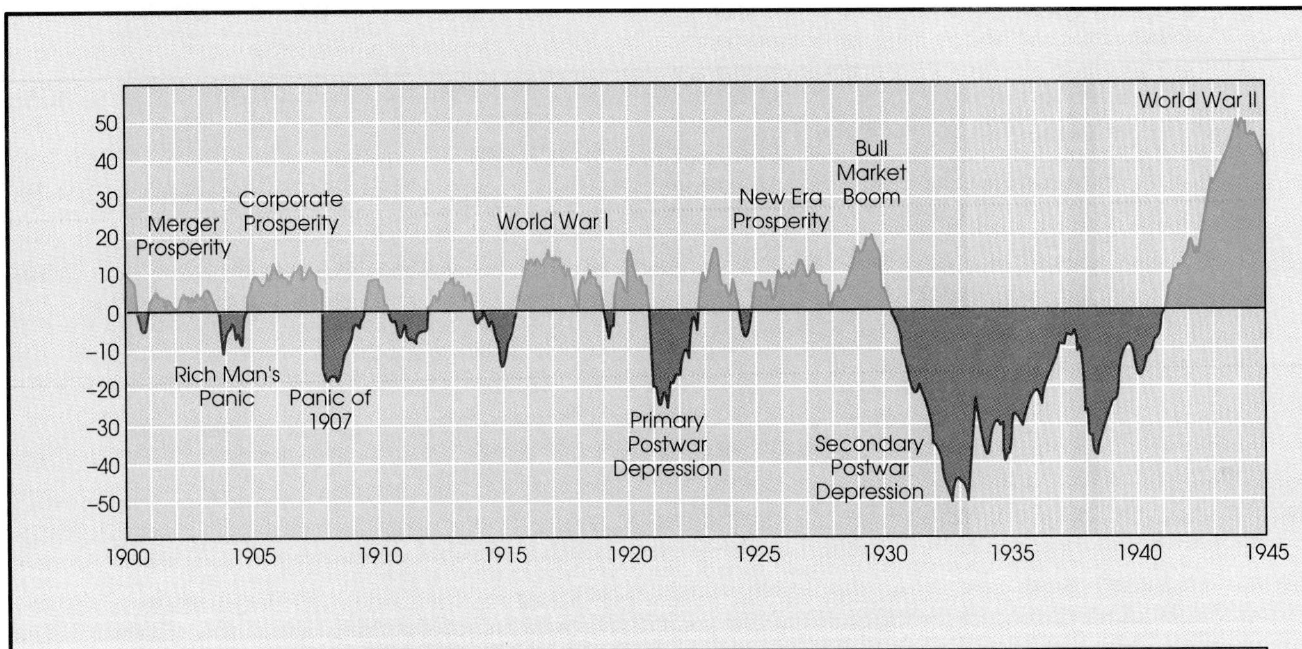


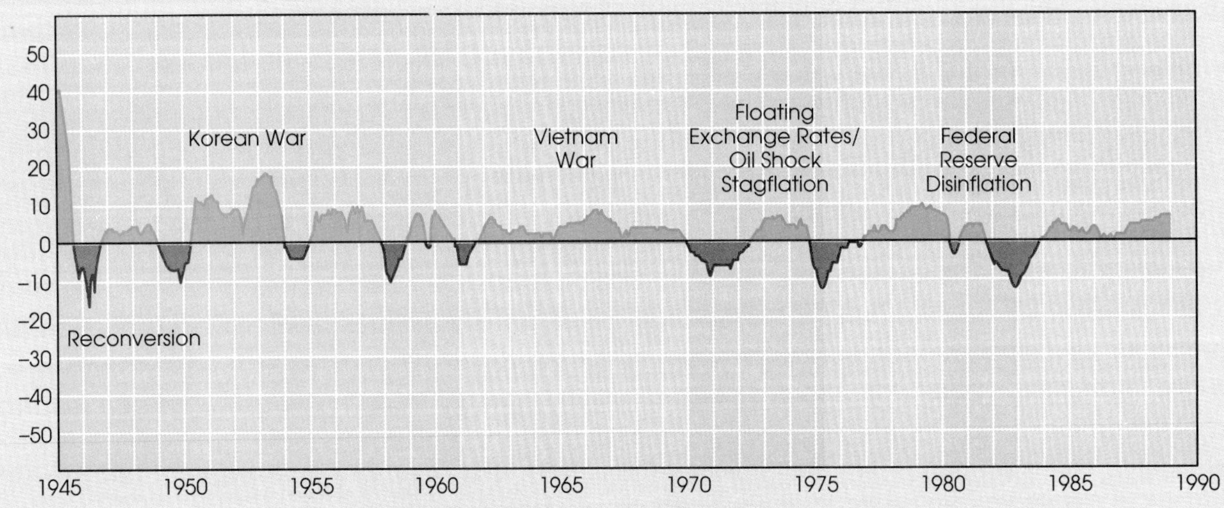

# 4

# Other Applications of the Derivative



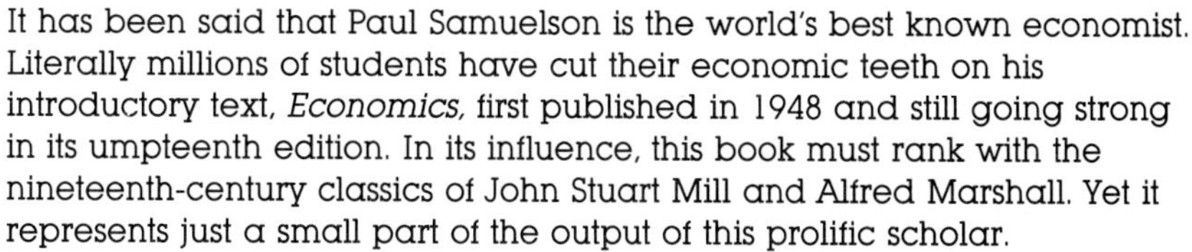

It has been said that Paul Samuelson is the world's best known economist. Literally millions of students have cut their economic teeth on his introductory text, *Economics,* first published in 1948 and still going strong in its umpteenth edition. In its influence, this book must rank with the nineteenth-century classics of John Stuart Mill and Alfred Marshall. Yet it represents just a small part of the output of this prolific scholar.

Paul Anthony Samuelson
(1915– )

Born in Gary, Indiana, Samuelson went to the University of Chicago for his B.A. (1935) and then to Harvard for a Ph.D. (1941). His early essay on the interaction of the multiplier and the accelerator (originally composed as a term paper for a graduate course) demonstrated his originality and insight. His doctoral dissertation, largely written when he was 23 years old, became the momentous book *Foundations of Economic Analysis* and formed part of the basis for his award of the Nobel prize in economic science in 1970.

Moving to Massachusetts Institute of Technology in 1941, Samuelson established himself as a scholar supremely able to apply the tools of mathematics to economic problems. The topic of maximization subject to constraints (a topic treated in its simplest setting in this chapter) is a favorite theme in many of his writings.

Samuelson's view of making mathematical models to explain economic data is expressed in his own words near the beginning of his elementary text: "Even if we had more and better data, it would still be necessary—as in every science—to *simplify,* to *abstract* from the infinite mass of detail. No mind can comprehend a bundle of unrelated facts. All analysis involves abstraction. It is always necessary to *idealize,* to omit detail, to set up simple hypotheses and patterns by which the facts can be related, to set up the right questions before going out to look at the world. Every theory, whether in the physical or biological or social sciences, distorts reality in that it oversimplifies. But if it is good theory, what is omitted is outweighed by the beam of illumination and understanding thrown over the diverse empirical data."

## 4.1
# OPTIMIZATION: GEOMETRIC PROBLEMS

We are all engaged in optimizing something. Students want to maximize their grades or perhaps minimize their study time. Professors aim to maximize their number of publications, the quality of their teaching, and the size of their salaries. These are complex problems and not easily subjected to mathematical analysis. There is, however, a host of practical optimization problems for which the tools of calculus are admirably suited. We begin with a particularly simple example.

### EXAMPLE A

Mark Farmer has 100 feet of fencing with which he plans to enclose a rectangular garden plot. What are the dimensions of the rectangle that will make the area of his garden as large as possible?

**Solution** Several possible rectangular plots are shown in Figure 1. They lend support to our intuitive feeling that the rectangle should be a square. Calculus can turn this feeling into an irrefutable fact.

Consider a general rectangle with sides of length $x$ and $y$, respectively (Figure 2.) We want to maximize the area $A = xy$, but unfortunately we as

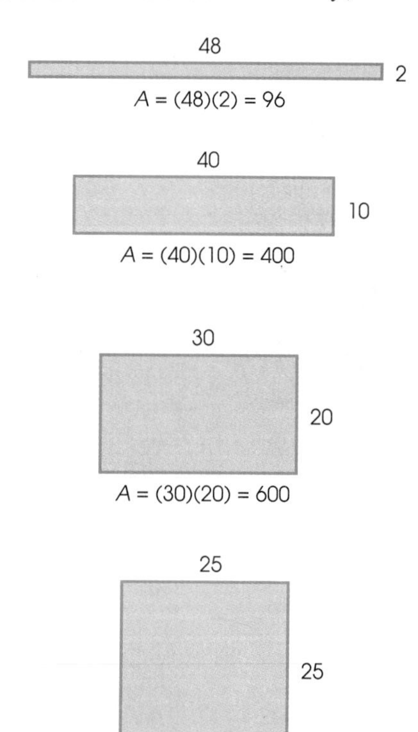


**FIGURE 1**

**POINT OF INTEREST**
In Example A, one could certainly argue that the allowable values for $x$ are $0 < x < 50$, claiming that the values $x = 0$ and $x = 50$ do not yield rectangles. If we adopt this view, we have only one critical value, $x = 25$. To see that it gives the maximum area, we use the First Derivative Test, noting that $dA/dx > 0$ for $x < 25$ and $dA/dx < 0$ for $x > 25$. See Example F for a specific illustration of this point of view.

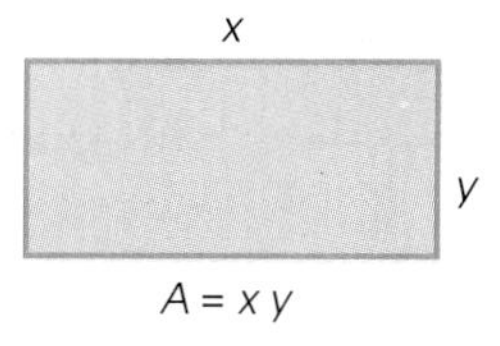


**FIGURE 2**

yet have no theory for maximizing a function of two variables. However, the fact that Mark has 100 feet of fencing places a *constraint* on $x$ and $y$, namely,

$$2x + 2y = 100$$

Happily, this allows us to solve for $y$ in terms of $x$ and thereby express $A$ in terms of $x$ alone.

$$\begin{aligned} x + y &= 50 \\ y &= 50 - x \\ A &= xy = x(50 - x) = 50x - x^2 \\ \frac{dA}{dx} &= 50 - 2x \end{aligned}$$

The allowable values for $x$ are $0 \le x \le 50$, so there are three critical points: the endpoints 0 and 50 and the point 25 where $dA/dx = 0$. At the endpoints, $A = 0$; at $x = 25$, $A = 50(25) - (25)^2 = 625$. Since a continuous function on a closed interval must attain a maximum (Max-Min Existence Theorem) and there are only the three indicated critical points, it must be that $x = 25$ yields the largest area. The rectangle of maximum area is a square.

The example just considered suggests a procedure to follow in solving practical optimization problems.

1. If possible, draw a picture that represents the problem.
2. Assign symbols to the significant variables and express the quantity $Q$ that is to be optimized in terms of them.
3. Determine the relations between the variables and use them to express $Q$ in terms of a single variable.
4. Use the techniques in Section 3.3 to locate the point that gives the optimum value.

## Variations on the Theme

When Mark Farmer thought about his problem, he realized that a bigger plot could be obtained by putting his garden next to a building.

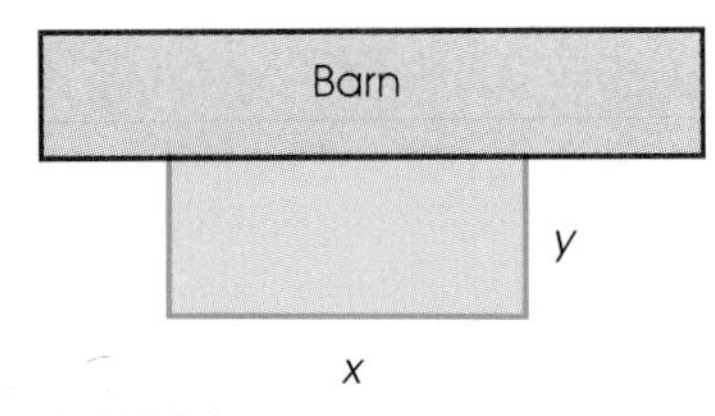


FIGURE 3

### EXAMPLE B

If Mark Farmer uses his 100 feet of fencing to fence a rectangular plot next to his long barn (Figure 3), what dimensions will make the garden as large as possible?

**Solution** The new feature is that the barn side of the garden does not need to be fenced. Following the suggested procedure, we have drawn a picture and assigned letters to the key variables. Again we want to maximize $A = xy$, but now the relation between $x$ and $y$ is $x + 2y = 100$. Thus,

$$y = 50 - \tfrac{1}{2}x$$

$$A = xy = x(50 - \tfrac{1}{2}x) = 50x - \tfrac{1}{2}x^2$$

$$\frac{dA}{dx} = 50 - x$$

The domain for $x$ is $0 \le x \le 100$, and so the critical points are the endpoints 0 and 100 and the point 50 where $dA/dx = 0$. The latter gives the maximum value of $A = 50(50) - \tfrac{1}{2}(50)^2 = 1250$. The rectangle has dimensions 50 by 25 feet. ■

### EXAMPLE C

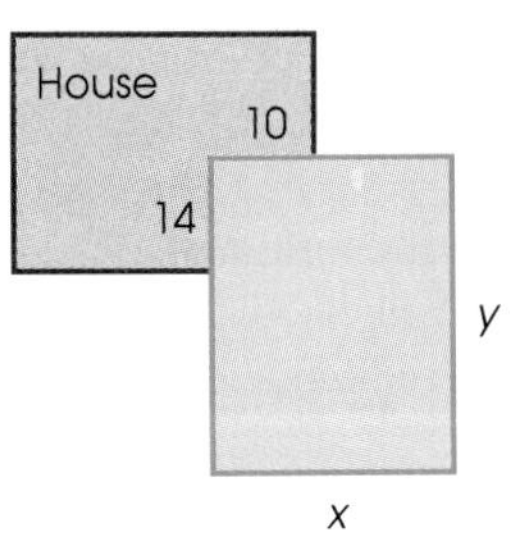


FIGURE 4

Could Mark Farmer do still better if he fitted his garden into the L-shaped corner next to his house (Figure 4)? What dimensions then give the maximum area and what is this area?

**Solution** All is as before except that now the relation between $x$ and $y$ is $x + y + (x - 10) + (y - 14) = 100$. Thus,

$$2x + 2y = 124$$

$$y = 62 - x$$

$$A = xy = x(62 - x) = 62x - x^2$$

$$\frac{dA}{dx} = 62 - 2x$$

Note that the smallest $y$ can be is 14, which corresponds to $x = 48$. We conclude that the domain for $x$ is $10 \le x \le 48$, and so the critical points are 10, 48, and 31. At these points, $A = 520$, 672, and 961, respectively. The dimensions of the plot of maximum area 961 are 31 by 31 feet. Farmer will do better by putting his garden next to the barn as in Example B. ■

## Best Containers

The next three examples illustrate how important it is to read a problem carefully before you try to solve it.

## EXAMPLE D

An open box is to be made from a square sheet of tin 12 centimeters on a side by cutting small squares from each of the corners and turning up the edges (Figure 5). What are the dimensions of the resulting box if its volume is to be a maximum?

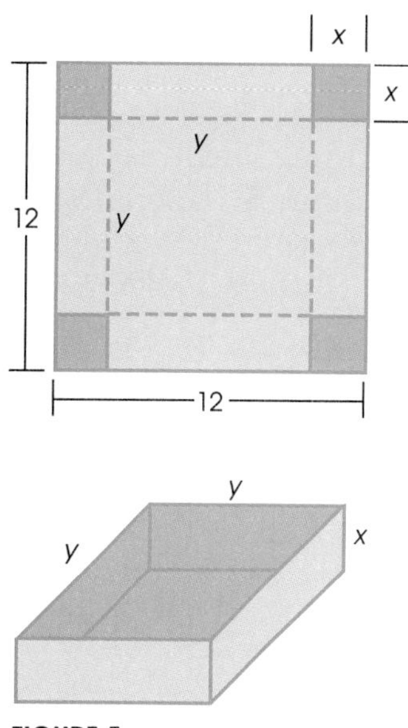


FIGURE 5

**Solution** The pictures have been drawn, and the significant variables labeled (Figure 5). We want to maximize the volume $V = xy^2$. Since $y = 12 - 2x$, we may write

$$V = xy^2 = x(12 - 2x)^2 = 144x - 48x^2 + 4x^3$$

$$\frac{dV}{dx} = 144 - 96x + 12x^2 = 12(12 - 8x + x^2) = 12(x - 6)(x - 2)$$

The domain for $x$ is $0 \le x \le 6$, so we have the three critical points 0, 6, and 2. Both $x = 0$ and $x = 6$ make $V = 0$. The value $x = 2$ yields the maximum volume $V = 2(8)^2 = 128$ cubic centimeters. The box of maximum volume is 2 centimeters deep with a base 8 centimeters by 8 centimeters. ■

## EXAMPLE E

An open box with square base is to have a total of 144 square centimeters of tin in its sides and base. What are the dimensions of the box if its volume is to be a maximum?

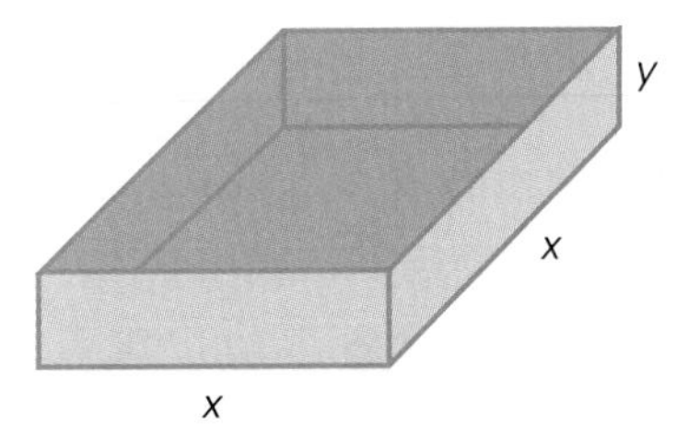


**FIGURE 6**

**Solution** A picture with labels appears in Figure 6. We want to maximize $V = x^2y$ with the constraint that the surface area be 144. There are four sides of area $xy$, and the base has area $x^2$. Thus,

$$4xy + x^2 = 144$$

$$4xy = 144 - x^2$$

$$y = \frac{144 - x^2}{4x}$$

$$V = x^2y = \frac{x^2(144 - x^2)}{4x} = 36x - \frac{1}{4}x^3$$

$$\frac{dV}{dx} = 36 - \frac{3}{4}x^2$$

The domain for $x$ is $0 \le x \le 12$. The endpoints 0 and 12 give volume 0. The other critical point is obtained by setting $dV/dx$ to 0 and solving for $x$.

$$36 - \tfrac{3}{4}x^2 = 0$$

$$36 = \tfrac{3}{4}x^2$$

$$48 = x^2$$

$$x = \sqrt{48} = 4\sqrt{3} \approx 6.928$$

The maximum volume is

$$V = x^2y = \frac{(48)(144 - 48)}{4(4\sqrt{3})} = (48)(2\sqrt{3}) = 96\sqrt{3} \approx 166.3$$

and the required dimensions are depth $2\sqrt{3}$ centimeters and base a square of side $4\sqrt{3}$ centimeters. ■

**EXAMPLE F**

A closed cylindrical can is to have a volume of 1000 cubic centimeters. Find the dimensions of the can that will minimize the amount of tin required to make it (assuming no waste).

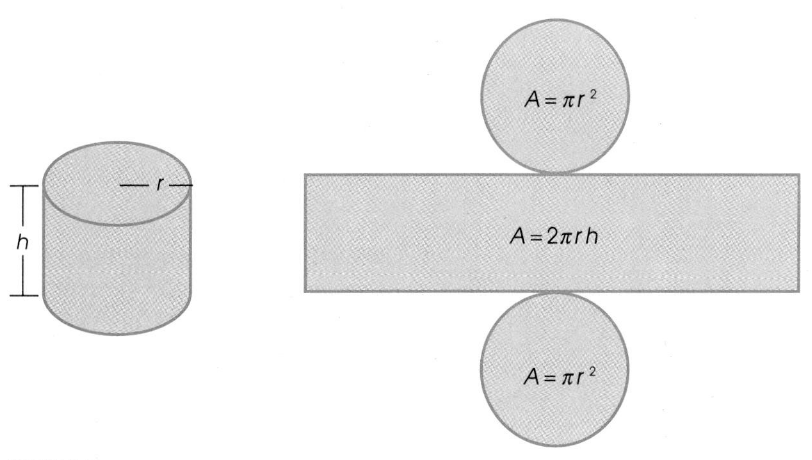


**FIGURE 7**

**Solution** Note the picture of the can together with the material needed to make it (Figure 7). We want to minimize the surface area $S = 2\pi r^2 + 2\pi rh$ given that $V = 1000 = \pi r^2 h$. We solve the latter equation for $h$ and substitute in the expression for $S$.

$$S = 2\pi r^2 + 2\pi r\frac{1000}{\pi r^2} = 2\pi r^2 + 2000r^{-1}$$

$$\frac{dS}{dr} = 4\pi r - \frac{2000}{r^2}$$

The domain for $r$ is $0 < r < \infty$. The only critical point is obtained by setting $dS/dr$ to 0.

$$4\pi r - \frac{2000}{r^2} = 0$$

$$4\pi r^3 - 2000 = 0$$

$$r^3 = \frac{2000}{4\pi}$$

$$r = \sqrt[3]{2000/4\pi} = \sqrt[3]{500/\pi} \approx 5.419$$

We find that $dS/dr < 0$ to the left of $\sqrt[3]{500/\pi}$ and $dS/dr > 0$ to the right of this point. This implies that $r = 5.419$ centimeters and $h = 1000/\pi r^2 \approx 10.839$ centimeters give the cylinder of smallest surface area. ■

## Best Routes

Our final examples are geometric in nature but also involve the familiar formula $D = RT$ (distance = rate × time) in its equivalent form $T = D/R$.

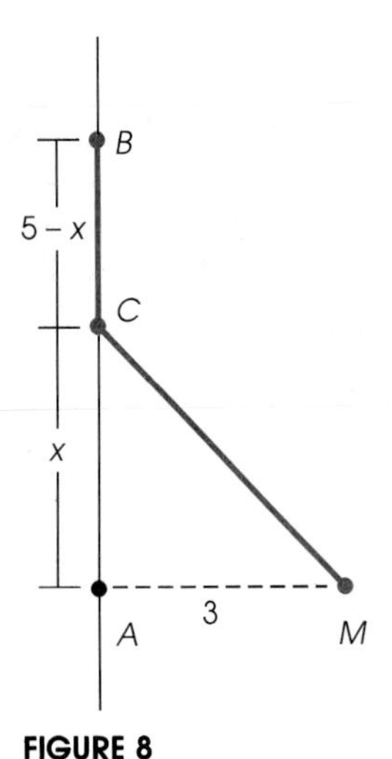


FIGURE 8

## EXAMPLE G

Mary Drinkwater is in a boat 3 miles from the nearest point $A$ on a straight shoreline. Down the shore 5 miles from $A$ is her home at $B$. If Mary can row 6 miles per hour and run 10 miles per hour, where should she land her boat to arrive home in the least amount of time?

**Solution** We think the problem is to find $x$ in Figure 8 so that the traveling time on the red path is a minimum. Note that $0 \leq x \leq 5$.

The distance $MC$ is $\sqrt{9 + x^2}$, and the time required to row it is $\sqrt{9 + x^2}/6$. The time needed to run $CB$ is $(5 - x)/10$. Mary's task is to minimize $T$, where

$$T = \frac{\sqrt{9 + x^2}}{6} + \frac{5 - x}{10}$$

Now

$$\frac{dT}{dx} = \frac{1}{6} \cdot \frac{1}{2}(9 + x^2)^{-1/2} \cdot 2x - \frac{1}{10}$$

$$= \frac{x}{6\sqrt{9 + x^2}} - \frac{1}{10}$$

When we set $dT/dx$ to 0 and solve, we obtain

$$\frac{x}{6\sqrt{9 + x^2}} = \frac{1}{10}$$

$$10x = 6\sqrt{9 + x^2}$$

$$100x^2 = 36(9 + x^2)$$

$$64x^2 = 324$$

$$x = \pm\frac{18}{8} = \pm 2.25$$

For the domain $0 \leq x \leq 5$, there are only the three critical points 0, 2.25, and 5. The corresponding values of $T$ are 1, 0.9, and 0.972. Mary should head for the point 2.25 miles down the shore from $A$. ■

All problems considered so far have achieved their optimum value at a point where the derivative is 0. This is the typical but not the universal situation in applied problems. Our last example exhibits an exception.

> **CAUTION**
> We claim that the equation $dT/dx = 0$ in Example H has no real solution. Don't take our word for it.
>
> CHECK IT OUT
>
> You should always read a math book with paper and pencil handy for filling in the details of an argument.

**EXAMPLE H**

If Mary in Example G has a small motor which allows her boat to go 12 miles per hour, then where should she land?

**Solution**

$$T = \frac{\sqrt{9 + x^2}}{12} + \frac{5 - x}{10}$$

$$\frac{dT}{dx} = \frac{x}{12\sqrt{9 + x^2}} - \frac{1}{10}$$

When we try to solve $dT/dx = 0$, we discover it has no real solution. Thus, the only critical points are the endpoints 0 and 5 where $T$ has values 0.75 and 0.486, respectively. Mary should head straight for $B$. ■

## Problem Set 4.1

*In Problems 1–8, find the maximum value of z subject to the given condition. Assume throughout that $x \geq 0$ and $y \geq 0$. Note that in order to use the theory we have developed, you will first have to use the condition to express z in terms of a single variable. For example, in Problem 1, first note that $y = 200 - 2x$; then write $z = x(200 - 2x) = 200x - 2x^2$.*

1. $z = xy$; $2x + y = 200$
2. $z = xy$; $3x + 2y = 450$
3. $z = 3x - 2y^{-2}$; $xy = 1$
4. $z = -4x - 3y$; $xy = 48$
5. $z = x^2y$; $x^2 + 4xy = 48$
6. $z = x^2y$; $3x^2 + 4xy = 225$
7. $z = \pi x^2y$; $2x^2 + 2xy = 150$
8. $z = \pi x^2y$; $x^2 + 2xy = 48$
9. The rectangle in Figure 9 has two sides along the coordinate axes and one vertex $(x, y)$ in quadrant I on the line $x + 2y = 8$. What values of $x$ and $y$ make the area of the rectangle a maximum? What is the maximum area?

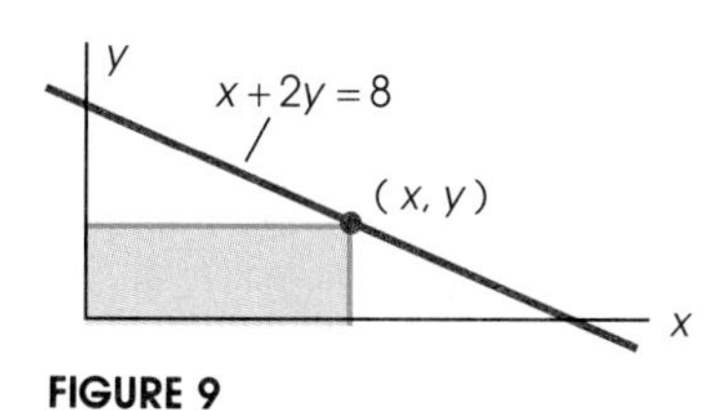


FIGURE 9

10. The rectangle in Figure 10 has two sides along the coordinate axes and one vertex $(x, y)$ in quadrant I on the parabola $y = 9 - x^2$. Find the dimensions of the rectangle for which its area is a maximum. What is the maximum area?

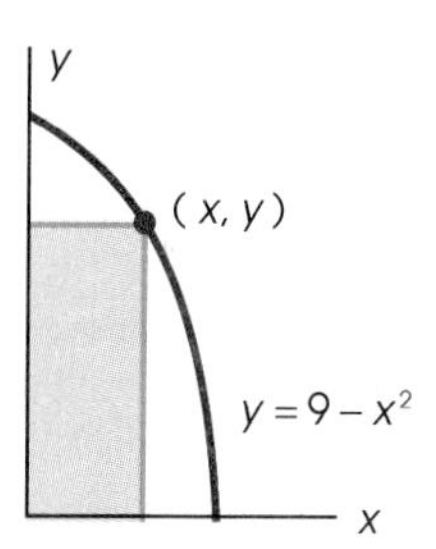


FIGURE 10

*For Problems 11–16, see Examples A, B, and C.*

11. Mary Wilson has 300 meters of fencing with which to enclose two adjacent lots as shown in Figure 11. Find the dimensions $x$ and $y$ that will make the total area a maximum. What is the maximum area?

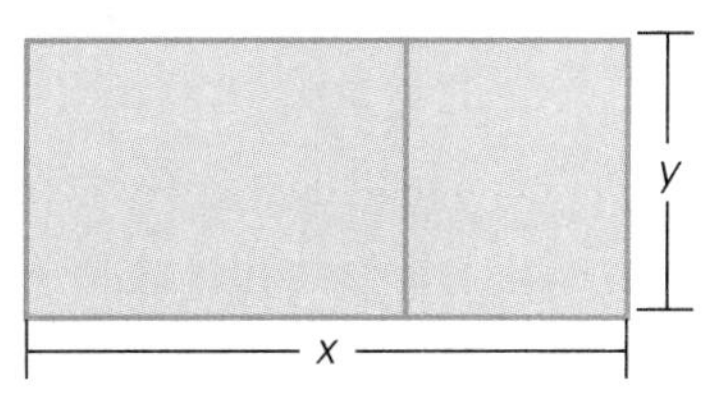


FIGURE 11

12. Suppose that Mary Wilson in Problem 11 chose to put the two lots along a river in such a way that the side labeled $x$ in Figure 11 does not require a fence. Find the dimensions $x$ and $y$ that make the total area a maximum. What is the maximum area?
13. Suppose that Mary Wilson in Problem 11 chose to enclose three rather than two adjacent lots with her 300 meters of fencing. What dimensions $x$ and $y$ would then make the total area a maximum? What is the maximum area?
14. A 3000-foot race track encloses a rectangle and its adjoining semicircular ends (Figure 12). Find the dimensions $x$ and $y$ of the rectangle of maximum area so enclosed.

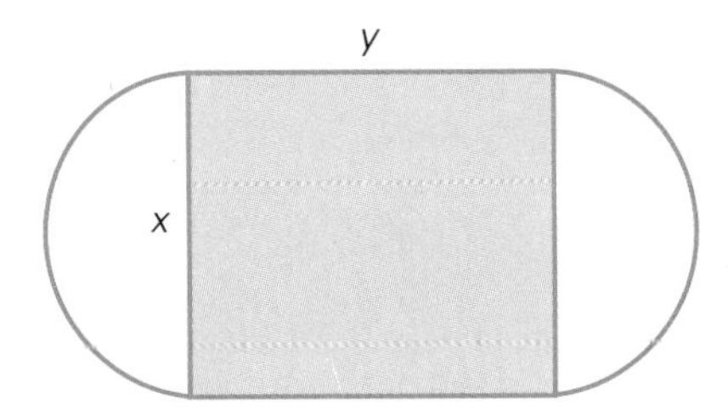


**FIGURE 12**

15. A gardener wishes to enclose two identical rectangular plots each of area 1200 square feet as in Figure 13. The outer boundary requires a heavy fence costing \$6 per foot, but the fence for the partition costs only \$4 per foot. For what dimensions $x$ and $y$ will the total cost of fencing be a minimum?
16. Christine Smith and Karen White have adjoining backyards. They have agreed to fence in adjoining garden plots (as in Figure 13) each of size 1000 square feet, sharing the cost of the partition fence on the common boundary. Determine the dimensions $x$ and $y$ that will minimize each woman's total cost. Assume that the same type of fence is used for the outer boundary and the partition.

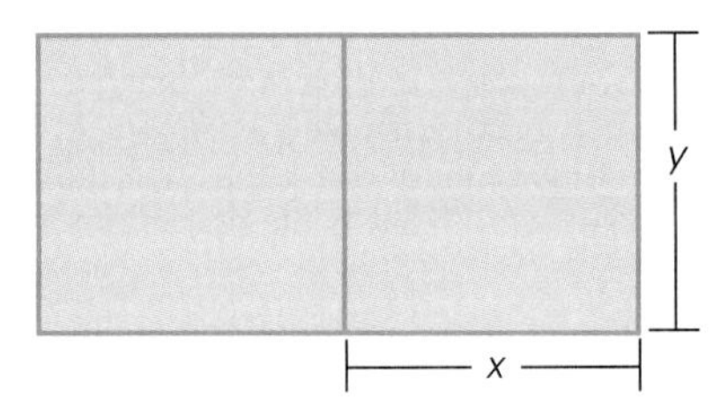


**FIGURE 13**

*For Problems 17–20, see Examples D, E, and F.*

17. A tray is to be made from a 9-inch by 24-inch sheet of copper by cutting identical squares from the corners and folding up the flaps. Find the dimensions of the tray of maximum volume.
18. A closed cedar chest with a rectangular base twice as long as wide is to be built to have a volume of $\frac{64}{3}$ cubic feet. Determine the dimensions of the chest so as to minimize its surface area.
19. A cistern with a square base is to hold 16,000 cubic feet of water. The metal top costs three times as much per square foot as the concrete sides and base. What are the dimensions of the most economical cistern?
20. A closed box with a square base is to have a volume of 24 cubic feet. The material for the sides, top, and base costs 25¢, 50¢, and \$1 per square foot, respectively. Find the dimensions of the most economical box.
21. A powerhouse is located at $P$ on one bank of a straight river that is 2000 feet wide. On the opposite bank and 10,000 feet downstream is a factory at $F$. A cable, costing twice as much for installation underwater as on land, is to be laid from $P$ to $B$ to $F$ as shown in Figure 14. Determine $x$ so as to minimize the cost of laying the cable.

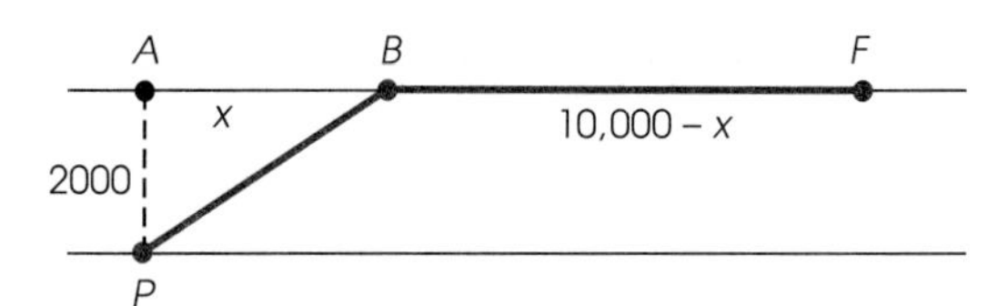


**FIGURE 14**

22. Do Problem 21 again, this time assuming that underwater installation is three times as costly as on-land installation.

---

23. Mary has been asked to make a copper plaque that will contain 150 square inches of printed matter. It is to have 3-inch margins at the top and bottom and 2-inch margins at the sides. What are the outside dimensions of the plaque if it is to require the least amount of copper?
24. The Maxwell Company ships specialty fruit in rectangular boxes with square cross sections. The shipping firm requires that the sum of the length of the package and the perimeter of a cross section not exceed 100 inches. Find the dimensions of the box of largest volume that the company can use.
25. If the strength of a rectangular beam is proportional to the product of its width and the square of its depth ($S = kwd^2$), find the dimensions of the cross section of the strongest beam that can be cut from a circular log of radius 1 foot (see Figure 15).

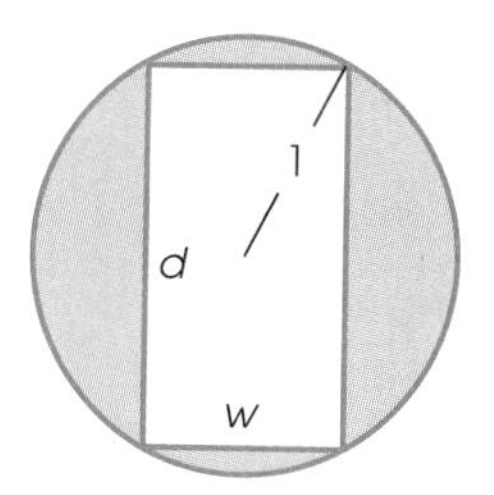


FIGURE 15

26. Split Rock is located in a lake 3 miles from the nearest point $P$ on a straight shoreline. Ole has been fishing all day anchored to this rock, but now it is time to head for home, located 12 miles down the shore from $P$. Ole will row at 4 miles per hour toward a point $A$ on the shore and then jog at 6 miles per hour toward home. How far down the shore from $P$ is $A$ if Ole chooses the fastest way home?
27. A wire 40 centimeters in length is to be cut into two pieces. One piece is to be bent to form a circle and the other a square, thus enclosing two areas. How long should the shorter piece be if the sum of the two areas is to be (a) a minimum, (b) a maximum? (Allow the possibility of no cut.)
28. A window in the shape of a rectangle surmounted by an equilateral triangle is to be cut through a wall but is subject to the condition that its perimeter cannot exceed 24 feet. What are the dimensions of the rectangle if the window is to admit the maximum amount of light?
29. Find the dimensions of the right circular cylinder of maximum volume that can be inscribed in a sphere of radius 12 inches (see Figure 16).

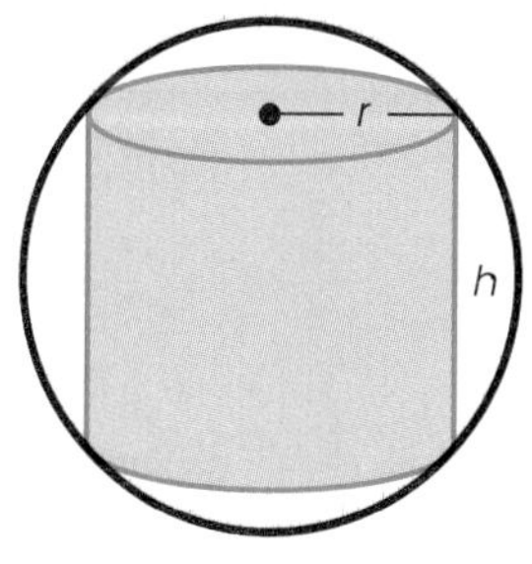


FIGURE 16

30. Find the dimensions of the right circular cone of maximum volume that can be inscribed in a sphere of radius 12 inches.
31. Find the points on the graph of the parabola $y = x^2 - 2x + 1$ that are closest to the point $(1, \frac{5}{2})$. *Hint:* Let $(x, y)$ be any point on the parabola. You are to minimize the distance between this point and the given point. An equivalent and easier problem is to minimize the square of this distance.
32. Consider a triangle formed by the coordinate axes and a line through the point (4, 2) as in Figure 17. Find the $x$- and $y$-intercepts of the line so that the area of the triangle is a minimum.

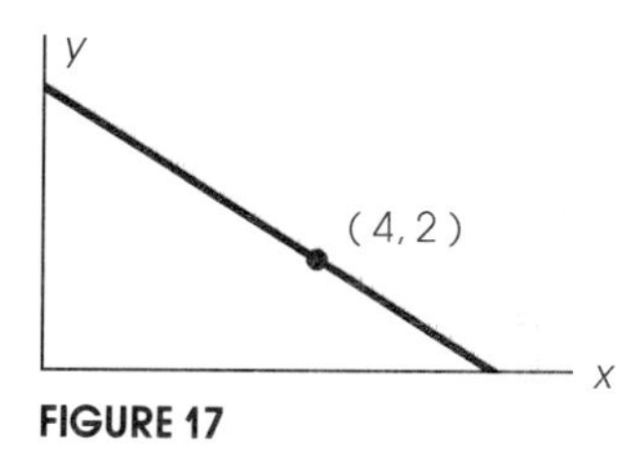


FIGURE 17

33. Find the $x$- and $y$-intercepts of the line in Problem 32 for which the hypotenuse has minimum length.
34. A paper company makes closed boxes from 5 feet by 8 feet sheets of cardboard (see Figure 18). To do this it cuts out the shaded parts shown in the figure and then folds on the dotted lines. Determine the dimensions $x$, $y$, and $z$ which will give the box of maximum volume.

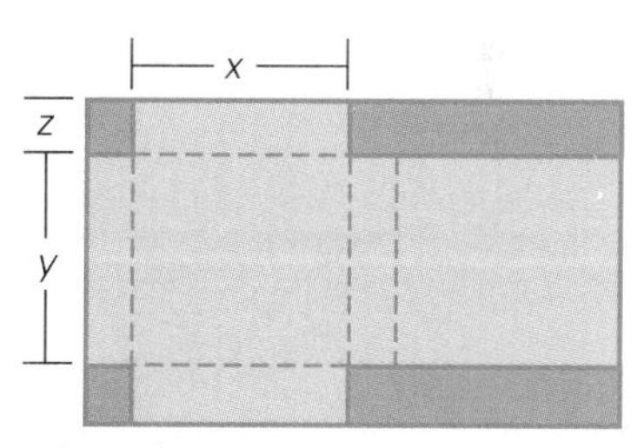


FIGURE 18

## 4.2 OPTIMIZATION: BUSINESS PROBLEMS

In Sections 1.4 and 1.6, we introduced a number of concepts used in business and economics. Of critical importance to a firm engaged in making and selling a product are three functions, always considered for some fixed time period (usually a quarter or a year).

$C(x)$: the cost of producing $x$ units of the product

$R(x)$: the revenue obtained from selling these $x$ units

$P(x) = R(x) - C(x)$: the profit generated by these $x$ units.

**POINT OF INTEREST**
Because many real-world phenomena are discrete in nature, mathematicians are giving increasing attention to a field called Discrete Mathematics. Calculus, however, is part of Continuous Mathematics.

In most cases, the product (refrigerators, bars of soap, reams of paper) will be in discrete units. Thus, in reality, the functions $C(x)$, $R(x)$, and $P(x)$ are defined only for nonnegative integers $x = 0, 1, 2, \ldots$ and consequently have graphs consisting of discrete points (Figure 19). However, it is conventional to connect these points with a smooth curve (Figure 20) and thereby pretend that $C(x)$, $R(x)$, and $P(x)$ are nice differentiable functions. The reason for doing this is quite obvious; it makes the powerful tools of the calculus available. Progress in analyzing real phenomena is often possible only when such simplifying assumptions are made. Making appropriate assumptions is a significant aspect of mathematical modeling.

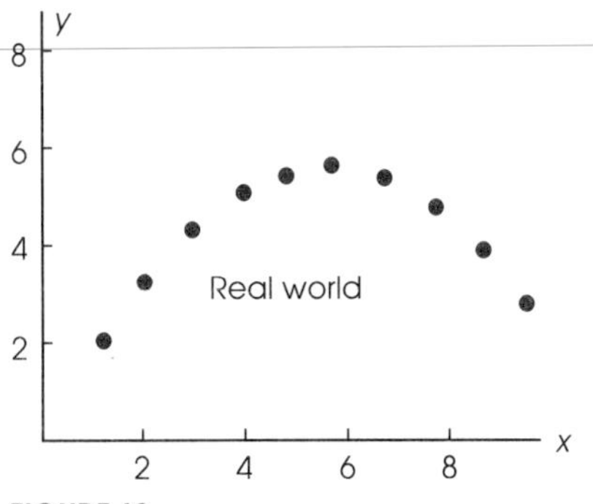

FIGURE 19

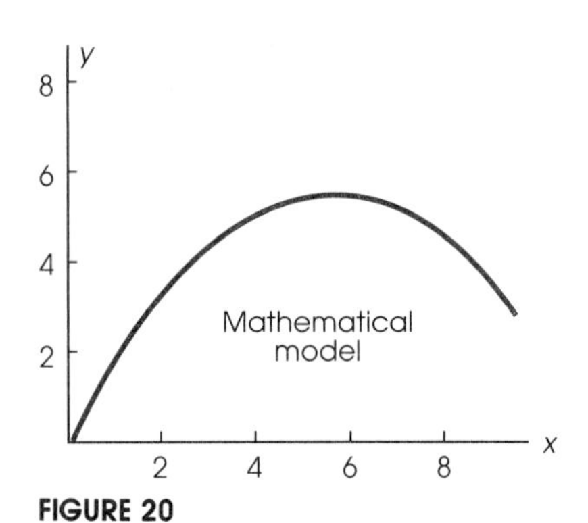

FIGURE 20

## Maximizing Profit

A firm may be concerned about many things—the pollution of its factories, the health of its employees, the reliability of its product. But in the final analysis, it will devote most of its energies to solving one problem—maximizing profit. Let's begin by looking at this problem from a general point of view.

We can expect revenue $R(x)$ to increase with $x$, though at a slower and slower rate. As the market becomes flooded, $R(x)$ may even decrease because of a lowering of the unit price that may be forced on the seller. As for cost $C(x)$, we can anticipate it to rise rapidly for small $x$ (start-up costs, inefficiency of low production), continue to increase but more slowly as production becomes more efficient, and then rise rapidly as production nears capacity (overtime pay, breakdowns). A typical situation is shown graphically in Figure 21A and B.

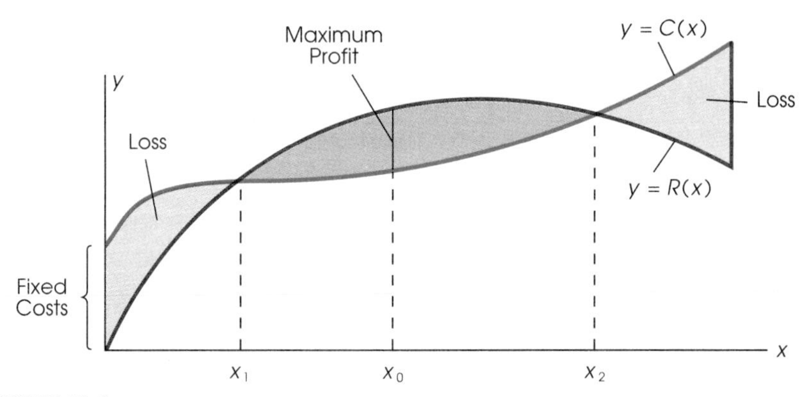

FIGURE 21 A

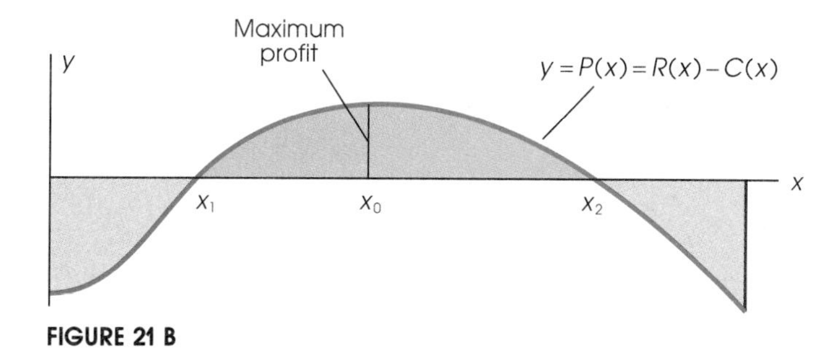


**FIGURE 21 B**

> **POINT OF INTEREST**
> At alternative statement of the economic principle is: A firm has attained maximum profit when it costs as much to produce an additional unit as can be obtained in additional revenue from that unit.

Note that the maximum profit occurs at the production level $x_0$ where $P'(x_0) = R'(x_0) - C'(x_0) = 0$, that is, where $R'(x_0) = C'(x_0)$. Thus, we have one of the fundamental principles of economics.

| Maximum profit occurs when marginal revenue equals marginal cost. |
|---|

To obtain appropriate cost and revenue functions for a particular firm is not a simple matter. Sometimes these functions are found by studying the history of the firm, sometimes they can be inferred from basic assumptions (see Example B), and occasionally they appear to be pulled out of a hat. Our first example is in the latter category.

### EXAMPLE A

The Dresswell Company makes and sells specialty dresses. Its daily cost function and revenue function (in dollars) are

$$C(x) = 500 + 360x - 9x^2 + x^3$$
$$R(x) = 480x$$

What production schedule (value for $x$) yields the maximum daily profit? Assume that peak production is 14 dresses per day.

**Solution**

$$\begin{aligned} P(x) &= R(x) - C(x) \\ &= 480x - 500 - 360x + 9x^2 - x^3 \\ &= -x^3 + 9x^2 + 120x - 500 \end{aligned}$$

Thus,

$$\begin{aligned} P'(x) &= -3x^2 + 18x + 120 \\ &= -3(x^2 - 6x - 40) \\ &= -3(x - 10)(x + 4) \end{aligned}$$

The only nonnegative solution to $P'(x) = 0$ is $x = 10$. This point together with the endpoints 0 and 14 must be checked for the maximum. Since $P(0) = -500$, $P(10) = 600$, and $P(14) = 200$, we conclude that a production schedule of 10 dresses per day will yield the maximum daily profit of \$600. ■

### EXAMPLE B

A company estimates that it can sell 1000 units per week if it sets the unit price at \$5.00, but that its weekly sales will rise by 100 units for each \$0.10 decrease in price. The company has fixed costs (overhead) each week of \$1050.00 and the costs in labor and materials to make a unit is \$1.10.

(a) Find expressions for $R(x)$ and $C(x)$.
(b) Then determine the production level $x$ that maximizes the profit $P(x)$ assuming $1000 \leq x \leq 3000$.

#### Solution

(a) Let $p(x)$ denote the unit price at production level $x$. Then

$$x = 1000 + 100\frac{5 - p(x)}{0.10}$$

This equation (called the *demand equation*) can be solved for $p(x)$, yielding

$$p(x) = 6 - 0.001x$$

Revenue is, of course, just the number of units sold times the price received for each; that is,

$$R(x) = xp(x) = x(6 - 0.001x) = 6x - 0.001x^2$$

Total cost $C(x)$ is the sum of the *fixed cost* \$1050.00 and the *variable cost* $1.10x$; that is,

$$C(x) = 1050 + 1.10x$$

(b) $P(x) = R(x) - C(x) = 6x - 0.001x^2 - 1050 - 1.10x$

$$= -0.001x^2 + 4.9x - 1050$$

and

$$P'(x) = -0.002x + 4.9$$

Setting $P'(x) = 0$ and solving gives the critical point $x = 2450$. Since $P(1000) = 2850$, $P(2450) = 4952.5$, and $P(3000) = 4650$, we conclude that a weekly production level of 2450 gives a maximum weekly profit of \$4952.50. ■

## Minimizing Inventory Costs

To be successful, a retail store must pay attention to the size of its inventory. Overstocking leads to extra interest costs, excessive warehouse rental,

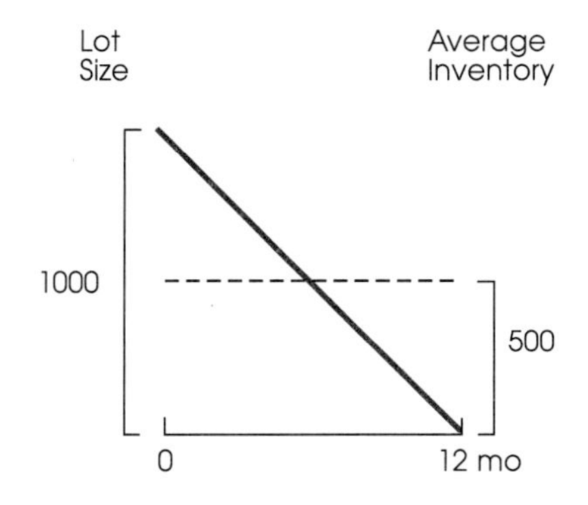


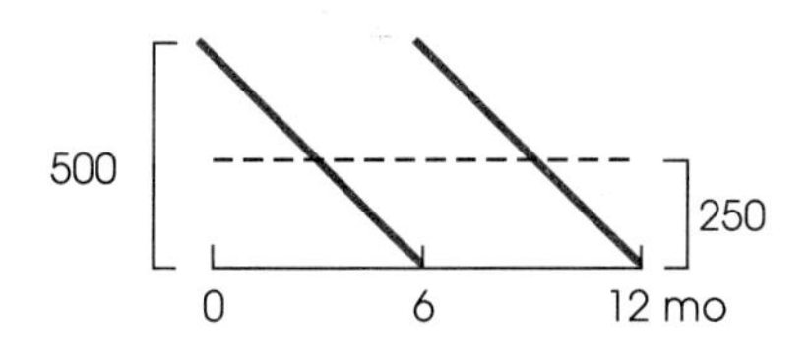


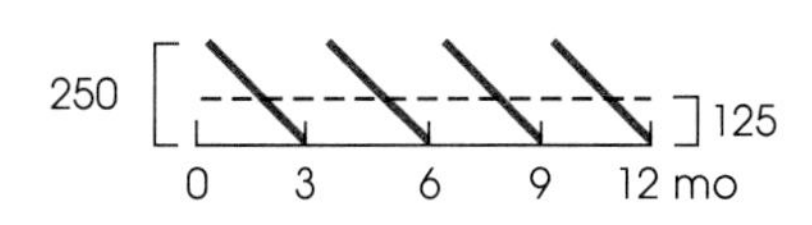


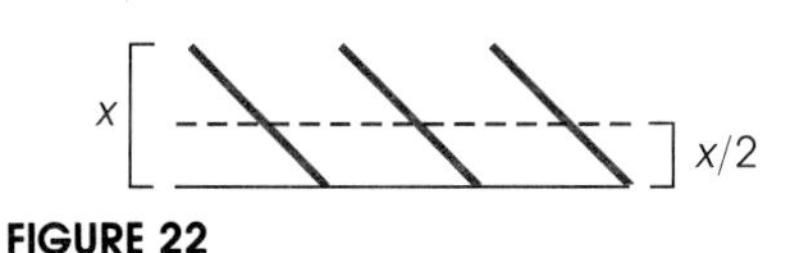


**FIGURE 22**

and the danger of obsolescence or deterioriation. Too small an inventory involves more paper work in reordering and extra delivery charges and increases the likelihood of running out of stock.

To illustrate, suppose the Sewbest Store expects to sell 1000 sewing machines each year. It could order in lots of size 1000, 500, 250 or in general lots of size $x$. On the average, the store will have $x/2$ sewing machines in stock on which it must pay inventory costs (Figure 22).

### EXAMPLE C

Suppose that it costs the Sewbest Store \$20 to hold a sewing machine for a year. To reorder a lot of size $x$ costs \$200 plus \$3 for each sewing machine. What lot size will result in the smallest inventory cost?

**Solution** Let $I(x)$ denote the yearly inventory cost corresponding to ordering in lots of size $x$.

$$\begin{aligned} I(x) &= \text{carrying cost} + \text{reordering cost} \\ &= \begin{pmatrix}\text{carrying cost}\\ \text{per unit}\end{pmatrix}\begin{pmatrix}\text{average}\\ \text{inventory}\end{pmatrix} + \begin{pmatrix}\text{cost of}\\ \text{each order}\end{pmatrix}\begin{pmatrix}\text{number}\\ \text{of orders}\end{pmatrix} \\ &= (20)\left(\frac{x}{2}\right) + (200 + 3x)\left(\frac{1000}{x}\right) \\ &= 10x + 200{,}000x^{-1} + 3000 \end{aligned}$$

$$I'(x) = 10 - 200{,}000x^{-2} = 10 - \frac{200{,}000}{x^2} = \frac{10(x^2 - 20{,}000)}{x^2}$$

When we set $I'(x) = 0$ and solve for $x$, we obtain $x = 141.42$. The First Derivative Test shows that this value of $x$ minimizes $I$. Sewbest should order in lots of size about 141. ■

## A Banking Problem

A bank naturally wants to increase its deposits and can do this by raising the interest it pays to savers. But raising the interest rate it pays lowers the net return it earns by reinvesting its deposits. Which is better: fewer deposits with a high net return or more deposits with a lower net return?

> **CAUTION**
> Think of percent as meaning per hundred. Thus, in Example D, a rate of 16.5% is really a rate of 0.165. In almost all problems involving percents, it is advisable to change percents to their equivalent decimal form before attempting the solution.

### EXAMPLE D

The ABC Saving and Loan Association plans to encourage the opening of new time-deposit accounts by setting an attractive interest rate. It has heard that the amount of money people will deposit is proportional to the square of the interest rate they will receive. If the association can earn 16.5% simple interest on the money it raises, what simple interest rate should the association offer to savers in order to maximize its profit?

**Solution** Let $x$ denote the interest rate the association will pay, and $M$ the corresponding amount of money people will put in time deposits. Then

$$M = kx^2$$

with $k$ being a positive constant. The association will earn on this money an amount

$$0.165M = 0.165kx^2$$

and will pay

$$xM = x(kx^2) = kx^3$$

resulting in an annual profit $P(x)$ given by

$$P(x) = 0.165kx^2 - kx^3$$

Setting $P'(x) = 0.33kx - 3kx^2 = 0$ gives the critical points $x = 0$ and $x = 0.33/3 = 0.11$. The association should offer an interest rate of 11%. ■

## Problem Set 4.2

*In Problems 1–4, C(x) and R(x) represent the weekly cost function and the weekly revenue function (both in dollars) for manufacturing and selling x units of a certain product. Find (a) the weekly profit function P(x), (b) the marginal profit function P'(x), and (c) the value of x that maximizes the weekly profit.*

1. $C(x) = 24x + 1000$, $R(x) = (4000x - x^2)/125$; $100 \le x \le 900$
2. $C(x) = 56x + 6000$, $R(x) = 140x - 0.1x^2$; $100 \le x \le 500$
3. $C(x) = 1000 + 20x - 0.05x^2$, $R(x) = 50x - 0.10x^2$; $50 \le x \le 350$
4. $C(x) = 400 + 3.5x - \frac{1}{300}x^2$, $R(x) = 7x - \frac{1}{200}x^2$; $200 \le x \le 1200$
5. In Problem 1, what is the unit price $p$ for which the weekly profit is a maximum?
6. In Problem 2, what is the unit price $p$ for which the weekly profit is a maximum?
7. The Radteck Company makes and sells radios. Its daily cost function and revenue function (both in dollars) are given by

   $$C(x) = 80 + 8x - 1.65x^2 + 0.1x^3 \qquad R(x) = 32x$$

   Assuming that the company can produce at most 30 radios per day, what production level $x$ will yield the maximum daily profit?
8. Blizzard Blowers makes and sells snowblowers. At a weekly production level of $x$ blowers $(0 \le x \le 30)$, its weekly cost function and weekly revenue function (both in dollars) are given by

   $$C(x) = 2000 + 40x - 33x^2 + 2x^3 \qquad R(x) = 400x$$

   Determine $x$ to give the maximum weekly profit.
9. A manufacturer of lamps estimates its daily production costs in producing $x$ lamps as $C(x) = 600 + 8x + 0.01x^3$ dollars with $0 \le x \le 40$.
   (a) How many lamps (an integer) should be produced each day to minimize the average cost per lamp, $C(x)/x$?
   (b) What is this minimum average cost?
10. A furniture company estimates that the profit $P$ in thousands of dollars resulting from spending $x$ thousands of dollars per year in advertising is given by $P = 280 + 40x - \frac{1}{4}x^2$. How much would you recommend the company spend on advertising each year?
11. A manufacturer of pen and pencil sets can sell $x$ sets per day if it sets the price per set at $p(x) = 4 - 0.002x$ dollars $(0 \le x \le 1200)$.
    (a) What production level $x$ and price $p$ will maximize daily revenue?
    (b) If the corresponding cost function is $C(x) = 200 + 1.5x$ dollars, what production level $x$ will maximize daily profit?

12. A maker of file cabinets with a capacity of 2500 units per month believes that it can sell $x$ units if it sets the price at $p(x) = 75 - 0.025x$ dollars. Its monthly cost function in dollars is given by $C(x) = 3000 + 30x - 0.015x^2$. How many cabinets should it make each month to maximize its profit?
13. The ABC Company, a manufacturer of metric tool sets, has fixed monthly costs of \$10,000 and direct costs of \$120 for each set produced. The company estimates that 200 sets can be sold each month if the unit price is \$200, and that 15 more units can be sold for each decrease of \$10 in the price. Assume $x$, the number of units sold per month, satisfies $200 \leq x \leq 400$.? See Example B.
    (a) Determine the monthly cost function $C(x)$.
    (b) Determine the monthly revenue function $R(x)$.
    (c) What number of tool sets should be produced to maximize monthly profit?
14. A manufacturer of hand-held calculators is confident that it can sell 3000 calculators per week at a price of \$12 each and believes that reducing the price by 20¢ each will increase its weekly sales by 200.
    (a) Find the weekly revenue $R(x)$ if the manufacturer produces $x$ calculators.
    (b) How many calculators should the manufacturer produce each week to maximize revenue? Assume a capacity of 10,000 calculators per week.
15. A stereo manufacturer estimates that in order to sell $x$ stereos per year, the unit price must be $p(x) = 1200 - x$ dollars. The annual cost of producing $x$ stereos is $C(x) = 30{,}000 + 100x$ dollars, and the annual capacity is 800 units.
    (a) Determine the annual revenue $R(x)$.
    (b) How many units should be produced to maximize the profit?
    (c) What is the maximum profit?
    (d) What price per unit maximizes the profit?
    (e) Find the value of the marginal revenue when $x = 200$.
    (f) Find (approximately) the added revenue gained by increasing production from 200 to 201.
16. Suppose the manufacturer in Problem 14 has weekly fixed costs of \$2000 and unit costs of \$2. How many calculators should it make each week to maximize its weekly profit?
17. The XYZ Company expects to sell 1200 power mowers per year. It costs the company \$40 to hold a mower for a year. Reordering a lot of size $x$ costs \$350 plus \$5 for each mower. What lot size will result in the smallest inventory cost for the year? See Example C.
18. Ace Sports Company can sell 24,000 bowling balls per year. It costs the company \$5 to hold a bowling ball for a year, and reordering a lot of size $x$ costs \$250 plus \$2 for each ball. In what lot size should bowling balls be reordered to have the least annual inventory cost?
19. Eastwood Publishing Company expects to sell 100,000 copies of a new hardcover novel per year. If setting up each printing run costs \$8000 and if storing a book for a year costs \$0.40, what size printing run will minimize these costs?
20. Quick Start Battery Company expects to make and sell 20,000 Super A batteries per year. Setting up a production run costs \$1200, and storing a battery for a year costs \$3. What size run will minimize these costs?
21. To encourage time deposits, Peabody Savings and Loan Association plans to offer an attractive interest rate. Its research indicates that a simple interest rate $x$ will result in deposits proportional to $x^{3/2}$. What interest rate should the association offer to maximize its profits if it can earn a 14% annual return on the money it raises? See Example D.
22. Suppose in Problem 21 that the amount people will deposit is proportional to $x^{5/2}$ but that the association can earn only a 12% return on the money it receives. What interest rate should be offered to maximize the association's profits?
23. A large state bank believes that the amount of money the public will deposit with it is proportional to the square of the increase in interest rate over 7%. If the bank expects to earn an 11% rate on the money it receives, what rate should it offer to investors to maximize profits? Assume that the rates are simple interest rates.
24. A research organization has told North Star Investments that it can expect deposits in proportion to the square of the increase in its interest rate beyond 4%. What rate should North Star offer to maximize profits if it can earn 10% on the money it receives?

---

25. An apartment complex has 80 units. When the rent is \$400 per month, all of the units are rented. For each \$10 increase in rent, one apartment unit becomes vacant. What rent should be charged to produce the maximum revenue?
26. A factory has 15 machines each capable of automatically assembling 40 toy trucks per hour. It costs \$60.00 to set up each machine and \$3.00 per hour to run it. One operator can oversee all 15 machines, and he earns \$19.20 per hour. How many machines should be set up to produce 6000 toy trucks in order to minimize costs?
27. Susan Blackwood is raising a steer that now weighs 600 pounds and if sold today would bring 60¢ per pound. She estimates that the steer is gaining 30 pounds per week but that the price per pound will

likely drop by 2¢ a week over the next several weeks. After how many weeks should she market her steer to maximize the amount she will get for it?

28. Problem 27 took no account of the cost of feeding a steer. Suppose it costs \$2.40 per week to feed a steer. After how many weeks should Susan sell her steer to maximize her profit?
29. Refer again to Problem 27. Suppose it costs 25¢ for the feed required to produce a 1-pound increase in the weight of the the steer. After how many weeks should Susan sell her steer if she wishes to maximize her profit?
30. John Adams estimates that his farm would yield 1000 bushels of sugar beets if harvested right now. Today's beet price is \$2.50 per bushel, but it will drop \$0.05 a bushel each day from now on. On the other hand, John guesses that the crop is still growing at 30 bushels per day. In how many days from now should John harvest and sell his crop to obtain maximum revenue (and presumably maximum profit)?
31. Northwood Independent Bank presently has \$1 million in savings deposits on which it is paying the equivalent of 7% = 0.07 simple interest. To increase deposits, it plans to increase the rate it pays by an amount $x$ and estimates that it will thereby gain new deposits $M$ proportional to $x^2$ (that is, $M = kx^2$). To keep its present deposits, it will pay the new rate on all deposits—old as well as new. The bank can earn an annual rate of 11% on all this money.
    (a) Show that the bank's annual profit $P(x)$ in millions of dollars on savings deposits under the new rate is given by

    $$P(x) = 0.04 - x + 0.04kx^2 - kx^3$$

    (b) One critical point is $x = 0$ (no increase). Show that the others are

    $$x = \frac{0.08 \pm \sqrt{0.0064 - 12/k}}{6}.$$

    C (c) Assuming that $k = 10{,}000$, determine the interest rate that maximizes $P(x)$.
    C (d) What is the maximum value of $P$ in dollars?

## 4.3* MORE APPLICATIONS

Differential calculus achieved its earliest successes in the fields of astronomy and physics. It is, in fact, in the physical sciences that the concepts of calculus find their most obvious uses. Yet, more and more, the life sciences and the social sciences are turning to mathematics (including calculus) to find tools with which to solve their problems. We saw in the previous section how economists use calculus to minimize costs and maximize profits. Here we offer a broad spectrum of other applications.

### Optimum Density

It is easy to be too greedy. Plant too many corn stalks on an acre of ground and you will actually reduce your yield; move tables too close together in a restaurant and diners will stay away in droves; put too many secretaries in the same office and their output will decrease. Calculus can help us analyze the problem of crowding.

#### EXAMPLE A

An apple grower estimates that each tree will produce 35 baskets of apples when there are 40 trees per acre, but that for each additional tree planted the yield per tree will go down by three-fourths of a basket. What density of trees will result in the largest apple harvest?

**Solution** The total yield $y$ in baskets per acre can be expressed in terms of $t$, the number of trees per acre, by the formula

$$\text{total yield} = (\text{number of trees})(\text{yield per tree})$$
$$y = t[35 - \tfrac{3}{4}(t - 40)]$$
$$= t(35 - \tfrac{3}{4}t + 30)$$
$$= 65t - \tfrac{3}{4}t^2$$

Thus

$$\frac{dy}{dt} = 65 - \tfrac{3}{2}t$$

and when set equal to 0, this gives

$$t = (65)(\tfrac{2}{3}) \approx 43.33$$

It is a simple matter to check that this value of $t$ makes $y$ as large as possible. Our apple grower should plant about 43 trees per acre. ■

## Best Truck Speed

Here is a problem faced by all firms that deliver their products by truck. Drivers are paid by the hour, which suggests that they should be asked to drive as fast as possible. On the other hand, truck operating costs tend to increase with speed; in particular, gas mileage goes down as speed increases. At what speed should a truck be driven to minimize the total cost?

**POINT OF INTEREST**
In connection with Example B, someone is sure to complain that 69 miles per hour exceeds the speed limit. Fair enough. Study the graph of cost $C$ against speed $s$ shown below. Conclude that to minimize cost a trucker should drive at the speed limit provided this limit is 69 miles per hour or less.

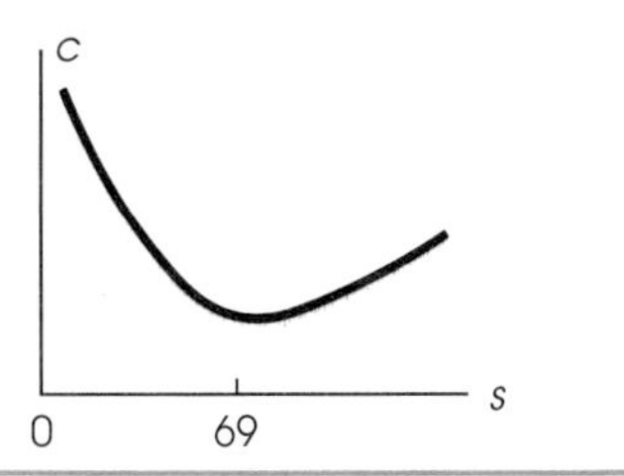


**EXAMPLE B**

The cost of operating a certain truck (gas, oil, and so on) is $40 + s/4$ cents per mile when driven at $s$ miles per hour. The driver gets $12 per hour. At what speed should the truck be driven to minimize the total cost of making a delivery to a distant city?

**Solution** Consider a trip of $k$ miles. The total cost $C$ (in cents) is given by

$$\text{total cost} = \text{truck cost} + \text{driver cost}$$

$$C = k\left(40 + \frac{s}{4}\right) + \frac{k}{s}(1200)$$

$$= 40k + \frac{ks}{4} + \frac{1200k}{s}$$

$$\frac{dC}{ds} = 0 + \frac{k}{4} - \frac{1200k}{s^2}$$

When we set $dC/ds$ equal to 0 and solve for $s$, we obtain

$$\frac{k}{4} = \frac{1200k}{s^2}$$

$$s^2 = 1200k\left(\frac{4}{k}\right) = (1200)(4)$$

$$s \approx 69 \text{ miles per hour}$$

Note that the constant $k$ canceled out at the end. The answer of 69 miles per hour is independent of the length of the trip. ∎

## Drug Dosage

Doctors need to understand the characteristics of the drugs they prescribe. For a given drug, one important item is the maximum allowable dosage $C$. Another is the so-called **sensitivity** of the body to the drug. If $x$ measures the dosage and $R(x)$ the strength of the reaction (change in body temperature or blood pressure or some other measurable body function), then $R'(x)$ measures sensitivity. It tells the doctor (approximately) how much change to expect in $R$ as the result of a unit change in the dosage. Generally, doctors like to prescribe a dosage of maximum sensitivity.

### EXAMPLE C

A common mathematical model for drug reactions is

$$R(x) = x^2\left(\frac{C}{2} - \frac{x}{3}\right) = \frac{Cx^2}{2} - \frac{x^3}{3} \qquad 0 \le x \le C$$

Show that the dosage of maximum sensitivity is $C/2$.

**Solution** Note that we want to maximize $R'(x)$ which we do by finding where $R''(x) = 0$.

$$R'(x) = Cx - x^2$$

$$R''(x) = C - 2x$$

$$R'''(x) = -2$$

Setting $R''(x) = 0$ yields $x = C/2$. Since $R'''(x)$ is negative, we can be certain that $R'(C/2)$ is at least a local maximum value for $R'(x)$. To check that it is the global maximum value, we calculate $R'$ at the three critical points: 0, $C/2$, and $C$. We obtain $R'(0) = R'(C) = 0$, while $R'(C/2) = C^2/4$. ∎

## Maximum Sustainable Harvest

We pointed out the danger of being too greedy earlier in this section. If one is in the business of harvesting fish, fur-bearing animals, or trees, restraint is a prerequisite to long-term success. Overharvesting will obviously interfere with the reproductive capabilities of the population.

To model this situation, let us suppose that a population of size $x$ will grow to size $f(x)$ in 1 year if unmolested by human beings. Then to keep from depleting the population, the yearly harvest should not exceed $h(x) = f(x) - x$. If our goal is to secure the largest possible harvest year after year, we should allow the population to grow to the size which makes $h(x)$ a maximum.

### EXAMPLE D

Each year, many states allow hunters to shoot deer during a limited open season whose length is carefully chosen to ensure a harvest that is sustainable year after year. For one state, the yearly growth curve for the deer population is estimated to be

$$f(x) = 1.4x - 0.0004x^2$$

where $x$ is measured in thousands. What is the optimal population size and what is the yearly kill that it will sustain?

**Solution**

$$h(x) = f(x) - x = 0.4x - 0.0004x^2$$

$$h'(x) = 0.4 - 0.0008x$$

$$h''(x) = -0.0008 < 0$$

The maximum value of $h$ is attained when $h'(x) = 0$, that is, at $x = 0.4/0.0008 = 500$. Thus, the state authorities should allow the deer population to climb to 500,000 which will then sustain a yearly harvest of

$$h(500) = 0.4(500) - 0.0004(500)^2 = 100$$

that is, 100,000 deer per year. ■

## Learning Theory

There are many phenomena in real life whose behavior over time is modeled by a curve of the type $y = P(t)$ shown in Figure 23. Here are some examples.

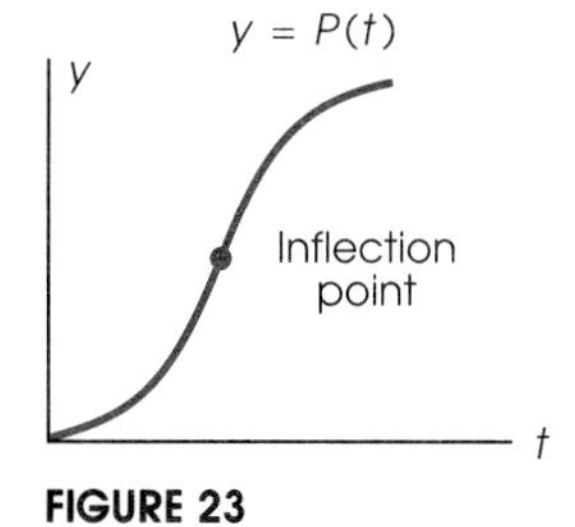

**FIGURE 23**

(a) $P(t)$ is the percentage of the population infected by a contagious disease during the first $t$ days of an epidemic.
(b) $P(t)$ is the number of people who have heard a rumor $t$ days after the rumor was started.
(c) $P(t)$ measures the amount of learning of a skill that takes place during the first $t$ hours of practice.

In each case $P$ grows slowly at first, then more and more rapidly, but eventually slower and slower. The time of most interest to doctors in case (a), to sociologists in case (b), and to psychologists in case (c) is the moment of fastest growth, that is, the moment when $P'(t)$ is largest. This is the time when $P''(t) = 0$ and corresponds to the inflection point on the curve in Figure 23.

**EXAMPLE E**

Typing skill is usually measured by typing speed in words per minute. For a class of 12-year-old boys, a psychologist discovered that on the average they could achieve a speed of $P(t)$ words per minute after $t$ hours of practice where

$$P(t) = 0.015t^2 - 0.0001t^3 \qquad (0 \le t \le 100)$$

After how many hours of practice are the boys learning most rapidly?

**Solution**

$$P'(t) = 0.03t - 0.0003t^2$$

$$P''(t) = 0.03 - 0.0006t$$

Setting $P''(t) = 0$ gives $t = 50$. The boys are learning most rapidly after 50 hours of practice. ■

## Problem Set 4.3

1. A commercial orange grower used to plant 30 trees per acre and normally harvested about 20 bushels of oranges per tree. A research study indicates that planting trees closer together may improve the total harvest per acre; in particular, the study estimates that for each additional tree (up to 15) beyond 30 that is planted on an acre, the yield per tree is reduced only by $\frac{1}{2}$ bushel. How many trees per acre should be planted to maximize the orange yield per acre?

2. A nut farmer estimates that a walnut tree will yield 60 pounds of nuts when there are 40 trees per acre and that for each additional tree per acre the yield per tree is reduced by 1 pound. If so, how many trees per acre should be planted to maximize yield?

3. Experience convinced the farmer in Problem 2 that the yield is reduced only $\frac{3}{4}$ pound for each additional tree beyond 40 (up to the maximum of 55 trees per acre permitted by government regulations).
   (a) Determine the yield per acre when there are 52 trees per acre.
   (b) What density of trees will maximize the per acre yield?

4. Acme Car Rental is able to rent out 120 cars per day when it charges \$12 per day (there is an additional mileage charge that just covers expenses). For each 25¢ increase in the daily rate, Acme thinks one less car would be rented.
   (a) Find the net daily income when the rate is set at \$15.
   (b) What daily rate would maximize the net daily income?

5. A downtown movie theater has had an average attendance of 150 people with an admission charge of \$2.50. Assume that it will consider only a charge that is a multiple of \$0.10 and that it believes it will lose three customers for each \$0.10 increase in the admission charge. What admission charge will maximize revenue and what revenue does this produce for each showing?

6. The Mark Twain Showboat advertises excursions to groups of 300 at \$15 per ticket but promises a reduction of 20¢ on all tickets for each group of 10 passengers in excess of 300 (assume that a group of 9 additional passengers produces no reduction in the ticket price). Exactly what (whole) number of passengers produces the maximum revenue for the showboat and what is this revenue?

7. Portage Transportation Company estimates the cost of operating a moving van is $40 + \frac{1}{4}x$ cents per mile

when driven at $x$ miles per hour. In addition, the driver gets \$10 per hour.

(a) At what speed should the van be driven to minimize the total cost of a 400-mile run?

(b) Find the minimum total cost.

8. Suppose that a truck averages $400/(x + 2)$ miles per gallon when driven at a speed of $x$ miles per hour $(40 \le x \le 65)$. Suppose further that gasoline costs \$1.20 per gallon, that the driver gets \$10.00 per hour, and that other costs are ignored. What speed will minimize the cost of a 200-mile run?

9. A long-distance trucker finds that when he maintains a speed of $x$ miles per hour, the truck consumes $V = 3x^{-1} + 0.0015x$ gallons of diesel fuel per mile.

   (a) If diesel fuel costs \$1.30 per gallon, what speed will result in the lowest fuel cost for a 400-mile trip?

   (b) Find the minimum fuel cost.

10. Suppose the driver in Problem 9 gets \$12 per hour of driving time. Ignoring costs other than fuel and wages, what speed will minimize the cost of the 400-mile trip?

11. Suppose that the human reaction $R$ to a certain drug is measured as

$$R(x) = \frac{Cx^2}{3} - \frac{x^3}{4}$$

where $x$ is the dosage and $C$ is the maximum allowable dosage.

   (a) Find the dosage that maximizes the reaction $R(x)$.

   (b) Find the dosage that maximizes the sensitivity $R'(x)$.

12. The reaction $R$ to an experimental drug is given by

$$R(x) = \frac{Cx^2}{a} - \frac{x^3}{a}$$

where $a$ is a constant and $C$ is the maximum allowable dosage. Show that the dosage that maximizes sensitivity to the drug is one-half the dosage that maximizes the reaction.

13. Blackhawk County annually attracts a large number of hunters for its fall hunting season and would like to attract even more. It is claimed that if there were no hunting season in the county, the pheasant population would grow from size $x$ (in thousands) to size $f(x) = 2.5x - 0.02x^2$ in 1 year. To what size should the county let the pheasant population grow to be able to maintain the largest yearly kill? What would this yearly kill then be?

14. In one of our western states, it was estimated that, if no hunting were allowed, an elk population of size $x$ (in thousands) would grow in 1 year to a population of size $f(x) = 1.6x - 0.0005x^2$. What population size allows for the maximum yearly kill year after year? What is the maximum kill?

15. A psychologist has found that if a 17-year-old girl is allowed to study $t$ hours for a certain test, her score $S$ can be predicted from the formula $S(t) = 30t^2 - 3t^3$ $(0 \le t \le 7)$.

   (a) How many hours of study produce the highest score?

   (b) At what time is learning occurring most rapidly?

16. A government study deals with the spread of a certain strain of flu. The study says that in a city of 10,000, the number of people $N$ who will contract the disease within the first $t$ weeks after the first reported case can be predicted from the formula

$$N(t) = 54t^2 - 2t^3 \qquad (0 \le t \le 18)$$

   (a) How many people from the city will get the disease?

   (b) When will the disease be spreading most rapidly?

---

17. Arne Carlson, a fruit grower in British Columbia, knows from experience that when he plants 24 cherry trees per acre, he can expect a yield per tree of 44 pounds. The local expert from the agricultural college claims that for each additional tree that Arne plants per acre, his yield per tree will drop by only 1 pound. Assume that Arne believes the expert.

   (a) How many additional trees (beyond 24) should Arne plant on each acre to achieve maximum yield per acre?

   (b) What is the maximum yield per acre?

18. A certain toll bridge averages 80,000 cars a day when the toll is \$1.50. The Bridge Authority is considering increasing the toll but, because there is a free bridge only 10 miles away, estimates that 1000 fewer cars will use its bridge for each \$0.05 increase in the toll. What toll (a multiple of \$0.05) will maximize the daily revenue?

19. At \$80 a night per room, the 120-room Royal Plaza Hotel is full to capacity. Suppose that each 80¢ increase in the room rate results in an additional vacant room and that it costs \$5 to service a rented room. What should the management charge per room to maximize its profit from room rentals? Assume that the number of rented rooms must be an integer.

20. Lake Serene has to be treated periodically to control the growth of harmful bacteria. If $t$ days after a treatment the concentration of bacteria per cubic centimeter is given by

$$C(t) = 18t^2 - 216t + 900 \qquad (0 \le t \le 12)$$

when would you recommend that your children go

swimming and what is the concentration of bacteria at this time?

21. When the treatment facility overflows and raw sewage is dumped into a lake, the oxygen content decreases until it reaches a minimum and then begins to increase. Suppose that $t$ days after raw sewage was dumped into Lake Wobegon, the concentration $C$ of oxygen (measured in appropriate units) in the lake was given by

$$C(t) = 600\left[1 - \frac{8}{t+4} + \frac{64}{(t+4)^2}\right]$$

(a) After how many days will the oxygen concentration be lowest?
(b) When will the oxygen concentration be increasing most rapidly?

22. Mark Farmer has found that when he uses $x$ bags of fertilizer per acre, his soybean yield in bushels per acre is

$$V(x) = \frac{32x}{x+8} + 50$$

If soybeans sell at $10 per bushel and fertilizer costs $30 per bag, how many bags per acre should he use to maximize his profit?

C 23. A school is to be built along a road between factories A and B which are 8 miles apart. Both factories emit air pollutants. At a point $x$ miles from factory A, the concentration of pollutants is measured as

$$p(x) = \frac{27c}{x} + \frac{8c}{8-x} \qquad (0 < x < 8)$$

where $c$ is a constant. How far from factory A should the school be built to minimize the air pollution problem?

24. The Hoodoo University Dragons are anxious to maximize the revenue from their football games. For several years they have charged $5.00 per ticket and have averaged 60,000 at their games. A typical fan spends $2.25 on concessions. A consulting firm says that for each $1.00 increase in the admission charge, there will be a loss of 6000 fans and on this basis recommends a new ticket price. What was their recommendation?

## 4.4 IMPLICIT DIFFERENTIATION AND RELATED RATES

The terms *explicit* and *implicit* are pretty standard English words. For example, we speak of explicit and implicit assumptions. An explicit assumption is openly stated and clearly formulated. In contrast, an implicit assumption may be hidden or vaguely expressed, but it is an assumption just the same. An important task for politicians, philosophers, and ordinary people is to make implicit assumptions explicit.

In mathematics, we use these two words to indicate two ways of determining a function. A function (with domain variable $x$) is described explicitly if it is determined by an equation of the type $y = f(x)$. For example, the equation

$$y = \frac{2+x}{3-x}$$

gives an explicit description of a function. In contrast, the equation

$$3y - xy = 2 + x$$

also determines a function (actually the same function), but now this fact is somewhat disguised. We must do some work (solve for $y$) to make the description explicit.

Sometimes it is easy to turn an implicit description into an explicit one; sometimes it is difficult; and sometimes it is impossible.

The equation $3y - xy = 2 + x$ is readily solved for $y$. First factor $y$ from the left side; then divide by $3 - x$:

$$3y - xy = 2 + x$$
$$y(3 - x) = 2 + x$$
$$y = \frac{2 + x}{3 - x}$$

Not many of us would want to try solving the equation

$$y^9 + 2xy + \log xy = 3y$$

for $y$; we believe this to be impossible. Here is an example of intermediate difficulty.

**EXAMPLE A**

Sove $x^2 + y^2 + 2y = 3$ for $y$.

**Solution** We begin by adding 1 to both sides, thus completing the square on the $y$'s.

$$x^2 + y^2 + 2y + 1 = 3 + 1$$
$$x^2 + (y + 1)^2 = 4$$
$$(y + 1)^2 = 4 - x^2$$
$$y + 1 = \pm\sqrt{4 - x^2}$$
$$y = -1 \pm \sqrt{4 - x^2}$$

The interesting fact discovered here is that our original equation determines two functions:

$$y = f(x) = -1 + \sqrt{4 - x^2} \quad \text{and} \quad y = g(x) = -1 - \sqrt{4 - x^2}$$ ■

## Implicit Differentiation

Now we ask a question. Is it possible to find the derivative $dy/dx$ without first having $y$ expressed explicitly in terms of $x$? The surprising answer is yes. Consider the equation

$$3y - xy = 2 + x$$

with which we started our discussion. Take the derivative of both sides with respect to $x$ and equate the results. Keep in mind that $y$ is a function of $x$;

> **CAUTION**
> A very common error is to leave out the parentheses around
>
> $$x\frac{dy}{dx} + y$$
>
> in the first step. This is fatal because it results in the wrong sign on $y$.

thus, we will have to use the Product Rule on the $xy$ term. We obtain

$$3\frac{dy}{dx} - \left(x\frac{dy}{dx} + y \cdot 1\right) = 0 + 1$$

$$3\frac{dy}{dx} - x\frac{dy}{dx} - y = 1$$

Next move $y$ to the right side; then factor $dy/dx$ from the left side.

$$\frac{dy}{dx}(3 - x) = 1 + y$$

$$\frac{dy}{dx} = \frac{1 + y}{3 - x}$$

Thc only troublesome feature of our answer is that it involves both $y$ and $x$ (we would prefer that it involved just $x$), but in many situations this is a minor problem.

An inquisitive reader is sure to ask whether the answer we just got for $dy/dx$ is equivalent to the one we would have gotten if we had first solved the original equation for $y$ and then taken the derivative. The answer is yes, though we shall not take the space to demonstrate this.

The method of finding the derivative $dy/dx$ without first solving the given equation explicitly for $y$ in terms of $x$ is called the method of **implicit differentiation.**

## EXAMPLE B

Find the slope of the tangent line to the circle

$$x^2 + y^2 = 25$$

at the point $(3, -4)$ using the method of implicit differentiation (see Figure 24).

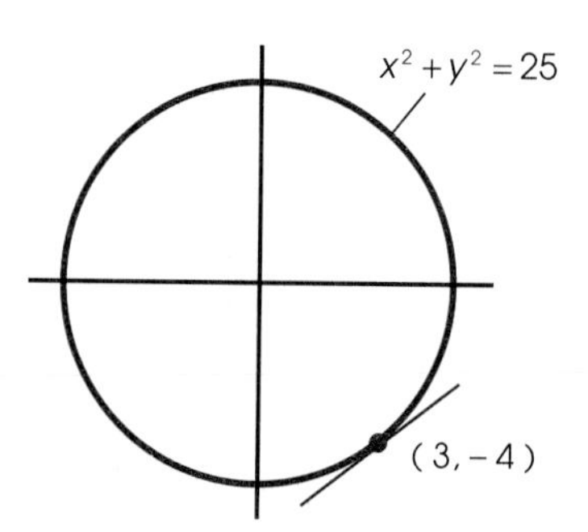


**FIGURE 24**

**Solution** We take the derivative of both sides with respect to $x$ and equate the results. Keep in mind that $y$ is a function of $x$; thus, we must use the General Power Rule on the term $y^2$.

$$2x + 2y\frac{dy}{dx} = 0$$

$$2y\frac{dy}{dx} = -2x$$

$$\frac{dy}{dx} = \frac{-2x}{2y} = \frac{-x}{y}$$

At $(3, -4)$, $dy/dx = -3/-4 = \frac{3}{4}$. The slope of the tangent line at this point is $\frac{3}{4}$. ■

**EXAMPLE C**

Find the equation of the tangent line to

$$xy^2 + y^3 = 12x$$

at the point $(1, 2)$.

**Solution** You should first check to see that $(1, 2)$ really does satisfy the equation. Then take the derivative of both sides with respect to $x$. On the term $xy^2$, you will have to use the Product Rule together with the General Power Rule; on the $y^3$ term, you will need just the General Power Rule.

$$x \cdot 2y\frac{dy}{dx} + y^2 \cdot 1 + 3y^2\frac{dy}{dx} = 12$$

$$\frac{dy}{dx}(2xy + 3y^2) = 12 - y^2$$

$$\frac{dy}{dx} = \frac{12 - y^2}{2xy + 3y^2}$$

At $(1, 2)$, $dy/dx$ has the value $(12 - 4)/(4 + 12) = \frac{1}{2}$. We conclude that the tangent line has the equation

$$y - 2 = \tfrac{1}{2}(x - 1)$$ ■

## Related Rates

Here is a common problem arising in the world of experience. Two variables, say $x$ and $y$, are both functions of a third variable $t$ though these functions are not known explicitly. An equation connecting $x$ and $y$ is given. Can we find $dy/dt$ if we know $dx/dt$? An example will clarify the question and provide a positive answer.

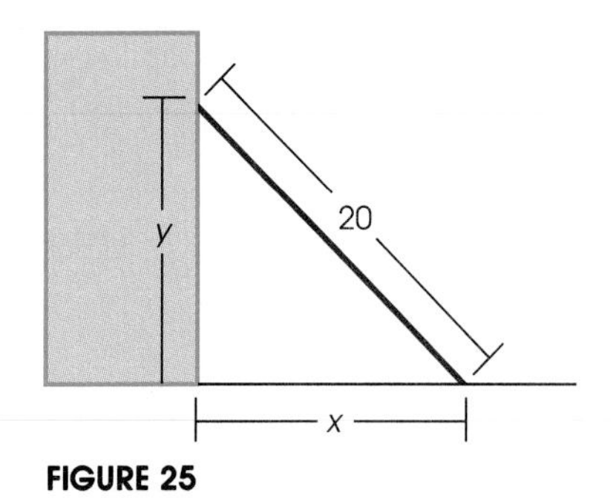


FIGURE 25

RECALL

In a right triangle,

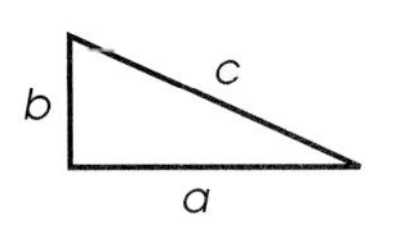


the three sides are related by the equation

$$a^2 + b^2 = c^2$$

This result is known as the Pythagorean Theorem.

## EXAMPLE D

A 20-foot ladder leans against a vertical wall (Figure 25). If the bottom of the ladder is pulled away from the wall at 3 feet per second, how fast is the top of the ladder sliding down the wall when it is 12 feet above the ground?

**Solution** Variables $x$ and $y$ have been assigned in Figure 25. Note that both $x$ and $y$ are functions of time $t$. We are given that $dx/dt = 3$, and the Pythagorean Theorem implies that

$$x^2 + y^2 = 400$$

We are asked to find $dy/dt$ when $y = 12$.

We use the method of implicit differentiation; that is, we take the derivative of both sides with respect to $t$, equating the results. Since both $x$ and $y$ are functions of $t$, we must apply the General Power Rule to both the $x^2$ and $y^2$ terms. We obtain

$$2x\frac{dx}{dt} + 2y\frac{dy}{dt} = 0$$

or since $dx/dt$ is given to be 3,

$$2x(3) + 2y\frac{dy}{dt} = 0$$

$$\frac{dy}{dt} = -\frac{6x}{2y} = -\frac{3x}{y}$$

This expression for $dy/dt$ is to be evaluated when $y = 12$. Unfortunately, we also need to know $x$. Once again the Pythagorean Theorem comes to our aid.

$$x^2 + (12)^2 = 400$$

$$x^2 = 256$$

$$x = 16$$

We conclude that when $y = 12$, $x = 16$ and

$$\frac{dy}{dt} = -\frac{3x}{y} = -\frac{48}{12} = -4 \text{ feet per second}$$

The negative sign is no surprise, since $y$ is decreasing with time. ■

> **CAUTION**
> In Example E, we are asked for the rate of increase in profit with respect to time, that is, for $dP/dt$. Be sure to distinguish this from marginal profit which is $dP/dx$, the rate of change in profit with respect to the number of units produced.

### EXAMPLE E

A manufacturer of hand-held calculators has weekly cost and revenue functions given by

$$C = 6000 + 5x$$

$$R = 10x - 0.0001x^2$$

where $x$ is the number of calculators produced each week. If $x$ is increasing at the rate of 400 calculators per week, find the rate of increase in profit $P$ when weekly production reaches 4000.

**Solution** We are given that $dx/dt = 400$ and want to find $dP/dt$ when $x = 4000$. Now

$$P = R - C = 10x - 0.0001x^2 - 6000 - 5x$$

that is,

$$P = -0.0001x^2 + 5x - 6000$$

We differentiate both sides of this equation with respect to $t$, keeping in mind that $x$ depends on $t$. We obtain

$$\frac{dP}{dt} = (-0.0001)(2x)\frac{dx}{dt} + 5\frac{dx}{dt} - 0$$

When $x = 4000$ and $dx/dt = 400$,

$$\frac{dP}{dt} = (-0.0001)(8000)(400) + 5(400)$$

$$= -320 + 2000 = 1680$$

Profit is increasing at $1680 per week. ■

## Problem Set 4.4

*The equations in Problems 1–12 determine y in terms of x but do so implicitly. Solve for y, thereby giving y explicitly in terms of x (see Example A and the discussion preceding it).*

1. $x^2 + 2y = 3$
2. $2xy - 5x = 6$
3. $4y + 3xy = 2 - x$
4. $x^2y - 3y = 2x + 1$
5. $x^2 + xy = y + x + 3$
6. $x^2y + 9 = 5x + 2y$
7. $(y - 4)(x^2 + x) = 3x + 2$
8. $(y - 2x)(x + 2) = 1 - x^2$
9. $y^2 + 2x = x^2 + 4$
10. $xy^2 + x^3 = 4$
11. $y^2 - x^2 - 8y + 11 = 0$
12. $y^2 + 2y - 2x + 4 = 0$
13. Find $dy/dx$ for Problem 5 in two ways (by differentiating the explicit relation you found there and by implicit differentiation). Show that your answers are equivalent.
14. Find $dy/dx$ for Problem 6 in two different ways.

*In Problems 15–24, find dy/dx by implicit differentiation (see the discussion preceding Example B as well as Examples B and C).*

15. $2x^2y = 7$
16. $3x^3y = 4x + 1$

17. $x^2y + 5 = x^2 - 4y$
18. $3xy - 2x + 4 = x^2 + 2y$
19. $y^3 + 3xy = 27$
20. $2y^2 - 3x^2y = 15$
21. $x^2 + 2xy + 3y^2 = 36$
22. $xy^2 - yx^2 = 4y + 1$
23. $(2x^2 + y^2)^4 = 20$
24. $(3 + y^2)^5 = 100x^2$

*In Problems 25–30, find the equation of the tangent line to the given curve at the specified point (see Examples B and C). Don't forget that you must evaluate $dy/dx$ to get the slope.*

25. $x^2 + y^2 = 169$; $(-12, 5)$
26. $x^2 - y^2 = 16$; $(5, 3)$
27. $x^2 + 4y^2 = 4y + 2x$; $(2, 1)$
28. $x^2 - y^3 = 34 + y$; $(2, -3)$
29. $x^2 - 2xy + 5y^2 = 13$; $(4, 1)$
30. $y^3 - 2xy = 5$; $(-2, 1)$
31. Suppose that the point $P$ is moving along the curve $y = x^2$ in such a way that $dx/dt$ is always 3 units per second. Find $dy/dt$ when $x = 2$. See Example D.
32. Suppose that the point $P$ moves along the curve $x = y^3$ so that $dx/dt$ is always 4 units per second. Find $dy/dt$ when $y = 2$.
33. A 50-foot ladder leaning against a vertical wall is beginning to slide. At the instant when the top of the ladder is 30 feet from the ground, the top is sliding down at the rate of 2 feet per second. How fast is the bottom of the ladder moving away from the wall at that instant?
34. A man who is 6 feet tall is walking directly away from a street light that is 15 feet above the ground. If he walks at 5 feet per second, how fast is the tip of his shadow moving when he is 100 feet from the base of the light post? *Hint:* In Figure 26 you are to find $dx/dt$ when $y = 100$.

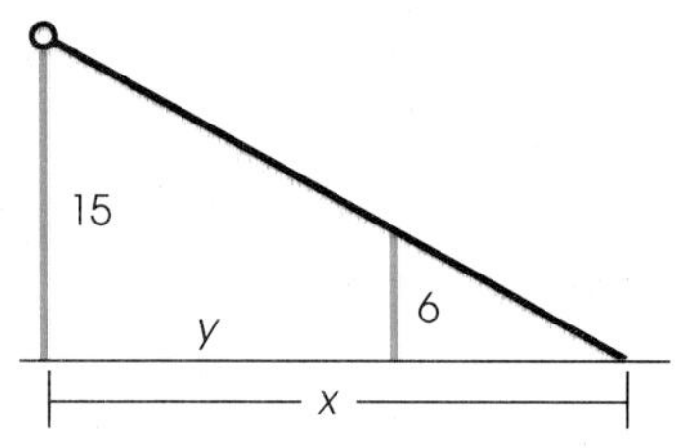

FIGURE 26

35. Suppose the point $P$ in Figure 27 is moving to the right along the line $y = 3$ at the rate of 4 units per second.
(a) How fast is the distance $\overline{OP}$ increasing at the instant when $\overline{AP} = 6$?

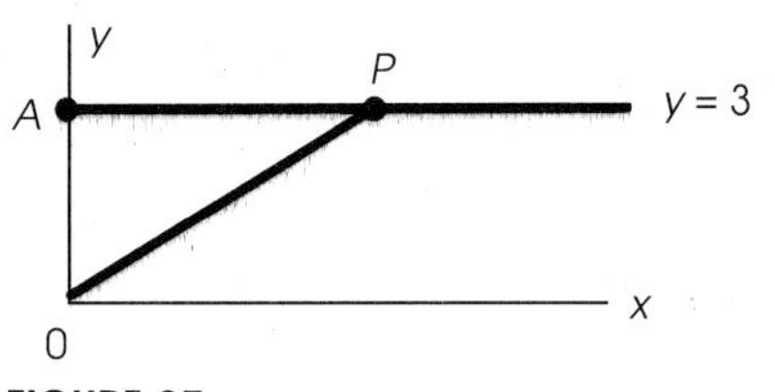

FIGURE 27

(b) How fast is the area of the triangle $OAP$ increasing at this instant?

36. The wind causes a girl's kite to move in a horizontal plane 300 feet above the ground at 30 feet per second. If she stands still, holding her end of the cord 5 feet above the ground, how fast is the cord slipping through her fingers when 500 feet of cord are out?
37. A manufacturer of bulldozers has a weekly revenue of $R$ thousand dollars when $x$ bulldozers are produced and sold per week, where

$$R = 200\sqrt{x} + 180x - x^2$$

If 100 bulldozers are now being produced per week and production is being increased at the rate of 3 units per week, at what rate is the weekly revenue changing?
38. A wholesale fruit company supplies a chain of grocery stores with $x$ thousand crates of peaches per day at $p$ dollars per crate. Demand $x$ is thought to be related to price $p$ (in dollars) by the equation

$$px + 8p - 3x - 240 = 0$$

Today's demand is 9000 crates but is expected to decrease at the rate of 180 crates per day. At what rate is the price changing right now?
39. In a discount supermarket, the weekly demand for a certain brand of tea satisfies the equation $px + 12p = 600$, where $x$ is the weekly demand in pounds and $p$ is the price in dollars per pound. Right now the price is \$6 per pound but is decreasing at the rate of 8¢ per week.
(a) At what rate is the demand changing?
(b) At what rate is the weekly revenue changing?
40. A fast-food restaurant has determined the cost $C$ (in dollars) of producing $x$ hamburgers per month is $C = 3000 + 0.48x$ and that the monthly demand satisfies $p = (40{,}000 - x)/15{,}000$, where $p$ is the price (in dollars). Right now the restaurant charges \$1.50 for a hamburger but plans to steadily increase the price at the rate of \$0.02 per month.
(a) Find the rate of change in the demand $x$.
(b) Find the rate of change in monthly revenue when the price of a hamburger reaches \$1.75.
(c) Find the rate at which the monthly profit is changing when the price of a hamburger reaches \$1.75.

41. Find the equation of the tangent line to $(y^2 + x - 1)^4 = 3x^2 + 4$ at the point (2, 1).
42. A squirrel is running along a telephone wire which has the shape of a parabola with equation $y = 15 + 0.01x^2$ $(-20 \le x \le 20)$, where $x$ and $y$ are in feet. Suppose that the horizontal component of the speed $dx/dt$ is 4 feet per second. Find the vertical component of the speed $dy/dt$ at the instant when the squirrel is at the point (10, 16).
43. Suppose that the legs $\overline{AB}$ and $\overline{BC}$ of the right triangle in Figure 28 are increasing at the constant rates of 6 centimeters per minute and 2 centimeters per minute, respectively. Find each of the following rates at the instant when $x = 12$ centimeters and $y = 5$ centimeters.
(a) The rate at which $\overline{AC}$ is increasing.
(b) The rate at which the perimeter is increasing.
(c) The rate at which the area is increasing.

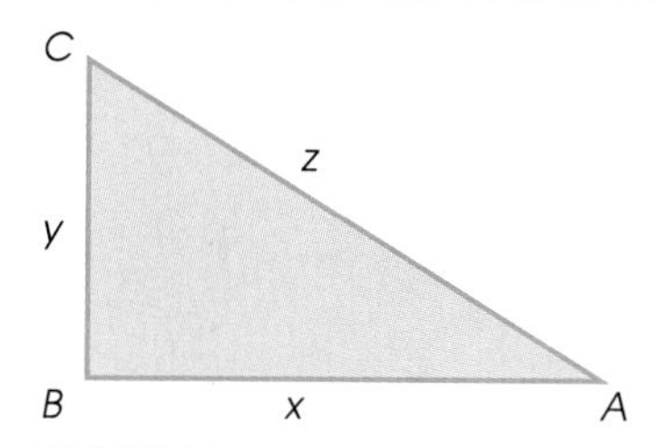

FIGURE 28

44. Suppose the radius of the cylinder in Figure 29 is decreasing at the rate of 0.5 centimeters per minute and the height $h$ is increasing at 1.5 centimeters per minute. Find the rate of change in volume with respect to time at the instant when $r = 6$ centimeters and $h = 12$ centimeters.

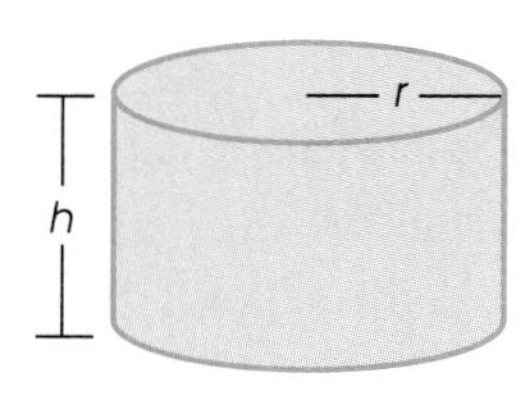

FIGURE 29

45. The elephant at the local zoo has a (spherical) tumor whose radius is growing at the alarming rate of 0.2 centimeters per day. Today its radius measures 8 centimeters. How fast is the volume of this tumor growing today?
46. Water is being poured into the hemispherical tank in Figure 30 at the constant rate of 30 cubic feet per minute. Find the rate at which the height of the water in the tank is rising when the height is 4 feet. *Hint:* The volume of the spherical segment of height $h$ is $V = \pi h^2(24 - h)/3$.

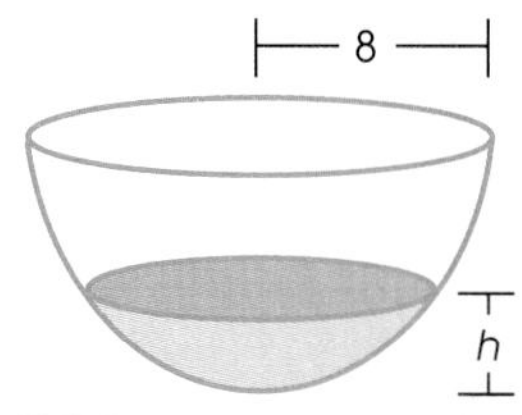

FIGURE 30

47. A manufacturer of 10-speed bicycles estimates that the yearly demand $x$ (in hundred thousands) for its bicycles will decrease in the future in such a way that $t$ years from now

$$x = \frac{50}{t + 4} + 3$$

On the other hand, its annual profit $P$ (in hundred thousands of dollars) is related to the number of bicycles it sells by the equation

$$3P - 2x^2 = 10$$

At what rate will the annual profit be changing 6 years from now?
48. A manufacturer of home computers estimates that the demand $x$ (in thousands of units per year) for its computers is related to the price $p$ in dollars by the equation $x = 4500 - 1.2p$. At present the price is \$1800 but is falling at the rate of \$150 per year. At what rate is the total annual revenue changing?

## 4.5* APPROXIMATIONS AND THE DIFFERENTIAL

As usual, let $y$ be a function of $x$; that is, $y = f(x)$. We are interested in what happens to $y$ when we change $x$ by a small amount $\Delta x$ (the change in $x$ is also denoted by $dx$). We denote the corresponding change in $y$ by the

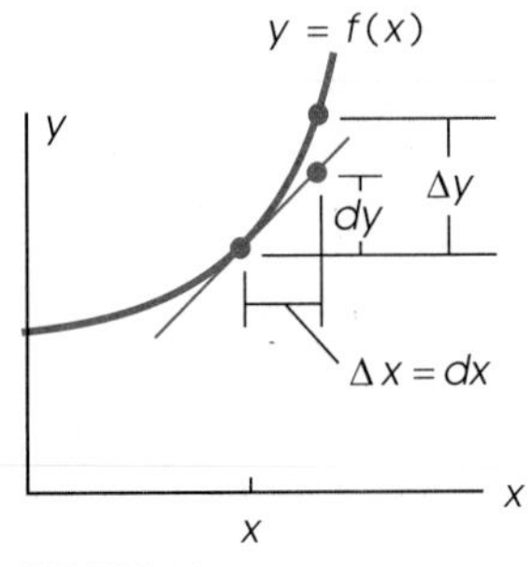

FIGURE 31

symbol $\Delta y$, a quantity that is often difficult to calculate. We propose to approximate $\Delta y$ by another quantity $dy$; called the differential of $y$. The **differential** of $y$ is the corresponding change on the tangent line and is therefore defined by

$$dy = f'(x)\,\Delta x = f'(x)\,dx$$

To understand what we have just said, refer to Figure 31. Note that $dy$ is a rather good approximation to $\Delta y$ and that this approximation will get better and better the smaller we make $\Delta x$.

We emphasize that $\Delta x$ and $dx$ are used interchangeably to indicate a small change in $x$, but that $\Delta y$ and $dy$ denote different quantitities: $\Delta y$ is the exact change in $y$, whereas $dy$ is the corresponding change on the tangent line (refer again to Figure 31).

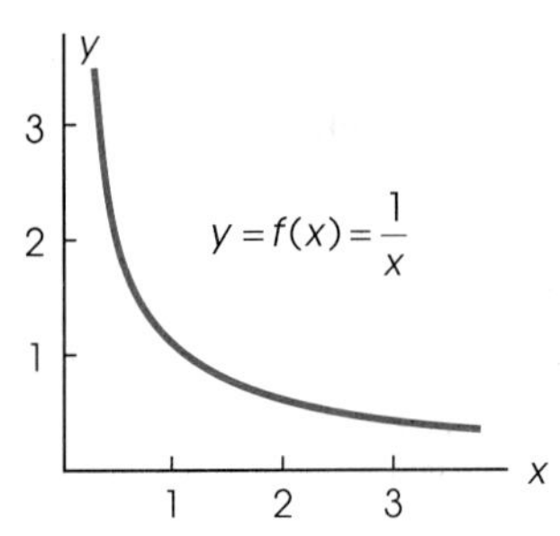

FIGURE 32

## EXAMPLE A

Let $f(x) = 1/x$. Suppose $x$ changes from 2 to 2.15, so that $dx = \Delta x = 0.15$. Find the corresponding exact change $\Delta y$ and its differential approximation $dy$ (see Figure 32).

### Solution

$$\Delta y = f(2.15) - f(2) = \frac{1}{2.15} - \frac{1}{2} = -0.03488372$$

To calculate $dy$, we first find $f'(x) = -1/x^2$ so that $f'(2) = -\frac{1}{4}$. Thus, $dy$ (evaluated when $x = 2$ and $\Delta x = 0.15$) is given by

$$dy = f'(x)\,\Delta x = (-\tfrac{1}{4})(0.15) = -0.0375$$

Clearly, $dy$ is not equal to $\Delta y$, but it does approximate it. ■

## EXAMPLE B

Suppose we need to know $\sqrt{81.3}$ and don't have a calculator handy. Use a differential approximation to find this number.

**Solution** We know that $\sqrt{81} = 9$. Let $y = f(x) = \sqrt{x}$ and consider what happens to $y$ when $x$ changes from 81 to 81.3. The corresponding approximate change in $y$ is given by the differential $dy$ evaluated when $x = 81$ and $\Delta x = 0.3$. Now $f'(x) = \frac{1}{2}x^{-1/2} = 1/2\sqrt{x}$, so $f'(81) = 1/2\sqrt{81} = \frac{1}{18}$. Thus,

$$dy = f'(x)\,\Delta x = (\tfrac{1}{18})(0.3) = 0.016666667$$

We conclude that

$$\sqrt{81.3} \approx \sqrt{81} + dy = 9 + 0.016666667 = 9.016666667$$

For comparison purposes, we mention that a direct calculation of $\sqrt{81.3}$ on our calculator gives 9.016651263. ■

## Estimating Errors Due to Inaccurate Measurement

Whenever we try to measure something (radius of a cylinder, strength of an electrical current, acidity of a solution), we are faced with a troublesome fact of life. Measurements cannot be made exactly. Thus, all scientists report measurements with an indication of the accuracy. For example, a chemist reports the mass of a sample as $2.0136 \pm 0.0003$ grams, an anthropologist reports the age of a bone specimen as $1230 \pm 25$ years, and a sociologist states that the percentage of one-parent homes in a certain state is $36 \pm 2$ percent. Any further calculation based on an inaccurate measurement of the basic data will be contaminated by these measurement errors. We pose the problem: Can we give a good estimate of the possible error in a calculated result if we have an estimate of the error in the data? The answer is yes, and it is based on the use of the differential approximation discussed above. We illustrate with a simple example.

### EXAMPLE C

A scientist reported that she had measured the radius of a sphere as $5.34 \pm 0.13$ centimeters. Calculate the volume of this sphere together with an estimate of the possible error in the answer.

**Solution** The formula for the volume of a sphere is $V = f(r) = \frac{4}{3}\pi r^3$. If 5.34 were the exact radius, we would calculate the volume as

$$V = \tfrac{4}{3}\pi(5.34)^3 = 637.8409242$$

The possible error of 0.13 in the measurement of the radius leads to an error in the calculated volume which we approximate by $dV$. Now,

$$dV = f'(r)\,\Delta r = 4\pi r^2\,\Delta r = 4\pi(5.34)^2(0.13) \approx 46.58$$

It would be reasonable to report the volume of the sphere as $638 \pm 47$ cubic centimeters. ■

## Relative Errors

An error of 1 foot in measuring the radius of the moon is of much less significance than is an error of 1 foot in measuring the height of a man. Often, it is the size of the error relative to the size of the quantity being measured that is of interest. Thus, we define **relative error:**

$$\text{relative error} = \frac{\text{size of error}}{\text{size of quantity being measured}}$$

Sometimes, the relative error is expressed as a percentage and then it is called the **percentage error.**

In Example C above, the relative error in the measurement of the radius is $0.13/5.34 \approx 0.024$, while the (approximate) relative error in the calculated volume is $47/638 \approx 0.074$. In alternative language, the percentage error in the measurement of the radius is 2.4%, while the (approximate) percentage error in the volume is 7.4%.

### EXAMPLE D

The side of a square is measured as $15.6 \pm 0.2$ feet. Use differentials to find the (approximate) error and relative error in the calculated area of the square.

**Solution** The formula for the area of a square is $A = f(x) = x^2$. Thus,

$$dA = f'(x)\,\Delta x = 2x\,\Delta x$$

The calculated area is

$$A = (15.6)^2 = 243.36$$

with an error of

$$dA = 2(15.6)(0.2) = 6.24$$

A sensible way to report the answer would be $A = 243.36 \pm 6.24$. The relative error is $6.24/243.36 \approx 0.026$ or 2.6%. ■

## The Economic Concept of Elasticity

A major concern of a manufacturer or a business person is to know how a change in the price of goods will affect the total revenue. To analyze the responsiveness of changes in demand (and of revenue) to changes in price, economists have introduced the concept of elasticity.

To begin, we suppose a manufacturer or business person is engaged in selling one type of goods. The **demand function** $x = f(p)$ relates the number of units $x$ that can be sold to the price $p$ charged and is always a decreasing function (the higher the price, the fewer units people will buy; see Figure 33). As a measure of responsiveness, it seems reasonable to calculate

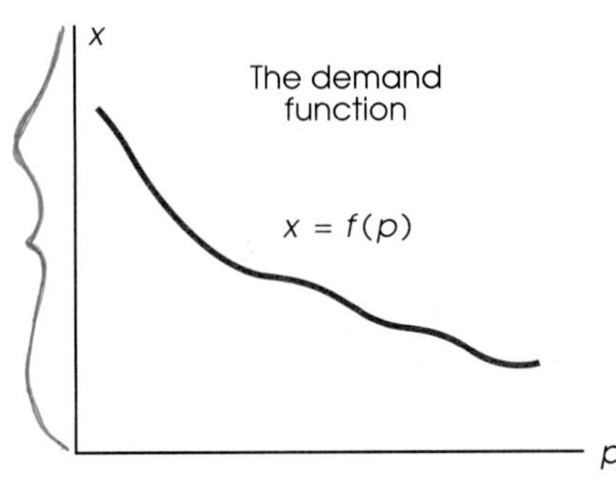


FIGURE 33

the ratio of the relative change in demand to the relative change in price but (in a calculus course) we should do this as the change in price tends to 0. Thus, consider

$$\lim_{\Delta p \to 0} \frac{\Delta x/x}{\Delta p/p} = \lim_{\Delta p \to 0} \frac{p\ \Delta x/\Delta p}{x} = \frac{p}{x} \lim_{\Delta p \to 0} \frac{\Delta x}{\Delta p} = \frac{p\ dx/dp}{x} = \frac{pf'(p)}{f(p)}$$

Now $f'(p)$ will be negative, whereas $p$ and $x$ are positive, so the measure above is negative. Since economists prefer to work with positive numbers, they define the **elasticity of demand** $E(p)$ to be the negative of the above expression.

**DEFINITION (ELASTICITY OF DEMAND)**
Let $x = f(p)$ be the demand function; that is, let $x$ denote the quantity demanded at price $p$ per unit. Then the **elasticity of demand** $E(p)$ is given by

$$E(p) = -\frac{p\ dx/dp}{x} = -\frac{pf'(p)}{f(p)}$$

To appreciate the significance of this concept, let us consider how the revenue responds to changes in price. Recall that

$$\text{revenue} = (\text{quantity sold})(\text{price per unit})$$

that is,

$$R(p) = x \cdot p = f(p) \cdot p$$

By the Product Rule for differentiation,

$$\begin{aligned} \frac{dR}{dp} &= \frac{d}{dp}[f(p) \cdot p] = f(p) \cdot 1 + p \cdot f'(p) \\ &= f(p)\left[1 + \frac{pf'(p)}{f(p)}\right] \\ &= f(p)[1 - E(p)] \end{aligned}$$

Suppose that $E(p) > 1$ (the so-called **elastic** case); then $dR/dp$ is negative and therefore $R$ is decreasing with $p$. This means that an increase in price will decrease revenue or, alternatively, that a cut in price will increase revenue. Similarly, if $E(p) < 1$ (the so-called **inelastic** case), then $dR/dp$ is positive and $R$ is increasing with $p$. This means that an increase in price will increase revenue (alternatively, that a cut in price will decrease revenue). In summary,

*a cut in price will increase revenue in the elastic case, but an increase in price will increase revenue in the inelastic case.*

### EXAMPLE E

A sales expert has estimated the monthly sales $x$ of a certain book to be related to its price according to the formula $x = f(p) = 10{,}000/\sqrt{p} = 10{,}000p^{-1/2}$, where $p$ is in dollars (see Figure 34). Presently the book is selling at \$9. Determine whether the demand is elastic or inelastic at that price level and advise the sales manager as to whether the price should be increased or decreased (to increase revenue).

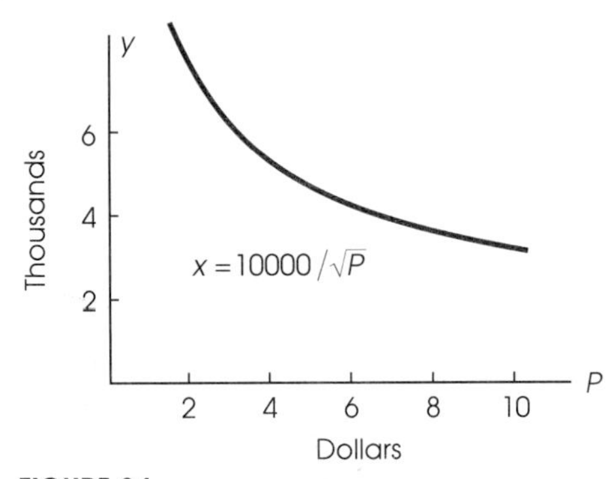


**FIGURE 34**

**Solution** First note that $f(9) = 10{,}000/3 \approx 3{,}333$. Also, $f'(p) = -5000p^{-3/2}$, and so $f'(9) = -5000(9)^{-3/2} \approx -185$. Thus,

$$E(9) = \frac{-9f'(9)}{f(9)} \approx \frac{-9(-185)}{3333} \approx 0.5$$

Demand is inelastic; the sales manager is advised to increase the price. ■

A good question to ask is, What value of the elasticity $E(p)$ will make the revenue a maximum? The maximum value of $R$ occurs when its derivative $dR/dp$ is 0. Referring to the formula $dR/dp = f(p)[1 - E(p)]$ derived above, we see that this happens when $E(p) = 1$, the so-called **unit elasticity** case.

## Problem Set 4.5

*In Problems 1–6, find the differential dy in terms of x and dx.*

1. $y = 2x^{3/2}$
2. $y = 3/x^2$
3. $y = 4\sqrt{x} + 5x^{-1}$
4. $y = 2/\sqrt{x} + x$
5. $y = (4 - 3x)^5$
6. $y = \sqrt{x^2 + 3x + 2}$

C *In Problems 7–12, find the exact change $\Delta y$ and the differential approximation dy corresponding to the change in x from $x_1$ to $x_2$ (see Example A).*

7. $y = 3x^2 - x + 3$; $x_1 = 2$, $x_2 = 2.1$
8. $y = 4 + 2x - x^2$; $x_1 = 1$, $x_2 = 1.2$
9. $y = \sqrt{x}$; $x_1 = 4$, $x_2 = 4.02$
10. $y = 4/x^2$; $x_1 = 2$, $x_2 = 2.03$
11. $y = \sqrt{2x^3 + 9}$; $x_1 = 2$, $x_2 = 1.95$
12. $y = (3 + 2\sqrt{x})^3$; $x_1 = 1$, $x_2 = 0.98$

*In Problems 13–18, use a differential approximation to find N (see Example B).*

13. $N = \sqrt{9.04}$
14. $N = \sqrt{25.06}$
15. $N = \sqrt[3]{26.75}$
16. $N = \sqrt[3]{63.8}$
17. $N = 4(1.98)^3 - 3/1.98$
18. $N = [2\sqrt{1.01} + 3(1.01)^2]^3$

19. A large cube of ice melted so that the length of each edge decreased from 25 inches to 24.5 inches. Approximate the corresponding change in (a) the volume, (b) the surface area.
20. One model for growing children says that the pulse rate $R$ in beats per minute is related to height $x$ in inches by $R = 610/\sqrt{x}$. About how much will the pulse rate of a child decrease as his or her height increases from 49 to 51 inches?
C 21. One leg of a right triangle is exactly twice as long as the other. The shorter leg is measured as $48.62 \pm 0.21$. Calculate the length of the hypotenuse and give an estimate for the possible error in the answer. See Example C.
C 22. The edge of a cube is measured as $24.6 \pm 0.4$. Calculate the volume and surface area of the cube, giving an estimate of the possible error in each case.
23. The height of a right circular cone is exactly four times the radius of the base. Recall that the volume of such a cone is $V = \pi r^2 h/3 = 4\pi r^3/3$. Suppose that the radius of the base can be measured with a possible error of 0.3 inch. Find (approximately) the corresponding relative error in the calculated volume if the radius is measured as (a) 8.4 inches, (b) 80.4 inches. Give your answers in percent.
24. Give (in percent) the size of the relative error in the calculated values in volume and surface area in Problem 22.
25. The radius of a sphere was measured as 10.5 inches with a percentage error of 1%. Determine (approximately) the percentage error in the calculated value of the volume. *Hint*: $V = 4\pi r^3/3$.
26. Recall that for a sphere the surface area is given by the formula $S = 4\pi r^2$. Deduce that $dS/S = 2\,dr/r$. Now use the latter fact to decide how big a percentage error we can allow in measuring $r$ if we insist that the percentage error in $S$ be less than 6%.
27. A manufacturer is presently selling its line of women's earrings at \$12 per pair and has reason to think that weekly demand $x$ is related to price by the equation $x = f(p) = 500 - 1.5p$.
    (a) Is the demand elastic or inelastic at this price?
    (b) Should the price be increased or decreased to increase revenue?
    (c) If the price is increased from \$12.00 to \$12.10, about how much will the weekly revenue change. *Hint:* From the formula $dR/dp = f(p)[1 - E(p)]$, which was derived just before Example E, we obtain $dR = f(p)[1 - E(p)]\,dp$.
28. A distributer of deluxe ballpoint pens is presently charging \$3.25 for each pen. A research study suggests that it could sell $x$ pens per week if it set the price at $p = 4 - 0.002x$. Note that this is equivalent to $x = f(p) = (4 - p)/0.002 = 2000 - 500p$.
    (a) Is the demand elastic or inelastic at this price?
    (b) Should the distributer increase or decrease the price to increase revenue?
    (c) If the price per pen is increased from \$3.25 to \$3.35, approximate the corresponding change in weekly revenue.
29. A maker of hand-held calculators knows that it can sell 3000 calculators per week at \$12 each and estimates that, for each 20¢ decrease in price, sales each week would increase by 200 calculators.
    (a) Write the demand equation relating the weekly demand $x$ to the price $p$.
    (b) Determine the elasticity of demand $E(p)$.
    (c) For what interval of $p$ values is the demand elastic?
30. Suppose the demand equation is $x = f(p) = 5625 - 3p^2$.
    (a) Find the elasticity of demand $E(p)$.
    (b) Determine the values of $p$ for which the demand is elastic.
    (c) If the present price is \$10, what percentage change in demand is caused by a 10% increase in price?

---

31. Find the differential $dy$ in each case.
    (a) $y = 3/x + x^2$ (b) $y = \sqrt{x^2 + 9}$
    (c) $y = (3t + 1)^4$ (d) $y = (2s + 4)/(s - 2)$
32. Suppose that $y = (x^2 + 2x - 2)^3$ and that $x$ increases from 3 to 3.1.
    (a) Find the corresponding approximate change $dy$ in $y$.
    (b) Find the corresponding exact change $\Delta y$ in $y$.
33. If $f(x) = \sqrt[3]{x} + 5x^2 - 280$, use differentials to approximate $f(8.01)$.
34. A cubical block of cheese with sides 18 inches long is to be covered with a coat of wax 0.15 inch thick. Use differentials to approximate the volume of wax needed.
35. The demand equation for a certain product (in units per month) is given by $x = 400(30 - \sqrt{p})$.
    (a) Find the elasticity of demand.
    (b) For what values of the price is the demand elastic?
    (c) If the price is increased from \$625 to \$630, find the approximate change in the monthly revenue.
36. A stereo distributor can sell $x$ stereos annually when the unit price is $p = 1200 - x$ dollars. Its cost is estimated at $C(x) = 3000 + 400x$.
    (a) Use differentials to find the approximate change in the annual profit if $x$ is increased from 400 to 410.
    (b) Find the approximate change in the average cost per unit, $C(x)/x$, when $x$ is increased from 400 to 410.
    (c) For what values of $p$ is the demand elastic?

## CHAPTER 4 REVIEW PROBLEM SET

1. Find the maximum value of $z = x^2y$ subject to the constraints $5x + 2y = 120$ and $x \geq 0$.
2. Find the maximum value of $z = xy + \frac{1}{2}x^2$ subject to the constraints $3x + 4y = 60$ and $x \geq 0$.
3. Find the points on the parabola $x = y^2$ that are closest to the point $(10, 0)$. *Hint:* Minimize the square of a distance.
4. Let $x$ and $y$ be positive numbers satisfying $x + 2y = 12$. Determine these numbers if their product $xy$ is a maximum.
5. A window is to have the shape of a rectangle topped by a semicircle as shown in Figure 35. If the perimeter is to be 20 feet, what are the dimensions $x$ and $y$ of the window admitting the most light?

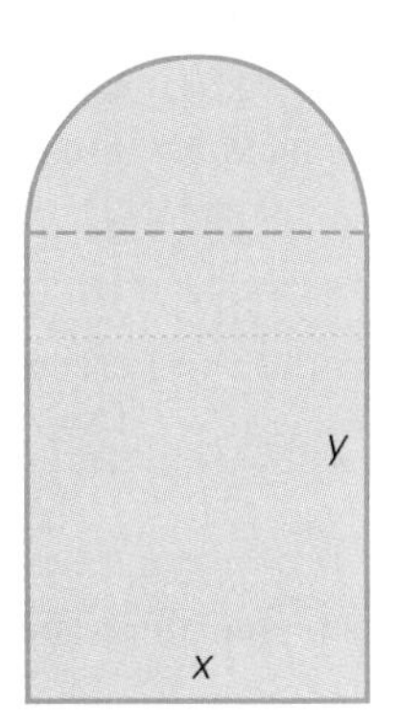

**FIGURE 35**

6. Suppose that the semicircular part of the window in problem 5 admits one-half as much light per square foot as does the rectangular part. What are the dimensions $x$ and $y$ of the window admitting the most light?
7. A cistern with a rectangular base twice as long as wide is to hold 24,000 cubic feet of water. If the metal top costs twice as much per square foot as the concrete sides and bottom, what are the dimensions of the most economical cistern?
8. A rectangular handbill is to contain 54 square inches of printed matter with 3-inch margins at the top and bottom and 2-inch margins on each side. What dimensions for the handbill would use the least paper?
9. A manufacturer of picnic table sets estimates the weekly cost of producing and selling $x$ sets to be

$$C(x) = 1000 + 50x + 0.07x^2$$

with $0 \leq x \leq 200$.
(a) How many sets (an integer) should be produced each week to minimize the average cost per set, $C(x)/x$?
(b) What is the minimum average cost?
10. Wonder Builders makes bird houses. The company estimates the weekly cost in dollars of producing $x$ bird houses to be $C(x) = 1200 + 60x + 0.08x^2$.
(a) How many houses (an integer) should be produced each week to minimize the average cost per house?
(b) What is this minimum average cost per house?
11. Suppose that the manufacturer in Problem 9 can sell picnic tables at a price of \$100 each.
(a) How many sets (an integer) should be produced to maximize weekly profit?
(b) What is the maximum weekly profit?
12. Suppose the company of Problem 10 can sell its bird houses at \$105 each.
(a) How many houses (an integer) should be produced each week to maximize profit?
(b) What is this maximum weekly profit?
13. An apartment complex with 220 rental units can fill all units when the monthly rent is set at \$480. It is estimated that for each \$15 per month increase in rent, 5 units will become vacant. The complex has fixed monthly costs of \$60,000 and a maintenance cost of \$60 for each rented apartment. What monthly rent should be charged to maximize the total profit of the complex?
14. A grapefruit orchard yields an average of 32 bushels of grapefruit per tree when 24 trees are planted on an acre of land. For each additional tree planted per acre, the yield of a tree is decreased by 0.75 bushels. How many trees per acre should be planted to get the highest yield?
15. The Royal Appliance Company expects to sell 6000 clock radios per year. It costs the company \$2 to hold a radio in stock for a year, and the cost of reordering a lot of size $x$ is \$100 plus \$1 for each radio. In what lot size should radios be reordered to minimize inventory costs?
16. The Hi-Life Sports Company expects to sell 8000 tennis rackets per year. It cost the company \$1.50 to keep a racket in stock for a year and the cost of reordering a lot of size $x$ is \$125 plus \$1.50 for each racket. In what lot size should tennis rackets be reordered to minimize inventory costs?
17. Martha Johnson, an apple grower, has 600 bushels of apples which she will sell to a single wholesaler. Today she could get \$10.00 per bushel, and the price is going up \$0.25 per day. On the other hand, she can count on spoilage of about 8 bushels per day. When

should she sell the apples in order to maximize the total revenue?

18. A local club has 120 members, the annual dues per member being \$36. A recent survey suggests that for each dollar decrease in the dues, 2 additional members will join the club. At what level should the dues be set to maximize the total revenue from dues?

19. Long Haul Movers estimates the vehicle cost of operating a moving van at $50 + \frac{1}{3}x$ cents per mile when driven at $x$ miles per hour. It pays its drivers \$12.50 per hour. At what speed should a van be driven to minimize the total operating cost per mile of one of its vans?

20. Suppose that a milk truck averages $420/(x + 3)$ miles per gallon when driven at an average speed of $x$ miles per hour, $40 \le x \le 70$. If gasoline costs \$1.25 per gallon and the driver gets \$12.50 per hour, what speed will minimize the cost of a 300-mile run? Calculate this minimum cost.

21. Find an equation for the tangent line to the curve $xy^3 + x^2y = 3x + 7$ at the point $(1, 2)$.

22. Find an equation for the tangent line to the curve $2xy^2 - x^3 = 4y - 11$ at the point in the first quadrant where $x = 3$.

23. A tank has the shape of a right circular cone 12 feet deep and 12 feet across at the top (Figure 36). Water is running into the tank at the rate of 120 cubic feet per minute. How fast is the level of water rising when this level reaches (a) 3 feet, (b) 9 feet? Recall that $V = \frac{1}{3}\pi r^2 h$ for a cone.

24. An airplane flying west at 400 miles per hour goes over a certain town at 12:00 noon. A second plane at the same altitude, flying south at 560 miles per hour, goes over the town at 12:30 P.M. How fast are the planes separating at 1:30 P.M.?

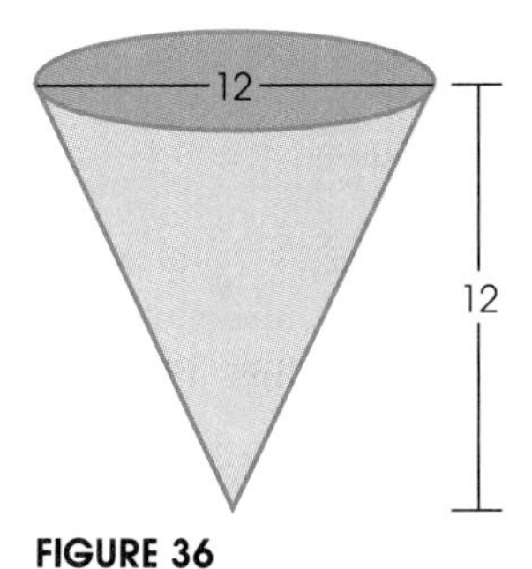

FIGURE 36

25. Refer to Figure 36 and assume that the height of the cone is exactly twice the radius of the top. Suppose the radius of the top is measured to be 6 feet with a possible error of 0.02 feet. Calculate the volume of the cone and use differentials to give an estimate of the possible error in this value.

26. A point is moving along the curve $y = x^2 + x + 4$ so that $dx/dt = 3$. Find the rate at which $y$ is changing with respect to $t$ when the point is at $(2, 10)$.

27. A wholesale store is presently charging \$6.50 for a pen and pencil set. It estimates that it can sell $x$ sets per week when the price per set is $p(x) = 8 - 0.005x$ dollars.
    (a) Is the demand elastic or inelastic?
    (b) Should the store increase or decrease the price to increase total revenue?
    (c) Approximate the change in weekly revenue when the price is increased from \$6.50 to \$7.00.

28. It is claimed in Boothill County that if there were no hunting season, the grouse population would grow from size $x$ (in thousands) to size $f(x) = 2.8x - 0.03x^2$ in one year. To what size should the grouse population be allowed to grow to maintain the largest yearly kill? What would this yearly kill be?

29. Consider $f(x) = x^2 + 4x^{-1}$.
    (a) Find the slope of the tangent line at $x = 1$ and at $x = 2$.
    (b) Show that the graph of $f$ is concave up on $(0, \infty)$.
    (c) Find the global minimum value of $f(x)$ on $(0, \infty)$.
    (d) Find the global maximum value of $f(x)$ on $[\frac{1}{2}, \frac{5}{2}]$.
    (e) Sketch the graph of $y = f(x)$ on $(0, 4]$.

30. Consider $f(x) = (x + 2)^2/x$ on the domain $\{x: x > 0\}$.
    (a) Find the slope of the tangent line at $x = 1/2$.
    (b) Where is the graph of $f$ concave up?
    (c) Show that $y = x + 4$ is an asymptote of the graph.
    (d) Find the global minimum value of $f$.
    (e) Sketch the graph of $f$.

31. Sketch the graph of a function with domain $[0, 8]$ having all of the following properties.
    (a) $f$ is continuous everywhere on $[0, 8]$.
    (b) $f$ fails to have a derivative at $x = 4$.
    (c) $f'(x) > 0$ on $(2, 4) \cup (4, 8)$.
    (d) $f''(x) > 0$ on $(0, 4)$ and $f''(x) < 0$ on $(4, 8)$.

32. Sketch the graph of a function with domain $[0, 8]$ having all of the following properties.
    (a) $f(0) = f(3) = 3$ and $f(x) = 5$ on $[6, 8]$
    (b) $f$ is continuous on $[0, 8]$ except at $x = 3$.
    (c) $f'(x) > 0$ on $(0, 3) \cup (3, 6)$.
    (d) $f''(x) < 0$ on $(0, 3)$ and $f''(x) > 0$ on $(3, 6)$.

33. Evaluate $\lim_{x \to \sqrt{2}} \dfrac{(2x - 3)(x - \sqrt{2})}{x^2 - 2}$.

34. Evaluate $\lim\limits_{x \to 2} \dfrac{(x - 2)^2(2x - 1)}{(x^3 - 8)^2}$.

35. Evaluate $f'(2)$, given that $f(x) = (3x^2 - \sqrt{x + 2})^4$.

36. Evaluate $f'(2)$ if $f(x) = \left[2\sqrt{x + 7} - \left(\dfrac{2}{x - 1}\right)^2\right]^4$.

*No picture better symbolizes the modern field of genetics than the double-helix model of DNA (deoxyribonucleic acid). DNA carries the genetic code; a tiny amount in a fertilized egg cell determines the physical characteristics of the fully developed animal. Unraveling this code is a complex mathematical task.*

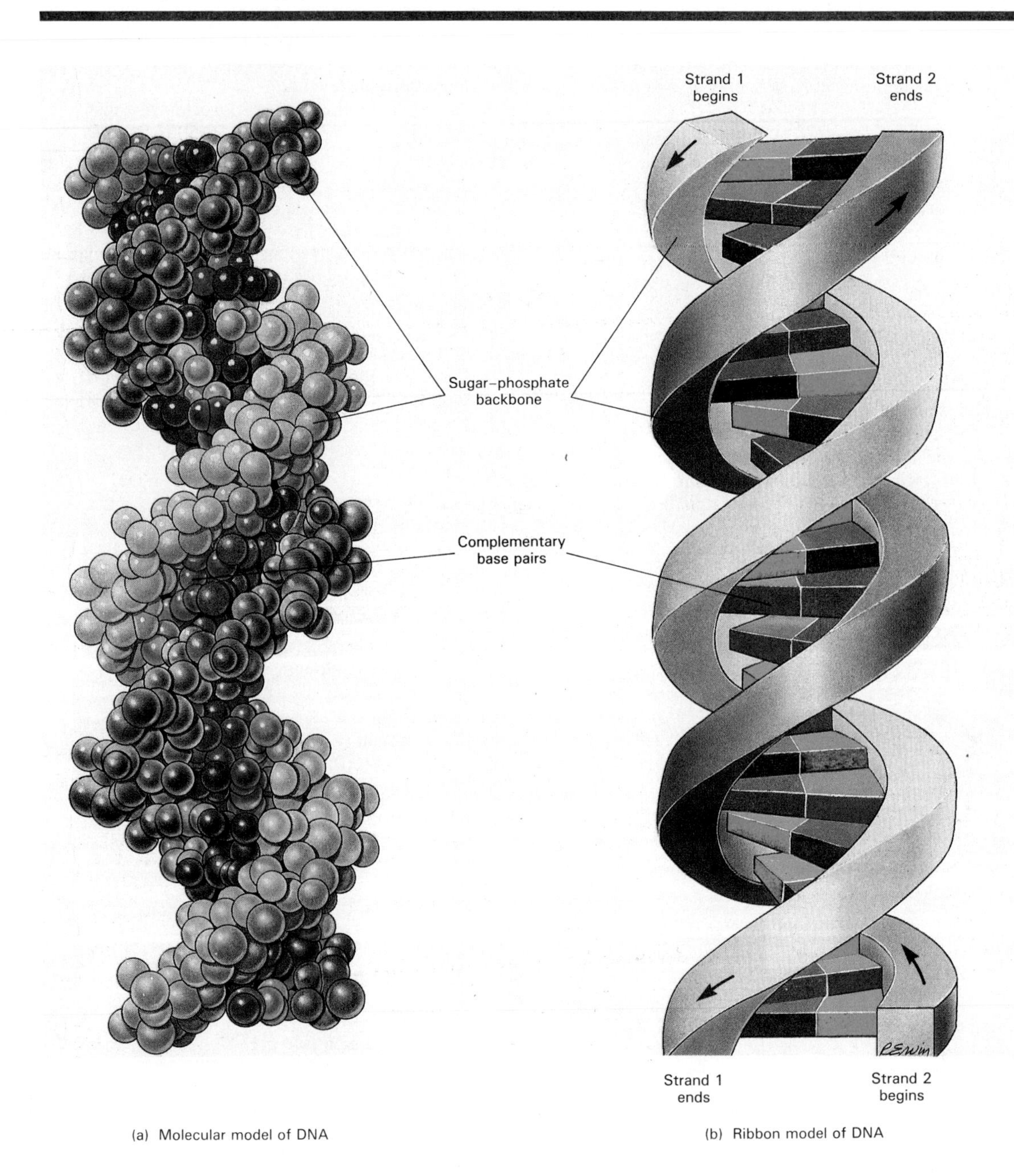


(a) Molecular model of DNA

(b) Ribbon model of DNA

# 5

# Exponential and Logarithmic Functions

Sir Ronald Aylmer Fisher
(1890–1962)

Ronald Fisher was the creator of much of modern statistical theory and was a pioneer in the mathematical theory of genetics. Born in London, he was educated at Cambridge University where he specialized in mathematics and physics. He suffered from extreme myopia, but this seems only to have strengthened his innate geometrical sense and forced him to solve problems entirely mentally.

In 1919, Fisher was appointed research statistician at the Rothamsted Experimental Station in Hertfordshire. His job was to analyze and reinterpret the records of 66 years of agricultural experiments at this station. Here he began the work that was to revolutionize statistical techniques. His 1925 book, *Statistical Methods for Research Workers,* became the bible of a whole generation of statisticians and still wields considerable influence. Its central theme is hypothesis testing and includes such notions as use of controls, randomization, and analysis of variance.

Fisher's second book, *The Genetical Theory of Natural Selection* (1930), is a classic in population genetics. It is a book of considerable mathematical depth and attempts to reconcile Mendel's and Darwin's theories of genetics. Two more books, *The Design of Experiments* (1935) and *Statistical Tables for Biological, Agricultural and Medical Research* (1938), have gone through many editions and had a profound influence.

In 1933, Fisher accepted the position of Galton Professor of Genetics at University College, London, and later the Balfour Chair of Genetics at Cambridge University. He was knighted in 1952.

Fisher's association with this chapter is appropriate, since both the exponential and logarithmic functions appear prominently in his work. The most important distribution in statistical theory is the standard normal distribution whose density function is given by

$$f(x) = (2\pi)^{-1/2}e^{-x^2/2}$$

## 5.1 EXPONENTIAL AND LOGARITHMIC FUNCTIONS

It is quite clear by now that the concept of function is basic to all that we do in calculus. However, if you review what we have done, you will realize that we have restricted our study to a fairly narrow class of functions, namely, the class of **algebraic functions.** These are the functions whose rule of correspondence can be obtained by using the real numbers, the symbol $x$, and the five algebraic operations of addition, subtraction, multiplication, division, and root taking. Examples are

$$f(x) = 3xxx - 5xx + 7x - 9 = 3x^3 - 5x^2 + 7x - 9$$

$$g(x) = \frac{3x^3 - 5x^2 + 7x - 9}{4x^5 - 3}$$

$$h(x) = 2\sqrt{x} + \frac{3}{\sqrt{x}} = 2x^{1/2} + 3x^{-1/2}$$

$$k(x) = (3x^2 - 1)^{2/3}(5x + 1)^{-5/6}$$

Note that though we have used exponents a great deal, in each case the exponent was a constant. Never have we allowed an exponent to be a variable. Removing this restriction is the key notion in this chapter. And doing this will open the door to a whole new world, the world of so-called **transcendental functions,** which is just a fancy term for nonalgebraic functions. Among these are the exponential functions, logarithmic functions, and trigonometric functions.

### The Exponential Functions

By an exponential function, we mean a function of the form

$$f(x) = b^x$$

where $b$ is a positive constant, called the **base** of the exponential function. While such functions have not previously appeared in this book, we feel confident that you have met them in earlier mathematics courses. To refresh your memory, we begin with a simple example.

#### EXAMPLE A

Sketch the graph of $f(x) = 2^x$.

**Solution** A table of values (partially obtained on a hand-held calculator) and the corresponding curve are shown in Figure 1. Note that the domain for this function consists of all real numbers, while its range consists of the positive real numbers. Perhaps the most significant feature of this function is its growth, which becomes more and more rapid as $x$ increases. ■

| $x$ | $y = 2^x$ |
|---|---|
| −4 | 0.0625 |
| −3 | 0.125 |
| −2 | 0.25 |
| −1 | 0.5 |
| 0 | 1 |
| 0.5 | 1.4142 |
| 1 | 2 |
| 1.3 | 2.4623 |
| 2 | 4 |
| 3 | 8 |
| 4 | 16 |

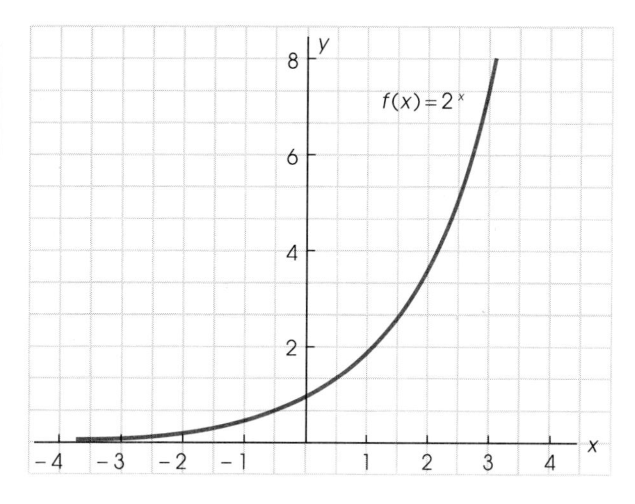


**FIGURE 1**

The graphs of a number of other exponential functions are shown in Figure 2. Note that if the base $b > 1$, then the graph rises to the right—more and more steeply the larger the base. In contrast, if $0 < b < 1$, the graph falls to the right. Also, the graphs of $f(x) = b^x$ and $g(x) = (1/b)^x = b^{-x}$ are symmetric to each other in the $y$-axis.

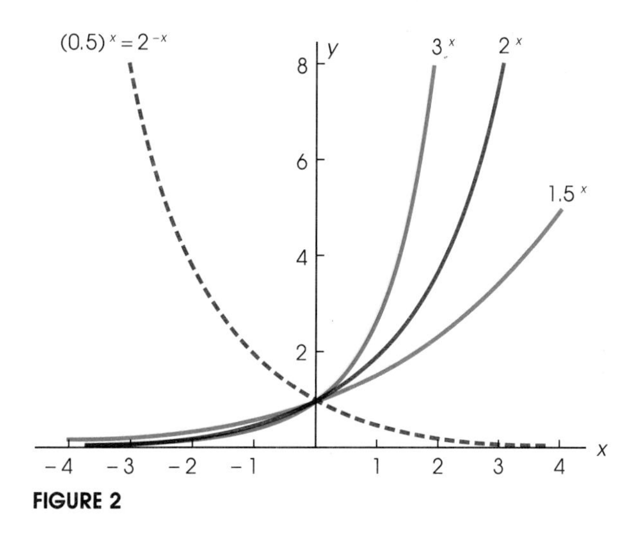


**FIGURE 2**

The properties of exponents were stated at the end of Section 2.2 and now are restated as they apply to exponential functions.

**PROPERTIES OF EXPONENTS**

For any real $x$ and $y$ and any $b > 0$,

1. $b^x b^y = b^{x+y}$
2. $b^x / b^y = b^{x-y}$
3. $(b^x)^y = b^{xy}$

EXAMPLE B

Simplify each of the following using the properties of exponents.

(a) $\left(\dfrac{b^3 b^4}{b^5}\right)^3$

(b) $\left(\dfrac{b^5 b^{-2}}{b^3 b^{-1/2}}\right)^{1/3}$

**Solution**

(a) $\left(\dfrac{b^3 b^4}{b^5}\right)^3 = \left(\dfrac{b^7}{b^5}\right)^3 = (b^2)^3 = b^6$

(b) $\left(\dfrac{b^5 b^{-2}}{b^3 b^{-1/2}}\right)^{1/3} = \left(\dfrac{b^3}{b^{5/2}}\right)^{1/3} = (b^{1/2})^{1/3} = b^{1/6}$ ■

## The Logarithmic Functions

The functions to be introduced next are closely related to the exponential functions. We begin with a definition.

**POINT OF INTEREST**

John Napier (1550–1617), farmer, inventor, and theologian, is credited with inventing logarithms. Father of 10 children, this Scotsman amused himself by studying mathematics and science. He sought to simplify astronomical calculations by replacing multiplications with additions—and succeeded.

**DEFINITION**

Assume that $b > 0$ and $b \neq 1$. The logarithm of $x$ to the base $b$ (written $\log_b x$) is the exponent to which $b$ must be raised to yield $x$, that is,

$$b^{\log_b x} = x$$

Another way to state the above definition is to say that

$$y = \log_b x \quad \text{if and only if} \quad b^y = x$$

Let's see if we can apply this definition in an example.

EXAMPLE C

Find each of the following logarithms.

(a) $\log_2 16$, (b) $\log_2 \frac{1}{8}$, (c) $\log_4 8$, (d) $\log_4 1$, (e) $\log_{10} 0.0001$

### Solution

(a) $\log_2 16 = 4$ since $2^4 = 16$
(b) $\log_2 \frac{1}{8} = -3$ since $2^{-3} = 1/2^3 = \frac{1}{8}$
(c) $\log_4 8 = \frac{3}{2}$ since $4^{3/2} = (\sqrt{4}4)^3 = 8$
(d) $\log_4 1 = 0$ since $4^0 = 1$
(e) $\log_{10} 0.0001 = -4$ since $10^{-4} = 0.0001$ ■

The properties of logarithms follow from those for exponents, as we shall show in a moment.

**PROPERTIES OF LOGARITHMS**

For any positive numbers $u$ and $v$ and any real number $r$,

1. $\log_b (uv) = \log_b u + \log_b v$
2. $\log_b (u/v) = \log_b u - \log_b v$
3. $\log_b (u^r) = r \log_b u$

To see why Property 1 is true, let $\log_b u = x$ and $\log_b v = y$. Then $b^x = u$ and $b^y = v$. It follows from the first property of exponents that

$$uv = b^x b^y = b^{x+y}$$

But this is precisely what it means to say that

$$\log_b(uv) = x + y = \log_b u + \log_b v$$

The other two properties of logarithms follow in a similar way.

**CAUTION**
Perhaps the most common error in the use of logarithms is to confuse $\log (x/y)$ and $\log x/\log y$ (see Example D).

### EXAMPLE D

Given that $\log_b 3 = 0.477$ and $\log_b 5 = 0.699$, find each of the following.

(a) $\log_b (3b^2/125)$ (b) $(\log_b 3b^2)/\log_b 125$

### Solution

(a)
$$\log_b \frac{3b^2}{125} = \log_b 3b^2 - \log_b 125$$
$$= \log_b 3 + \log_b b^2 - \log_b 5^3$$
$$= \log_b 3 + 2 \log_b b - 3 \log_b 5$$
$$= 0.477 + 2(1) - 3(0.699) = 0.380$$

(b) Be sure to distinguish this problem from the one just completed.

$$\frac{\log_b 3b^2}{\log_b 125} = \frac{\log_b 3 + 2\log_b b}{3\log_b 5}$$
$$= (0.477 + 2)/3(0.699)$$
$$= 2.477/2.097 = 1.181$$ ■

### EXAMPLE E

Write each of the following as a single logarithm by using the properties of logarithms.

(a) $\log_b x + \log_b y - \log_b z$
(b) $3\log_b x - \frac{1}{2}\log_b y$

**Solution**

(a) $\log_b x + \log_b y - \log_b z = \log_b xy - \log_b z = \log_b \frac{xy}{z}$
(b) $3\log_b x - \frac{1}{2}\log_b y = \log_b x^3 - \log_b y^{1/2}$
$$= \log_b \frac{x^3}{y^{1/2}} = \log_b \frac{x^3}{\sqrt{y}}$$ ■

## Graphs of Logarithmic Functions

Let us begin by noting that the functions $f(x) = b^x$ and $g(x) = \log_b x$ are **inverse functions.** By this we mean that they undo each other; that is,

$$f(g(x)) = x \quad \text{and} \quad g(f(x)) = x$$

This amounts to saying that

$$b^{\log_b x} = x \quad \text{and} \quad \log_b b^x = x$$

> **RECALL**
> If $f$ and $g$ are inverse functions, then $y = f(x)$ and $x = g(y)$ are equivalent and have the same graph. However, $x = g(y)$ and $y = g(x)$ have the roles of $x$ and $y$ interchanged, which graphically corresponds to a reflection across the line $y = x$.

both of which are direct consequences of the definition of a logarithm. From this it follows that the graphs of $y = b^x$ and $y = \log_b x$ are reflections of each other across the line $y = x$. We illustrate this fact for the case $b = 2$ in Figure 3.

Figure 3 illustrates, for the case $b = 2$, several important facts about the exponential function $f(x) = b^x$ and its inverse, $g(x) = \log_b x$. The domain of $f$ is the set of all real numbers; its range is the set of positive real numbers. For the logarithmic function $g$, exactly the opposite is true. Its domain is the set of positive real numbers, and its range is the set of all real numbers. In particular, *negative numbers and 0 do not have logarithms,* but every real number is the logarithm of something. Finally, note from Figure 3 how rapidly $2^x$ grows as $x$ becomes large, but conversely how slowly $\log_2 x$ grows. These opposite behaviors are characteristic of $b^x$ and $\log_b x$ provided $b > 1$.

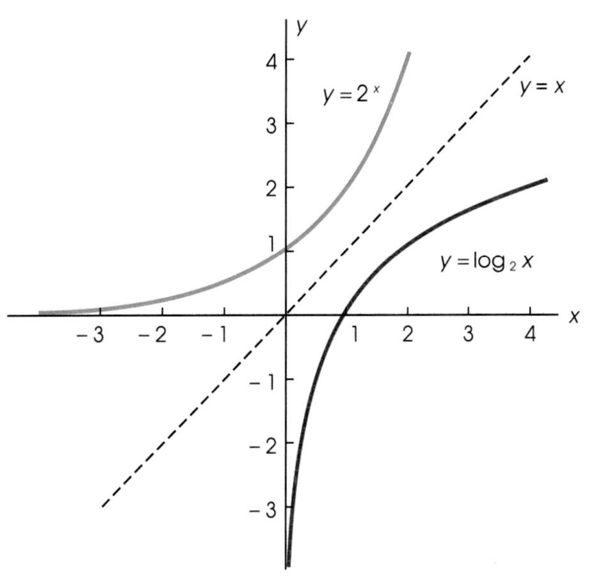


FIGURE 3

## Problem Set 5.1

*Evaluate each of the numerical expressions in Problems 1–8.*

1. $16^{3/2}$
2. $27^{2/3}$
3. $5^2 5^{-1}$
4. $3^{4/3}3^{2/3}3^{-2}$
5. $(2^{-3}2^4)^{-1}$
6. $(4^{-1/4}4^{1/3})^{12}$
7. [C] $(1.7)^{2.3}$
8. [C] $\pi^{1.234}$
9. [C] $(5.12)^{-2.3}(1.23)^2$
10. [C] $(2.23)^{4.12}/(3.21)^3$

*In Problems 11–14, sketch the graphs of each set of functions using the same axes (see Example A).*

11. $f(x) = 4^x$, $g(x) = (\frac{1}{4})^x$
12. $f(x) = 5^x$, $g(x) = 5^{-x}$
13. [C] $f(x) = (0.8)^x$, $g(x) = (1.25)^x$, $h(x) = (2.25)^x$
14. [C] $f(x) = (1.1)^x$, $g(x) = (1.8)^x$, $h(x) = (2.3)^x$

*Simplify the expressions in Problems 15–24, leaving no negative exponents in your answers (see Example B).*

15. $(b^{-3}b^{7/2})^2$
16. $(b^{-2/3}b^2)^3$
17. $(c^3c^{-4}c^{1/2})^4$
18. $(c^{-2/3}c^2c^{-1})^3$
19. $[b^4/(b^{-2}b^5)]^3$
20. $(b^{-7}b^3/b^{-8})^2$
21. $(\pi^8\pi^{-4})^{-2}$
22. $(\pi^{4/3}\pi^0\pi^{2/3})^{-3/2}$
23. $(a + 1)^{2/3}(a + 1)^{-1/2}$
24. $(b - 1)^{-3/4}(b - 1)^{5/3}$

*Evaluate each of the logarithms in Problems 25–38 (see Example C). Note that on many calculators the $\log_{10}$ key is denoted simply by log.*

25. $\log_3 27$
26. $\log_4 16$
27. $\log_{10} 100{,}000$
28. $\log_{10} 0.0001$
29. $\log_9 27$
30. $\log_{16} 64$
31. $\log_b b^{-3/2}$
32. $\log_b (1/b^2)$
33. [C] $\log_{10} 0.4321$
34. [C] $\log_{10} 43.21$
35. [C] $\log_{10} (5.14/6.32)$
36. [C] $(\log_{10} 5.14)/(\log_{10} 6.32)$
37. $3^{\log_3 (4.32)}$
38. $\log_{2.2}(2.2^{3.3})$

*Find each of the logarithms in Problems 39–46 given that $\log_a 2 = 0.693$, $\log_a 3 = 1.099$, and $\log_a 5 = 1.609$. For example, in Problem 39 use the fact that $12 = 2^2 3$ and proceed as in Example D.*

39. $\log_a 12$
40. $\log_a \frac{6}{5}$
41. $\log_a \frac{27}{4}$
42. $\log_a (5\sqrt{2})$
43. $\log_a (40/a)$
44. $\log_a (30a^2)$
45. $[\log_a (5a^2)]/\log_a 81$
46. $\log_a (5a^2/81)$

*In problems 47–52, write each expression as a single logarithm by using the properties of logarithms (see Example E).*

47. $\log_b x - 2\log_b 8$
48. $\log_b x + \log_b y - \log_b z$
49. $5\log_b x - 4\log_b y$
50. $2\log_b (x+3) + 4\log_b (x-3)$
51. $3[\log_b x - \log_b y + 2\log_b z]$
52. $\frac{1}{2}[\log_b x - 2(\log_b x - \log_b z)]$

*In Problems 53–56, sketch the graphs of each pair of functions using the same coordinate axes.*

53. $f(x) = 3^x$, $g(x) = \log_3 x$
54. $f(x) = \log_2 x$, $g(x) = \log_2 x^3$
55. $f(x) = \log_3 x$, $g(x) = \log_3 (x-2)$
56. $f(x) = \log_2 x$, $g(x) = \log_2 (x+3)$

---

*Solve for x in the equations in Problems 57–66.*

57. $\log_2 16 = x$
58. $\log_3 x = -2$
59. $\log_x \frac{1}{8} = -3$
60. $\log_3 3^{\pi} = x$
61. $5^{\log_5 9} = x$
62. $3^{3\log_3 2} = x$
63. $\log_5 (x+2) = 2$
64. $\log_4 (x-1) = 1.5$
65. $3^{2x-1} = 27$
66. $2^{3x+2} = 64$

67. Which is larger, $10^{1000}$ or $10{,}000^{200}$?
68. Which is larger, $\log_5 117$ or $\log_6 117$?
69. [C] Which is larger, $10^{10000}$ or $13^{9000}$? *Hint:* Take logs.
70. [C] Think of a large sheet of paper being folded 40 times (this is possible only in our imagination). Calculate the height of the resulting stack of paper assuming 1000 thicknesses of paper to the inch. Express your answer in miles.
71. [C] One of the authors (Walter Fleming) remembers seeing the following problem on a streetcar transfer (long ago when American cities had streetcars and ordinary people knew about horses). "A rich man went to the blacksmith to have his horse shoed. It takes 28 nails to shoe a horse (7 for each shoe). The man was charged 1¢ for the first nail, 2¢ for the second nail, 4¢ for the third, 8¢ for the fourth, and so on. How much did he have to pay?"
72. [C] An old legend says that an Eastern prince was so fond of playing chess that he offered the inventor of the game a reward of his choice. The man called for a chessboard and asked for one grain of rice for the first square, two grains for the second square, four grains for the third, and so on. The prince exploded with laughter at such a modest request. Show that the inventor had the last laugh by calculating the number of kilograms of rice that would be needed. You will need to know that a chess board has 64 squares and that a grain of rice weighs about 0.02 gram.

## 5.2 THE NATURAL EXPONENTIAL AND LOGARITHMIC FUNCTIONS

Historically, the most common base used for the exponential and logarithmic functions was base 10. However, in calculus and all of advanced mathematics another base plays a fundamental role. This base is the number $e$, a number we first met back in the introduction to Section 2.1. The number $e$ is defined by a limit as follows.

$$e = \lim_{u \to 0} (1+u)^{1/u}$$

| $u$ | $(1+u)^{1/u}$ |
|---|---|
| .1 | 2.59374 |
| .01 | 2.70481 |
| .001 | 2.71692 |
| .0001 | 2.71815 |
| .00001 | 2.71827 |
| .000001 | 2.718280 |
| .0000001 | 2.718282 |
| .00000001 | 2.718282 |

**FIGURE 4**

Figure 4 gives the results of some calculations done on a hand-held calculator and suggests an approximate value for $e$.

$$e \approx 2.71828$$

**POINT OF INTEREST**

To Leonhard Euler (1701–1783) we owe the modern definition of logarithms and recognition of the importance of the number $e$. This Swiss mathematician is said to be the most prolific mathematical writer of all time (856 titles), all the more remarkable since he was totally blind during a good part of his career.

Actually, $e$ is an irrational number whose decimal expansion is known to hundreds of thousands of places; however, the approximation above is good enough for any calculations that we will need to make.

The functions $f(x) = e^x$ and $g(x) = \log_e x$ are called the **natural exponential function** and the **natural logarithmic function,** respectively. We use the abbreviation $\ln x$ in place of the more cumbersome $\log_e x$. The two functions $e^x$ and $\ln x$ are inverse functions, and so

$$e^{\ln x} = x \quad \text{and} \quad \ln e^x = x$$

Thus, the graphs of $y = e^x$ and $y = \ln x$ are reflections of each other across the line $y = x$ as we show in Figure 5.

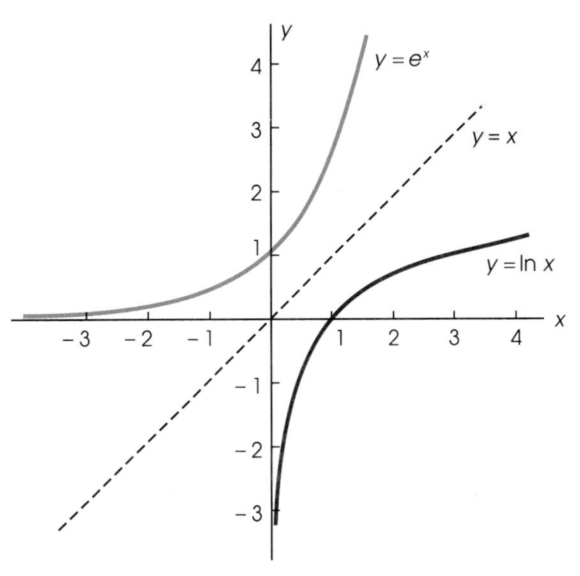


**FIGURE 5**

Note that the domain of $e^x$ is the set of all real numbers, whereas for $\ln x$ it is the set of positive real numbers. Our graphs suggest that both of these functions should have derivatives. We will investigate that subject soon.

## Some Calculations with $\ln x$ and $e^x$

The appendix includes tables of $\ln x$ and $e^x$. However, most of our readers will own a scientific calculator which allows calculations with these functions at the touch of a button. (On some calculators $e^x$ is obtained as inv $\ln x$.) To gain experience in using these functions, we suggest you work through Example A on your calculator.

### EXAMPLE A

Calculate each of the following using a scientific calculator. (a) $\ln(2.0123)$, (b) $\ln(0.00542)$, (c) $\ln(-1.2373)$, (d) $e^{0.01234}$, (e) $e^{3.765}$, (f) $e^{-2.3672}$, (g) $[\ln(2.367 + e^{-5.43})]/e^{0.154}$

**Solution**

> **CAUTION**
> Only positive numbers have logarithms. For example, $\ln(-1.2373)$ is undefined. The reason is simple. There is no number $x$ for which
>
> $$e^x = -1.2373$$

(a) $\ln(2.0123) = 0.6992783465$
(b) $\ln(0.00542) = -5.217659464$
(c) $\ln(-1.2373) =$ error; negative numbers do not have logarithms.
(d) $e^{0.01234} = 1.012416452$
(e) $e^{3.765} = 43.16370586$
(f) $e^{-2.3672} = 0.0937428391$
(g) $[\ln(2.367 + e^{-5.43})]/e^{0.154} = 0.7402315607$ ■

### EXAMPLE B

Use natural logarithms to solve the equation

$$3^{2x+1} = 4.9$$

> **RECALL**
> For natural logarithms, the three properties of logarithms take the form:
>
> 1. $\ln uv = \ln u + \ln v$
> 2. $\ln \frac{u}{v} = \ln u - \ln v$
> 3. $\ln u^r = r \ln u$

**Solution** When the unknown appears in an exponent, it is a good idea to begin by taking logarithms of both sides (we will use natural logarithms, though logarithms to other bases could be used). Note our use of the third property of logarithms.

$$\begin{aligned} \ln(3^{2x+1}) &= \ln 4.9 \\ (2x+1)\ln 3 &= \ln 4.9 \\ 2x + 1 &= (\ln 4.9)/(\ln 3) = 1.446584224 \\ 2x &= 0.446584224 \\ x &= 0.223292112 \end{aligned}$$ ■

## The Derivative of $f(x) = \ln x$

Our approach is very basic; we go back to the definition of the derivative. In the derivation that follows, we use the properties of logarithms and then at a key step we replace $h/x$ by $u$ (note that if $h \to 0$, then $u \to 0$).

$$\begin{aligned} f'(x) &= \lim_{h\to 0} \frac{f(x+h) - f(x)}{h} = \lim_{h\to 0} \frac{\ln(x+h) - \ln x}{h} \\ &= \lim_{h\to 0} \frac{1}{h} \ln\left(\frac{x+h}{x}\right) = \lim_{h\to 0} \frac{1}{x}\frac{x}{h} \ln\left(1 + \frac{h}{x}\right) \\ &= \frac{1}{x} \lim_{h\to 0} \ln\left(1 + \frac{h}{x}\right)^{x/h} = \frac{1}{x} \lim_{u\to 0} \ln(1+u)^{1/u} \\ &= \frac{1}{x} \ln e = \frac{1}{x} \log_e e = \frac{1}{x} \end{aligned}$$

We have shown that

$$\frac{d(\ln x)}{dx} = \frac{1}{x}$$

This is a fundamental result. When combined with the fact that $d^2(\ln x)/dx^2 = -1/x^2$, it yields the conclusion that the graph of $\ln x$ is always rising and concave down, agreeing with the picture in Figure 5. Now let us use the boxed result in two simple examples.

### EXAMPLE C

Find the equation of the tangent line to $y = \ln x$ at the point $(1, 0)$.

**Solution** At $x = 1$, the derivative $1/x$ has a value of 1, which means that the slope of the tangent line is 1. The equation of this line is $y - 0 = 1(x - 1)$; that is, $y = x - 1$. ■

### EXAMPLE D

Find $f'(x)$ if (a) $f(x) = x^3 \ln x$, (b) $f(x) = (\ln x)^5$.

**Solution**

(a) We use the Product Rule.

$$f'(x) = (x^3)\frac{1}{x} + (\ln x)(3x^2) = x^2 + 3x^2 \ln x = x^2(1 + 3 \ln x)$$

(b) We use the General Power Rule.

$$f'(x) = 5(\ln x)^4 \frac{1}{x}$$

■

## The Derivative of $f(x) = e^x$

Once more, we go back to the definition of the derivative.

$$f'(x) = \lim_{h \to 0} \frac{f(x + h) - f(x)}{h} = \lim_{h \to 0} \frac{e^{x+h} - e^x}{h}$$

$$= \lim_{h \to 0} \frac{e^x(e^h - 1)}{h} = e^x \lim_{h \to 0} \frac{e^h - 1}{h}$$

| $h$ | $(e^h - 1)/h$ |
|---|---|
| .1 | 1.05171 |
| .01 | 1.00502 |
| .001 | 1.00050 |
| .0001 | 1.00005 |
| .00001 | 1.00000 |

**FIGURE 6**

Our problem thus reduces to evaluating the last limit. The calculations in Figure 6 were done on a hand-held calculator. They suggest that

$$\lim_{h \to 0} \frac{e^h - 1}{h} = 1$$

Taking this for granted, we conclude that

$$\frac{d(e^x)}{dx} = e^x$$

> **POINT OF INTEREST**
> The only functions with the remarkable property that they are their own derivatives are the multiples of $e^x$, that is, functions of the form
>
> $$f(x) = ce^x$$

another fundamental fact. It says that the derivative of $e^x$ is itself. You have a right to be a little skeptical of the demonstration above, since it rests on guessing at a limit based on the numbers in Figure 6. However, we promise another and very different demonstration of the boxed result in the next section.

From the fact that $d(e^x)/dx = e^x$, we infer that $d^2(e^x)/dx^2 = e^x$, which means that both the first and second derivatives of $e^x$ are always positive. This implies that the graph of $y = e^x$ is always rising and concave up, confirming the shape of the graph in Figure 5.

## EXAMPLE E

Find $g'(x)$ if (a) $g(x) = e^x(x^2 + 3x)$ (b) $g(x) = xe^{-x} = x/e^x$
(c) $g(x) = \sqrt{e^x + 5x} = (e^x + 5x)^{1/2}$

**Solution**

(a) We use the Product Rule.

$$g'(x) = e^x(2x + 3) + (x^2 + 3x)e^x = e^x(x^2 + 5x + 3)$$

(b) We use the Quotient Rule.

$$g'(x) = \frac{(e^x)(1) - (x)(e^x)}{(e^x)^2} = \frac{e^x(1 - x)}{(e^x)^2} = \frac{1 - x}{e^x}$$

(c) This calls for the General Power Rule.

$$g'(x) = \tfrac{1}{2}(e^x + 5x)^{-1/2}(e^x + 5)$$

■

## EXAMPLE F

Find the maximum value of $f(x) = x/e^x$ on the interval $[0, \infty)$. Also determine where the graph of $f(x)$ is concave up and where it is concave down. Finally, sketch the graph of $f(x)$.

**Solution** From Example E, we know that $f'(x) = (1 - x)/e^x$. Thus, the critical points are the endpoint 0 and the point 1, where the derivative is 0. Since $f(0) = 0$ while $f(1) = 1/e$, we tentatively accept $f(1) = 1/e \approx 0.36788$ as the maximum value. To confirm this, we note that $f'(x) > 0$ when $x < 1$, and $f'(x) < 0$ for $x > 1$.

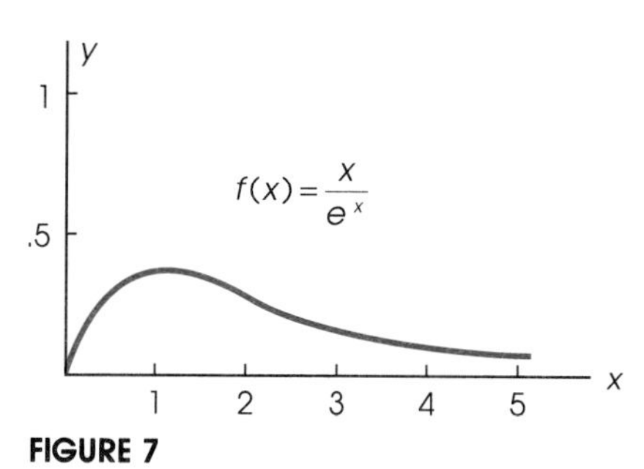

FIGURE 7

To analyze concavity, we find the second derivative.

$$f''(x) = \frac{e^x(-1) - (1 - x)e^x}{(e^x)^2} = \frac{-2e^x + xe^x}{e^{2x}}$$

$$= \frac{e^x(x - 2)}{e^{2x}} = \frac{x - 2}{e^x}$$

From this we see that $f''(x) < 0$ for $x < 2$, and $f''(x) > 0$ for $x > 2$. We conclude that the graph of $f(x)$ is concave down for $x < 2$ and concave up for $x > 2$. The graph is shown in Figure 7. ■

## Problem Set 5.2

C *Use your calculator to evaluate each of the expressions in Problems 1–14.*

1. $\ln 14.6$
2. $\ln 3961$
3. $\ln 0.0096$
3. $\ln 0.109$
5. $\ln (9.82)^3$
6. $3 \ln 9.82$
7. $\ln \sqrt[3]{61.9}$
8. $\ln \sqrt[4]{4.63}$
9. $e^{3.18}$
10. $e^{0.023}$
11. $e^{-\pi}$
12. $\pi^e$
13. $(2 \ln 14.1 + 19)/(e^{2.1} - 14)$
14. $e^{3.2} (\ln 61 + e^{-1.2})^3$

C *Use natural logarithms and your calculator to solve the equations in Problems 18–24 (see Example B).*

15. $3^x = 15$
16. $5^x = 0.41$
17. $2^{3x-1} = 20.5$
18. $4^{x-3} = 9.82$
19. $(\pi^2 + 1)^x = 2$
20. $(\sqrt{3} + 2)^x = 144$
21. $4^{\sqrt{x}} = 10$
22. $10^{1/x} = 4.214$
23. $2^{x^2-x} = 5$
24. $3^{x^2+x} = 7$
25. Find the slope of the tangent line to the curve $y = \ln x$ at $x = 0.5, 1, 3$, and $100$.
26. Find the slope of the tangent line to the curve $y = e^x$ at $x = 0.5, 1, 3$, and $100$.

*In Problems 27–30, find the equation of the tangent line to the given curve at the indicated value of x (see Example C).*

27. $y = 3 \ln x$ at $x = e$
28. $y = 2 \ln x + x$ at $x = 1$
29. $y = (3 + \ln x)^2$ at $x = 1/e$
30. $y = (x^2 + \ln x)^3$ at $x = 1$
31. At what point on the curve $y = 4 + 3 \ln x$ is the tangent line parallel to the line $5y - x = 10$?
32. For what values of $x$ is the slope of the curve $y = x - 2 \ln x$ less than $0.3$?

*Find f'(x) for each of the functions given in Problems 33–46 (see Examples D and E).*

33. $f(x) = x^2 \ln x$
34. $f(x) = (\ln x)/x^3$
35. $f(x) = 3(\ln x)^2$
36. $f(x) = \sqrt{2 + \ln x}$
37. $f(x) = 4/(\ln x)^3$
38. $f(x) = (x^2 + 2 \ln x)/x$
39. $f(x) = (x^2 + 2 \ln x)^3$
40. $f(x) = (3x + \ln x)^{1.2}$
41. $f(x) = (2x + 1)e^x$
42. $f(x) = 3x^2e^x$
43. $f(x) = x^2/(e^x + 3)$
44. $f(x) = 2e^x/(3x - 1)$
45. $f(x) = \sqrt[5]{5e^x + x^5}$
46. $f(x) = (e^x + 2x)^3$
47. Consider (as in Example F) the function $f(x) = xe^x$.
    (a) Where is this function increasing and where is it decreasing?
    (b) Find its minimum value on $(-\infty,\infty)$.
    (c) Determine where the graph of $f$ is concave up and where it is concave down.
    (d) Sketch the graph of $f$.
48. Follow the instructions in Problem 47 for the function $f(x) = x^2/e^x$ except that in part (b) find the maximum value.

---

C 49. Calculate.
    (a) $(e^{3.1} + \ln 43.2)^{1/3}$
    (b) $[\ln (\pi^3 + \sqrt{e})]/[e^{\sqrt{\pi}} - \ln 2]$

C 50. Solve for $x$.
    (a) $(3.2)^x = (2.8)^{x+1}$ (b) $e^{x^2-x-6} = 1$
    (c) $\ln (3x + 1) = 2.966$ (d) $\ln (x^2 + x) = 1$

51. Write $2 \ln 3x + \ln (x^2 - 1) - \ln (x - 1) - \ln x$ as a single logarithm and simplify.

52. Simplify each of the following.
    (a) $(e^{\ln x})^2$
    (b) $e^{3 \ln 2}$
    (c) $e^{(0.5) \ln (x^2)}$
    (d) $\dfrac{\ln (x^4)}{\ln (x^2)}$
53. Find $g'(0)$ for each of the following.
    (a) $g(x) = (3e^x + 5x)^2$
    (b) $g(x) = (x^2 + 1)/\sqrt{4 + 5e^x}$
54. Find $g''(x)$ for each of the following. Simplify your answer.
    (a) $g(x) = x^3 e^x$
    (b) $g(x) = x(x + \ln x)^3$
55. Consider the function $f(x) = e^x/x$ on $(0, \infty)$.
    (a) Where is this function increasing and where is it decreasing?
    (b) Determine its minimum value.
    (c) Determine where its graph is concave up and where it is concave down.
    (d) Sketch its graph.

## 5.3 THE CHAIN RULE

In the previous section, we learned how to find the derivatives of two important functions, namely, $e^x$ and $\ln x$. Now we want to consider how to find the derivative of some complicated combinations of these functions, functions like

$$F(x) = e^{x^3+4x-9}$$

and

$$G(x) = \ln \frac{x^2 + 5}{x - 1}$$

Both of the displayed functions can be thought of as *composite functions* (see Section 1.1). For example,

$$F(x) = f(g(x)) \quad \text{where} \quad f(x) = e^x \quad \text{and} \quad g(x) = x^3 + 4x - 9$$

$$G(x) = h(k(x)) \quad \text{where} \quad h(x) = \ln x \quad \text{and} \quad k(x) = \frac{x^2 + 5}{x - 1}$$

Our problem boils down to this. How can we find the derivative of a composite function given that we know the derivatives of the constituent parts? The answer is given by the Chain Rule, certainly one of the most useful results in calculus.

### The Chain Rule Stated

We begin by asking the question about the derivative of a composite function in a slightly different way. Suppose that

$$y = f(u) \qquad \text{where} \quad u = g(x)$$

We ask the question: How can we find $dy/dx$ given that we know $dy/du$ and $du/dx$? Stated this way, there is a simple intuitive answer. Remember that derivatives represent rates of change. If $y$ changes $r$ times as fast as $u$, and $u$ changes $s$ times as fast as $x$, then $y$ ought to change $rs$ times as fast as $x$. For example, if Renee can type three times as fast as Susan, and Susan can type twice as fast as Tammy, then Renee can type (3)(2) = 6 times as fast as Tammy. This suggests the following pattern.

Left variable → $y = f(u)$; Middle variable → $u$ (in $f(u)$) and $u$ (in $u = g(x)$); Right variable → $x$

$$y = f(u) \quad \text{and} \quad u = g(x)$$

$$\begin{bmatrix}\text{Derivative of}\\ \text{left variable}\\ \text{with respect to}\\ \text{right variable}\end{bmatrix} = \begin{bmatrix}\text{Derivative of}\\ \text{left variable}\\ \text{with respect to}\\ \text{middle variable}\end{bmatrix} \cdot \begin{bmatrix}\text{Derivative of}\\ \text{middle variable}\\ \text{with respect to}\\ \text{right variable}\end{bmatrix}$$

$$\frac{dy}{dx} = \frac{dy}{du} \cdot \frac{du}{dx}$$

Here is a formal statement of the Chain Rule. We postpone its proof until the end of the section.

**CHAIN RULE**

Let $y = f(u)$ and $u = g(x)$ determine the composite function $y = f(g(x))$. If both $dy/du$ and $du/dx$ exist, then $dy/dx$ exists and is given by the formula

$$\frac{dy}{dx} = \frac{dy}{du} \cdot \frac{du}{dx}$$

## The Chain Rule Illustrated

**CAUTION**

It is a truism that a student who fails to master the Chain Rule will never succeed in calculus. This rule is chief among our growing supply of tools for finding derivatives.

We illustrate first with the two functions mentioned in the introduction to this section.

### EXAMPLE A

Find $dy/dx$ for each of the following.

(a) $y = e^{x^3+4x-9}$ (b) $y = \ln[(x^2 + 5)/(x - 1)]$

**Solution**

(a) Think of this as $y = e^u$, where $u = x^3 + 4x - 9$. Then by the Chain Rule,

$$\frac{dy}{dx} = \frac{dy}{du}\frac{du}{dx} = e^u(3x^2 + 4) = e^{x^3+4x-9}(3x^2 + 4)$$

Note that the final answer is stated in terms of $x$, the variable in the original problem.

(b) Think of this as $y = \ln u$, where $u = (x^2 + 5)/(x - 1)$. Then by the Chain Rule followed by the Quotient Rule,

$$\frac{dy}{dx} = \frac{dy}{du}\frac{du}{dx} = \frac{1}{u}\frac{(x-1)(2x) - (x^2+5)(1)}{(x-1)^2}$$

$$= \frac{x-1}{x^2+5}\,\frac{x^2-2x-5}{(x-1)^2} = \frac{x^2-2x-5}{(x^2+5)(x-1)} \qquad \blacksquare$$

**EXAMPLE B**

Find $f'(x)$ if (a) $f(x) = xe^{-x}$, (b) $f(x) = [\ln (x^2 + 1)]/x$.

**Solution**

(a) We apply the Product Rule first.

$$f'(x) = x\frac{d(e^{-x})}{dx} + e^{-x}(1)$$

To find the derivative of $e^{-x}$, we let $u = -x$ and apply the Chain Rule, obtaining $e^u(-1) = e^{-x}(-1)$. We conclude that

$$f'(x) = x(-e^{-x}) + e^{-x} = e^{-x}(-x + 1)$$

(b) We begin with the Quotient Rule and then apply the Chain Rule to the derivative of $\ln (x^2 + 1)$.

$$f'(x) = \frac{x\dfrac{d[\ln (x^2+1)]}{dx} - \ln (x^2+1)\dfrac{d(x)}{dx}}{x^2}$$

$$= \frac{x\dfrac{1}{x^2+1}2x - \ln (x^2+1)}{x^2}$$

$$= \frac{2x^2 - (x^2+1)\ln (x^2+1)}{x^2(x^2+1)} \qquad \blacksquare$$

**EXAMPLE C**

For $y = \ln (e^{-x} + x^2)$, find $dy/dx$.

**Solution** Initially, we let $u = e^{-x} + x^2$ and apply the Chain Rule to obtain

$$\frac{dy}{dx} = \frac{dy}{du}\frac{du}{dx} = \frac{1}{u}\frac{du}{dx} = \frac{1}{e^{-x} + x^2}\frac{du}{dx}$$

Calculating $du/dx$ will involve a second use of the Chain Rule in finding the derivative of $e^{-x}$ (we saw how this was done in Example B). We get $du/dx = -e^{-x} + 2x$, which yields the final result

$$\frac{dy}{dx} = \frac{-e^{-x} + 2x}{e^{-x} + x^2}$$ ■

## The General Power Rule Revisited

We have been using one instance of the Chain Rule all along. Recall that if $u = u(x)$ and if $y = u^r$, then $dy/dx = ru^{r-1}(du/dx)$. In other words, to find the derivative of $y = u^r$, where $u$ is a function of $x$, we first take the derivative of $y$ with respect to $u$ (which is $ru^{r-1}$) and multiply the result by $du/dx$. This means that when we prove the Chain Rule (coming shortly), we will have automatically proven the General Power Rule, a matter we left dangling back in Chapter 2.

The three uses of the Chain Rule so far observed can be summarized as follows.

1. $\dfrac{d(u^r)}{dx} = ru^{r-1}\dfrac{du}{dx}$
2. $\dfrac{d(e^u)}{dx} = e^u\dfrac{du}{dx}$
3. $\dfrac{d(\ln u)}{dx} = \dfrac{1}{u}\dfrac{du}{dx}$

We must develop skill in using these three results, so much so that we can use them mechanically. Here is another three-part example.

### EXAMPLE D

Find $dy/dx$ for (a) $y = [\ln(3x + 1)]^3$, (b) $y = (e^{\sqrt{x}} + 2)^4$, (c) $y = \ln[\ln(3x^2)]$.

**Solution** Each of these involves a double use of the Chain Rule. See if you can make the mental substitutions required to obtain the following answers.

(a) $\dfrac{dy}{dx} = 3[\ln(3x + 1)]^2\dfrac{1}{3x + 1}3$

(b) $\dfrac{dy}{dx} = 4(e^{\sqrt{x}} + 2)^3(e^{\sqrt{x}})\dfrac{1}{2}x^{-1/2}$

(c) $\dfrac{dy}{dx} = \dfrac{1}{\ln(3x^2)}\dfrac{1}{3x^2}6x$ ■

Our next example will play an important role in Chapter 7.

**EXAMPLE E**

Find $dy/dx$ if $y = \ln|x|$.

**Solution** We suppose that $x \neq 0$, since $\ln 0$ is undefined. For $x > 0$, $|x| = x$, and so in this case $\dfrac{d(\ln|x|)}{dx} = 1/x$. For $x < 0$, $|x| = -x$, and so (using the Chain Rule),

$$\frac{d(\ln|x|)}{dx} = \frac{d[\ln(-x)]}{dx} = \frac{1}{-x}(-1) = \frac{1}{x}$$

We conclude that whether $x > 0$ or $x < 0$,

$$\frac{d(\ln|x|)}{dx} = \frac{1}{x} \qquad x \neq 0$$

■

## The Chain Rule Proved

Refer to the statement of the Chain Rule near the beginning of this section. Let

$$\Delta u = g(x + \Delta x) - g(x) \quad \text{and} \quad \Delta y = f(u + \Delta u) - f(u)$$

Note that if $\Delta x \to 0$, then $\Delta u \to 0$. Thus,

$$\frac{dy}{dx} = \lim_{\Delta x \to 0} \frac{\Delta y}{\Delta x} = \lim_{\Delta x \to 0} \frac{\Delta y}{\Delta u}\frac{\Delta u}{\Delta x}$$

$$= \lim_{\Delta u \to 0} \frac{\Delta y}{\Delta u} \lim_{\Delta x \to 0} \frac{\Delta u}{\Delta x} = \frac{dy}{du} \cdot \frac{du}{dx}$$

**SHARP CURVE**
Our proof of the Chain Rule ultimately collapses under the weight of the maxim—Thou shalt not divide by zero. To construct a rigorous proof requires a complicated maneuver best left for more advanced books.

This proof seems almost too simple and, in fact, there is a subtle flaw. The division and multiplication by $\Delta u$ in the first step would be illegal if $\Delta u = 0$. This would happen, for example, if $g$ were a constant function. The Chain Rule turns out to be valid even in this case, though our proof breaks down. For a complete proof of the Chain Rule which covers all possible cases, see *Calculus With Analytic Geometry,* 5th edition, by Purcell and Varberg, p. 827.

## Another Demonstration That $d(e^x)/dx = e^x$

Whenever we know the derivative of a function, we can use the Chain Rule to find the derivative of its inverse function. We illustrate with the familiar function $\ln x$. Assume we know that $d(\ln x)/dx = 1/x$. Let $u = e^x$ so that $\ln u = x$. In this last equation, take the derivative of both sides with respect to $x$, keeping in mind that $u$ is a function of $x$. On the left we use the Chain Rule. We obtain the following in straightforward steps.

$$\ln u = x$$

$$\frac{1}{u}\frac{du}{dx} = 1$$

$$\frac{du}{dx} = u$$

$$\frac{d(e^x)}{dx} = e^x$$

## Problem Set 5.3

*In Problems 1–16, find $dy/dx$.*

1. $y = 4e^{3x}$
2. $y = 2e^{-3x}$
3. $y = e^{x^2} = e^{(x^2)}$
4. $y = 3e^{x^3}$
5. $y = 4 \ln (2x + 5)$
6. $y = 3 \ln (1 - x^2)$
7. $y = \ln (5x + 1/x)$
8. $y = \ln (\sqrt{x} + x^{-2})$
9. $y = 3e^{4\sqrt{x}}$
10. $y = -2e^{1/x}$
11. $y = \ln [(3x + 1)/(x^2 + 2)]$
12. $y = \ln [x^2/(2x - 1)]$
13. $y = e^{2x/(x+3)}$
14. $y = -e^{(1-x)/x^2}$
15. $y = \ln [(2x^2 + 5)(3x - 2)]$
16. $y = \ln [(5x - 3)(x^3 + 2x)]$

*Find $f'(x)$ in Problems 17–24 (see Example B).*

17. $f(x) = x^2e^{3x}$
18. $f(x) = (2x + 1) \ln 5x$
19. $f(x) = (2x + 1)/e^{2x}$
20. $f(x) = [\ln (4x + 1)]/3x$
21. $f(x) = (1 + 3/x) \ln (x^2 + 4)$
22. $f(x) = 5\sqrt{x}e^{2/x}$
23. $f(x) = e^{x^2} \ln (x^2)$
24. $f(x) = [\ln (2x + 1)]/e^{2x+1}$

*Find $g'(t)$ in Problems 25–32 (see Example C).*

25. $g(t) = \ln (2t + e^{5t})$
26. $g(t) = 3 \ln (t^2 - e^{2t})$
27. $g(t) = e^{t-\ln (t^2+1)}$
28. $g(t) = e^{3t+\ln (t^3-2)}$
29. $g(t) = \ln [\ln (t)]$
30. $g(t) = e^{e^t}$
31. $g(t) = \ln [\ln (\ln t)]$
32. $g(t) = \ln [\ln (t^2 - t)]$

*In Problems 33–44, find $dy/dx$ (see Example D).*

33. $y = (3e^{2x} - 1)^3$
34. $y = (2x + e^{2x})^{1/2}$
35. $y = [\ln (x^2 + 4)]^4$
36. $y = (2x + 3 \ln x)^5$
37. $y = 4(e^x + \ln x)^{-2}$
38. $y = (2xe^x + 4)^{3/2}$
39. $y = x^2(e^x + 2)^3$
40. $y = (x + \ln x)^2/e^x$
41. $y = e^{(2x+1)^2}$
42. $y = (e^{2x+1})^2$
43. $y = [\ln (x + 4 \ln x)]^3$
44. $y = [\ln (x \ln x)]^3$

---

45. Find $dy/dt$.
(a) $y = 2e^{3t} - 4e^{-t}$
(b) $y = 3 \ln (t^2 - 4)$
(c) $y = t^2(e^t - \ln t)$
(d) $y = e^{2t}/(t^2 - 5)$
(e) $y = e^{(2t+1)^2}$
(f) $y = \ln [(2t + \sqrt{t})^4]$
(g) $y = [\ln (2t + \sqrt{t})]^4$
(h) $y = (4t - e^{2/t})^6$
46. Find $g'(s)$.
(a) $g(s) = s^2 - 5e^{2s}$
(b) $g(s) = 2 \ln (s + s^2 + s^3)$
(c) $g(s) = (e^s - s)/e^{2s}$
(d) $g(s) = 3s \ln (s^2 + 1)$
(e) $g(s) = \ln (\sqrt{3 + s^2} + 2)$
(f) $g(s) = 3e^{(2-s)^4}$
(g) $g(s) = [(5/s) + e^{3s}]^4$
(h) $g(s) = [s^2 \ln (1 + 5s)]^3$
47. If $f(x) = x^2e^{-x^2}$, find $f''(x)$.
48. Find the equation of the tangent line to the curve $y = [2 + \ln (1 + x)]^3$ at the point where $x = 1$.
49. Find the equation of the tangent line to the curve $y = \sqrt{x^2 + 1} \ln (e^{2x} + x)$ at the point where $x = 0$.
50. [C] Suppose that after $t$ weeks of practice, a typical typing student is able to type $f(t) = 80(1 - e^{-0.25t})$ words per minute.
(a) How fast can such a student type after 4 weeks?
(b) When will the student be able to type 75 words per minute?
(c) How fast is the student's typing speed increasing after 3 weeks?
51. [C] If $P$ dollars is invested at rate $r$ *compounded continuously*, its accumulated value after $t$ years is given by

$A = Pe^{rt}$ (a subject to be discussed fully in Section 6.3). Taking this for granted, answer the following questions under the assumption that $r = 9\% = 0.09$ and $P = 10{,}000$.
(a) What is the accumulated value of this investment after 5.5 years?
(b) At what rate in dollars per year is this investment growing at the end of 10 years?
(c) How long will it take this investment to double in value?

C 52. To test retention, a group of calculus students was given a final exam and then equivalent forms of the exam at 1-month intervals thereafter. Students were told not to study between exams. The average score on these exams after $t$ months was found to be $S(t) = 72 - 15 \ln (t + 1)$.
(a) Determine the average score after 12 months.
(b) At what rate was the average score changing after 4 months?
(c) When will the average score drop to three-fourths its original value?

C 53. A company believes that it will be able to sell $n = n(x)$ units of its product each year if it spends $x$ thousand dollars on advertising, where $n = 2500 + 400 \ln x$. Further, its annual profit $P$ (ignoring advertising costs) in thousands of dollars is related to the number of units sold by the formula $P = -30 + 0.22n - 0.000002n^2$.
(a) Determine the profit when \$40,000 is spent on advertising.
(b) Calculate the rate of change in profit with respect to advertising when \$40,000 is spent on advertising.
(c) Interpret your answer to part (b).

54. Sketch the graph of $y = e^{-(x-1)^2}$, showing the coordinates of the maximum point and the inflection points.

## 5.4* GENERAL LOGARITHMIC AND EXPONENTIAL FUNCTIONS

We have said that it is the natural logarithmic and exponential functions ($\ln x$ and $e^x$) that are important in calculus and its applications. Even the most superficial examination of the literature will support this statement. It is also true that any problem involving other logarithmic or exponential functions can easily be transformed into a problem about the corresponding natural functions. We demonstrate this now.

We begin with the equation that actually defines the general logarithm $\log_b x$, namely,

$$b^{\log_b x} = x$$

When we take natural logarithms of both sides of this equation and equate the results, we get

$$(\log_b x)(\ln b) = \ln x$$

or

$$\boxed{\log_b x = \frac{\ln x}{\ln b}}$$

This makes it obvious that any question involving $\log_b x$ can be rephrased as a question about $\ln x$.

In a similar manner, recall that for any $u$, $e^{\ln u} = u$. In particular

$$e^{\ln b^x} = b^x$$

or by Property 3 of logarithms,

$$e^{x \ln b} = b^x$$

We may rewrite this as

$$b^x = e^{kx} \qquad \text{where } k = \ln b$$

EXAMPLE A

Using the $\ln x$ and $e^x$ keys on a calculator, evaluate (a) $\log_3 14$, (b)$3^{2.24}$

**Solution**

(a) $\log_3 14 = (\ln 14)/(\ln 3) = 2.63905733/1.098612289$
$= 2.402173503$

(b) $3^{2.24} = e^{(\ln 3)(2.24)} = e^{(1.098612289)(2.24)}$
$= e^{2.460891527} = 11.71525135$

Of course $3^{2.24}$ can be obtained on your calculator using the $y^x$ key, but internally the calculator probably made the calculation as we have shown it. ■

## The Derivatives of $\log_b x$ and $b^x$

Using the boxed results above, it is a simple matter to find formulas for the required derivatives. After noting that $\ln b$ is a constant, we write

$$\frac{d(\log_b x)}{dx} = \frac{1}{\ln b}\frac{d(\ln x)}{dx} = \frac{1}{\ln b} \cdot \frac{1}{x}$$

That is,

$$\frac{d(\log_b x)}{dx} = \frac{1}{x \ln b}$$

Similarly, a simple use of the Chain Rule gives

$$\frac{d(b^x)}{dx} = \frac{d(e^{kx})}{dx} = ke^{kx} = (\ln b)e^{kx}$$

That is,

$$\frac{d(b^x)}{dx} = b^x \ln b$$

### EXAMPLE B

Find $dy/dx$ if (a) $y = \log_3 x$, (b) $y = \log_2(x^2 + 3)$, (c) $y = 2^x$, (d) $y = 3^{e^x - e^{-x}}$

**Solution**

(a) $\dfrac{dy}{dx} = \dfrac{1}{x \ln 3}$

(b) We apply the Chain Rule with $u = x^2 + 3$ to obtain

$$\frac{dy}{dx} = \frac{1}{u \ln 2}\frac{du}{dx} = \frac{2x}{(x^2 + 3)\ln 2}$$

(c) $\dfrac{dy}{dx} = 2^x \ln 2$

(d) $\dfrac{dy}{dx} = 3^{e^x - e^{-x}}(\ln 3)(e^x + e^{-x})$ ■

## The Derivative of $(g(x))^{f(x)}$

> **CAUTION**
> Be sure to distinguish between the four functions $x^e$, $e^x$, $x^x$, and $e^e$. Their derivatives are:
>
> $$\frac{d(x^e)}{dx} = ex^{e-1}$$
>
> $$\frac{d(e^x)}{dx} = e^x$$
>
> $$\frac{d(x^x)}{dx} = x^x(1 + \ln x)$$
>
> $$\frac{d(e^e)}{dx} = 0$$

We now know how to find the derivative of $[f(x)]^c$ and $c^{f(x)}$, $c$ being a constant. The first is handled by the General Power Rule; the second is handled by the method illustrated in part (d) of Example B above. Thus,

$$\frac{d[(x^2 - 5)^6]}{dx} = 6(x^2 - 5)^5(2x)$$

while

$$\frac{d(6^{x^2-5})}{dx} = 6^{x^2-5}(\ln 6)(2x)$$

But what about the derivative of a function raised to an exponent which is itself a function? For example, how does one find the derivative of $(x^2 - 5)^{3x}$? Rather than state a new rule, we are going to suggest a procedure that goes by the name **logarithmic differentiation.** It amounts to this: Before differentiating $y = f(x)$, take natural logarithms of both sides.

### EXAMPLE C

Find $dy/dx$ for $y = (x^2 + 5)^{3x}$.

**Solution** As suggested, we begin by taking natural logarithms, obtaining

$$\ln y = 3x \ln(x^2 + 5)$$

Then we take the derivative of both sides. On the left, we must keep in mind that $y$ is a function of $x$; on the right side, we use the Product Rule.

We get

$$\frac{1}{y}\frac{dy}{dx} = 3x\frac{d[\ln(x^2 + 5)]}{dx} + [\ln(x^2 + 5)](3)$$

$$= 3x\frac{1}{x^2 + 5}2x + 3\ln(x^2 + 5)$$

$$= \frac{6x^2}{x^2 + 5} + 3\ln(x^2 + 5)$$

Next, we solve for $dy/dx$ (multiplying by $y$). We get

$$\frac{dy}{dx} = y\left[\frac{6x^2}{x^2 + 5} + 3\ln(x^2 + 5)\right]$$

$$= (x^2 + 5)^{3x}\left[\frac{6x^2}{x^2 + 5} + 3\ln(x^2 + 5)\right]$$

■

The method of logarithmic differentiation is sometimes helpful in finding the derivative of complicated expressions which involve products and quotients of powers. We illustrate.

## EXAMPLE D

Find $dy/dx$ if

$$y = \frac{(2x - 3)^4(3x + 5)^5}{(5x + 4)^6}$$

**Solution** First, we take logarithms of both sides and simplify using the properties of logarithms; then we take the derivative of both sides with respect to $x$; finally, we solve for $dy/dx$.

$$\ln y = 4\ln(2x - 3) + 5\ln(3x + 5) - 6\ln(5x + 4)$$

$$\frac{1}{y}\frac{dy}{dx} = 4\frac{2}{2x - 3} + 5\frac{3}{3x + 5} - 6\frac{5}{5x + 4}$$

$$\frac{dy}{dx} = \frac{(2x - 3)^4(3x + 5)^5}{(5x + 4)^6}\left(\frac{8}{2x - 3} + \frac{15}{3x + 5} - \frac{30}{5x + 4}\right)$$

■

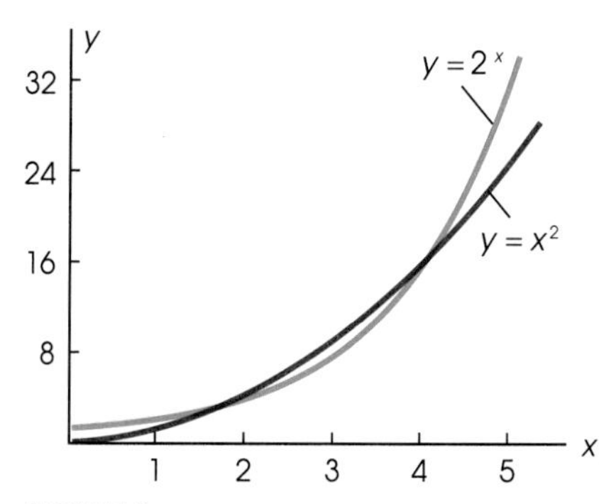


**FIGURE 8**

## Exponential Functions Versus Power Functions

Once again, we emphasize the difference between the exponential function $f(x) = 2^x$ and the power function $g(x) = x^2$. Both functions are increasing for $x > 0$ (Figure 8), but eventually $f(x)$ grows incomparably faster. When trying to model a real-life situation, one often is faced with choosing between an exponential model and a power model. How can one make this decision in a sensible manner?

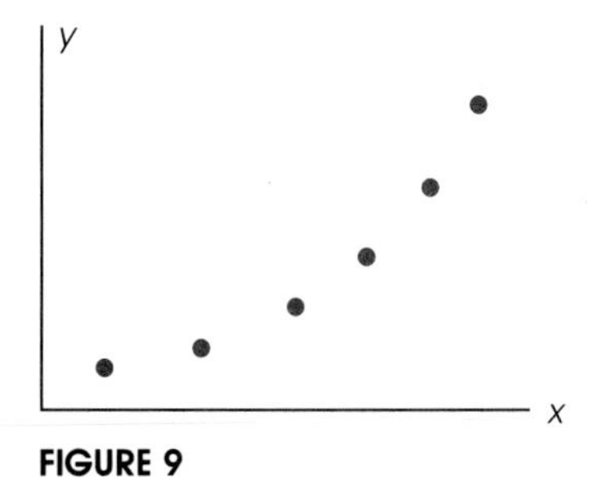

FIGURE 9

Let us state the problem more generally. Suppose you have a set of $(x, y)$ data where $y$ increases rather rapidly as $x$ increases as in Figure 9. You would like to fit a curve to these data. Consider for $c > 0$ and $b > 1$ the following two models.

| Exponential model | Power model |
|---|---|
| $y = cb^x$ | $y = cx^b$ |

Is there a definite way of distinguishing between them and of choosing $c$ and $b$? Yes, there is, though we shall not go into all the details. The basic idea is to take logarithms to obtain, respectively,

$$\ln y = \ln c + x \ln b \qquad \ln y = \ln c + b \ln x$$

$$y = C + Bx \qquad Y = C + bX$$

Here we have made the substitutions $Y = \ln y$, $C = \ln c$, $B = \ln b$, and $X = \ln x$. The important thing to notice is that in both cases, we get a linear equation (with a straight-line graph). But in the first case, we get the straight-line graph when we plot $\ln y$ against $x$; in the second, we get the straight line when we plot $\ln y$ against $\ln x$. Two kinds of graph paper are available to make this procedure easy. One is called semilog graph paper and has a log scale on the vertical axis; the other is called log-log graph paper and has logarithmic scales on both axes. If plotting data on semilog paper leads to a straight line, the exponential model is appropriate; on the other hand, if plotting data on log-log paper gives a straight line, the power model is indicated.

## Problem Set 5.4

C *In Problems 1–8, first write each expression in terms of e and ln. Then evaluate the result on your calculator. See Example A.*

1. $\log_3 29$
2. $\log_5 14$
3. $4^{2.63}$
4. $3^{1.98}$
5. $\log_6 0.432$
6. $\log_5 0.098$
7. $(3.2)^{2.4}$
8. $(5.06)^{0.31}$

C 9. Write $(4.71)^t$ in the form $e^{kt}$.
C 10. Write $(0.987)^t$ in the form $e^{kt}$.
C 11. Write $(8.12)^{-2t}$ in the form $e^{kt}$.
C 12. Write $\log_3(t)$ in the form $k \ln t$.
C 13. Write $\log_6(t^3)$ in the form $k \ln t$.
C 14. Write $\log_\pi(t^{\sqrt{2}})$ in the form $k \ln t$.

*Find dy/dx in Problems 15–28 (see Example B).*

15. $y = \log_5 x$
16. $y = \log_9 x$
17. $y = 4^x$
18. $y = 10^x$
19. $y = \log_3 (2x + 1)$
20. $y = \log_5 (1 - 3x)$
21. $y = 6^{4-5x^2}$
22. $y = 3^{x^2+2x}$
23. $y = 3 \log_4 (3x^2 + 1)$
24. $y = 10 \log_2 (x^3 - 2x^2)$
25. $y = 5^{4e^x+3}$
26. $y = 2^{x+e^{1/x}}$
27. $y = 2^x 3^{-2x}$
28. $y = 3^{2x} \log_2 (3x)$

*In Problems 29–36, use logarithmic differentiation to find dy/dx (see Example C).*

29. $y = (2x + 1)^x$
30. $y = (3x + 2)^x$
31. $y = (x^2 + 5x)^{2x}$
32. $y = (3x - 2x^2)^{3x}$
33. $y = (5x)^{x^2-x^{-1}}$
34. $y = (2x + 4)^{4e^{-x}}$
35. $y = (e^x + 4)^{\ln x}$
36. $y = (3 \ln x + 2)^{e^x}$

*Use logarithmic differentiation to find dy/dx in Problems 37–44 (see Example D).*

37. $y = x^{10}(2x - 1)^6$
38. $y = (3x + 1)^4/x^8$

39. $y = (x + 1)^3(x + 2)^4/(x + 3)^5$
40. $y = (2x - 1)^5(x + 3)^6(2 - x)^3$
41. $y = (x^2 + 1)^8(2x + 5)^5$
42. $y = (3x + 5)^4/(x^2 - x + 9)^3$
43. $y = \sqrt{x^2 + 2}(x + 2)^4/\sqrt[3]{3x - 5}$
44. $y = (3x + 2)^{5/3}(x - 1)^9/e^{2x}$

---

C 45. Calculate $\log_b 100$ for $b = 2$, $e$, 10, and 200.

C 46. Write each of the following in the form $e^{kt}$: $10^{-t}$, $3^{-t}$, $1.5^t$, and $9^t$.

47. Find $dy/dx$.
(a) $\log_3 (\pi^2 + e^3)$ (b) $\log_3 (e^{2x} + x)$
(c) $y = 5^{3e^{2x}}$ (d) $y = (4 + 3x^2)^{x^2}$

48. Find the equation of the tangent line to the curve $y = 4^{x^2-3}$ at the point where $x = 2$.

49. Use logarithmic differentiation to find $f'(0)$, given that

$$f(x) = (\sqrt{x + 1} + 1)^4[2(x + 1)^{-1} - 1]^{10}$$

50. Show that $f(x) = \log_5(\log_2 x)$ can be rewritten as $f(x) = [\ln (\ln x) - \ln (\ln 2)]/\ln 5$ and use this to find $f(e)$ and $f'(e)$.

51. Transform the equation $xy^2 = e^{-5}$ to a linear equation by an appropriate change in variables. Then find the slope and the "y-intercept" of the resulting equation.

C 52. For the following data sets, decide whether $y = cb^x$ or $y = cx^b$ is the better model. One way to do this (suggested in the text) is to plot $\ln y$ against $x$ and $\ln y$ against $\ln x$ to see which more closely approximates a line.

(a)

| $x$ | 1 | 2 | 3 | 4 |
|---|---|---|---|---|
| $y$ | 96 | 145 | 216 | 325 |

(b)

| $x$ | 0 | 1 | 2 | 4 |
|---|---|---|---|---|
| $y$ | 243 | 162 | 108 | 48 |

(c)

| $x$ | 1 | 2 | 3 | 5 |
|---|---|---|---|---|
| $y$ | 12 | 190 | 975 | 7490 |

(d)

| $x$ | 1 | 4 | 9 |
|---|---|---|---|
| $y$ | 16 | 128 | 432 |

C 53. Suppose that the number of bacteria in a certain culture $t$ hours from now will be $400(1.6)^t$.
(a) Determine the bacteria count 4.5 hours from now.
(b) When will the count reach 4000?

54. The magnitude $R$ of an earthquake on the *Richter scale* is given by $R = \log_{10} (I/I_0)$, where $I$ is the actual intensity of the earthquake and $I_0$ is a standard intensity used for comparison. A recent quake measured 4.2 on this scale, whereas the 1906 San Francisco quake measured 8.2. How much more intense was the San Francisco quake than the recent one?

55. The Weber-Fechner law in psychology says that people react to external stimuli on a logarithmic scale. Because of this, the loudness $L$ of sounds is usually measured in decibels using the formula $L = 10 \log_{10} (I/I_0)$. Here $I$ is the actual intensity of the sound in watts per square meter and $I_0$ is a constant (the least intense sound that can be heard). A certain conversation seemed to be five times louder than a whisper. If so, how did the actual intensity of the sound compare with that of a whisper?

## CHAPTER 5 REVIEW PROBLEM SET

1. Evaluate each of the following without the help of a calculator.
(a) $9^{-3/2}$ (b) $(5^{-3/2}5^{2/3}5^{3/4})^{12}$
(c) $\log_5 125$ (d) $\log_8 2$
(e) $\log_8 16$ (f) $\log_5 5^{\pi+1}$

2. Evaluate each of the following without using a calculator.
(a) $8^{-4/3}$ (b) $(3^{-5/2}3^{4/3}3^{11/12})^{12}$
(c) $\log_{10} 10{,}000$ (d) $\log_{16} 4$
(e) $\log_4 4^{3.2}$ (f) $\log_4 32$

C 3. Use your calculator to evaluate each of the following.
(a) $(3\pi)^{2.13}$ (b) $(4.11)^{2.1}/(1.79)^{1.4}$
(c) $\ln [(14.2)^3 - 962]$ (d) $e^{\pi^2/2}$
(e) $(e^{-2.5} + 2 \ln 1.08) \ln 142$
(f) $\log_7 72.45$

C 4. Evaluate using a calculator.
(a) $(2e + 1)^5$ (b) $e^{\sqrt{3}+2}$
(c) $\ln(6.17^2 + 12.3^3)$ (d) $e^{(3-\sqrt{2})^2} \ln 13.68$
(e) $(14.2^2 - 5 \ln 14.6)/\sqrt[4]{98.1}$
(f) $\log_8 (e^3 + 192.4)$

C 5. Sketch the graphs of $f(x) = e^{-0.1x}$ and $g(x) = e^{0.4x}$ using the same axes.

C 6. Sketch the graphs of $f(x) = e^{0.6x}$ and $g(x) = e^{-0.3x}$ using the same axes.

7. If $\log_a 2 = 0.356$, $\log_a 3 = 0.565$, and $\log_a 5 = 0.827$, find each value.
(a) $\log_a 40$ (b) $\log_a \frac{12}{25}$ (c) $\log_a (\sqrt{15}/a)$

8. If $\log_a 2 = 0.431$, $\log_a 3 = 0.683$, and $\log_a 7 = 1.209$, find each value.
(a) $\log_a 54$ (b) $\log_a (\sqrt{7}/6)$ (c) $\log_a (1.5a\sqrt{a})$

9. Express $2 \ln x + 3 \ln 5 - 5 \ln y$ as a single logarithm.

10. Express $3 \ln x - 2 \ln y + \frac{1}{2} \ln 15$ as a single logarithm.

C 11. Solve for $x$.
(a) $\ln (3x + 1) = 4$ (b) $e^{2+3x} = \pi^x$
(c) $\ln (x^2 - x) - \ln 3 = 1$ (d) $e^{x^2-3} = \ln 140$

12. Solve for $x$.
(a) $\ln(5x - 2) = 3$ (b) $3^{2-x} = \pi^2$
(c) $\ln(2x + 1) - \ln(3 + 4x) = 2$
(d) $e^{2x^2+1} = \ln 2000$

13. Find $f'(x)$ in each case.
(a) $f(x) = 2e^{x^2-3}$ (b) $f(x) = 5 \ln (4 + 3/x)$
(c) $f(x) = e^{x/(2x+1)}$
(d) $f(x) = \ln [(x^2 + 3)^2 + 1]$
(e) $f(x) = (x^2 + 4) \ln (2x + 1)$
(f) $f(x) = e^{2x}/(x + 2)$
(g) $f(x) = (3 + 2x \ln x)^4$
(h) $f(x) = (xe^{-2x} - 1)^{3/2}$

14. Find $f'(x)$ in each case.
(a) $f(x) = \dfrac{1}{2}e^{x^2-2x+10}$ (b) $f(x) = 12 \ln\left(x - \dfrac{4}{x}\right)$
(c) $f(x) = e^{(3x-2)(3x+2)}$
(d) $f(x) = \ln(\sqrt{x^2 + 5} - 3)$
(e) $f(x) = (2x - 1)^5 \ln(4 - 2x)$
(f) $f(x) = e^{2+\sqrt{x}}/(4 - 3x)$
(g) $f(x) = (x^2 + 2 \ln x)^5$
(h) $f(x) = (x^2e^{4x} + 4)^{5/2}$

15. Consider $f(x) = (x - 2)e^x$.
(a) Where is $f$ increasing? Decreasing?
(b) Where is the graph of $f$ concave up? Concave down?
(c) Determine the minimum value of $f$.
(d) Sketch the graph of $f$.

16. Consider $g(x) = (x - 3)e^{-x}$.
(a) Where is $g$ increasing? Decreasing?
(b) Where is the graph of $g$ concave up? Concave down?
(c) Determine the maximum value of $g$.
(d) Sketch the graph of $g$.

17. Where does the graph of $f(x) = e^{-x^2/2}$ change the direction of its concavity?

18. Where does the graph of $y = \ln(x^2 + 4)$ change its concavity?

C 19. To test retention of European history, a class was given equivalent forms of a final exam at 1-month intervals after the class had ended. A careful study of the results showed that the average exam score $S(t)$ at the end of $t$ months could be modeled by the formula

$$S(t) = 76 - 16 \ln (t + 1).$$

(a) Determine the average score at the end of 4 months.
(b) At what rate was the average score changing at the end of 4 months?
(c) When did the average score first fall below 40?

20. The number $N$ of bacteria in a certain culture $t$ hours from now is predicted to be

$$N = 1000 + 20{,}000 \ln(1 + 0.8t)$$

(a) What is the size of the present population?
(b) At what rate will the population be increasing 10 hours from now?
(c) When will the population reach 16,000?

C 21. Write $(2\pi)^t$ in the form $e^{kt}$.

C 22. Write $(\sqrt{3} + 2)^t$ in the form $e^{kt}$.

C 23. Write $\log_5 t$ in the form $k \ln t$.

C 24. Write $\log_8 t^3$ in the form $k \ln t$.

25. Find $dy/dx$.
(a) $y = 10^{4\sqrt{x}}$ (b) $y = 6 \log_4(3x^2 + 5)$

26. Find $du/dt$.
(a) $u = 7^{\sqrt{4+t^2}}$ (b) $u = \log_5(t^2 + 2t - 4)$

27. Find $f'(1)$ using logarithmic differentiation.
(a) $f(x) = (x^2 + 1)^{x^3+1}$
(b) $f(x) = (3x^2 - 2)^6(2x - 1)^8/\sqrt{2x + 1}$

28. Find $g'(2)$ using logarithmic differentiation.
(a) $g(x) = (2x - 3)^{\sqrt{2x-3}}$
(b) $g(x) = (3x^2 - 11)^5(3 - x)^4/\sqrt{x + 7}$

29. Transform $y = 4e^{-3x+4}$ to a linear equation by an appropriate change in variables. Then find the slope and "y-intercept" of the resulting equation.

30. Transform $y = 3e^{\pi x+2}$ to a linear equation by a suitable change of variables. Then find the slope and "y-intercept" of the resulting equation.

31. Find the equation of the tangent line to the curve $x \ln y + y^2 = 5x - 9$ at the point (2, 1).

32. Find the equation of the tangent line to the curve

$$y^2 \ln x + 2ye^{x-1} = 3y - 3$$

at the point (1, 3).

33. A point is moving along the curve $y = 6e^{x^2/4}$ in such a way that the horizontal component $dx/dt$ of its velocity is always 3. Find $dy/dt$ at the instant when $x = 2$.

34. A point is moving along the curve $y = 3e^{2(x-1)}$ in such a way that the horizontal component of its velocity is always 4.
   (a) Find $dy/dt$ at the instant when $x = 1$.
   (b) Find the rate (with respect to time) at which the distance from the origin is increasing when $x = 1$.

35. Evaluate

$$\lim_{h\to 0} \frac{[\ln(5 + h)]^2 - [\ln 5]^2}{h}$$

by recognizing it as a value of a certain derivative.

36. Evaluate $\lim_{h\to 0} \frac{e^{3(2+h)^2} - e^{12}}{h}$ by recognizing it as a value of a certain derivative.

37. Let $f(x) = x^\pi - \pi^x$.
   (a) Is $f(e)$ positive or negative?
   (b) Is $f'(e)$ positive or negative?

38. Let $f(x) = \ln(\ln(\ln x))$.
   (a) Determine the domain of $f$.
   (b) Evaluate $f[e^{(e^e)}]$.
   (c) Evaluate $f'(e^4)$.

39. Let $f(x) = (\ln x)^2 - \ln x$ on $(0, \infty)$.
   (a) Where is $f(x) > 0$?
   (b) Where is $f'(x) > 0$?
   (c) Where is $f''(x) > 0$?

40. Let $f(x) = \ln x - \sqrt{\ln x}$ on $(1, \infty)$.
   (a) Where is $f(x) > 0$?
   (b) Where is $f'(x) > 0$?
   (c) Evaluate $f''(e)$.

*Differential equations provide mathematical models for complicated physical phenomena such as the flow of particles past airplane wings. Solving such equations is a tremendous challenge and often can only be accomplished by numerical methods using supercomputers.*

# 6

# Applications of Exponential and Logarithmic Functions

Ol'ga Arsen'evna Oleinik
(1925– )

Ol'ga Arsen'evna Oleinik, a superb Soviet mathematician and head of the Department of Differential Equations at Moscow State University, was born in 1925 near Kiev, Russia. After completing her undergraduate work with honors in mathematics, she pursued graduate study under I. G. Petrovskii and in 1954 successfully defended her doctoral dissertation. It was on a topic in differential equations, a subject in which she had already obtained important new results as a student and to which she was to be a major contributor throughout her career.

Her rise through the ranks at Moscow State University has been rapid, culminating in her appointment as head of the Department of Differential Equations in 1973. Besides administering this department, she regularly teaches a course on differential equations in which she uses and constantly revises her own textbook, has supervised over 40 dissertations, serves as editor-in-chief of *Transactions of the Moscow Mathematical Society*, is a member of the editorial board of several other journals, participates in international congresses and conferences, and is invited to give lectures at prestigious institutions around the world. But what is still more amazing is that by the time she celebrated her sixtieth birthday in 1985, she had published 233 research papers. One can only wonder at such energy, such productivity, such scholarship.

This is not the place to attempt a description of her research except to say that it covers three main areas: differential equations (a subject introduced in this chapter), algebraic geometry, and mathematical physics. In terms of applications, she has studied filtration of fluids in porous media, heat transfer, boundary layer problems, inelasticity theory, and the mechanics of inhomogeneous media.

Oleinik's work has been recognized in many ways. She has won the Chebotarev and the Lomonosov prizes in her own country, as well as the medals of the College de France and of Charles University. She has been elected to several foreign mathematical societies and holds an honorary doctorate from the University of Rome. Her contributions to encouraging world peace resulted in a medal from the World Peace Council. She is said to be a person of integrity, kindliness, and personal charm. Ol'ga Arsen'evna Oleinik is a truly remarkable woman.

## 6.1 EXPONENTIAL GROWTH AND DECAY

We are interested in a whole class of real-life problems where we are given the time rate of change in a quantity $y$ and want to determine explicitly how $y$ depends on time. In particular in this section, we treat the case where the time rate of change in $y$ is proportional to the amount present, that is, where $dy/dt = ky$, $k$ being a constant. This is an example of a **differential equation,** an equation involving derivatives in which the unknown is a function. An important task is to solve such equations, that is, to determine the unknown function. Methods for solving differential equations are discussed in Section 7.4. Here we merely state the result we need and apply it in practical problems.

> **THEOREM**
>
> The solution $y = f(t)$ to the differential equation
>
> $$\frac{dy}{dt} = ky$$
>
> subject to $y = y_0$ at $t = 0$ is
>
> $$y = y_0 e^{kt}$$

It is easy to verify that $y = y_0e^{kt}$ is a solution to the given differential equation, a matter we ask you to investigate in Problems 1 and 2. What is not so clear is how this solution was found or why it is the only solution. These are matters that will be clarified in Section 7.4.

The significance of the theorem is that any physical phenomenon satisfying $dy/dt = ky$ will have an exponential growth (or decay) curve. It will grow exponentially if the constant $k$ is positive; it will decay exponentially if $k$ is negative (Figure 1).

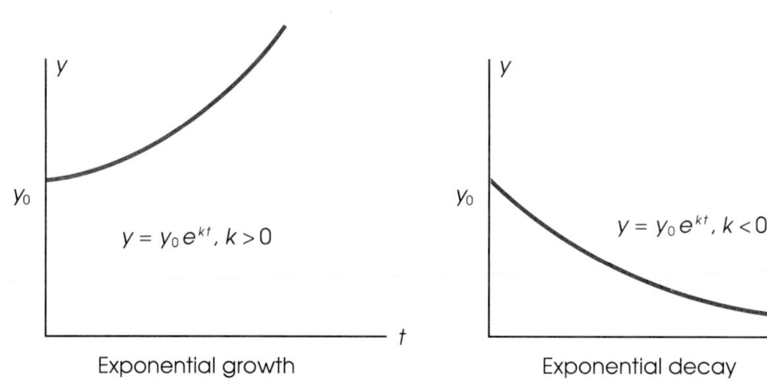


FIGURE 1

**EXAMPLE A**

Solve the differential equation $dy/dt = 0.04y$, given that $y_0 = 15$, and determine $y$ when $t = 6$.

**Solution** We simply appeal to the theorem. The solution is

$$y = 15e^{0.04t}$$

When $t = 6$,

$$y = 15e^{0.04(6)} \approx 19.0687$$

■

## Biological Growth

Consider a biological population, such as a bacteria culture, which is growing in the presence of adequate food supplies. It seems reasonable to conjecture that the rate of growth of such a population is proportional to the size of the population. (The number of offspring per time period is proportional to the size of the population.) Experience confirms this conjecture.

**EXAMPLE B**

The cells in a certain bacterial culture divide on the average every 2.5 hours. If there are 500 cells initially, how many cells would we expect to find after 12 hours?

**Solution** Let $y$ denote the number of cells present at time $t$. The appropriate differential equation is $dy/dt = ky$ with $y_0 = 500$, and this has the solution

$$y = 500e^{kt}$$

To determine $k$, we note that the population will double in size every 2.5 hours, and so

$$1000 = 500e^{k(2.5)}$$

or on dividing by 500,

$$2 = e^{2.5k}$$

Since the unknown $k$ is in the exponent, we solve by taking natural logarithms of both sides.

$$\ln 2 = 2.5k$$

$$k = (\ln 2)/2.5 \approx 0.277259$$

We conclude that

$$y = 500e^{0.277259t}$$

a result valid for all $t$. Finding $y$ when $t = 12$ is simply a matter of substituting $t = 12$ in this equation.

POINT OF INTEREST

The time required for a quantity to double in size is called the **doubling time.** It is a characteristic of exponential growth that if a quantity doubles in an initial time interval of length $T$, it will double in any interval of length $T$.

$$y = 500e^{0.277259(12)} \approx 13{,}929$$

It would probably be better to report an answer of 13,900 or even 14,000, since our differential equation is surely only an approximate model for the real situation. ■

The differential equation $dy/dt = ky$ has been used in describing the world population of human beings. History indicates that the population $y$ of the world (during the last 200 years) has been growing approximately according to this model with growth constant $k = 0.0198$.

### EXAMPLE C

The population of the world was about 4 billion on January 1, 1975. Assuming that the exponential growth model with $k = 0.0198$ is appropriate, when will the world's population reach 10 billion?

**Solution** Let $t$ denote the number of years after January 1, 1975, and $y$ the population in billions at time $t$. Then $y_0 = 4$, and so

$$y = 4e^{0.0198t}$$

We wish to determine when $y = 10$; that is, we wish to solve the equation

$$10 = 4e^{0.0198t}$$

for $t$. Dividing by 4 and taking natural logarithms yields

$$\ln 2.5 = 0.0198t$$

$$t = (\ln 2.5)/0.0198 \approx 46.28$$

According to this model, the world's population will reach 10 billion in $1975 + 46.28$ or during the year 2021. ■

It seems clear that the world's population cannot continue to grow indefinitely according to the model above. Food supplies cannot continue to keep up with this kind of population growth. We will consider other and perhaps more realistic models for human population growth in the next section.

> **POINT OF INTEREST**
>
> A Famous Quotation
>
> "I said that population, when unchecked, increased in a geometrical ratio [exponentially], and subsistence for man in an arithmetical ratio [linearly]."
>
> Thomas Malthus, 1798

## Radioactive Decay

It is an experimental fact that radioactive elements (uranium, radium, and so on) decay according to the differential equation $dy/dt = ky$ with $k$ being a negative constant associated with the corresponding element. The decay constant $k$ is usually not specified. Rather, scientists give the **half-life** of the element, that is, the time required for a given amount of the element to decay to half its size.

**EXAMPLE D**

Radium has a half-life of 1690 years. Determine the decay constant $k$ for radium and then calculate how much of 10 grams of radium will be left after 2400 years.

**Solution** From the given information, we conclude that 2 units of radium will decay to 1 unit in 1690 years. Thus,

$$1 = 2e^{k(1690)}$$

or

$$0.5 = e^{k(1690)}$$

Following our standard procedure, we take natural logarithms of both sides and solve for $k$.

$$\ln 0.5 = k(1690)$$

$$k = (\ln 0.5)/1690 = -0.000410146$$

We conclude that the amount $y$ of radium present after $t$ years is $y = y_0 e^{-0.000410146t}$, where $y_0$ is the initial amount. When $t = 2400$, then 10 grams of radium will have decayed to

$$y = 10e^{-0.000410146(2400)} \approx 3.737 \text{ grams}$$

■

## Carbon Dating

The ratio of the radioactive isotope carbon 14 to the stable isotope carbon 12 in the atmosphere remains constant, since the amount of carbon 14 that decays is exactly balanced by the formation of new carbon 14 as a result of cosmic rays hitting the atmosphere. All living things contain these two isotopes of carbon in the same ratio as in the atmosphere. However, when a plant or animal dies, it ceases to absorb carbon from the atmosphere, and so the proportion of carbon 14 in dead organic material decreases over time as a result of radioactive decay. This allows scientists to date old organic objects, as we now illustrate.

**EXAMPLE E**

Human hair from a grave in Africa proved to have only 74% of the carbon 14 found in living tissue. Given that carbon 14 has a half-life of 5570 years, determine when the body died.

**Solution** We first determine the decay constant $k$ from the equation

$$0.5 = e^{k(5570)}$$

$$\ln 0.5 = k(5570)$$

$$k = (\ln 0.5)/5570 = -0.000124443$$

Thus, if $y$ denotes the fraction of the original amount of carbon 14 present at time $t$,

$$y = e^{-0.000124443t}$$

and so we must solve the equation

$$0.74 = e^{-0.000124443t}$$

for $t$. This we do by first taking natural logarithms.

$$\ln 0.74 = -0.000124443t$$

$$t = (\ln 0.74)/(-0.000124443) \approx 2420$$

The body died about 2420 years ago. ■

## Problem Set 6.1

1. In each case, differentiate the given $y$ to show that it satisfies the accompanying differential equation.
   (a) $y = 5e^{0.03t}$; $dy/dt = 0.03y$
   (b) $y = 20e^{-0.6t}$; $dy/dt = -0.6y$
2. Differentiate $y = y_0e^{kt}$ to show that it satisfies the differential equation $dy/dt = ky$.

C *In Problems 3–8, solve the given differential equation with the given value of $y_0$. Then determine the value of $y$ when $t = 8$.*

3. $dy/dt = 0.8y$; $y_0 = 16$
4. $dy/dt = 0.09y$; $y_0 = 12$
5. $dy/dt = -0.05y$; $y_0 = 20$
6. $dy/dt = -0.006y$; $y_0 = 40$
7. $dy/dt = 0.193y$; $y_0 = 3.21$
8. $dy/dt = -2.1y$; $y_0 = 612$
9. A colony of bacteria has 100 members now and will double every 70 minutes. How many members will it have
   (a) after 210 minutes?
   C (b) after $t$ minutes?
   C (c) after 120 minutes?
10. A population of bacteria numbers 600 now and will double every 3.25 hours. How many bacteria will there be
    (a) after 13 hours?
    C (b) after $t$ hours?
    C (c) after 9.7 hours?

C 11. An exponentially growing culture had 16,000 cells initially and grew to 60,000 cells in 12 hours.
   (a) How many cells were there after $t$ hours?
   (b) How long did it take to reach 80,000 cells?

C 12. Suppose that initially there were 800 bacteria in a certain culture and that now, after 6 hours, there are 3500.
   (a) How many bacteria will be present 4 hours from now?
   (b) How many hours from now will the number of bacteria reach 10,000?

C 13. Assuming that the growth model for world population given in Example C is correct ($k = 0.0198$), how long does it take the world's population to double?

C 14. Population figures for the world are unreliable, and some statisticians give a considerably lower figure for $k$ than that reported in Example C. How long does it take the world's population to double if $k = 0.0183$?

C 15. The population of a certain country was 10.5 million on January 1, 1978, and 14.6 million 10 years later. Assume that the exponential growth model is appropriate.
   (a) Find the formula for the population $t$ years after January 1, 1978.
   (b) What will the population be on January 1, 2000?

C 16. The population of the world was 4 billion on January 1, 1975, and a newspaper reported it to be 5 billion on April 1, 1985 (approximate figures). This information allows a determination of the value for $k$ in the exponential growth model. Find this $k$.

C 17. Uranium 238 has a half-life of 4.51 billion years. Determine the decay constant $k$ and then calculate how much of 20 grams of this element will be left after 1 billion years. See Example D.

C 18. The dangerously radioactive element, radon 220, has a half-life of 3.92 seconds. How much of 5 grams of radon 220 will be left after 2.5 seconds?

C 19. Polonium 214 disintegrates according to the differential equation $dy/dt = -4621y$, where $y$ is the number of grams present after $t$ seconds. Find the half-life of this element.

C 20. Uranium 237 satisfies the differential equation $dy/dt = -0.1022y$, where $y$ is the number of grams present after $t$ days. Find the half-life of this element.

C 21. A skeleton found in a farmer's field has only 62% of the amount of carbon 14 found in living human bones. How old is the skeleton? See Example E.

C 22. A fragment of wood found during the excavation of an old building site has 51% of the amount of carbon 14 found in living wood of the same type. About how many years ago was the building erected?

C 23. An anthropologist claimed that a human skull unearthed in southern Europe was 24,500 years old. If carbon dating was used to make this claim, what percentage of the amount of carbon 14 found in living human bones was in that skeleton?

C 24. When an ancient burial site was excavated, the amount of carbon 14 present in the skeletons ranged from 71 to 75% of that in living bones. About how many years had this burial site been used?

---

25. A certain population is growing exponentially so that it doubles every 25 years. If the population was 10,000 in 1980, what will it be in 2080?

26. A radioactive element decays with a half-life of 20 days. If I have 64 grams of this element now, how much will I have after 120 days?

27. A fossilized leaf contains 25% of the carbon 14 of a living leaf. How old is this fossil?

28. Twenty-five percent of a radioactive substance remained after 10 years. What is its half-life?

C 29. Solve the differential equation $dy/dt = ky$, given that $y = 10$ when $t = 3$ and $y = 24$ when $t = 4$. *Hint:* The solution is $y = y_0e^{kt}$, but you must determine both $y_0$ and $k$.

C 30. The rabbit population in a certain area is said to be growing exponentially. This population was estimated at 48,000 in 1975 and at 62,000 in 1985. How large will this population be in 2005?

C 31. To diagnose a certain disease, a tracer dye is injected into the pancreas. A healthy person will secrete 5% of the dye remaining in the pancreas each minute. Thirty minutes after injecting 0.3 gram of this dye into a patient, 0.1 gram remained. How much would have remained in a healthy individual?

C 32. Atmospheric pressure $p$ (in millibars) is related to altitude $h$ above sea level (in meters) by the differential equation $dp/dh = kp$. At sea level, $p = 1013.5$, and at $h = 1500$ meters, $p = 845.6$.
(a) What is the pressure at the top of Mount Everest which is 8848 meters above sea level?
(b) How high is Mount A if the pressure at its peak measures 627.3 millibars?

C 33. Colony A and colony B of bacteria grow according to the differential equations $dN/dt = 0.4N$ and $dN/dt = 1.2N$, respectively, where $t$ is measured in hours. Right now colony A is 10 times as large as colony B. When will the two colonies have the same population?

C 34. The wolf population in a certain state grew exponentially from 1950 to 1960 with growth constant $k = 0.035$ but then began a long exponential decline with decay constant $k = -0.015$. How many wolves did the state have in 1985 if there were 6400 in 1950?

## 6.2 OTHER GROWTH MODELS

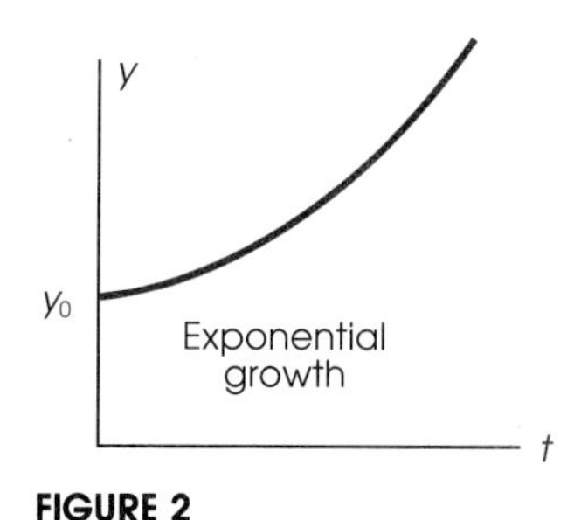

FIGURE 2

There are many situations where the exponential growth model in the last section is inappropriate. The equation $y = y_0e^{kt}$, with $k > 0$, describes processes which continue to grow—and at a faster and faster rate (Figure 2). Clearly, the world population of humans cannot continue to grow in this way indefinitely, since the amount of food the earth can produce is limited. There may be an upper bound $L$ beyond which the population of the world cannot grow.

To take another, simpler, example, consider what happens when a cold object is put in an oven whose temperature is $L$. The temperature of the object will increase, but not indefinitely. It will never exceed the upper bound $L$. Our task through most of this section is to consider growth models where there is an upper limit beyond which growth is impossible.

## Simple Bounded Growth

Many processes in nature can be modeled by the differential equation

$$\frac{dy}{dt} = k(L - y) \qquad (k > 0)$$

This says that the rate of increase in $y$ is large when $y$ is small, but that this rate tends to zero as $y$ gets closer and closer to $L$. We will show how to solve this differential equation in a later section (Section 7.4), but for now we ask you to accept the following result.

**THEOREM**
The solution to the differential equation

$$\frac{dy}{dt} = k(L - y)$$

subject to $y = y_0$ at $t = 0$, is

$$y = L - (L - y_0)e^{-kt} = L(1 - e^{-kt}) + y_0 e^{-kt}$$

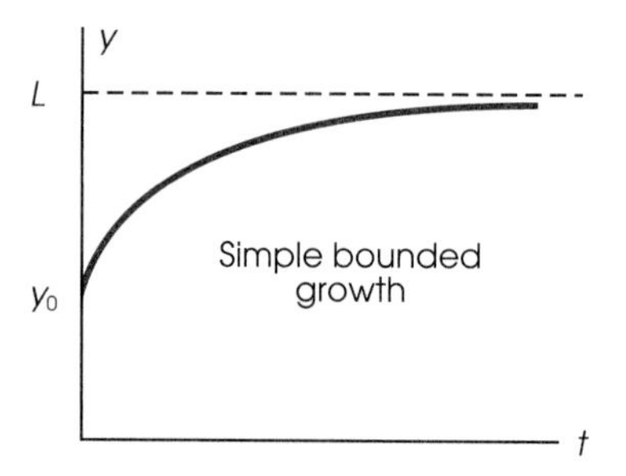


**FIGURE 3**

Note in the solution that as $t \to \infty$, $e^{-kt} \to 0$, and so $y \to L$. The graph of the solution is shown in Figure 3.

### EXAMPLE A

Solve the differential equation

$$\frac{dy}{dt} = 0.4(10 - y)$$

given that $y_0 = 3$. Then calculate $y$ when $t = 6$.

**Solution** Since $k = 0.4$, $L = 10$, and $y_0 = 3$, the theorem says that

$$y = 10 - (10 - 3)e^{-0.4t} = 10 - 7e^{-0.4t}$$

At $t = 6$,

$$y = 10 - 7e^{-0.4(6)} \approx 9.36$$

■

Sociologists have found that important news tends to be diffused in a population according to the model above. At first the news spreads very rapidly, then more and more slowly as the number of people who have heard it nears the total population.

## EXAMPLE B (SPREAD OF INFORMATION)

The news that the mayor of a certain city had been killed was announced at noon, and in 3 hours it was thought that 75% of the people in the city had heard it. How long will it take for 99% of the people to hear it?

**Solution** Let $y$ denote the percentage (written as a decimal) of the people who hear the news within $t$ hours and note that $y_0 = 0$ and that the bound $L$ is 1 (corresponding to 100%). Using these results and appealing to the theorem (second form of the solution), we obtain

$$y = 1 - e^{-kt}$$

We can determine $k$ from the fact that $y = 0.75$ at $t = 3$; that is,

$$0.75 = 1 - e^{-k(3)}$$

or

$$e^{-3k} = 0.25$$

Taking natural logarithms yields

$$\begin{aligned} -3k &= \ln 0.25 \\ k &= (\ln 0.25)/(-3) \approx 0.462098 \end{aligned}$$

The value for $y$ at any time $t$ is thus

$$y = 1 - e^{-0.462098t}$$

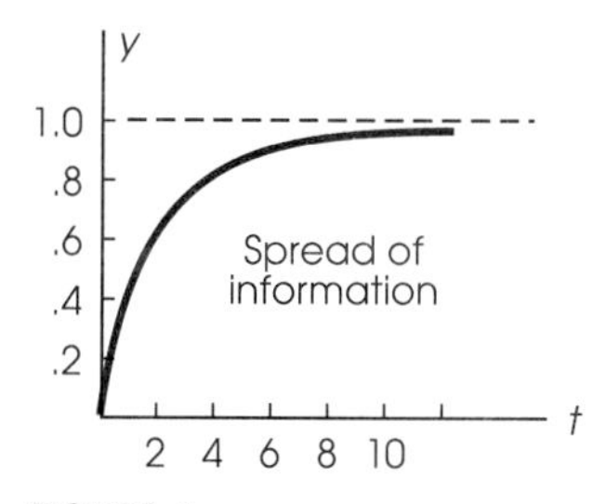


FIGURE 4

an equation graphed in Figure 4. We want to know $t$ when $y = 0.99$; that is, we want to solve

$$0.99 = 1 - e^{-0.462098t}$$

for $t$. Simplifying and taking logarithms gives

$$\begin{aligned} e^{-0.462098t} &= 0.01 \\ -0.462098t &= \ln 0.01 \\ t &= (\ln 0.01)/(-0.462098) \approx 9.97 \end{aligned}$$

In just under 10 hours, 99% of the people in the city will have heard the news. ■

## EXAMPLE C (VON BERTALANFFY FISH GROWTH MODEL)

The classical model for fish growth assumes that the rate of change in fish length is proportional to the difference between a theoretical maximum length $L$ and the actual length $y$. A certain variety of fish is hatched at length 0.5 inches and never grows beyond 12 inches. If a typical such fish has a length of 6 inches in 20 weeks, how long will it be after 50 weeks?

**Solution** If $y$ denotes length at time $t$, then $dy/dt = k(12 - y)$ and $y_0 = 0.5$. Thus, by our theorem,

$$y = 12 - (12 - 0.5)e^{-kt} = 12 - 11.5e^{-kt}$$

Substituting $y = 6$ when $t = 20$ gives

$$6 = 12 - 11.5e^{-k(20)}$$

or

$$\begin{aligned} 11.5e^{-20k} &= 6 \\ e^{-20k} &= 6/11.5 \approx 0.521739 \\ -20k &= \ln 0.521739 \\ k &= (\ln 0.521739)/(-20) \approx 0.032529 \end{aligned}$$

We conclude that

$$y = 12 - 11.5e^{-0.032529t}$$

At $t = 50$,

$$y = 12 - 11.5e^{-0.032529(50)} \approx 9.74 \text{ inches}$$

■

## Logistic Growth

If the food supply (or space) is limited, a better model for population growth than either exponential growth or bounded growth is a compromise between them called **logistic growth.** We suppose that the rate of change in $y$ is proportional to both $y$ and $L - y$; that is,

$$\frac{dy}{dt} = ky(L - y) \qquad (k > 0)$$

A population growing according to this model will appear to exhibit exponential growth initially (when $y$ is small and food is plentiful) and then bounded growth for large $t$ (when $y$ nears $L$). The relevant theorem, proved in Section 7.4, is

**THEOREM**

The solution to the differential equation

$$\frac{dy}{dt} = ky(L - y)$$

subject to $y = y_0$ at $t = 0$, is

$$\begin{aligned} y &= \frac{Ly_0}{y_0 + (L - y_0)e^{-Lkt}} \\ &= \frac{L}{1 + Be^{-Lkt}} \qquad [B = (L - y_0)/y_0] \end{aligned}$$

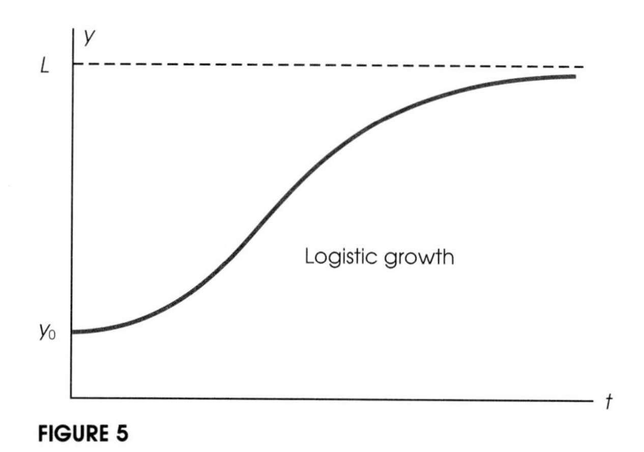


**FIGURE 5**

The graph of a typical logistic growth curve is the S-shaped curve shown in Figure 5.

## EXAMPLE D

A fruit fly population, initially numbering 20, is growing in a laboratory container. After 3 days, there are 56 flies, and experience indicates that not more than 240 flies can live in such a confined space. Assuming the logistic growth model is appropriate, how many flies can we expect after 10 days?

**Solution** Let $y$ denote the number of fruit flies at time $t$. Then, $y_0 = 20$ and $L = 240$, so $B = (240 - 20)/20 = 11$. Thus,

$$y = \frac{L}{1 + Be^{-Lkt}} = \frac{240}{1 + 11e^{-240kt}}$$

Substituting $y = 56$ when $t = 3$ gives

$$56 = \frac{240}{1 + 11e^{-720k}}$$

or

$$\begin{aligned} 1 + 11e^{-720k} &= 240/56 = 4.28571 \\ e^{-720k} &= 3.28571/11 = 0.29870 \\ -720k &= \ln 0.29870 \\ k &= (\ln 0.29870)/(-720) = 0.0016782 \end{aligned}$$

Thus,

$$y = \frac{240}{1 + 11e^{-240(0.0016782)t}} = \frac{240}{1 + 11e^{-0.40277t}}$$

When we substitute $t = 10$ in the last equation, we get $y = 200.67$. There will be approximately 200 flies after 10 days. ■

## C Problem Set 6.2

*In Problems 1–4, solve the given differential equation with the specified initial value $y_0$. Then calculate $y$ for the indicated value of $t$.*

1. $dy/dt = 0.09(8 - y)$; $y_0 = 2$; $t = 6$
2. $dy/dt = 0.3(10 - y)$; $y_0 = 4$; $t = 8$
3. $dy/dt = 0.45(20 - y)$; $y_0 = 12$; $t = 4$
4. $dy/dt = 0.08(25 - t)$; $y_0 = 12$; $t = 5$
5. It was announced on the 9:00 A.M. news that movie star Golden Eddie was coming to visit his home town of Blooming Meadow. By 12:00 noon of that day, 60% of the people in the town had heard the exciting news. See Example B.
   (a) What percentage of the townspeople had heard the news by 3:00 P.M.?
   (b) How long did it take for 98% of the people to hear it?
6. Central Auto Parts recently began an advertising campaign on a local TV station offering huge discounts in a closeout sale. After 10 days, a poll showed that 40% of the viewership of 24,000 had seen the advertisement.
   (a) How many viewers will see the advertisement within the first $t$ days?
   (b) How long must the store continue the advertising to reach 80% of the station's viewers?
7. Newton's Law of Cooling says that an object at temperature $T$ when placed in a medium of temperature $T_m$ with $T_m < T$, will cool at a time rate that is proportional to the difference $T - T_m$.
   (a) Write the differential equation that models this situation.
   (b) A kettle of soup at a temperature of 180°F is placed on a table in a room where the temperature is 70°F. If the temperature drops to 140°F after 10 minutes, what will its temperature be after another 10 minutes?
8. A ham at room temperature 27°C was placed in a refrigerator set at 3°C. In 20 minutes, the temperature of the ham dropped to 15°C. How long will it take to drop to 5°C?
9. Let $W = W(t)$ denote the number of words per minute that an average person can type after $t$ weeks of practice. Suppose that $W$ satisfies the differential equation $dW/dt = 0.6(72 - W)$.
   (a) Find the formula for $W$.
   (b) What is the maximum typing speed an average person can hope to attain?
   (c) How long does it take an average person to achieve a typing speed of 68 words per minute?
10. Based on one model, the yield $y$ (in bushels) of corn per acre should increase in relation to the number $x$ (in pounds) of fertilizer used per acre according to the differential equation $dy/dx = k(240 - y)$. When no fertilizer was used, the yield was 80 bushels per acre, and when 100 pounds of fertilizer per acre was applied, the yield was 140 bushels per acre. What yield can one expect if 150 pounds per acre are applied?
11. A variety of bass, hatched at a length of 2 centimeters, never grows beyond a length of 100 centimeters. Assuming that the Von Bertalanffy model applies, how long will a bass be after 20 weeks that was 8 centimeters long after 7 weeks? See Example C.
12. A type of worm increases in length according to the Bertalanffy model and never grows beyond 20 centimeters long. A worm with an initial length of 1 centimeter grew to 6 centimeters in 1 month.
    (a) What was its length after 4 months?
    (b) How long did it take to reach 18 centimeters?
13. An initial bacteria population of 2000 was growing in a container that can support a maximum population of 36,000. After 1.5 hours, this population numbered 10,000. Assuming the logistic model for growth, how large was it after 5 hours? See Example D.
14. Suppose that the bacteria in a certain culture had an initial density of $2 \times 10^7$ cells per milliliter and that the limiting density is $2 \times 10^8$ cells per milliliter. If this density increased to $5 \times 10^7$ after 3 hours, what will it be after 6 hours? Assume the logistic growth model.
15. A very contagious type of Asian flu is spreading through a city of 120,000 at a rate proportional to both the number already infected and the number not infected. When first diagnosed, 30 people had the disease, and 10 days later 3000 were infected. Assuming that everyone is susceptible to the disease, how long will it take until 100,000 have been infected? Use the logistic growth model.
16. A newly created lake has been stocked with 500 adult (1 pound and bigger) walleyed pike. It is estimated that the lake can support only 40,000 adult walleyes. No fishing is to be allowed until the count of adult walleyes reaches 10,000. How long will this take if a survey at the end of 1 year indicated a count of 2000 walleyes? Use the logistic growth model.

---

17. Suppose that the total revenue $R$ (in thousands of dollars) from sales of a certain model of home freezer during the first $t$ years it has been on the market satisfies the differential equation $dR/dt = k(3500 - R)$ and that the revenue during the first year was \$1500 thousand. When will the revenue reach \$3000 thousand?

18. Let $P = P(x)$ be the number of Tinker Builder sets that a new worker on a production line can assemble on the $x$th day on the job and suppose that $P$ can be modeled by the differential equation $dP/dx = k(40 - P)$. If a worker assembled 25 sets on the fifth day, how many sets can we expect this worker to assemble on the eighth day?
19. An experimental psychologist has discovered that the number of new French words a typical language student learns in $x$ hours of a special tutoring course fits the bounded growth model. If learning 1500 vocabulary words is the most that can be expected and if it took Samantha 4 hours to learn 100 words, how many hours of special tutoring would she require to reach the 1000-word level?
20. Several years ago a service organization began a drive to increase its membership, with a goal of reaching 600 members. Two years ago, the membership was 80, and today it is 250. Assuming the logistic growth model is appropriate, what can the organization expect its membership to be 4 years from now?
21. A suburb of a certain city has enough space for 20,000 single-family homes. When incorporated, there were only 100 such homes. Five years later, there were 800 of these homes. No multiple-family dwellings are planned for this suburb. Assuming the logistic growth model for this suburb and that (on the average) each home will have 1.2 school-age children, how large a school system will be needed after 20 years?
22. Refer to Problem 21. If the average home in this suburb has 4.2 members and this is expected to continue to be true, what will the rate of increase in population be at the end of 20 years?
23. Let us suppose that the yield $y$ of tomatoes in bushels per acre increases, in relation to the number $x$ pounds of fertilizer applied, according to the differential equation $dy/dx = k(480 - y)$. When no fertilizer is applied, the yield is 100 bushels per acre, and when 200 pounds per acre is applied, the yield is 280 bushels per acre. If tomatoes sell for \$6.00 per bushel and fertilizer costs \$0.80 per pound, how many pounds of fertilizer per acre should be applied to yield the maximum profit per acre?
24. Refer to Problem 23. Let us make the model a little more realistic by supposing that it costs 50¢ a bushel to pick and market a bushel of tomatoes. What level of fertilizer application will then yield the maximum profit per acre?

## 6.3 MATHEMATICS OF FINANCE

In the western world, people expect money that is invested to grow in value. The simplest and surest way of achieving such growth is to put money in a savings account at a bank. Suppose that a **principal** of $P$ dollars is put in a savings account today and left there for $t$ years. Then at periodic times, the bank will add amounts called **interest** to the original amount $P$, thus causing the value of the account to grow. There are a number of different rules for specifying how the interest is figured, and it is important that we understand them. We do not exaggerate in saying that this understanding is fundamental to success in the business world.

> **POINT OF INTEREST**
> The concept of money earning interest is very old.
> "Then you ought to have put my money on deposit and on my return I should have got it back with interest."
> Parable of the talents
> Matthew 25:27

If only the original principal $P$ draws interest, we say that the bank is paying **simple interest.** Suppose, for example, that Mary puts \$100 in the bank today in an account that pays 9% simple interest annually. What will the account be worth at the end of 6 years? At the end of each year, the bank will add 9% of \$100 [that is, $0.09(100) = \$9$] to the account. In 6 years, the bank will add a total of $6(9) = \$54$ to the account, bringing its value to \$154.

In general, if $P$ dollars is invested in an account drawing simple interest at rate $r$ (written as a decimal), then after $t$ years this account will be worth

$$F = P + Prt$$

We use the letter $F$ to indicate that it represents the final or future value of the original $P$ dollars.

### EXAMPLE A

If \$500 is invested at 12% simple interest, what will its value be after 9 years?

**Solution**

$$F = P + Prt = 500 + 500(0.12)(9) = \$1040$$ ■

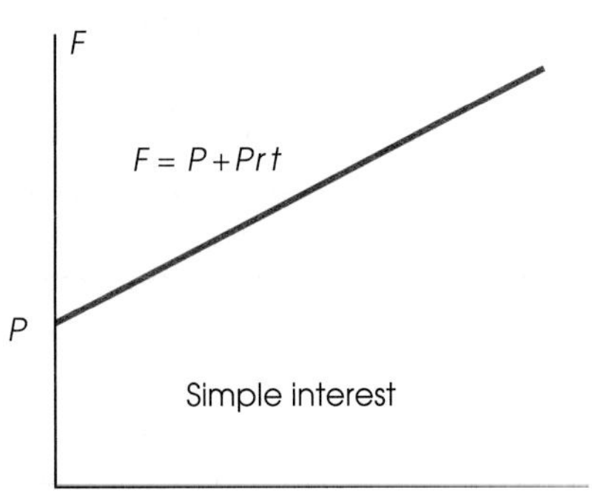


**FIGURE 6**

If we graph $F = P + Prt$ (note that $P$ and $r$ are fixed), we get a line with slope $Pr$ (Figure 6). Actually, this graph is not a true reflection of what is in the account except at year ends, since it is only at these times that the bank adds interest to the account. However, a straight-line graph is a good conceptual model for thinking about simple interest and is to be contrasted with the curved graphs used to model compound interest, the subject we consider next.

## Compound Interest (With Annual Compounding)

Of far more significance than simple interest in today's world is compound interest, since most banks and investment firms offer this type of interest. If at periodic intervals interest is converted to principal so that the interest also earns interest, then we say that the account is earning **compound interest.** Suppose that Mary puts her \$100.00 in an account earning 9% compound interest with interest compounded (converted to principal) annually. Then how much will she have at the end of 6 years? At the end of 1 year, the bank will add \$9.00 to the account just as in the simple interest problem treated earlier. But now \$109.00 becomes the principal during the second year and will therefore earn $0.09(109) = \$9.81$ in interest during that year. The principal for the third year is thus $109 + 9.81 = \$118.81$. And so on.

Rather than carry out this six-step process completely, let us see if we can develop a formula that will simplify our work. Note that the value at the end of 1 year is

$$100 + 0.09(100) = 100(1 + 0.09) = 100(1.09)$$

Thus, to get the value at the end of a year, we multiply the principal at the beginning of the year by 1.09. To get the value at the end of the second year, we take the principal at the beginning of that year, namely, 100(1.09), and multiply it by 1.09, obtaining

$$100(1.09)(1.09) = 100(1.09)^2$$

You can check, using your calculator, that this is 118.81, the number we got earlier. At the end of the third year, the value will be

$$100(1.09)^2(1.09) = 100(1.09)^3 = \$129.50$$

and at the end of 6 years, it will be

$$100(1.09)^6 = \$167.71$$

As expected, this is considerably more than the \$154.00 Mary received at 9% simple interest.

The general situation is similar. If $P$ dollars is invested at a rate $r$ (written as a decimal) with interest compounded annually, then after $t$ years the value will be

$$F = P(1 + r)^t$$

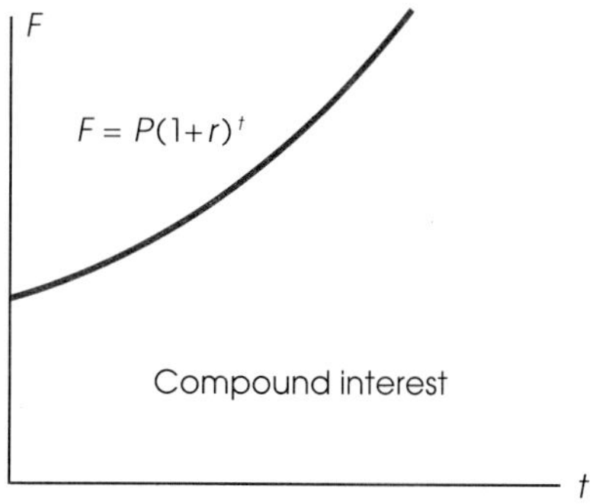

FIGURE 7

The graph of this equation is a curve, an exponential curve (Figure 7). It rises more and more rapidly as $t$ gets bigger. Money invested at compound interest eventually grows very rapidly in value, a fact you should never forget.

**EXAMPLE B**

If \$500 is invested at 12% interest compounded annually, what will it be worth after 9 years?

**Solution**

$$F = P(1 + r)^t = 500(1.12)^9 = \$1386.54$$ ■

## Compounding $k$ Times Per Year

Interest rates are always stated as annual rates (also called **nominal rates**). However, compounding may be done annually, quarterly, monthly, or at any stated time intervals (called periods) of equal length. Suppose interest is compounded $k$ times per year. Then a nominal rate $r$ corresponds to a rate of $i = r/k$ per period. For example, a nominal rate of 12% corresponds to a period rate of $i = 0.12/12 = 0.01$ if interest is compounded monthly. If $P$ dollars is invested at rate $r$ compounded $k$ times per year, then its value at the end of one period is $P(1 + i)$, at the end of two periods $P(1 + i)^2$, at the end of three periods $P(1 + i)^3$, and so on. In $t$ years, there is a total of $n = kt$ periods, so at the end of $t$ years the value is

$$F = P(1 + i)^{kt} = P(1 + i)^n$$

Here $i = r/k$ is the interest rate per period, and $n$ is the total number of periods.

## EXAMPLE C

Find the value of \$500 at the end of 9 years if it is invested at 12% interest compounded (a) annually, (b) quarterly, (c) monthly, (d) daily.

### Solution

(a) We worked this out in Example B; the answer is \$1386.54.
(b) $i = 0.12/4 = 0.03$ and $n = 4(9) = 36$, so

$$F = 500(1.03)^{36} = \$1449.14$$

(c) $i = 0.12/12 = 0.01$ and $n = 12(9) = 108$, so

$$F = 500(1.01)^{108} = \$1464.46$$

(d) $i = 0.12/365 = 0.00032876712$ and $n = 365(9) = 3285$, so

$$F = 500(1.00032876712)^{3285} = \$1472.08$$

Several things are to be noted about these calculations. First, we give our answers to the nearest penny. This requires that we take a rather accurate value for $i$ (using as many digits as our calculator allows). Second, we ignore leap years in making the calculation in part (d). Third, the final value $F$ increases as we compound more and more frequently. You expected this didn't you? ■

## Continuous Compounding

In recent years, most banks have offered to compound interest daily, at least on some of their savings accounts. A few banks have gone even further; they offer continuous compounding of interest. What does this mean? Imagine what would happen to $F$ if interest were compounded every day, every minute, every second, and so on. That is, consider what happens to $F$ as we let the period of compounding grow shorter and shorter. Appealing to the formula above, we are asking for an evaluation of

$$\lim_{k\to\infty} P\left(1 + \frac{r}{k}\right)^{kt}$$

With a little algebra, we can rewrite this limit in a more manageable form.

$$\begin{aligned}
\lim_{k\to\infty} P\left(1 + \frac{r}{k}\right)^{kt} &= \lim_{k\to\infty} P\left[\left(1 + \frac{r}{k}\right)^{k/r}\right]^{rt} \\
&= P\left[\lim_{k\to\infty}\left(1 + \frac{r}{k}\right)^{k/r}\right]^{rt} \\
&= P\left[\lim_{u\to 0}(1 + u)^{1/u}\right]^{rt} \qquad \text{(replacing } r/k \text{ by } u\text{)} \\
&= Pe^{rt} \qquad \text{(using Section 5.2)}
\end{aligned}$$

**SHARP CURVE**
The argument involved two subtleties. First, we used a property of exponents, namely,

$$(x^a)^b = x^{ab}$$

Second, we replaced $r/k$ by $u$ and noted that $k \to \infty$ is equivalent to $u \to 0$.

Here is what we have learned. If $P$ dollars is invested at rate $r$ compounded continuously, then after $t$ years its value will be

$$F = Pe^{rt}$$

**EXAMPLE D**

If \$500 is invested at 12% interest compounded continuously what will its value be at the end of 9 years?

**Solution** Comparing with Example C, we expect the answer to be larger than that obtained by daily compounding, that is, larger than \$1472.08.

$$F = Pe^{rt} = 500e^{0.12(9)} = \$1472.34$$

is larger, but not by much. The difference between daily compounding and continuous compounding is a matter of pennies unless applied to very large principals. ■

## Effective Rates

Which is the better interest rate—9% compounded daily or 9.1% compounded monthly? Because most people do not know how to answer this question, the government now requires banks to state not only their nominal interest rate but also something called the effective rate. For a given compound rate, the corresponding **effective rate** of interest is the rate which compounded annually yields the same return as the given rate. Here is a simpler way to say the same thing. The effective rate is just the interest earned on \$1 during 1 year at the given rate. Thus, the effective rate $r_e$ corresponding to the nominal rate $r$ compounded $k$ times per year is

$$r_e = \left(1 + \frac{r}{k}\right)^k - 1$$

**EXAMPLE E**

Which yields the better return: (a) 9% compounded daily or (b) 9.1% compounded monthly?

**Solution** To answer, we calculate the corresponding effective rates.

(a) $r_e = (1 + 0.09/365)^{365} - 1 \approx 0.094162$
(b) $r_e = (1 + 0.091/12)^{12} - 1 \approx 0.094893$

The second rate is somewhat better. ■

## Doubling Times

How long will it take money to double in value at a given interest rate? The answer, called the **doubling time** for the interest rate, is of considerable interest to investors.

### EXAMPLE F

How long does it take money to double if the interest rate is 10% compounded monthly?

**Solution** Let $n$ be the number of months required for \$1 to grow to \$2. Then

$$2 = 1(1 + 0.10/12)^n = (1.008333333)^n$$

an equation we wish to solve for $n$. Since the unknown is in the exponent, our first step is to take natural logarithms of both sides. We obtain

$$\ln 2 = n \ln 1.008333333$$

$$n = (\ln 2)/(\ln 1.008333333) = 83.52$$

The doubling time is about 84 months, or 7 years. ■

## C Problem Set 6.3

*In Problems 1–6, find the value F after t years if P dollars is invested at the rate r simple interest.*

1. $P = 1000$, $r = 0.08$, $t = 4$
2. $P = 800$, $r = 0.09$, $t = 3$
3. $P = 1200$, $r = 0.095$, $t = 4.5$
4. $P = 1600$, $r = 0.075$, $t = 5.25$
5. $P = 1463$, $r = 0.093$, $t = 16.75$
6. $P = 2182$, $r = 0.0821$, $t = 14.5$
7. If \$1000 is invested at the rate $r = 8\% = 0.08$ compounded annually, what will it be worth at the end of 4 years?
8. If \$800 is invested at 9% compounded annually, what will it be worth after 3 years?
9. If \$1250 is invested at 9.5% compounded annually, what will it be worth after 11 years?
10. If \$1800 is invested at 7.75% compounded annually, what will it be worth after 8 years?
11. Find the value of \$1000 at the end of 10 years if it is invested at 9% compounded (a) annually, (b) semiannually, (c) quarterly, (d) monthly, and (e) daily.
12. Do Problem 11 assuming the interest rate is 8%.
13. Henry Baker borrowed \$3650 from Jimmy Hawks at 11% compounded quarterly and paid off his debt in a lump sum 6 years and 6 months later. What was this lump sum?
14. On January 1, 1990, Linda borrowed \$4200 from her mother, promising to repay this sum together with interest at 8% compounded semiannually on July 1, 1995. How much will she have to pay her mother then?
15. If \$1000 is invested at 9% interest compounded continuously, what will it be worth at the end of 10 years? Be sure to compare this value with the results for Problem 11.
16. If \$1000 is invested at 8% interest compounded continuously, what will it be worth at the end of 10 years?
17. Find the accumulated value at the end of 16 years and 9 months if \$1463 is invested at 9.3% compounded continuously.
18. If \$2182 is invested at 8.21% compounded continuously, what will its value be at the end of 14 years and 6 months?

*In Problems 19–22, find the effective rate of interest for the given compound rate and period of compounding.*

19. 10%, quarterly
20. 9%, monthly
21. 8.9%, daily
22. 12%, daily

*In Problems 23–26, determine which of the two rates gives the better return (see Example E).*

23. Eight percent compounded monthly or 8.2% compounded semiannually.
24. Ten percent compounded quarterly or 9.9% compounded monthly.
25. Ten percent compounded annually or 9.6% compounded monthly.
26. Ten percent compounded annually or 9.51% compounded daily.
27. How long does it take money to double in value at 11% compounded monthly?
28. How long does it take money to double in value at 10% compounded daily? See Example F.
29. How long does it take money to double in value at 9.5% compounded continuously?
30. How long does it take money to triple in value at 6.36% compounded daily?

---

31. If $2400 is invested today, find its value at the end of 8 years if the interest rate is 9.6% (a) simple interest, (b) compounded monthly, (c) compounded daily, (d) compounded continuously.
32. How much money should be invested today at 12% compounded monthly in order to have $4000 at the end of 5 years? *Hint:* Solve $4000 = P(1.01)^{60}$ for $P$.
33. What principal $P$ invested at 8.4% compounded quarterly will accumulate to a value of $4800 at the end of 12 years?
34. What principal $P$ invested at 8.4% simple interest will accumulate to a value of $4800 at the end of 12 years?
35. How long will it take $1000 to grow to $4800 if it is invested at 10% compounded daily?
36. How long will it take $1000 to grow to $4800 if it is invested at 10% compounded continuously?
37. At what rate of interest compounded annually will an investment of $2100 accumulate to $3400 by the end of 6 years?
38. At what rate of interest compounded monthly will an investment of $1000 grow to $3000 in 15 years?
39. At what rate of interest compounded continuously will an investment of $1000 grow to $3000 in 15 years?
40. Jack McDuff borrowed $5000 from Jim Slye, promising to pay back $7000 at the end of 2 years. What rate compounded monthly was Jack paying?
41. Amy Wise invested $2500 in a savings account. Find its value at the end of 8 years if it earned 7.3% compounded daily during the first 4 years and 7.9% compounded daily during the last 4 years.
42. When Lisa was born, her father put $1000 in a savings account for her with an initial interest rate of 10% compounded monthly. At the end of each 4-year period, the interest went up by $\frac{1}{2}$%. How much was in her account when she went away to college at age 18?

## 6.4 MORE ON THE MATHEMATICS OF FINANCE

The basic principles on which the mathematics of finance rests were enunciated in the previous section. Here our goal is to flesh out some details and to take into account the important factor of inflation, that is, the continuing decrease in the purchasing power of money.

We begin with the fundamental notion of a time line, a line extending indefinitely into both the future and the past. Because money earns interest, a given amount of money with value $P$ right now will have increased value at a future time but had less value at a time in the past. We imagine that associated with each point on the time line is the value of $P$ at that time. For example, suppose that Mary has $100 right now and that interest is at a rate

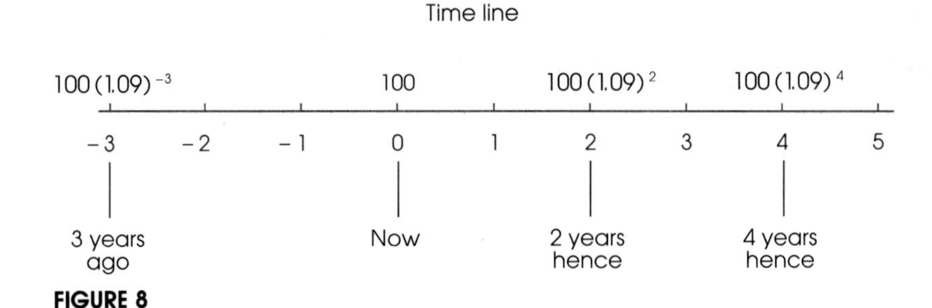


FIGURE 8

of 9% compounded annually. The value of the \$100 at various times is shown on the time line in Figure 8.

More generally, if Mary has $P$ dollars right now and interest is compounded at a rate of $i$ per period, then the value of $P$ at a time $n$ periods in the future will be $P(1 + i)^n$ and its value at a time $m$ periods into the past was $P(1 + i)^{-m}$. We **accumulate** money $n$ periods into the future by multiplying by $(1 + i)^n$; we **discount** money $m$ periods into the past by multiplying by $(1 + i)^{-m}$. The corresponding time line diagram is shown in Figure 9.

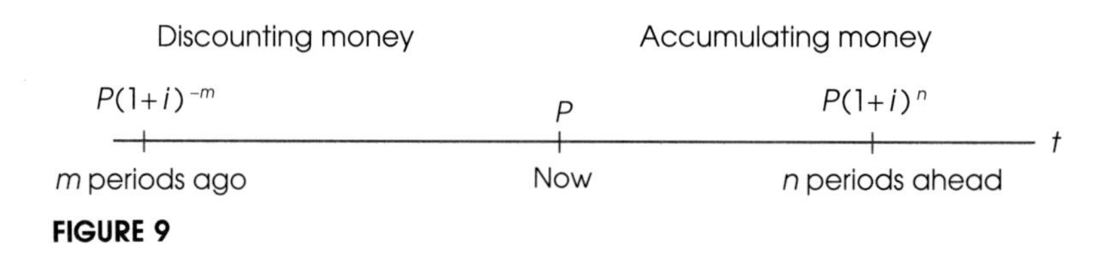


FIGURE 9

**EXAMPLE A**

Zero-coupon U.S. Treasury bonds are popular with conservative investors. These bonds carry no interest coupons but rather are bought at huge discounts. How much should Celia be willing to pay for a zero-coupon bond with a face value of \$10,000 due in 15 years if interest is at 13% compounded annually?

**Solution** This is simply a matter of discounting \$10,000 back 15 years. The answer is

$$10{,}000(1.13)^{-15} = \$1598.91$$

■

## Equivalent Values

A common financial problem is to calculate the value of a set of money payments at some specified time.

### EXAMPLE B

Chung Tai expects to receive royalty payments of \$2500 one year from now, \$3000 two years from now, and \$3500 three years from now. What is the total value of these payments right now? That is, what is their total present value? Assume that interest is 12% compounded monthly.

**Solution** We must discount the three payments by the appropriate number of periods, namely, 12 months, 24 months, and 36 months, respectively. The interest rate per period is $i = 0.12/12 = 0.01$. We conclude that the total present value $T$ of the three payments is

$$\begin{aligned} T &= 2500(1.01)^{-12} + 3000(1.01)^{-24} + 3500(1.01)^{-36} \\ &= 2218.62 + 2362.70 + 2446.24 = \$7027.56 \end{aligned}$$

■

### EXAMPLE C

Marcia has an obligation of \$5000 which is due 1 year from now. She is negotiating with her creditor to make instead three equal payments at the end of 2, 3, and 4 years. If interest is assumed to be at 14% compounded annually, what should be the size of the three payments?

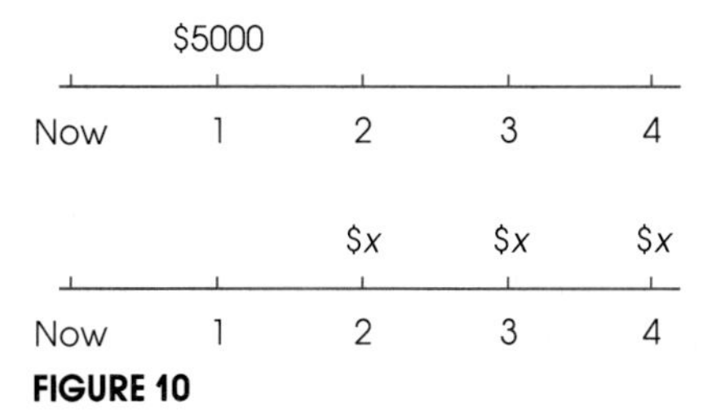

FIGURE 10

**Solution** In terms of time line diagrams, Marcia wants to replace the top diagram in Figure 10 by the bottom one. We choose to determine the values of the payments shown in the two diagrams at the common date of 1 year from now (any date can be chosen as the comparison date in stating the equivalence of two sets of values). We obtain

$$\begin{aligned} 5000 &= x(1.14)^{-1} + x(1.14)^{-2} + x(1.14)^{-3} \\ &= x[(1.14)^{-1} + (1.14)^{-2} + (1.14)^{-3}] \\ &= x(2.321632027) \\ x &= 5000/2.321632027 = \$2153.66 \end{aligned}$$

■

## The Effect of Inflation

While it is true that money invested in a savings account will earn interest and so grow in value, part of this increased value will be eroded away be-

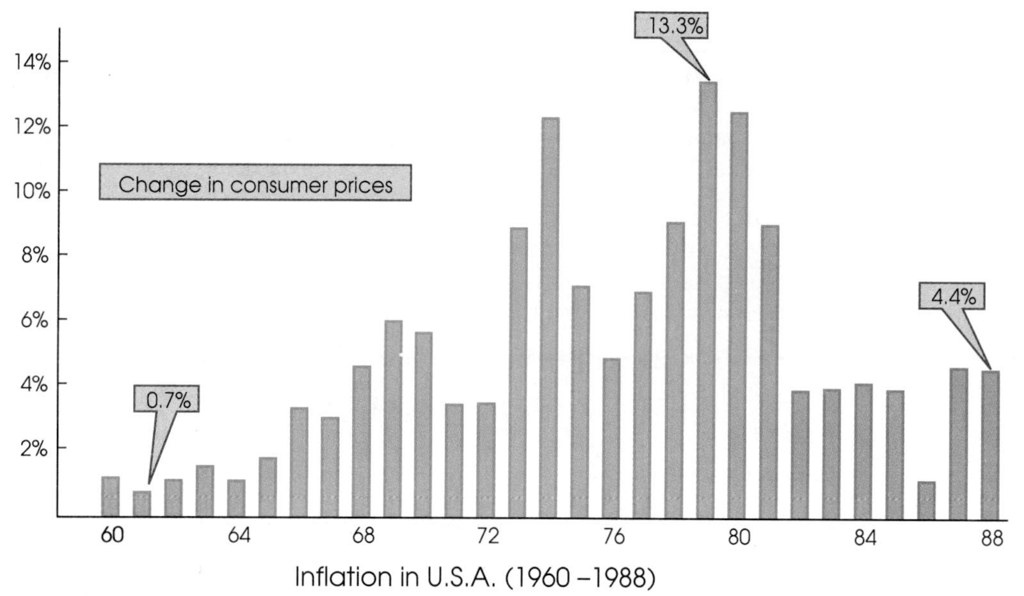


**FIGURE 11**

cause (at least in recent years) the purchasing power of the dollar is continually decreasing. Since 1960, the so-called **inflation rate** has varied from a low of 0.7% in 1961 to a high of 13.3% in 1979 (Figure 11). If we want to know what is happening to money in terms of its true purchasing power, we must correct for inflation.

For purposes of illustration, let us suppose that the interest rate per period is $i$ and that the inflation rate per period is $j$. It is natural to guess that the true growth rate $g$ of money is $i - j$, but it turns out that this is wrong though it is approximately correct. Let us find the exact value of $g$. Because it is earning interest at rate $i$, $P$ dollars will grow to $P(1 + i)^n$ after $n$ periods. But this value must be discounted because of the inflation rate $j$ to find its true value (that is, its value in constant dollars). The true value is

$$P(1 + i)^n(1 + j)^{-n}$$

We have called the true growth rate $g$. Thus,

$$P(1 + i)^n(1 + j)^{-n} = P(1 + g)^n$$

In other words,

$$\frac{(1 + i)^n}{(1 + j)^n} = (1 + g)^n$$

or

$$\frac{1 + i}{1 + j} = 1 + g$$

**POINT OF INTEREST**

If both interest and inflation are assumed to be compounded continuously at rates $i$ and $j$, respectively, then the true value of $P$ dollars after $t$ years is

$$Pe^{it}e^{-jt} = Pe^{(i-j)t}$$

This means that the true growth rate $g$ considered as a continuous rate is $g = i - j$.

or

$$g = \frac{1 + i}{1 + j} - 1 = \frac{i - j}{1 + j}$$

**EXAMPLE D**

In 1984, inflation was running at an annual rate of about 4%, while one could expect to earn interest in a savings account at an annual rate of about 8.5%. At that time, Rodriguez had $2500 in a savings account. He wanted to know how much he would have in 10 years in constant value dollars. Assuming the rates of interest and inflation to remain constant (a huge assumption), give Rodriguez an answer.

**Solution** The true growth rate of money under the given assumptions is

$$g = (0.085 - 0.04)/1.04 \approx 0.0432692$$

In 10 years, Rodriguez will have

$$2500(1.0432692)^{10} = \$3818.60$$

measured in constant value dollars. ■

## Depreciation

The Internal Revenue Service keeps changing the rules for figuring depreciation of machines, equipment, and real property. However, there are two basic theoretical models for depreciation. The first, straight-line depreciation, was discussed in Chapter 1. The second, exponential depreciation, allows for more rapid depreciation during the early years of service. One version of the latter is the *double-declining-balance* method. If this is used over $N$ years, the original value $C$ is depreciated each year by $100(2/N)$ percent of its value at the beginning of that year. Thus, the value $V$ after $n$ years is given by

$$V = C\left(1 - \frac{2}{N}\right)^n$$

**EXAMPLE E**

A piece of equipment costing $10,000 is to be depreciated by the double-declining-balance method over 15 years. What is its value after (a) 5 years, (b) 10 years, (c) 15 years?

**Solution** $1 - \frac{2}{N} = 1 - \frac{2}{15} = 0.8666666667.$

(a) $V = 10{,}000(0.8666666667)^5 = \$4889.46$
(b) $V = 10{,}000(0.8666666667)^{10} = \$2390.68$
(c) $V = 10{,}000(0.8666666667)^{15} = \$1168.91$ ■

## Problem Set 6.4

1. Accumulate \$2500 for 6 years at 8% compounded monthly.
2. Accumulate \$4200 for 8 years at 9% compounded quarterly.
3. Discount \$2500 for 10 years at 8% compounded monthly.
4. Discount \$4200 for 18 years at 9% compounded annually.
5. How much should a person be willing to pay for a zero-coupon bond with a face value of \$5000 due in 12 years if the interest rate is 8.5% compounded annually? See Example A.
6. Do Problem 5 assuming the interest rate is 10% compounded annually.
7. Helen intends to buy a zero-coupon bond with a face value of \$10,000 due in 14 years. What will she have to pay if the interest rate is 9% compounded monthly?
8. John is planning to buy a zero-coupon bond with a face value of \$5000 due 8 years from now. What should he be willing to pay if the prevailing interest rate is 8.4% compounded quarterly?

*In Problems 9–12, find the total present value of the given set of payments if the interest rate is 9% compounded annually (see Example B).*

9. \$2000 two years from now and \$3500 six years from now.
10. \$5000 three years from now and \$3200 four years from now.
11. \$2000 at the end of each of the next 4 years.
12. \$2400 at the end of 2 years, 4 years, and 6 years.
13. Do Problem 9 assuming interest is compounded monthly.
14. Do Problem 12 assuming interest is compounded monthly.
15. Romulo has an obligation of \$6000 due 4 years from now but has negotiated instead to pay \$3000 three years from now and then a payment of $x$ dollars 5 years after that. Determine $x$ if the interest rate is 8% compounded annually. See Example C.
16. To settle a debt, Mary had promised to pay Phyllis \$4500 three years from now and another \$2500 four years after that. Mary wants to renegotiate this contract and has asked to be allowed to make a single payment 8 years from now. How large should this single payment be if Phyllis insists on earning interest at 9% compounded quarterly?
17. Walter has an obligation to pay Margaret \$10,000 four years from now but has asked instead to make two equal payments at the ends of 3 and 6 years. If interest is figured at 9% compounded annually, what should be the size of these payments?
18. Tony, having borrowed money to go to college, owes his alma mater \$8000 right now. He has agreed to pay off this debt by making one payment 2 years from now and a second payment twice as large 4 years from now. Determine the size of the first payment if interest is at 6% compounded quarterly.
19. Maria has \$10,000 in a savings account earning 6.4% compounded annually, and the annual inflation rate is 3.5%. Assuming these rates hold steady, what will the account be worth in constant value dollars 5 years from now? What is the true growth rate of her money? See Example D.
20. Mark Farmer anticipates that he will need a new tractor 6 years from now. Assuming that the interest rate and the inflation rate will remain steady at 9% and 5%, respectively, how much money needs to be set aside to purchase a model that sells for \$60,000 today?
21. Which is better from the point of view of an investor: 8% interest with 4% inflation or 14% interest with 10% inflation? Explain.
22. Show that for the true growth rate to be at least one-half of the annual interest rate $i$, it is necessary that the annual inflation rate $j$ satisfy the inequality $j < i/(i + 2)$.
23. A \$50,000 tractor is to be depreciated by the double-declining-balance method over 10 years. What is its value after (a) 3 years, (b) 6 years, (c) 10 years? See Example E.

24. A bulldozer costing \$80,000 is to be depreciated over 12 years by the double-declining-balance method. What is its value after (a) 4 years, (b) 8 years, (c) 12 years?
25. A building valued at \$100,000 is to be depreciated by the double-declining-balance method over 18 years. How long does it take to lose half of its value?
26. How long did it take the bulldozer in Problem 24 to lose half of its value?

---

27. An obligation consisting of payments of \$2000 in 2 years and \$4000 in 5 years is to be replaced by two equal payments at the ends of 3 and 6 years, respectively. Determine the size of the latter payments if interest is at 9% compounded monthly.
28. John is supposed to pay his friend David \$2000 two years from now but has asked instead to make equal payments at the end of each of the next four years. How large should these payments be if interest is at 8% compounded quarterly?
29. To pay for a new car that she expects to cost \$16,000 three years from now, Betty determines to make three equal deposits in her savings account—now and at the end of the next 2 years. How large must these deposits be if the bank pays 7% compounded annually?
30. As a result of a divorce settlement, Alice is to receive payments of \$10,000 at the end of each of the next 4 years. To raise the down payment for a house, she has decided to sell these payments to a bank that has offered to buy them paying 6% interest compounded annually. How much will she receive as a lump sum?
31. On Monday morning, Rene bought a zero-coupon bond with a face value of \$20,000. It matures in 15 years and pays interest at 8% compounded annually.
    (a) How much did she pay for the bond?
    (b) On Monday afternoon of the same day, she had to sell the bond because of an emergency, but in that short time the prevailing interest rate had risen to 8.5% compounded annually. How much did she lose not counting broker fees?
32. Roberta bought a zero-coupon bond with a face value of \$10,000. It matures in 12 years and carries an interest rate of 9% compounded annually. One year later when Roberta sold the bond the prevailing interest rate had dropped to 7% compounded annually. How much more did Roberta get for the bond than she paid for it?
33. Howard bought a zero-coupon bond with a face value of \$25,000 for \$10,000. The bond matures in 10 years. What interest rate is he getting based on annual compounding?
34. In a certain Latin American country, the annual inflation rate is 80%. What annual interest rate would a lender need in order to achieve a true annual growth rate of 5%?
35. A road-paving company makes a practice of selling each of its pieces of heavy equipment when its value depreciates to one-fourth its original value. After how many years does it sell equipment if it uses the double-declining-balance method of depreciation over 12 years?

## CHAPTER 6 REVIEW PROBLEM SET

[C] 1. Solve the differential equation $dy/dt = -0.029y$, given that $y_0 = 80$. Then determine the value of $y$ when $t = 21$.

[C] 2. Solve the differential equation $dy/dt = 0.013y$ given that $y_0 = 1200$. Then determine the value of $y$ when $t = 15$.

[C] 3. Solve the differential equation $dy/dt = 0.07(12 - y)$, given that $y_0 = 3$. Then evaluate $y$ when $t = 5$.

[C] 4. Solve the differential equation $dy/dt = 0.045(20 - y)$, given that $y_0 = 8$. Then find the value of $y$ when $t = 11$.

[C] 5. Solve the differential equation $dy/dt = 0.05y(20 - y)$, given that $y_0 = 8$. Then evaluate $y$ when $t = 3.5$.

[C] 6. Solve the differential equation $dy/dt = 0.04y(30 - y)$, given that $y_0 = 10$. Then evaluate $y$ when $t = 6$.

[C] 7. Suppose that a bacteria population is growing exponentially, that it numbers 2400 now, and that it will number 12,000 after 12 hours.
    (a) Write a formula for the number $y$ of bacteria $t$ hours from now.
    (b) How many bacteria will there be at the end of 5 hours?
    (c) At what rate is the population growing at the end of 1 hour?
    (d) After how many hours will the population reach 16,000?

[C] 8. Suppose that the population of a certain country was 26 million on January 1, 1980, and 34.5 million 8

years later. Assume that the exponential growth model is appropriate.
(a) Find the formula for the population $t$ years after January 1, 1980.
(b) What will the population be on January 1, 1996?

C 9. Recall that carbon 14 has a half-life of 5570 years. The thigh bone of a prehistoric animal contains 6% of the carbon 14 found in living bones. How old is this fossil?

C 10. Polonium 214 disintegrates according to the differential equation $dy/dt = -4621y$, where $y$ is the number of grams present after $t$ seconds. Determine the half-life of this element.

C 11. The news that Morningvale's favorite son, Henry Fastman, had drowned in a boating accident in Sweden was first heard by 1000 people on the 8:00 A.M. news. By 11:00 A.M., two-thirds of Morningvale's population of 48,000 had learned the bad news. Assuming the bounded growth model applies, how many people will have heard the news by 3:00 P.M.?

C 12. The news that 10-year-old Lisa had been kidnapped was reported on the 9:00 A.M. news. Only 1500 of Porter City's population of 84,000 heard this report, but by noon the news had spread to three-fourths of the population. Assuming the bounded growth model is valid, determine how many people will have heard the news by 4:00 P.M.

C 13. When a certain national park opened, it had a deer population of 4000. Three years later, this population had risen to 4600. It is estimated that the park cannot support more than 9000 deer. How many deer will there be 20 years after the park opening, assuming the logistic growth model is appropriate?

C 14. A flu epidemic is spreading through a city of 50,000 at a rate proportional to both the number already infected and the number not yet infected. When the disease was first recognized, 60 people had it, and within 5 days a total of 4000 had been infected. Assuming everyone is subject to the disease, how long will it take until 40,000 have been infected?

C 15. Find the accumulated value at the end of 10 years of $3000 invested today at:
(a) 8% simple interest,
(b) 8% compounded annually,
(c) 8% compounded monthly,
(d) 8% compounded continuously.

C 16. If $4500 is invested today, find its accumulated value at the end of 8 years if interest is at:
(a) 9% simple interest;
(b) 9% compounded annually;
(c) 9% compounded quarterly;
(d) 9% compounded continuously.

C 17. How long does it take money to double at 8.5% compounded (a) semiannually, (b) continuously?

C 18. How long does it take an investment of $3200 to accumulate to $5000 if interest is at 9.5% compounded: (a) monthly, (b) continuously?

C 19. Find the effective rate corresponding to a rate of 9.4% compounded monthly.

C 20. Which rate of interest gives a lender a better return on his money: 9.2% compounded monthly or 9.15% compounded continuously?

C 21. What is the present value of a set of four equal annual payments of $560 each, the first one occurring 1 year from now, if the interest is at the rate of 10% compounded annually?

C 22. Joanne will discharge a debt by making 6 annual payments of $1000, the first one occurring one year from now. If interest is figured at 8.5% compounded annually, what is her present indebtedness?

C 23. Luanne Fisher bought a zero-coupon bond with a face value of $20,000 for $11,000. The bond matures at the end of 8 years. What interest rate is she getting based on annual compounding?

C 24. John paid $4500 for a zero-coupon bond which has a face value of $10,000 and matures at the end of 10 years. What interest rate is John realizing on his investment based on semiannual compounding?

C 25. Maurice has an obligation of $10,000 due 3 years from now. He has negotiated with his creditor to make instead one payment at the end of 2 years and another twice as large at the end of 5 years. Determine the size of these payments if the interest is at 7.5% compounded annually.

C 26. Felicia has promised to pay off her debt to her alma mater by making two payments of $5000 each, one at the end of 2 years, the other at the end of 5 years. She wishes to renegotiate her debt so that she will make one payment at the end of 1 year and another three times as large at the end of 6 years. Determine the size of these payments if interest is at 7.6% compounded annually.

C 27. A bus costing $120,000 is to be depreciated over 10 years by the double-declining-balance method.
(a) Determine its value after 5 years.
(b) When will its value first fall below $60,000?

C 28. How much money should be invested today at 9% compounded quarterly to have $4800 at the end of $8\frac{1}{2}$ years?

C 29. The wolf population in a certain state is believed to have grown exponentially from 1950 to 1980. If the population was 12,000 in 1970 and 13,500 in 1980, what was it in 1950?

C 30. A certain radioactive element decays exponentially so that 67% of a given amount will remain after 10 years. What percent of an initial amount will remain after 500 years?

[C] 31. A moving company knows that it must buy a new van 5 years from now. The interest rate and the inflation rate are expected to remain steady at 8.4% and 5.6%, respectively, each compounded annually. How much money should be set aside now to take care of this purchase if the required type of van sells for $60,000 now?

[C] 32. Mary invested $5000 in a fund that earned interest at the rate of 7.5% compounded annually during the first 8 years. Then the interest rate jumped to a higher rate for the next 6 years, again compounded annually. What was this higher interest rate if the accumulated amount at the end of the 14 years was $15,375?

33. Show by direct substitution that $y = L - (L - y_0)e^{-kt}$ satisfies the differential equation $dy/dt = k(L - y)$.

34. Show by substitution that $y = 100/(1 + e^{-100kt})$ satisfies the differential equation $dy/dt = ky(100 - y)$.

35. Show that $y = (A + Bt)e^{2t}$ satisfies the differential equation $y'' - 4y' + 4y = 0$.

36. Show that $y = A + (B + Ct)e^{3t}$ satisfies the differential equation $y''' - 6y'' + 9y' = 0$.

37. A population $y$ is said to be growing so that the rate of growth is proportional to the time elapsed and to the number present. Write down the corresponding differential equation. In the long run, will this growth be faster or slower than exponential growth?

38. An object is moving along a straight line so that its acceleration is inversely proportional to the square root of its velocity.
    (a) What differential equation does the velocity $v$ satisfy?
    (b) Find the formula for $v$ in terms of $t$, given that $v = 4$ when $t = 0$ and $v = 9$ when $t = 2$.

39. An object is moving along a line so that its acceleration is proportional to the square of its velocity. If $y$ is its distance from a fixed point, what differential equation does $y$ satisfy?

[C] 40. Bacteria populations $A$ and $B$ grow exponentially, doubling in size in 2 hours and 3 hours, respectively. If the present population of $A$ is 400 and that of $B$ is 600, after how many hours will the two populations be equal in size?

*Are there definite stages in a child's mental development? Do teachers sometimes introduce abstract concepts too early? Does manipulation of concrete objects improve geometric understanding? Would children learn mathematics better if teachers understood and used Piaget's learning theory model? The answers to all these questions appear to be yes.*

# 7

# The Antiderivative

Jean Piaget
(1896–1980)

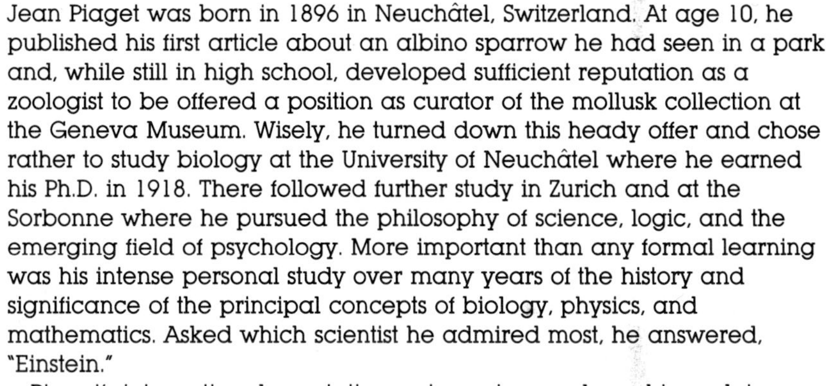

Jean Piaget was born in 1896 in Neuchâtel, Switzerland. At age 10, he published his first article about an albino sparrow he had seen in a park and, while still in high school, developed sufficient reputation as a zoologist to be offered a position as curator of the mollusk collection at the Geneva Museum. Wisely, he turned down this heady offer and chose rather to study biology at the University of Neuchâtel where he earned his Ph.D. in 1918. There followed further study in Zurich and at the Sorbonne where he pursued the philosophy of science, logic, and the emerging field of psychology. More important than any formal learning was his intense personal study over many years of the history and significance of the principal concepts of biology, physics, and mathematics. Asked which scientist he admired most, he answered, "Einstein."

Piaget's international reputation rests most securely on his work in developmental psychology. Careful observation of the learning patterns of his own children and later of large numbers of others led him to propose that children go through four definite stages of development: sensorimotor period (birth to 2 years), preoperational period (2 to 7 years), concrete operational period (7 to 11 years), and formal operational period (11 to 15 years). It is in the latter stage that children begin to think abstractly and logically—to do true mathematical thinking. More than any other person, Piaget has investigated how children learn the concepts and operations of arithmetic, geometry, algebra, and logic.

Piaget's psychological writings are studded with mathematical ideas and terminology. He is the author of a book on algebraic logic (1949) and another on groups and lattices (1952). When asked why mathematics is important in the study of the development of knowledge, he replied, "Because, along with formal logic, mathematics is the only entirely deductive discipline."

# 7.1 ANTIDERIVATIVES

Most of the operations that we perform in mathematics can be undone. The operation of adding 10 to a number can be undone by subtracting 10 from the result of the first operation. We call addition and subtraction inverse operations. Other examples of inverse operations are multiplication and division, squaring and taking square roots, and taking the natural logarithm of a number and exponentiating with the base $e$. Our task in this section is to study the operation that is inverse to differentiation (finding the derivative). We call this operation *antidifferentiation,* and the object obtained by this process is an *antiderivative*. Here is a formal definition.

DEFINITION

We call $F(x)$ an **antiderivative** of $f(x)$ on an interval if $F'(x) = f(x)$ for all $x$ in the interval, that is, if

$$\frac{d[F(x)]}{dx} = f(x)$$

EXAMPLE A

Find an antiderivative of $f(x) = x^2$.

**Solution** The function $F(x) = x^3$ occurs to us, but we immediately reject it because $d(x^3)/dx = 3x^2$. However, this does suggest $F(x) = x^3/3$, which is correct since

$$\frac{d(x^3/3)}{dx} = \frac{1}{3}\frac{d(x^3)}{dx} = \frac{1}{3}3x^2 = x^2$$

A little more thought suggests other possibilities. The functions $\frac{1}{3}x^3 + 7$ and $\frac{1}{3}x^3 - 129$ also have the right derivative. In fact, any function of the form $F(x) = \frac{1}{3}x^3 + C$ is an antiderivative of $f(x) = x^2$. ■

## The General Antiderivative of a Function

Example A above indicates that there are many antiderivatives for any given function. If you find one antiderivative, you may add any constant to it and have another antiderivative. Are all antiderivatives of a given function obtained this way? The answer is yes, and we now indicate why.

We begin with the intuitive fact that the only type of function whose derivative is identically 0 is a constant function. For if a function is not constant, then it will increase or decrease somewhere and this will make its derivative nonzero.

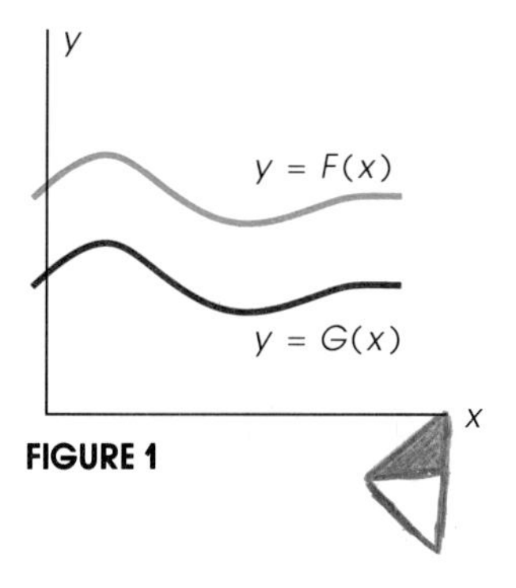

FIGURE 1

Next consider two functions $F(x)$ and $G(x)$ which have the same derivative everywhere on an interval (see Figure 1). Then the function $H(x) = F(x) - G(x)$ has a derivative which is identically 0, which means that it is a constant function; that is, $H(x) = C$ for some constant $C$. But this means that $F(x) - G(x) = C$ or, equivalently, $F(x) = G(x) + C$. We have shown that if two functions have the same derivative, then one can be obtained from the other by the addition of a constant. In the language of antiderivatives, we have shown that if we find one antiderivative of a given function $f(x)$, we can get any other antiderivative from it by the addition of a constant. The whole family of antiderivatives so obtained is called the **general antiderivative of** $f(x)$. In Example A, we would call $F(x) = \frac{1}{3}x^3 + C$ the general antiderivative of $f(x) = x^2$.

### EXAMPLE B

Find the general antiderivative of $f(x) = x^4$.

**Solution** After experimenting a little, we discover that $\frac{1}{5}x^5$ is one antiderivative. Thus $F(x) = \frac{1}{5}x^5 + C$ is the general antiderivative of $f(x) = x^4$. ■

From now on, when we ask you to find the antiderivative of a function, we mean that you should find the general antiderivative. Thus, your answers and ours will involve an arbitrary constant, traditionally denoted by the letter $C$.

We need a notation for antiderivatives and a systematic way of finding them. These are matters we take up next.

## Some Rules for Antiderivatives

First, we introduce a symbol that goes way back to one of the founders of the calculus, Gottfried Leibniz. To denote the antiderivative of $f(x)$, he used the symbol $\int f(x)\,dx$. The elongated ess, $\int$, is called the **integral sign,** the function $f(x)$ is called the **integrand,** and for now $dx$ is just a symbol that indicates that we are antidifferentiating with respect to $x$. In this notation, we indicate the results of Examples A and B as follows.

$$\int x^2\,dx = \tfrac{1}{3}x^3 + C \qquad \text{and} \qquad \int x^4\,dx = \tfrac{1}{5}x^5 + C$$

We remark that an antiderivative is also sometimes called an **indefinite integral.**

These two results suggest a general rule; it is called the Power Rule (for antiderivatives).

**POWER RULE**

If $r$ is any real number except $-1$, then

$$\int x^r\,dx = \frac{x^{r+1}}{r+1} + C$$

The proof of this rule is simple. Just take the derivative of the right side and note that it is the integrand on the left side. This is how you should remember this rule: *To find the antiderivative of a power, increase the exponent by 1 and divide by the new exponent.*

Be sure to note that the rule above has one exception, namely, when $r = -1$. However, we take care of this exception by using Example E from Section 5.3.

$$\int x^{-1}\,dx = \int \frac{1}{x}\,dx = \ln|x| + C \qquad x \neq 0$$

**EXAMPLE C**

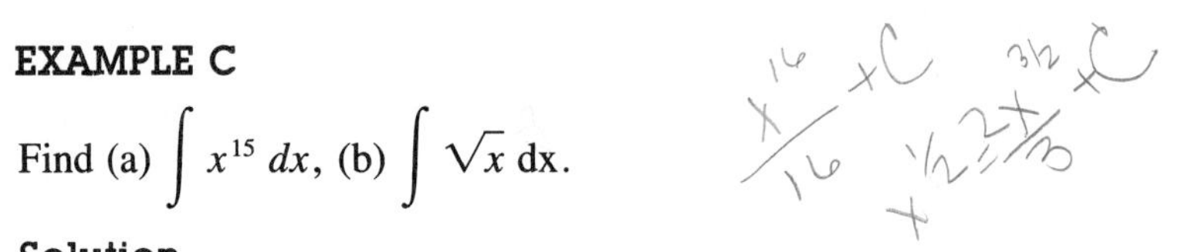

Find (a) $\int x^{15}\,dx$, (b) $\int \sqrt{x}\,\text{dx}$.

**Solution**

(a) $\displaystyle\int x^{15}\,dx = \frac{x^{16}}{16} + C$

(b) $\displaystyle\int \sqrt{x}\,dx = \int x^{1/2}\,dx = \frac{x^{3/2}}{\frac{3}{2}} + C = \frac{2}{3}x^{3/2} + C$ ■

As with derivatives, the antiderivative of a sum is the sum of the antiderivatives, and a constant may be moved past an integral sign. In symbols,

$$\int [f(x) + g(x)]\,dx = \int f(x)\,dx + \int g(x)\,dx$$

$$\int cf(x)\,dx = c\int f(x)\,dx$$

With these two properties in hand, we can find many more antiderivatives.

**RECALL**
It is clear that

$$\int 1\,dx = x + C$$

since the derivative of $x + C$ is 1. This is consistent with the Power Rule. Just remember that $1 = x^0$, apply the Power Rule, and you will get the boxed result.

**EXAMPLE D**

Find the antiderivative of $4x^2 - 6x + 5$.

**Solution**

$$\begin{aligned}\int [4x^2 - 6x + 5]\,dx &= \int 4x^2\,dx + \int (-6)x\,dx + \int 5(1)\,dx \\ &= 4\int x^2\,dx - 6\int x\,dx + 5\int 1\,dx \\ &= \frac{4x^3}{3} - \frac{6x^2}{2} + 5x + C \\ &= \tfrac{4}{3}x^3 - 3x^2 + 5x + C\end{aligned}$$ ■

**EXAMPLE E**

Find the antiderivative of $x^\pi + 3x^{-1/3} + 5x^{-1}$.

**Solution**

$$\begin{aligned}\int [x^\pi + 3x^{-1/3} + 5x^{-1}]\,dx &= \int x^\pi\,dx + 3\int x^{-1/3}\,dx + 5\int x^{-1}\,dx \\ &= \frac{x^{\pi+1}}{\pi + 1} + \frac{3x^{2/3}}{\frac{2}{3}} + 5\ln|x| + C \\ &= \frac{x^{\pi+1}}{\pi + 1} + \frac{9}{2}x^{2/3} + 5\ln|x| + C\end{aligned}$$ ■

We mention one more rule that will be used continuously from now on, the rule for antidifferentiating $e^x$.

$$\int e^x\,dx = e^x + C$$

As usual, the proof is simple. Just take the derivative of the right side and see that you get the integrand on the left.

**EXAMPLE F**

Find the antiderivative of $2e^x - 3$.

**Solution**

$$\int (2e^x - 3)\,dx = 2\int e^x\,dx - 3\int 1\,dx = 2e^x - 3x + C$$ ■

## Antiderivatives with Side Conditions

We have emphasized the fact that the general antiderivative of a given function involves an arbitrary constant and thus is a whole family of functions. Thus the general antiderivative of the function $f(x) = 2x$ is the family of functions $F(x) = x^2 + C$. This family is displayed in Figure 2.

Sometimes we are given a condition which picks out from a family of antiderivatives a particular function which satisfies that condition.

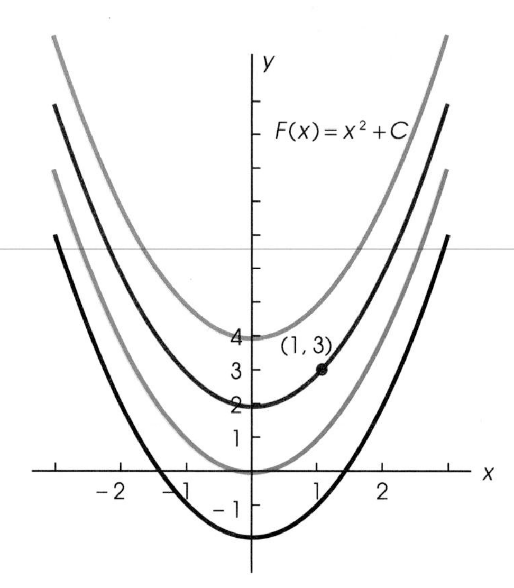


FIGURE 2

### EXAMPLE G

Find the antiderivative of $f(x) = 2x$ which satisfies the side condition $F(1) = 3$, that is, whose graph goes through the point $(1, 3)$.

**Solution** The general antiderivative of $f(x) = 2x$ is $F(x) = x^2 + C$. The condition $F(1) = 3$ means that

$$3 = F(1) = 1^2 + C$$

which requires that $C = 2$. We conclude that $F(x) = x^2 + 2$. Geometrically, the side condition has picked out the blue curve from the whole family of curves (Figure 2). ■

## Problem Set 7.1

*Fill in the blanks in Problems 1–6.*

1. If $G'(x) = g(x)$, we call $G(x)$ an Antideriv________ of $g(x)$.

2. If $G'(x) = H'(x)$ for all $x$, then $G(x) - H(x)$ is ________.

3. The derivative of the antiderivative of $f(x)$ is ________.
4. The general antiderivative of $f'(x)$ is ________.
5. $x^4$ is an antiderivative of ________.
6. The general antiderivative of $x^4$ is ________.

*In Problems 7–18, find the general antiderivative of the given function.*

7. $f(x) = x^3$
8. $f(x) = x^5$
9. $f(x) = x^{-3}$
10. $f(x) = x^{-5}$
11. $f(x) = x^{5/2}$
12. $f(x) = x^{3/4}$
13. $f(x) = x^{\pi}$
14. $f(x) = x^{\sqrt{2}}$
15. $f(x) = 1/x^4$
16. $f(x) = 1/x$
17. $f(x) = x^{-1/2}/x^{1/2}$
18. $f(x) = x^4/\sqrt{x}$

*In Problems 19–40, find the indicated antiderivatives.*

19. $\int (3x + 2)\, dx$
20. $\int (x^2 + 4x)\, dx$
21. $\int (5x^4 - 6x)\, dx$
22. $\int (3x^3 + 2x - 3)\, dx$
23. $\int (5x^{3/2} - 3x^{1/2})\, dx$
24. $\int (4x^{1/3} + 3)\, dx$
25. $\int (3\sqrt{x} + 5/x)\, dx$
26. $\int (4/x^3 + 2\sqrt[3]{x})\, dx$
27. $\int (5/x^2 - 2/x^3)\, dx$
28. $\int (2/x^5 - 3)\, dx$
29. $\int (5e^x - 4/x)\, dx$
30. $\int (x^{-2} + 4e^x)\, dx$
31. $\int (3\sqrt[3]{t} + 2t)\, dt$
32. $\int (4\sqrt[3]{s} - 1/s)\, ds$
33. $\int (2u^3 - 4/u^2)\, du$
34. $\int (5/v^6 + 2v^5)\, dv$
35. $\int (2x - 1)(x + 3)\, dx$
36. $\int (x + 2)(3x - 1)\, dx$
37. $\int (2t + 3)^2\, dt$
38. $\int (3u^2 - 1)^2\, du$
39. $\int (x^2 + 4x + 2)/x^2\, dx$
40. $\int (3x^2 - 5)/x^{3/2}\, dx$

*In Problems 41–48, find the particular antiderivative F(x) of f(x) that satisfies the given side condition (see Example G).*

41. $f(x) = 4x - 2;\ F(1) = 3$
42. $f(x) = 3 - 2x;\ F(2) = 7$
43. $f(x) = x^2 - 2x + 5;\ F(2) = \frac{23}{3}$
44. $f(x) = 3x^2 + x - 4;\ F(-1) = \frac{13}{2}$
45. $f(x) = 4e^x;\ F(0) = 2$
46. $f(x) = e^x - 2x;\ F(0) = 5$
47. $f(x) = 3/x + 6x^2;\ F(1) = 5$
48. $f(x) = x^2 - 2/x;\ F(1) = \frac{10}{3}$

---

49. Find the following antiderivatives.
    (a) $\int (2x - 4/x)\, dx$ (b) $\int (x + \sqrt{x})^2\, dx$
    (c) $\int (\sqrt{x} + 1)(3/\sqrt{x})\, dx$
    (d) $\int (3x^2 - 4)/x^3\, dx$
    (e) $\int (6 + 4e^s)\, ds$ (f) $\int (e^t + t^e)\, dt$
50. Suppose that the slope of the graph of $f$ at $x$ is given by $3x^2 - e^x + 5$ and that the graph passes through $(0, 4)$. Find $f(x)$.
51. Let $R(x)$ be the revenue in dollars from producing and selling $x$ camping trailers per year and note that $R(0) = 0$. If the marginal revenue is $3000 - 0.9x$, find the revenue in a year in which 1500 trailers were produced and sold.
52. Suppose that $g''(t) = 6/\sqrt{t}$, $g'(4) = 0$, and $g(1) = 10$. Find $g(t)$. *Hint:* Antidifferentiate twice.
53. Suppose that an object moves along a straight line in such a way that its velocity $v$ after $t$ seconds is given by $v = ds/dt = 6t^2 - 4t + 2$ feet per second. Find the position $s = s(t)$ of the object at time $t$, given that $s(1) = 14$.
54. Suppose that a car starting from rest accelerates at 8 feet per second. Find its velocity and how far it has traveled by the end of 5 seconds.
55. A rapidly growing city had a population of 10,000 in 1985. If its growth rate $t$ years after 1985 is estimated as $dP/dt = 250 + 300\sqrt{t}$, what will the population be in 2001?
56. [C] During the first 13 days of chemotherapy, the mass $M(t)$ of a malignant tumor decreases at the rate $M'(t) = -0.2t + 0.015t^2$ grams per day. What is the mass of such a tumor after 10 days of treatment if it started out weighing 180 grams?
57. [C] Let $C(x)$ denote the number of dollars it costs a publishing company to produce $x$ copies of a certain calculus book. If the marginal cost is $400/\sqrt{x}$ and if the

set-up and preparation costs are \$150,000, find (a) the total cost of producing 25,600 books and (b) the average cost per book at this production level.

58. A nursery sells a certain kind of pine tree when such a tree reaches a height of 2 feet and claims that it will have a growth rate $t$ years later given by $dh/dt = 4t^{-1/3}$, where $h$ is the height in feet. How tall will such a tree be 8 years later?

## 7.2 THE SUBSTITUTION RULE

The Chain Rule plays a fundamental role in finding derivatives. The analog of this rule for antiderivatives is the Substitution Rule. It is equally important and will be used in almost every section from here on.

Before we can state this important rule, we need to review an idea introduced for another purpose in Section 4.5. Let $u = g(x)$ and let $dx$ be a symbol called the **differential** of $x$. You may think of $dx$ as representing a small change in $x$ if you wish. Then $du$, the differential of $u$, is defined by

$$du = g'(x)\,dx$$

If you studied Section 4.5, you know that $du$ gives an approximation to the corresponding change in $u$, an approximation that gets better the smaller $dx$ is chosen. For right now, all that is important is that you remember the boxed formula. With it, we can state the Substitution Rule in a particularly convenient form.

**THE SUBSTITUTION RULE**

If $u = g(x)$, then $du = g'(x)\,dx$ and

$$\int f(g(x))g'(x)\,dx = \int f(u)\,du$$

To see why this rule is correct, let $F(u)$ be an antiderivative of $f(u)$, so that

$$F'(u) = f(u) \quad \text{and} \quad \int f(u)\,du = F(u) + C$$

Now by the Chain Rule,

$$\frac{d[F(u)]}{dx} = \frac{d[F(u)]}{du}\frac{du}{dx} = F'(u)g'(x)$$

That is,

$$\frac{d[F(g(x))]}{dx} = F'(g(x))g'(x) = f(g(x))g'(x)$$

In terms of antiderivatives, this says

$$F(u) + C = F(g(x)) + C = \int f(g(x))g'(x)\,dx$$

or, equivalently,

$$\int f(u)\,du = \int f(g(x))g'(x)\,dx$$

## The Substitution Rule Illustrated

In applying the Substitution Rule to an integral, we must be able to recognize the integrand as having the form $f(g(x))g'(x)\,dx$ so that the substitution $u = g(x)$ will change this expression to the form $f(u)\,du$. This requires insight and practice. Our first example can be done two ways—with or without the Substitution Rule.

### EXAMPLE A

Find $\int (2x^4)^3 8x^3\,dx$.

**Solution** Method 1 (using the Substitution Rule). Let $u = 2x^4$, so that $du = 8x^3\,dx$. Then

$$\int (2x^4)^3 8x^3\,dx = \int u^3\,du = \frac{u^4}{4} + C = \frac{(2x^4)^4}{4} + C = 4x^{16} + C$$

Method 2 (by first multiplying out the integrand).

$$\int (2x^4)^3 8x^3\,dx = \int 8x^{12} 8x^3\,dx = \int 64x^{15}\,dx$$

$$= \frac{64x^{16}}{16} + C = 4x^{16} + C$$

Naturally, we get the same answer both ways. ■

**SHARP CURVE**
Two methods of doing the same integration may lead to different-looking answers.

$$\int (2x + 3)2\,dx$$

$$= \frac{1}{2}(2x + 3)^2 + C$$

$$= 2x^2 + 6x + \frac{9}{2} + C$$

$$\int (2x + 3)2\,dx = \int (4x + 6)\,dx$$

$$= 2x^2 + 6x + K$$

However, these answers are equivalent since $\frac{9}{2} + C$ represents the same arbitrary constant in the first case that $K$ does in the second case.

### EXAMPLE B

Find $\int (x^2 + 3)^3 2x\,dx$.

**Solution** This example could be done by cubing $x^2 + 3$, multiplying by $2x$, and then finding the antiderivative; but that is far too much work. Rather, we make the substitution $u = x^2 + 3$, so $du = 2x\,dx$. This gives

$$\int (x^2 + 3)^3 2x\,dx = \int u^3\,du = \frac{u^4}{4} + C = \frac{(x^2 + 3)^4}{4} + C$$ ■

**EXAMPLE C**

Find $\int (x^4 - 5)^6 x^3\, dx$.

**Solution** The substitution $u = x^4 - 5$ occurs to us, but unfortunately, $du = 4x^3\, dx$, differing from what we have by a factor of 4. However, if we remember that constants can be shifted past an integral sign, we see what to do. Multiply and divide by 4.

$$\int (x^4 - 5)^6 x^3\, dx = \int (x^4 - 5)^6 \frac{1}{4} 4x^3\, dx$$

$$= \frac{1}{4}\int u^6\, du = \frac{1}{4}\frac{u^7}{7} + C$$

$$= \frac{1}{28}(x^4 - 5)^7 + C \quad \blacksquare$$

**EXAMPLE D**

Find $\int e^{x^2} x\, dx$.

**Solution** We try the substitution $u = x^2$, noting that $du = 2x\, dx$. The factor of 2 is missing in the integrand, but we can introduce this factor provided we compensate with a factor of $\frac{1}{2}$.

$$\int e^{x^2} x\, dx = \int e^{x^2} \frac{1}{2} 2x\, dx = \frac{1}{2}\int e^u\, du$$

$$= \frac{1}{2}e^u + C = \frac{1}{2}e^{x^2} + C \quad \blacksquare$$

**EXAMPLE E**

Find $\int \frac{x^2 + 2x}{x^3 + 3x^2 + 4}\, dx$.

**Solution** We make the substitution $u = x^3 + 3x^2 + 4$ so $du = (3x^2 + 6x)\, dx$. Clearly, we need an extra factor of 3 in the integrand. We may introduce this factor provided we compensate for it with a factor of $\frac{1}{3}$, a factor which may be passed in front of the integral sign.

$$\int \frac{x^2 + 2x}{x^3 + 3x^2 + 4}\, dx = \frac{1}{3}\int \frac{3(x^2 + 2x)}{x^3 + 3x^2 + 4}\, dx = \frac{1}{3}\int \frac{1}{u}\, du$$

$$= \frac{1}{3}\ln|u| + C = \frac{1}{3}\ln|x^3 + 3x^2 + 4| + C \quad \blacksquare$$

**EXAMPLE F**

Find $\int \frac{(\ln x)^4}{x}\, dx$

**Solution** Let $u = \ln x$ so $du = (1/x)\, dx$. Then

$$\int \frac{(\ln x)^4}{x}\, dx = \int u^4\, du = \frac{u^5}{5} + C = \frac{(\ln x)^5}{5} + C$$ ■

## A Miscellany of Examples

The examples so far considered are rather standard; they all use basically the same idea. However, we must be alert to a number of variations.

**EXAMPLE G**

Find $\int (x + 3)^{50} x\, dx$.

**Solution** The substitution $u = x + 3$ occurs to us but does not seem promising since $du = dx$. What shall we do with the extra factor $x$ that appears in the integrand? Note that if $u = x + 3$, then $x = u - 3$. If we replace all factors involving $x$ by their correspondents in terms of $u$, we obtain

$$\int (x + 3)^{50} x\, dx = \int u^{50}(u - 3)\, du$$

which still looks rather complicated. However, we have achieved something, because we can now multiply the factors of the integrand together. We get

$$\begin{aligned}\int u^{50}(u - 3)\, du &= \int (u^{51} - 3u^{50})\, du \\ &= \frac{u^{52}}{52} - \frac{3u^{51}}{51} + C \\ &= \frac{(x + 3)^{52}}{52} - \frac{(x + 3)^{51}}{17} + C\end{aligned}$$ ■

**EXAMPLE H**

Find $\int (x^3 - 3)^2\, dx$.

**Solution** No substitution will work here. However, the simplest possible idea does work: Expand the integrand.

$$\begin{aligned}\int (x^3 - 3)^2\, dx &= \int (x^6 - 6x^3 + 9)\, dx \\ &= \frac{x^7}{7} - \frac{6x^4}{4} + 9x + C\end{aligned}$$ ■

### EXAMPLE I

Find $\int (2x^4 - 3x^2 + 5)/5x\, dx$.

**Solution** No substitution will work. Here the trick is to break the integrand into three terms.

$$\int \frac{2x^4 - 3x^2 + 5}{5x}\, dx = \int \left[\frac{2}{5}x^3 - \frac{3}{5}x + \frac{1}{x}\right] dx$$

$$= \frac{2x^4}{20} - \frac{3x^2}{10} + \ln|x| + C$$

$$= \frac{x^4}{10} - \frac{3x^2}{10} + \ln|x| + C \qquad \blacksquare$$

### EXAMPLE J

Find $\int [x^3/(x^2 + 1)]\, dx$ and $\int (x^3 - 6x + 8)/(x - 2)\, dx$.

**Solution** Both integrands are examples of what are called improper fractions—rational functions in which the degree of the numerator is greater than or equal to that of the denominator. Our first step is to perform long division (see RECALL). In the first case, this gives

$$\frac{x^3}{x^2 + 1} = x - \frac{x}{x^2 + 1} = x - \frac{1}{2}\frac{2x}{x^2 + 1}$$

This combined with the mental substitution $u = x^2 + 1$ gives

$$\int \frac{x^3}{x^2 + 1}\, dx = \int x\, dx - \frac{1}{2}\int \frac{2x}{x^2 + 1}\, dx$$

$$= \frac{x^2}{2} - \frac{1}{2}\ln(x^2 + 1) + C$$

In the second case, we obtain

$$\int \frac{x^3 - 6x + 8}{x - 2}\, dx = \int (x^2 + 2x - 2)\, dx + 4\int \frac{1}{x - 2}\, dx$$

$$= \frac{x^3}{3} + x^2 - 2x + 4\ln|x - 2| + C$$

Here, we made the mental substitution $u = x - 2$ in the last step. $\blacksquare$

**RECALL**

Here we illustrate what we hope is a familiar process—long division.

$$\begin{array}{r} x^2 + 2x - 2 \\ x - 2\,\overline{)\,x^3 \qquad\quad - 6x + 8} \\ \underline{x^3 - 2x^2 \qquad\qquad\quad} \\ 2x^2 - 6x \qquad \\ \underline{2x^2 - 4x} \qquad \\ -2x + 8 \\ \underline{-2x + 4} \\ 4 \end{array}$$

Thus,

$$\frac{x^3 - 6x + 8}{x - 2} = x^2 + 2x - 2 + \frac{4}{x - 2}$$

## Problem Set 7.2

*In Problems 1–4, find the general antiderivative in two ways: by a substitution and by first multiplying out the integrand (see Examples A and B).*

1. $\int (2x^3)^4 6x^2\, dx$
2. $\int (2x^5)^3 10x^4\, dx$
3. $\int (x^2 + 3)^2 2x\, dx$
4. $\int (2x^2 - 1)^2 4x\, dx$

*In Problems 5–14, use an appropriate substitution to find the indicated antiderivative (see Examples B and C).*

5. $\int (x^3 - 4)^5 3x^2\,dx$

6. $\int (2x^3 + 1)^3 6x^2\,dx$

7. $\int (x^4 + 5)^6 x^3\,dx$

8. $\int (3x^2 - 4)^4 x\,dx$

9. $\int (x^2 + 6x + 4)^4 (2x + 6)\,dx$

10. $\int (2x^2 + 8x - 2)^3 (4x + 8)\,dx$

11. $\int (x^3 + 3x - 2)^{10} (x^2 + 1)\,dx$

12. $\int (x^3 + 3x^2 - 11)^7 (x^2 + 2x)\,dx$

13. $\int (\sqrt{x} + 1)^4 (1/\sqrt{x})\,dx$

14. $\int (3x^{-1} + 1)^3 (x^{-2})\,dx$

*Find each antiderivative in Problems 15–50 (see Examples D–J).*

15. $\int e^{3x}\,dx$

16. $\int e^{-2x}\,dx$

17. $\int 3e^{2t+1}\,dt$

18. $\int 4e^{3u-4}\,du$

19. $\int e^{2x^2+1} x\,dx$

20. $\int e^{4-3x^2} x\,dx$

21. $\int e^{\sqrt{x}} (1/\sqrt{x})\,dx$

22. $\int e^{4/x} (1/x^2)\,dx$

23. $\int \dfrac{2x + 5}{x^2 + 5x + 4}\,dx$

24. $\int \dfrac{3x^2 - 1}{x^3 - x + 7}\,dx$

25. $\int \dfrac{x^3 + x}{x^4 + 2x^2 - 2}\,dx$

26. $\int \dfrac{x^3 - 1}{2x^4 - 8x + 5}\,dx$

27. $\int \dfrac{e^x}{3e^x + 1}\,dx$

28. $\int \dfrac{e^x + 2}{4e^x + 8x}\,dx$

29. $\int (4 + \ln x)^{3/2} (1/x)\,dx$

30. $\int \sqrt{2 + 3\ln x}\,(1/x)\,dx$

31. $\int [\ln (2x + 1)]^4 / (2x + 1)\,dx$

32. $\int \int (x \ln x)^3 (1 + \ln x)\,dx$

33. $\int [\ln (x^2 + 1)][x/(x^2 + 1)]\,dx$

34. $\int [\ln (\ln x)]^4 [1/(x \ln x)]\,dx$

35. $\int (x - 4)^{14} x\,dx$

36. $\int (x - 2)^{10} x\,dx$

37. $\int (2x + 1)^8 4x\,dx$

38. $\int (1 - 3x)^7 x\,dx$

39. $\int (x^2 + 5)^2\,dx$

40. $\int (2x^2 - 3)^2\,dx$

41. $\int (\sqrt{x} + 3)^2\,dx$

42. $\int (3/x + 2)^2\,dx$

43. $\int \dfrac{3x^3 + x - 7}{x^2}\,dx$

44. $\int \dfrac{5x^4 - 2x^2 + 7}{x^3}\,dx$

45. $\int \dfrac{2\sqrt{x} + 5}{\sqrt{x}}\,dx$

46. $\int \dfrac{x^{5/2} + 2x^{3/2} + 1}{x^{3/2}}\,dx$

47. $\int \dfrac{x^2 + x + 1}{x - 1}\,dx$

48. $\int \dfrac{x^2 + 5x - 4}{x + 2}\,dx$

49. $\int \dfrac{x^3 + 2x}{x^2 - 1}\,dx$

50. $\int \dfrac{x^3 + 6x}{x^2 + 2}\,dx$

---

*Find the antiderivatives in Problems 51–62.*

51. $\int (2u^2 + 1)^3\,du$

52. $\int (2t^2 + 1)^{11} t\,dt$

53. $\int x^2 e^{x^3}\,dx$

54. $\int (e^{2x} - 3)^2\,dx$

55. $\int (x + 1)^{10} x\,dx$

56. $\int (\ln x)^3 x^{-1}\,dx$

57. $\int \dfrac{t^2 + 1}{t^3 + 3t - 4}\,dt$

58. $\int \left(t^2 + \dfrac{4}{t}\right)^2 dt$

59. $\int \dfrac{\sqrt{t} + 2t + 3}{t}\,dt$

60. $\int \dfrac{t^2 + 9t}{t + 3}\,dt$

61. $\int \dfrac{e^t}{e^t + 2} \ln (e^t + 2)\,dt$

62. $\int \dfrac{(t + 1)^3}{(t + 2)^5}\,dt$

63. Find the equation of the curve that passes through the point (0, 3) and whose tangent line at any point $(x, y)$ on the curve has slope $(2x + 3)/(x^2 + 3x + 1)$.

[C] 64. A commercial truck, purchased for $60,000, depreciates at a rate of $28{,}000e^{-t/2}$ dollars per year $t$ years after purchase (that is, $dV/dt = -28{,}000e^{-t/2}$). Find its value at the end of 10 years.

[C] 65. A manufacturer estimates that the marginal cost of producing $x$ picnic tables per week is $C'(x) =$

$50(1 + e^{-0.02x})$ and that the fixed costs per week are \$2500.

(a) How much will it cost to produce 80 tables per week?

(b) What is the average cost per table in this case?

C 66. A recreational lake is treated periodically with a chemical during the summer to reduce the bacteria level. The rate at which the concentration of bacteria is changing $t$ days after treatment is given by $dB/dt = -1500t^2/(1 + t^3)$, where $B = B(t)$ is the number of bacteria per milliliter of water. If the initial concentration of bacteria was 6000 per liter, find the bacteria concentration after 8 days.

C 67. A particle moves along a straight line in such a way that its velocity $v$ after $t$ minutes is $v = t(3 - 0.2t)^{1/2}$ feet per second $(0 \le t \le 15)$. If $s = s(t)$ denotes the coordinate of the particle at time $t$ and $s(10) = 10$, find (a) $s(t)$ and (b) the distance traveled during the interval $0 \le t \le 15$. *Hint:* Let $u = (3 - 0.2t)^{1/2}$, so that $u^2 = 3 - 0.2t$ and $2u\, du = -0.2\, dt$.

## 7.3 SOME INTEGRATION TECHNIQUES

We have not used the word *integration* before because the word *antidifferentiation* seems more descriptive. However, from now on, the two words will be used interchangeably. In like manner, the expression *integral* (technically *indefinite integral*) will often replace *antiderivative*.

Our main tool for integration is the *Substitution Rule* introduced in the previous section. There is another general rule that is sometimes useful. It is traditionally known as **integration by parts,** but the phrase *double substitution rule* is more descriptive.

### Integration by Parts

The Product Rule for derivatives says that

$$\frac{d[u(x)v(x)]}{dx} = u(x)v'(x) + v(x)u'(x)$$

When we integrate both sides of this equation, we obtain

$$u(x)v(x) = \int u(x)v'(x)\, dx + \int v(x)u'(x)\, dx$$

or, equivalently,

$$\int u(x)v'(x)\, dx = u(x)v(x) - \int v(x)u'(x)\, dx$$

In differential symbols, this takes the following form called Integration by Parts.

**RECALL**

Keep in mind that antidifferentiation and differentiation are inverse operations. Thus

$$\int \frac{d}{dx} f(x)\, dx = f(x) + C$$

In particular,

$$\int \frac{d}{dx}[u(x)v(x)]\, dx = u(x)v(x) + C$$

a fact that we have used in the derivation at the right. Why were we able to suppress the constant $C$ in this derivation?

**INTEGRATION BY PARTS**

Let $u = u(x)$ and $v = v(x)$ be differentiable functions. Then,

$$\int u\, dv = uv - \int v\, du$$

For this result to be of use in finding an integral $\int f(x)\, dx$, we must be able to write $f(x)\, dx$ in the form $u\, dv$. Even then, it has little value unless the integral $\int v\, du$ is something we know how to handle. An example will clarify what we mean.

## EXAMPLE A

Find $\displaystyle\int xe^x\, dx$.

**Solution** First, you should convince yourself that the ordinary Substitution Rule is of no help. Turning to integration by parts, make the double substitution

$$u = x \qquad dv = e^x\, dx$$

Then

$$du = dx \qquad v = \int e^x\, dx = e^x$$

and so,

$$\begin{aligned}\int xe^x\, dx &= \int u\, dv = uv - \int v\, du \\ &= xe^x - \int e^x\, dx \\ &= xe^x - e^x + C\end{aligned}$$

■

## EXAMPLE B

Find $\displaystyle\int xe^{x^2}\, dx$.

**Solution** Don't jump to integration by parts. Always begin by trying an ordinary substitution. This integral was found by the substitution $u = x^2$ in the previous section (Example D). ■

## EXAMPLE C

Find $\displaystyle\int x^2 e^x\, dx$.

**Solution** After a bit of thought, we decide that an ordinary substitution will not be fruitful so we try integration by parts. Let

$$u = x^2 \qquad dv = e^x\, dx$$

Then,

$$du = 2x\,dx \qquad v = e^x$$

and so,

$$\int \overbrace{x^2}^{u}\, \overbrace{e^x\,dx}^{dv} = \overbrace{x^2 e^x}^{u\,v} - \int \overbrace{e^x}^{v}\, \overbrace{2x\,dx}^{du} = x^2 e^x - 2\int xe^x\,dx$$

To find the last integral, we appeal to Example A (otherwise, we would have to give up or, if especially clever, apply integration by parts again). We obtain

$$\begin{aligned}\int x^2 e^x\,dx &= x^2 e^x - 2(xe^x - e^x + C)\\ &= x^2 e^x - 2xe^x + 2e^x + K\end{aligned}$$ ■

### EXAMPLE D

Find $\int x^2 \ln x\,dx$.

**Solution** This is an integration by parts problem, but what shall we substitute for $u$ and $dv$? If we let $dv = \ln x\,dx$, we are in trouble because we don't know how to integrate $\ln x$. As an alternative, let us try

$$u = \ln x \qquad dv = x^2\,dx$$

Then

$$du = \frac{1}{x}\,dx \qquad v = \frac{x^3}{3}$$

and

$$\begin{aligned}\int x^2 \ln x\,dx &= \frac{(\ln x)x^3}{3} - \int \frac{x^3}{3}\frac{1}{x}\,dx\\ &= \frac{1}{3}(x^3 \ln x) - \frac{1}{3}\int x^2\,dx\\ &= \frac{1}{3}(x^3 \ln x) - \frac{1}{3}\frac{x^3}{3} + C\\ &= \frac{1}{3}(x^3 \ln x) - \frac{1}{9}x^3 + C\end{aligned}$$ ■

## Integration Using Partial Fractions

It is a standard algebraic exercise to combine two or more simple fractions into one fraction. For example,

$$\frac{3}{x-1} - \frac{2}{x+2} = \frac{3x+6-2x+2}{(x-1)(x+2)} = \frac{x+8}{(x-1)(x+2)} = \frac{x+8}{x^2+x-2}$$

In integration problems, it will be to our advantage to go in the opposite direction since, while we do not know how to integrate the expression at the right, we are able to integrate the one at the left. We illustrate.

### EXAMPLE E

Find $\int \frac{x+8}{x^2+x-2}\,dx$.

**Solution**

$$\int \frac{x+8}{x^2+x-2}\,dx$$

$$= \int \left(\frac{3}{x-1} - \frac{2}{x+2}\right) dx = 3\int \frac{1}{x-1}\,dx - 2\int \frac{1}{x+2}\,dx$$

The first of these integrals is found by the substitution $u = x - 1$, and the second by the substitution $u = x + 2$. Note that in both cases $du = dx$. We obtain as our final answer

$$3 \ln|x-1| - 2\ln|x+2| + C$$

■

We pose now this question: Can a complicated fraction always be written as a sum of simpler fractions? More precisely, can a fraction of the form

$$\frac{f(x)}{(x-a)(x-b)(x-c)\cdots(x-p)} \qquad (a, b, c, \ldots, p \text{ all different})$$

be expressed as a sum

$$\frac{A}{x-a} + \frac{B}{x-b} + \frac{C}{x-c} + \cdots + \frac{P}{x-p}$$

The answer is yes provided the numerator $f(x)$ is a polynomial with degree less than that of the denominator. The process is called decomposing into partial fractions.

### EXAMPLE F

Decompose $4/(x^2 + 5x + 6)$ into partial fractions.

**Solution** First, we note that the denominator can be factored as the product $(x + 3)(x + 2)$. Thus, we expect a decomposition of the form

$$\frac{4}{x^2+5x+6} = \frac{A}{x+3} + \frac{B}{x+2}$$

After multiplying both sides by $(x + 3)(x + 2)$, this takes the form

$$4 = A(x + 2) + B(x + 3)$$

We want the latter to be true for all $x$, in particular for $x = -3$ and $x = -2$. Substituting these values gives

$$4 = A(-3 + 2)$$

and

$$4 = B(-2 + 3)$$

We conclude that $A = -4$ and $B = 4$. Thus,

$$\frac{4}{x^2 + 5x + 6} = \frac{-4}{x + 3} + \frac{4}{x + 2}$$ ■

**EXAMPLE G**

Find $\int 4/(x^2 + 5x + 6)\, dx$.

**Solution** We use the result from Example F to write the above integral as

$$-4 \int \frac{1}{x + 3}\, dx + 4 \int \frac{1}{x + 2}\, dx$$

When we make the substitution $u = x + 3$ in the first integral and $u = x + 2$ in the second, we obtain

$$-4 \ln |x + 3| + 4 \ln |x + 2| + C$$ ■

## Tables of Integrals

Faced with an integration problem, you now have a number of techniques to try. Suppose none of them work. What then? One possibility is to turn to a *table of integrals*. There are whole books consisting of integral formulas, carefully arranged by category. You can probably find one or more of them in your college library. A small such table of integrals is shown at the back of this book, and we urge you to familiarize yourself with it. Here is an example of its use.

**POINT OF INTEREST**
Some things are hard; others are impossible. It is impossible to represent $\pi$ as a quotient of two integers. But $\pi$ can be approximated by 22/7 or even better by 31416/10000. Similarly $\int e^{x^2}\, dx$ cannot be expressed in terms of elementary functions. It can, however, be approximated by such functions. This is a huge topic and one that we cannot delve into at this point.

**EXAMPLE H**

Find $\int \frac{1}{x^2 \sqrt{9 - x^2}}\, dx$

**Solution** Apply Formula 58 from the table of integrals at the back of the book to obtain

$$\int \frac{1}{x^2 \sqrt{9 - x^2}}\, dx = -\frac{\sqrt{9 - x^2}}{9x} + C$$ ■

We conclude by correcting a wrong impression that we may have given. Even with all of the known techniques and with the biggest table of integrals that has been devised, we are unable to perform many indefinite integrations. Some integrations are difficult, some are extremely difficult, and some are impossible. Yes, we do mean impossible. It has been shown that the indefinite integrals of many functions cannot be expressed in terms of the elementary functions of calculus. For example, $\int e^{x^2}\,dx$ is known to be such an indefinite integral.

## Problem Set 7.3

*Find the integrals in Problems 1–12 using integration by parts.*

1. $\int xe^{2x}\,dx$
2. $\int xe^{-3x}\,dx$
3. $\int x \ln x\,dx$
4. $\int \sqrt{x} \ln x\,dx$
5. $\int x^{-2} \ln x\,dx$
6. $\int x^{7/2} \ln x\,dx$
7. $\int x(1 + x)^6\,dx$
8. $\int x(2x + 1)^6\,dx$
9. $\int x^2e^{-2x}\,dx$
10. $\int x^2e^{4x}\,dx$
11. $\int x \ln (x + 2)\,dx$
12. $\int \ln (x - 2)\,dx$

*Decompose each of the expressions in Problems 13–20 into partial fractions as in Example F.*

13. $3/[(x + 1)(x + 2)]$
14. $24/[(x - 3)(x + 3)]$
15. $(5x - 9)/(x^2 - 3x)$
16. $(x + 11)/(x^2 + x - 2)$
17. $(6x + 24)/[(x + 1)(x + 2)(x + 3)]$
18. $(6x^2 - 8x - 4)/[(x - 2)(x - 1)(x + 2)]$
19. $(3x^2 - 11x - 30)/(x^3 + x^2 - 6x)$
20. $(x^2 - 26x + 9)/[(x^2 - 9)(x - 1)]$

21–28. Use the results obtained above to find the integrals of the expressions in Problems 13–20 as in Examples E and G.

*Find the integrals in Problems 29–40.*

29. $\int 4e^{2x}\,dx$
30. $\int 6e^{-3x}\,dx$
31. $\int (x + 2)e^{-5x}\,dx$
32. $\int (x - 3)e^{4x}\,dx$
33. $\int xe^{4x^2}\,dx$
34. $\int x^3e^{-x^2}\,dx$
35. $\int (\ln x)/\sqrt{x}\,dx$
36. $\int (\ln x)^4/x\,dx$
37. $\int \frac{2x - 1}{x^2 - x - 2}\,dx$
38. $\int \frac{2x + 3}{x^2 + 3x - 10}\,dx$
39. $\int \frac{2x + 7}{x^2 + 3x - 10}\,dx$
40. $\int \frac{3x - 9}{x^2 - x - 2}\,dx$
41. Find $\int \frac{7x^2 - 7x - 9}{(x + 1)^2(x - 4)}\,dx$

    *Hint:* The integrand can be decomposed in the form $\frac{A}{x + 1} + \frac{B}{(x + 1)^2} + \frac{C}{x - 4}$.
42. See the hint in Problem 41 to help you find $\int \frac{-10x^2 + 36x - 17}{(x - 2)^2(x + 1)}\,dx$
43. An object is moving along a line in such a way that its velocity at time $t$ seconds is $2te^{2t}$ feet per second. How far will the object travel between $t = 0$ and $t = 8$?
44. Find the function $f$ such that the tangent line to the graph of $f$ at the point $[x, f(x)]$ has slope $x^2 \ln x$, given that $f(1) = 2$.
45. [C] Let $P(t)$ denote the number of milligrams of tetracycline assimilated into the bloodstream of a person $t$ minutes after taking a pill containing the drug. Suppose the rate of assimilation is $3te^{-0.4t}$ milligrams per minute at time $t$. How much of the drug will be assimilated in 15 minutes?
46. [C] The rate of growth of a certain city is $10(12 - t)e^{t/4}$ people per year, for $0 \le t \le 12$, where $t$ denotes the number of years after 1990. What will the population of this city be in 2000 if it was 13,640 in 1990?

# 7.4* DIFFERENTIAL EQUATIONS

The phrase *differential equation* occurred earlier in the book (Sections 6.1 and 6.2). A **differential equation** is an equation involving derivatives (or differentials) in which the unknown is a function. A typical example is the differential equation of exponential growth and decay, namely, $dy/dt = ky$. In the earlier sections, we gave the solutions to three important differential equations, but we did it without indicating how these solutions were found. For example, we said that $dy/dt = ky$ has the solution $y = y_0 e^{kt}$. We propose in this section to introduce a fairly general method for solving simple differential equations, but we warn our readers that we will only scratch the surface of an immense subject.

Our discussion of antiderivatives (indefinite integrals) makes the solution of one kind of differential equation obvious. The solution to $dy/dx = f(x)$ is $y = \int f(x)\,dx$. For example, the solution to $dy/dx = 3x^2$ is $y = x^3 + C$. Actually, this is a whole family of solutions, the family shown graphically in Figure 3. In this section, we refer to this family of solutions as the **general solution** to the differential equation. Sometimes, we will pick a **particular solution** from this family by giving an initial condition. Thus, we may say: Solve the differential equation $dy/dx = 3x^2$ subject to the initial condition that $y_0 = 2$. This picks from the family the particular solution $y = x^3 + 2$ (see again Figure 3).

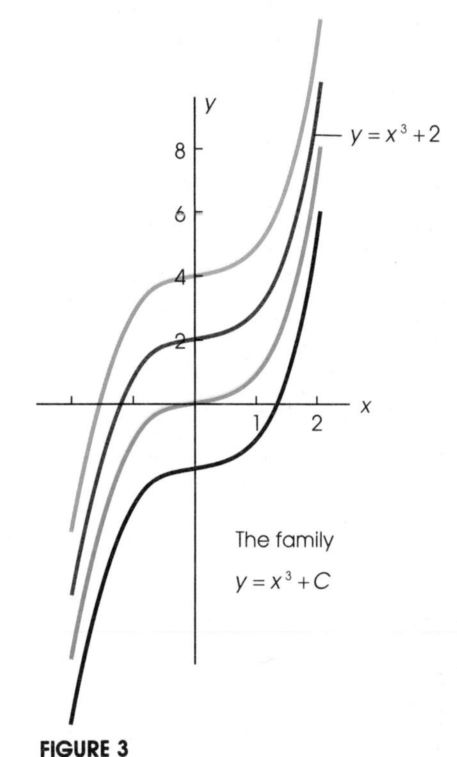


FIGURE 3

## Separation of Variables

We now formulate a new way of solving the differential equation $dy/dx = 3x^2$. Interpret $dy$ and $dx$ in this equation as differentials and multiply by $dx$ to obtain

$$dy = 3x^2\,dx$$

Then integrate both sides; that is, write successively

$$\int dy = \int 3x^2\,dx$$

$$y + C_1 = x^3 + C_2$$

$$y = x^3 + (C_2 - C_1)$$

Nothing is lost by replacing the constant $C_2 - C_1$ by $C$. Accordingly, from now on, we shall add a constant of integration to one side only. In this case, we obtain the same solution we got earlier, namely, $y = x^3 + C$.

What is significant about the method just illustrated is that it has wide applicability. It works whenever a simple algebraic manipulation of the differential equation results in a **separation of variables,** that is, a grouping of all the $y$'s and $dy$'s on one side of the equation and all the $x$'s and $dx$'s on the other side.

**EXAMPLE A**

Solve the differential equation

$$\frac{dy}{dx} = \frac{2x}{3y}$$

**Solution** We begin by multiplying both sides by $3y\,dx$ to separate variables and then integrate both sides.

$$3y\,dy = 2x\,dx$$

$$\int 3y\,dy = \int 2x\,dx$$

$$\frac{3y^2}{2} = x^2 + C$$

We could now solve this last equation for $y$ explicitly, but since it would involve both a plus and minus and a messy square root, we accept $3y^2/2 = x^2 + C$ as an adequate representation of the general solution. ■

**EXAMPLE B**

Solve the differential equation

$$\frac{dy}{dx} = 3\sqrt{xy}$$

subject to the condition that $y = 256$ when $x = 4$.

**Solution** We separate variables (multiply by $dx$ and divide by $\sqrt{y}$) and then integrate.

$$y^{-1/2}\,dy = 3x^{1/2}\,dx$$

$$\int y^{-1/2}\,dy = \int 3x^{1/2}\,dx$$

$$2y^{1/2} = 2x^{3/2} + C$$

Substituting the initial data yields

$$2\sqrt{256} = 2(4)^{3/2} + C$$

$$32 = 16 + C$$

$$16 = C$$

The required particular solution therefore satisfies

$$2y^{1/2} = 2x^{3/2} + 16$$

or

$$y^{1/2} = x^{3/2} + 8$$

or

$$y = (x^{3/2} + 8)^2 = x^3 + 16x^{3/2} + 64$$ ■

## Three Examples Related to Growth Problems

We consider now the three differential equations which were used to model various types of growth in Sections 6.1 and 6.2. In each case, $y$ represents a positive quantity whose rate of growth (or decay) $dy/dt$ has a known form. The results obtained here justify the theorems of the earlier sections.

### EXAMPLE C (EXPONENTIAL GROWTH MODEL)

Solve the differential equation $dy/dt = ky$ subject to the condition that $y = y_0$ when $t = 0$ ($k$ is a constant and $y_0 > 0$).

**Solution** To separate variables, we multiply by $dt$ and divide by $y$; then we integrate.

$$\frac{1}{y}\,dy = k\,dt$$

$$\int \frac{1}{y}\,dy = \int k\,dt$$

$$\ln y = kt + C$$

Changing the latter equation to its equivalent exponential form yields

$$y = e^{kt+C} = e^C e^{kt}$$

Substituting $y = y_0$ when $t = 0$ gives

$$y_0 = e^C$$

Thus, the desired solution is

$$y = y_0 e^{kt}$$ ■

### EXAMPLE D (BOUNDED GROWTH MODEL)

Solve the differential equation $dy/dt = k(L - y)$ subject to $y = y_0$ when $t = 0$ ($k$ and $L$ are constants and $0 \le y_0 < L$).

**Solution** We separate variables and integrate.

$$\frac{dy}{L - y} = k\,dt$$

$$\int \frac{dy}{L - y} = \int k\,dt$$

In the left integral, we make the substitution $u = L - y$ and $du = -dy$, leading to

$$-\int \frac{du}{u} = \int k\,dt$$

$$-\ln u = kt + C$$

$$\ln u = -kt - C$$

$$u = e^{-kt-C} = e^{-C}e^{-kt}$$

$$L - y = e^{-C}e^{-kt}$$

$$y = L - e^{-C}e^{-kt}$$

Substituting $y = y_0$ at $t = 0$ gives

$$y_0 = L - e^{-C} \quad \text{or} \quad e^{-C} = L - y_0$$

We conclude that

$$y = L - (L - y_0)e^{-kt}$$

Our derivation assumed that $y < L$ (so that $u = L - y > 0$), an assumption consistent with the bounded growth model. ■

**EXAMPLE E (LOGISTIC GROWTH MODEL)**

Solve the differential equation $dy/dt = ky(L - y)$ subject to $y = y_0$ when $t = 0$ ($k$ and $L$ are constants and $0 < y_0 < L$).

**Solution** We rewrite the differential equation as

$$\frac{1}{y(L - y)}\,dy = k\,dt$$

and then as

$$\left(\frac{1}{L}\frac{1}{y} + \frac{1}{L}\frac{1}{L - y}\right)dy = k\,dt$$

Thus,

$$\frac{1}{L}\int \frac{1}{y}\,dy + \frac{1}{L}\int \frac{1}{L - y}\,dy = \int k\,dt$$

As in Example D, these integrations lead to

$$\frac{1}{L}\ln y - \frac{1}{L}\ln (L - y) = kt + C$$

From here on, the details are somewhat messy, but they are of a straightforward algebraic nature (Problem 35). The final result is

$$y = \frac{Ly_0}{y_0 + (L - y_0)e^{-Lkt}}$$ ■

> **SHARP CURVE**
> Writing $1/y(L - y)$ as the sum of two simple fractions is an example of decomposing a rational expression into partial fractions, a subject treated in Section 7.3

## An Economic Application

In Section 4.5, we introduced the important economic concept of *elasticity*. If $x = f(p)$ represents the demand function for a certain manufactured item (that is, $x$ is the number of units that can be sold at price $p$ per unit in some specified length of time), then the elasticity of demand $E$ is given by

$$E = -\frac{p}{x}\frac{dx}{dp}$$

**EXAMPLE F**

If the elasticity of demand for a certain item is constant with value $\frac{1}{2}$, determine the demand function $x = f(p)$.

**Solution** We must solve the differential equation

$$-\frac{p}{x}\frac{dx}{dp} = \frac{1}{2}$$

On separation of variables, this takes the form

$$-2\frac{dx}{x} = \frac{dp}{p}$$

or after integration of both sides

$$-2 \ln x = \ln p + C$$

We can rewrite this successively as

$$\ln p + 2 \ln x = -C$$

$$\ln (px^2) = -C$$

$$px^2 = e^{-C}$$

$$x^2 = \frac{e^{-C}}{p}$$

$$x = \frac{K}{\sqrt{p}}$$

Here $K$ is a constant that cannot be determined without more information. ■

## Problem Set 7.4

*Find the general solution of the differential equations in Problems 1–12.*

1. $\dfrac{dy}{dx} = \dfrac{x}{y^2}$

2. $\dfrac{dy}{dx} = \dfrac{1}{xy}$

3. $\dfrac{dy}{dx} = \dfrac{x^2}{1 - 3y^2}$

4. $\dfrac{dy}{dx} = \dfrac{1 + x^2}{y^{3/2}}$

5. $\dfrac{dy}{dx} = \dfrac{xy^2}{1 + x^2}$

6. $\dfrac{dy}{dx} = (1 + 2x)e^{-y}$

7. $\frac{dy}{dx} = \frac{x^3(1 + y^2)}{y(1 + x^4)}$

8. $\frac{dy}{dx} = \left(\frac{y + 4}{2 - x}\right)^2$

9. $\frac{dy}{dx} = \frac{2y}{x^2 - x}$

10. $\frac{dy}{dx} = \frac{7x - 1}{(x^2 - 1)y}$

11. $\frac{dy}{dx} = \frac{3x^2}{ye^y}$

12. $\frac{dy}{dx} = \frac{\sqrt{x} \ln x}{\sqrt{y}}$

*In Problems 13–24, find the particular solution of the indicated differential equation that satisfies the given condition (see Examples B–E).*

13. $\frac{dy}{dx} = \frac{x}{2y}$; $y = 4$ when $x = 2$

14. $\frac{dy}{dx} = \sqrt[3]{x/y}$; $y = 8$ when $x = 1$

15. $\frac{dy}{dt} = -y^2t(t^2 + 2)^4$; $y = 1$ when $t = 0$

16. $\frac{dy}{dt} = \sqrt[3]{2t + 1}$; $y = 0$ when $t = 0$

17. $\frac{dy}{dt} = 0.02y$; $y = 4$ when $t = 0$

18. $\frac{dy}{dt} = -0.18y$; $y = 10$ when $t = 0$

19. $\frac{dy}{dt} = -0.36y$; $y = 5$ when $t = 1$

20. $\frac{dy}{dt} = 0.0125y$; $y = 20$ when $t = 4$

21. $\frac{dy}{dt} = 4(12 - y)$; $y = 8$ when $t = 0$

22. $\frac{dy}{dt} = 0.9(20 - y)$; $y = 13$ when $t = 0$

23. $\frac{dy}{dt} = 0.03y(12 - y)$; $y = 10$ when $t = 0$

24. $\frac{dy}{dt} = 0.04y(6 - y)$; $y = 5$ when $t = 0$

25. Suppose the elasticity of demand, $-(p/x)(dx/dp)$, is always equal to the constant $\frac{3}{4}$ and that $x = 10$ when $p = 16$. See Example F.
   (a) Express $x$ in terms of $p$.
   (b) Find $x$ when $p = 81$.

26. Suppose that the elasticity of demand (see Problem 25) is equal to $x$ and that $x = 1000$ when $p = 1$. Find $x$ when $p = 1.2$.

---

*In Problems 27–34, find the general solution of the given differential equation.*

27. $\frac{dy}{dt} = -4y$

28. $\frac{dy}{dx} = x^2y^3$

29. $\frac{dy}{dx} = \frac{e^x}{y}$

30. $\frac{dy}{dx} = \frac{y \ln x}{x}$

31. $\frac{dy}{dx} = xy^2e^{x^2}$

32. $\frac{dy}{dx} = xye^{-x}$

33. $\frac{dy}{dx} = \frac{8y}{x(x - 2)}$

34. $\frac{dy}{dx} = (y - 5)(10 - y)$

35. Fill in the missing steps in the solution to Example E; that is, show how to get from the line

$$\frac{1}{L} \ln y - \frac{1}{L} \ln (L - y) = kt + C$$

to the final result

$$y = \frac{Ly_0}{y_0 + (L - y_0)e^{-Lkt}}$$

36. Show that if the elasticity of demand for a certain item is always equal to 1, then the product of the price and the demand is a constant.

C 37. Suppose that the rabbit population $P = P(t)$ in a certain area $t$ months after January 1, 1980, was growing at a rate equal to 9% of $P$ (in rabbits per month). If $P(0)$ was 1200, what was the increase in population during the month of January 1981?

C 38. A company is advertising a new brand of bath soap on local television. Suppose that the rate at which new consumers become aware of the product is proportional to the number who are not yet aware of it. Let $p(t)$ denote the percentage of the total number of possible viewers who are aware of the product after $t$ weeks.
   (a) Write the differential equation that $p(t)$ must satisfy.
   (b) If no one knew about the new soap when the advertising campaign began and 20% were aware of it after 1 week, what percentage were aware of it at the end of 3 weeks?

C 39. A city of 5000 has been hit by a flu epidemic. The rate at which the epidemic is spreading is proportional to the product of $y$, the number of people already infected, and the number not yet infected.
   (a) Write the differential equation that $y$ must satisfy.
   (b) If 5 people were infected initially and 30 were infected after 1 week, how many were infected at the end of 4 weeks?

C 40. Let $P = P(t)$ denote the population of Wonderland $t$ years after 1980. The internal growth rate (that is, neglecting migration) in the population of this country is equal to 2% of $P$, but there is also a net migration into the country of 20,000 per year.
(a) Write the differential equation for $P$.
(b) If the population was 12 million in 1980, what will it be in the year 2000?

C 41. Suppose that in a certain country the population $P = P(t)$, $t$ years after 1990, satisfies the differential equation $P'(t) = 0.02P(t)$; and suppose also that the gross national product $G(t)$ satisfies the differential equation $G'(t) = 0.015G(t)$. If the 1990 per capita income $G/P$ was \$920, what will the per capita income be in the year 2000?

C 42. In a certain wilderness region, the bear population ranges from a low of 500 to a high of 2500 and presently numbers 1000. A biologist has suggested that the growth rate of this population is equal to $0.0002(P - 500)(2500 - P)$, where $P = P(t)$ is the population $t$ years from now. Estimate the population 15 years from now.

## CHAPTER 7 REVIEW PROBLEM SET

1. Determine the general antiderivative of each of the following functions.
(a) $f(x) = 8x^7 - 4x^3 + \pi$
(b) $f(x) = 2\pi x^{\pi-1} - 10x^{3/2}$
(c) $f(x) = -\dfrac{4}{x} + 2e^{4x}$
(d) $f(x) = 3x^2e^{x^3} - 2$

2. Determine the general antiderivative of each function.
(a) $f(x) = 15x^4 - 8x^3 + 2x - \sqrt{3}$
(b) $f(x) = 3\sqrt{2}\,x^{\sqrt{2}-1} - 14x^{5/2}$
(c) $f(x) = 4e^{2x} - \sqrt{x} + \dfrac{5}{x}$
(d) $f(x) = \dfrac{3x^2 + 1}{x^3 + x - 4} + \dfrac{1}{x^2}$

3. Find each indefinite integral.
(a) $\displaystyle\int (3x^2 - 12x + 5)\,dx$ (b) $\displaystyle\int \left(3\sqrt{x} + \frac{5}{x^2}\right) dx$
(c) $\displaystyle\int \frac{(2x + 1)^2}{x}\,dx$ (d) $\displaystyle\int \left(4e^{3x} - \frac{6}{\sqrt[3]{x}}\right) dx$

4. Find each indefinite integral.
(a) $\displaystyle\int (4x^3 - 6x^2 - 3)\,dx$ (b) $\displaystyle\int \frac{(2x^3 + 1)^2}{x^4}\,dx$
(c) $\displaystyle\int \left(5t\sqrt{t} + \frac{4}{\sqrt{t}}\right) dt$ (d) $\displaystyle\int (e^{2x} + 3x^e)\,dx$

5. Determine $F(x)$ if $F'(x) = 4/x^2 + 4 + x^2$ and $F(2) = 3$.

6. Find $F(x)$ if $F'(x) = x^3 + 6x + \dfrac{12}{x} + \dfrac{8}{x^3} + 4$ and $F(1) = 4$.

7. If $f''(t) = 12/t^5$, $f'(1) = 1$, and $f(1) = -7$, find $f(t)$.

8. If $f''(t) = 15\sqrt{t}$, $f'(1) = 12$, and $f(1) = 3$, find $f(t)$.

9. The Deluxe Apparel Company estimates the marginal cost of making and selling $x$ tuxedos per month to be $280 - 12x + 0.6x^2$ dollars. Find the formula for $C(x)$, the cost of making and selling $x$ tuxedos per month, given that $C(10) = \$3100$.

10. The Good Time Company estimates the marginal cost of making and selling $x$ alarm clocks per month to be $5 - 0.08x + 0.12x^2$ dollars. Find the formula for $C(x)$, the cost of making and selling $x$ alarm clocks per month, given that $C(100) = \$42{,}100$.

11. A proposed model for the deer population $P$ in a certain state is that its rate of growth satisfies $dP/dt = 2000/\sqrt{t}$, where $t$ is the number of years after January 1, 1980. If this model is a good one and if the deer population was 18,500 on January 1, 1980, what deer population is predicted for January 1, 2005?

12. Suppose that the population $P$ of a certain city, $t$ years after January 1, 1990, satisfies the differential equation $dP/dt = 150e^{-0.015t}$. If the population on January 1, 1990, was 25,000, find the population on January 1, 2000.

13. Find each indefinite integral.
(a) $\displaystyle\int (x^3 + 2x)^4(3x^2 + 2)\,dx$
(b) $\displaystyle\int (e^{2x} + 4)^3 e^{2x}\,dx$
(c) $\displaystyle\int \frac{2x + 1}{x^2 + x - 7}\,dx$
(d) $\displaystyle\int \frac{x^3 + x}{(x^4 + 2x^2 - 7)^3}\,dx$
(e) $\displaystyle\int \frac{1}{x(\ln x)^2}\,dx$ (f) $\displaystyle\int x(x + 3)^9\,dx$
(g) $\displaystyle\int \frac{x^2 + x - 15}{x - 3}\,dx$ (h) $\displaystyle\int (x^3 + 2)^2 x\,dx$

14. Find each indefinite integral.

(a) $\int (2x^3 - 5x + 4)^3(6x^2 - 5)\, dx$

(b) $\int (\ln t + 5)^4(1/t)\, dt$

(c) $\int \frac{4x - 1}{2x^2 - x + 8}\, dx$ (d) $\int \frac{6t^2 + 6}{(2t^3 + 6t - 5)^4}\, dt$

(e) $\int \frac{3\sqrt{\ln x}}{x}\, dx$ (f) $\int (x - 4)^{18} x\, dx$

(g) $\int \frac{4x^3 - 5x + 7}{2x - 1}\, dx$ (h) $\int (y^4 + 3)^2 y^2\, dy$

15. Find using integration by parts.

(a) $\int xe^{4x}\, dx$ (b) $\int \frac{\ln x}{x^{3/2}}\, dx$

16. Find by using integration by parts.

(a) $\int xe^{-3x}\, dx$ (b) $\int 49x^{5/2} \ln x\, dx$

17. Use a partial fraction decomposition to find each integral.

(a) $\int \frac{3x + 2}{x^2 - x - 12}\, dx$ (b) $\int \frac{2x^2 - 9x - 12}{x^3 + x^2 - 12x}\, dx$

18. Use a partial fraction decomposition to find each integral.

(a) $\int \frac{x - 16}{x^2 + 3x - 10}\, dx$ (b) $\int \frac{x^2 - 15x + 18}{x(x^2 - 9)}\, dx$

19. Find the general solution of the following differential equations.

(a) $\frac{dy}{dx} = 8x^3y^2$ (b) $t\,\frac{dy}{dt} = \sqrt{y} \ln t$

20. Find the general solution to each differential equation.

(a) $\frac{dy}{dx} = 6x^2\sqrt{y}$ (b) $(t + 1)e^s \frac{ds}{dt} = (e^s + 2)t$

21. Find the particular solution to the given differential equation with the associated condition.

(a) $\frac{dy}{dx} = \frac{1}{2}y^3e^x$; $y = \frac{1}{3}$ when $x = 0$.

(b) $\frac{dy}{dx} = -4y^2x\sqrt{5 + x^2}$; $y = \frac{1}{2}$ when $x = 2$.

22. Determine the indicated particular solution.

(a) $\frac{dy}{dx} = -y^2x^2(x^3 + 2)^3$; $y = \frac{1}{8}$ when $x = 1$.

(b) $(e^x + 2)\frac{dy}{dx} = e^{x-y}$; $y = 0$ when $x = 0$.

23. Find the equation of the curve that passes through the point $(1, 2)$ and whose slope at any point $(x, y)$ on the curve is $(1 - 2\sqrt{x})^3/\sqrt{x}$.

24. Find the equation of the curve that passes through the point $(1, 5)$ and whose slope at any point $(x, y)$ on the curve is $(9x^2\sqrt{x^3 + 3})/y$.

25. A strain of Asian flu has hit a city of 25,000. The rate at which the flu is spreading is proportional to $y$, the number of people already infected, and to the number not yet infected.
(a) Write the differential equation that $y$ satisfies.
(b) If 50 people were infected initially and a total of 750 by the end of 2 weeks, how many were infected by the end of 5 weeks?

26. Ten people started the rumor of a scandal in city government in Abbeville, a city of 50,000 people. Assume that the rate at which the rumor spread was proportional to $y$, the number who had heard the rumor, and to the number who had not heard the rumor.
(a) Write the differential equation satisfied by $y$.
(b) If 5,000 people had heard the rumor by the end of 5 days, how many had heard it by the end of 12 days?

27. Let $f$ have domain $(-3, \infty)$ and satisfy $f''(x) = -10/(x + 3)^3$, $f'(0) = \frac{5}{9}$, and $f(0) = \frac{-2}{3}$.
(a) Where is the graph of $f$ concave up?
(b) Where is $f$ increasing?
(c) Where is $f(x) > 0$?
(d) Find $\lim_{x\to\infty} f(x)$.
(e) Sketch the graph of $f$.

28. Let $f$ have domain $(-1, \infty)$ and satisfy $f''(x) = -10/(x + 1)^2$, $f'(0) = 1$, and $f(0) = 1$.
(a) Where is the graph of $f$ concave up?
(b) Where is $f$ increasing?
(c) Where is the slope of the graph equal to $1/3$?
(d) Where is $f(x) > 0$?
(e) Sketch the graph of $f$.

*Take a closed wire loop of arbitrary shape, dip it in liquid soap, and a soap film of smallest area spanning the loop will form. Richard Courant analyzed these minimal surfaces and was fascinated by them. Below, soap films are used to find minimal paths connecting four points and five points. Intersections always form at 120° angles.*

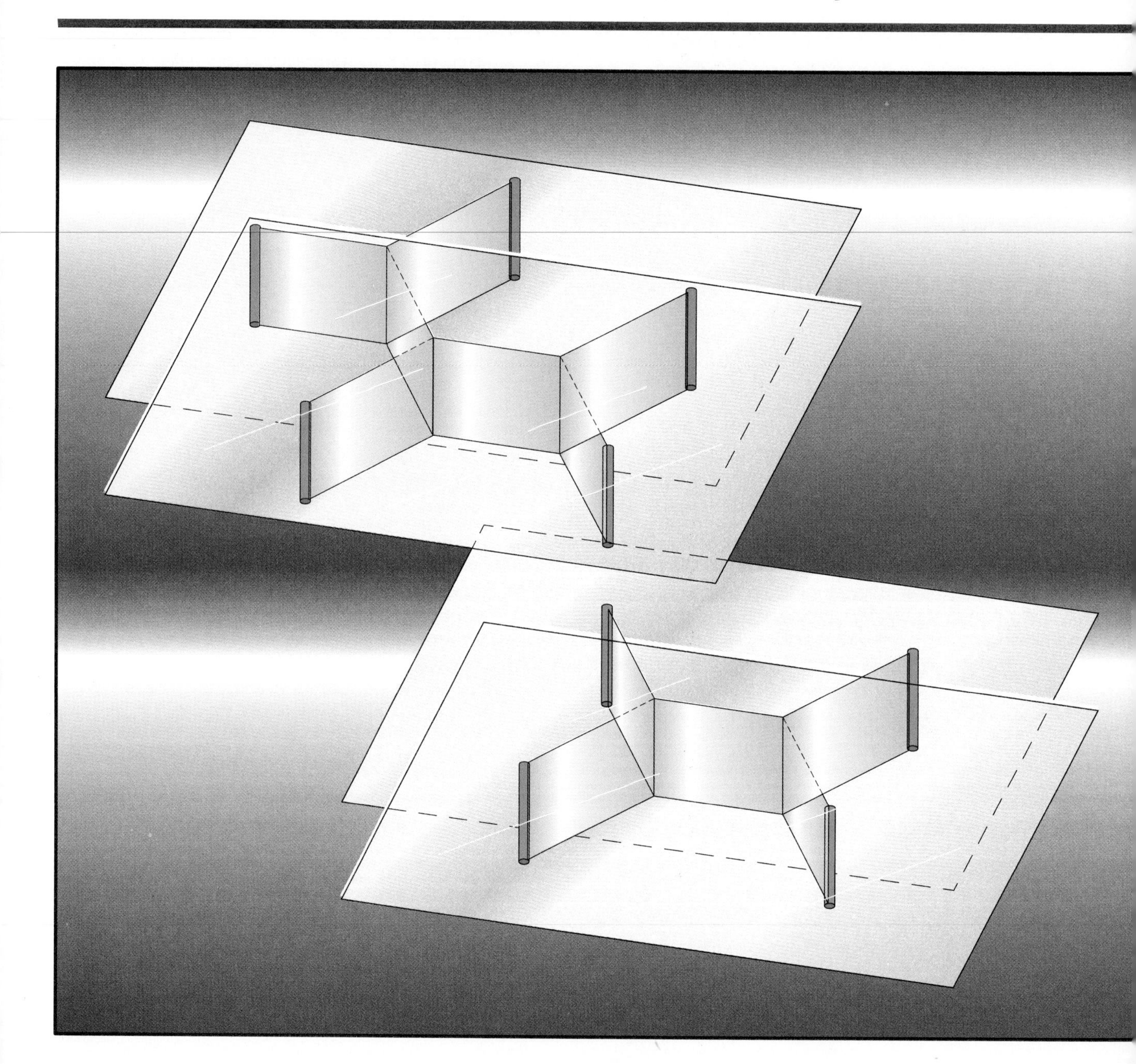

# 8

# The Definite Integral

Richard Courant
(1888–1972)

Richard Courant represents one of the richest traditions in applied mathematics. After his early training in mathematics, physics, and philosophy in Breslau, Courant went to Göttingen in central Germany to obtain his Ph.D. Since the days of the great Karl Gauss (c. 1800), the university in this small city had enjoyed the reputation of being the center of the world in mathematics and physics. It was *the* place to study if one had mathematical talent. There Courant came under the influence of the greatest mathematician of the day, David Hilbert. Hilbert recognized both the mathematical and administrative gifts of the young Courant, made him his assistant, and later helped to secure him the position of director of the mathematics institute.

As director, Courant was determined to continue and increase the influence of the institute, and he was successful in attracting a majority of the world's greatest mathematicians to either permanent or visiting positions. While Courant valued pure mathematics, he was ever concerned that mathematical research should both grow out of and help solve the problems of the real world. He felt that what distinguished the institute at Göttingen was its integration of mathematics and science.

Within months after he came to power, Adolf Hitler began to destroy what Courant and his predecessors had built because many of the brightest stars at the institute (including Courant) were Jewish. Fortunately, most of them were able to come to the United States. Before long, Courant was beginning to build at New York University a mathematical center (now called the Courant Institute of Mathematical Sciences) modeled after the one at Göttingen. Here the great tradition of Göttingen lives on.

While Courant is said to have been egotistical, self-serving, and often controversial, all agree that he was a shrewd judge of mathematical talent, an exceptional administrator, an able mathematician, and a superb mathematical writer. His books, always with an applied flavor, include *Differential and Integral Calculus, What Is Mathematics* (with Herbert Robbins), and *Methods of Mathematical Physics* (with David Hilbert).

## 8.1
## AREA AND THE DEFINITE INTEGRAL

We used the geometric notion of *slope of a curve* to introduce the first of the two most important concepts in calculus, the derivative. Another geometric idea, *area of a curved region,* will lead to the second of these concepts, the definite integral.

What is area? This is not a simple question, and the definition for curved regions must finally be based on the notion of limit just as was that of slope. We will get at this question in Section 8.4. However, for now we ask you to trust your intuition and accept the following properties of area.

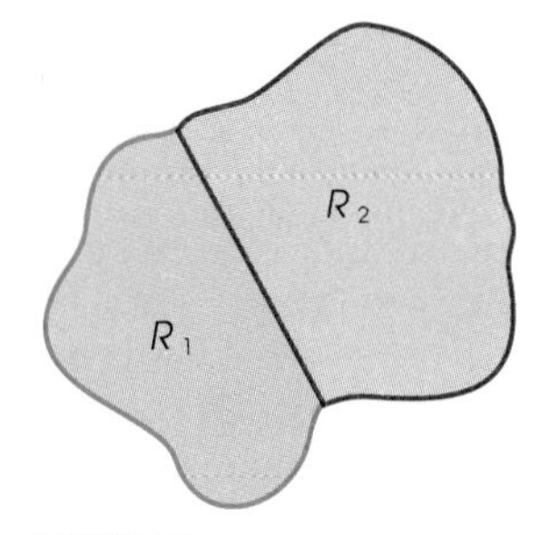

FIGURE 1

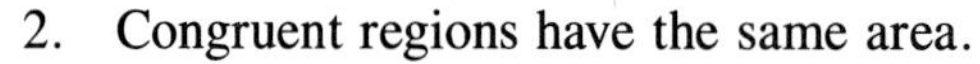
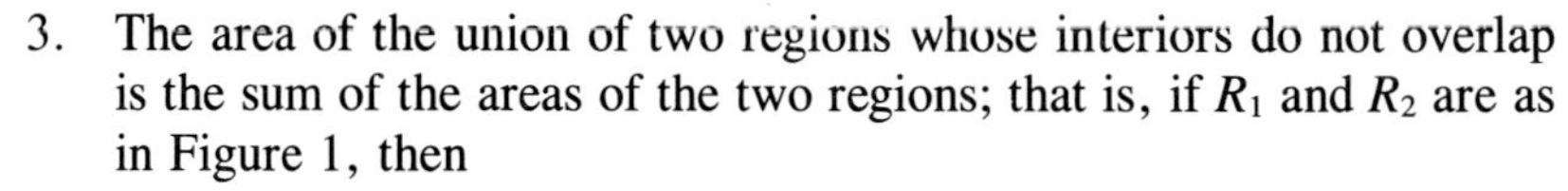

1. The area of a plane region is a nonnegative number.
2. Congruent regions have the same area.
3. The area of the union of two regions whose interiors do not overlap is the sum of the areas of the two regions; that is, if $R_1$ and $R_2$ are as in Figure 1, then

$$A(R_1 \cup R_2) = A(R_1) + A(R_2)$$

4. The area of a rectangle is the product of its length and breadth.

We remind you of some area formulas that you probably learned in grade school (Figure 2).

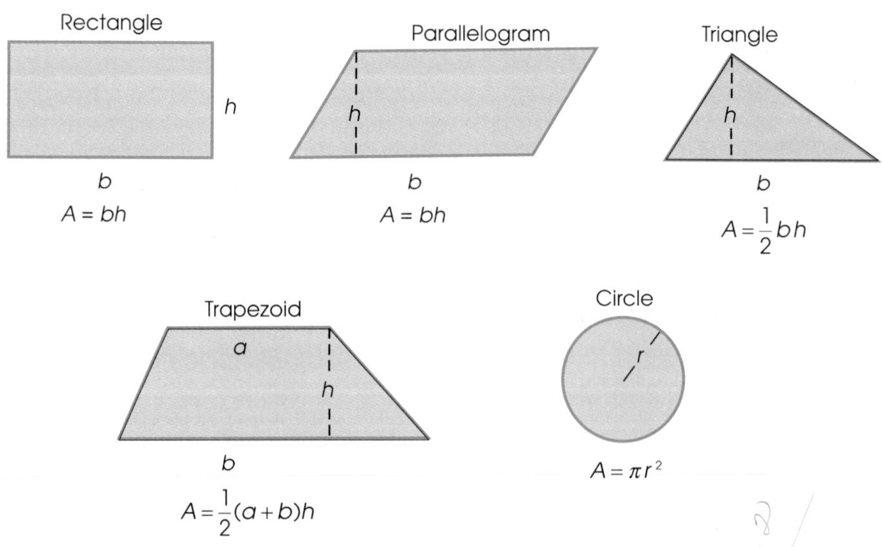

FIGURE 2

### EXAMPLE A

Find the area of the region $R = R_1 \cup R_2 \cup R_3$ shown in Figure 3.

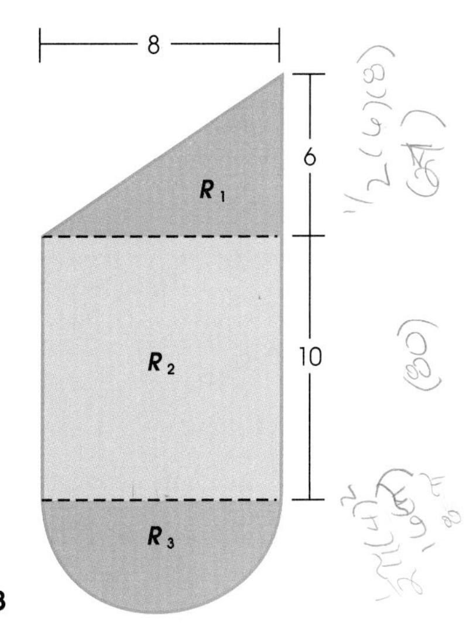


FIGURE 3

**Solution**

$$\begin{aligned} A(R) &= A(R_1) + A(R_2) + A(R_3) \\ &= (\tfrac{1}{2})(8)(6) + (8)(10) + (\tfrac{1}{2})(\pi)(4^2) \\ &= 24 + 80 + 8\pi \approx 129.13 \end{aligned}$$

■

## The Definite Integral

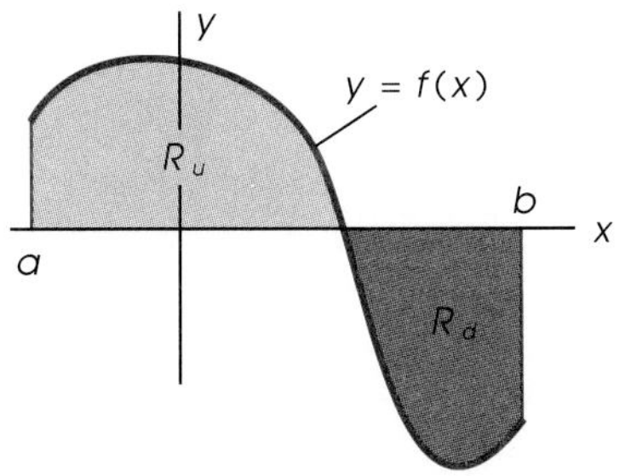

Consider a region $R$ in the $xy$-plane bounded by a curve $y = f(x)$ and the $x$-axis and lying between $x = a$ and $x = b$ (Figure 4). In general $R$ will consist of two parts: $R_u$ (the part above the $x$-axis and referred to as $R$ up) and $R_d$ (the part below the $x$-axis and referred to as $R$ down). We define the definite integral from $a$ to $b$ in terms of the areas of these two regions as follows.

FIGURE 4

> **DEFINITION**
> The symbol $\int_a^b f(x)\,dx$, called the **definite integral of $f(x)$ from $a$ to $b$,** is defined to be the area of $R_u$ minus the area of $R_d$. That is,
>
> $$\int_a^b f(x)\,dx = A(R_u) - A(R_d)$$

This innocent-sounding definition has far-reaching consequences that we will be exploring in both this chapter and the next. Our first job is to make sure that we are able to calculate some simple definite integrals.

### EXAMPLE B

Let $f(x) = -\frac{1}{2}x$ on the interval $-2$ to 4. Sketch its graph and then calculate $\int_{-2}^{4} f(x)\,dx$.

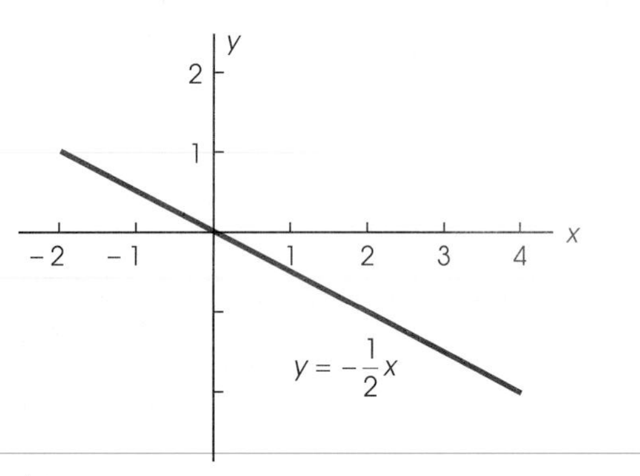


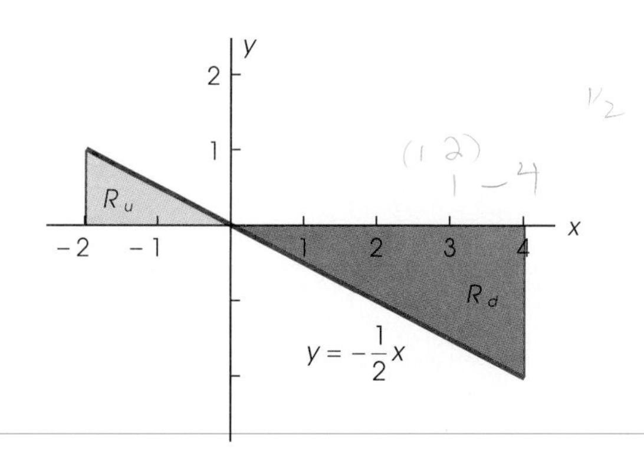


FIGURE 5

**Solution** The graph is shown on the left in Figure 5. The diagram on the right of this figure shows $R_u$ and $R_d$.

$$\int_{-2}^{4} f(x)\,dx = A(R_u) - A(R_d) = 1 - 4 = -3 \qquad \blacksquare$$

### EXAMPLE C

Let the graph of $f(x)$ be as shown in the left part of Figure 6. Calculate $\int_{-1}^{3} f(x)\,dx$.

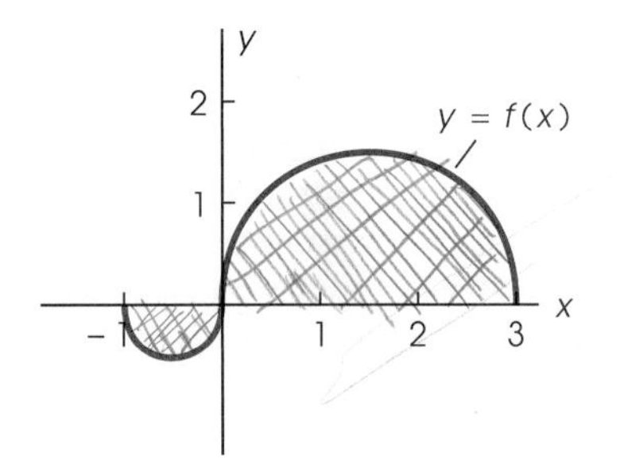


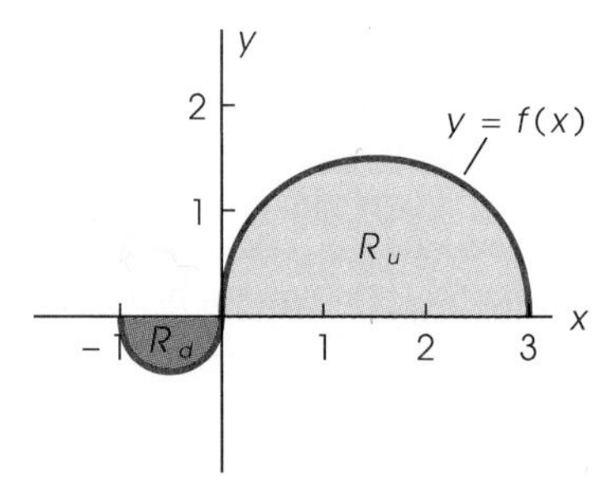


FIGURE 6

**Solution** The regions $R_u$ and $R_d$ are shown in the right part of Figure 6; both are semicircles. Thus,

$$\int_{-1}^{3} f(x)\,dx = A(R_u) - A(R_d) = \frac{\pi(1.5)^2}{2} - \frac{\pi(0.5)^2}{2} \approx 3.14 \qquad \blacksquare$$

### EXAMPLE D

Sketch the graph and calculate $\int_{-1}^{2} f(x)\,dx$ for

$$f(x) = \begin{cases} 3x & -1 \le x \le 1 \\ -x + 1 & 1 < x \le 2 \end{cases}$$

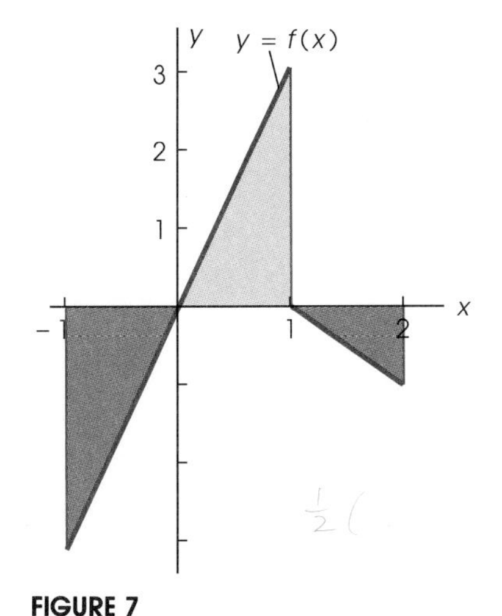

FIGURE 7

**Solution** The graph of $f$ and the regions $R_u$ and $R_d$ are shown in Figure 7. Note that between $-1$ and 1, the part below the $x$-axis and the part above the $x$-axis are the same size, and so their contribution to the integral is 0. The contribution on the interval 1 to 2 is $-\frac{1}{2}$. We conclude that

$$\int_{-1}^{2} f(x)\,dx = -\frac{1}{2}$$

■

## Use of Symmetry

Example D suggests a way to use symmetry as an aid in evaluating definite integrals. We explore this further now. Recall from Section 1.2 that if $f$ is an odd function [meaning that $f(-x) = -f(x)$], then its graph is symmetric with respect to the origin. Similarly, if $f$ is an even function [meaning that $f(-x) = f(x)$], then its graph is symmetric with respect to the $y$-axis. For example, $f(x) = x^3$ is an odd function, while $g(x) = x^2$ is an even function (Figure 8). On an interval $[-a, a]$ that is symmetric about the origin, the

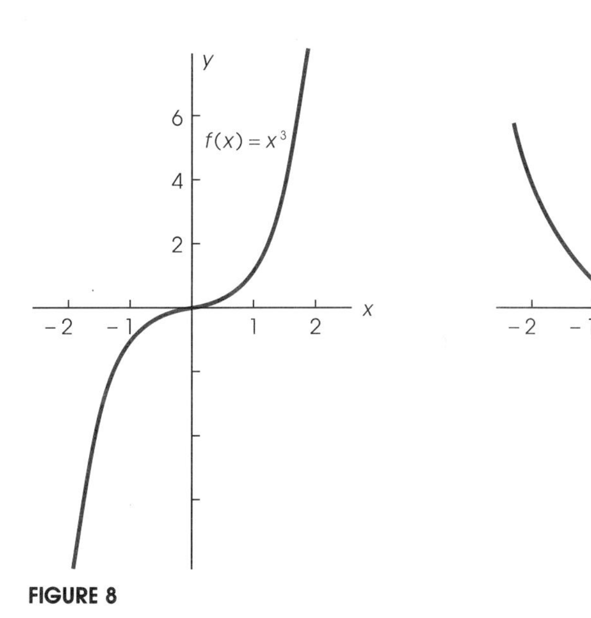

FIGURE 8

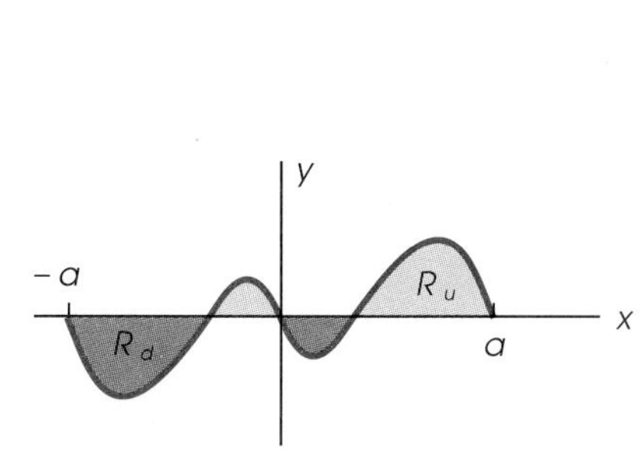

Odd function: $A(R_u) = A(R_d)$

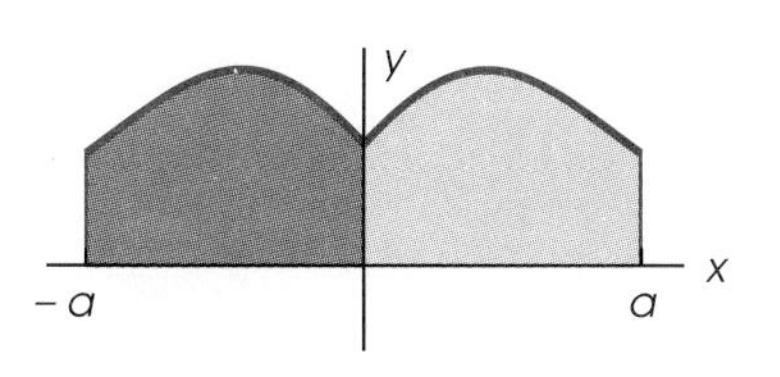

Even function: $A(R_\ell) = A(R_r)$

FIGURE 9

graph of an odd function and the $x$-axis will enclose a region with the same amount of area above the $x$-axis as below this axis (Figure 9). On the other hand, the region bounded by the graph of an even function and the $x$-axis on the interval $[-a, a]$ has exactly the same area to the left of the $y$-axis as to the right of this axis. These facts lead to the following conclusions.

1. If $f$ is an odd function, $\int_{-a}^{a} f(x)\,dx = 0$.
2. If $f$ is an even function, $\int_{-a}^{a} f(x)\,dx = 2\int_{0}^{a} f(x)\,dx$.

**EXAMPLE E**

In the next section, we will learn that

$$\int_0^a x^n\,dx = \frac{a^{n+1}}{n+1} \qquad n \neq -1$$

Use this fact and symmetry considerations to calculate (a) $\int_{-4}^{4} x^2\,dx$, (b) $\int_{-2}^{2} 5x^3\,dx$, (c) $\int_{-1}^{1} x^6\,dx$

**Solution**

(a) $$\int_{-4}^{4} x^2\,dx = 2\int_0^4 x^2\,dx = \frac{2(4^3)}{3} = \frac{128}{3}$$

(b) $$\int_{-2}^{2} 5x^3\,dx = 0, \text{ since } f(x) = 5x^3 \text{ is an odd function}$$

(c) $$\int_{-1}^{1} x^6\,dx = 2\int_0^1 x^6\,dx = \frac{2(1^7)}{7} = \frac{2}{7}$$ ■

## Problem Set 8.1

*In Problems 1–10, make use of standard area formulas to find the area of each shaded region.*

1. 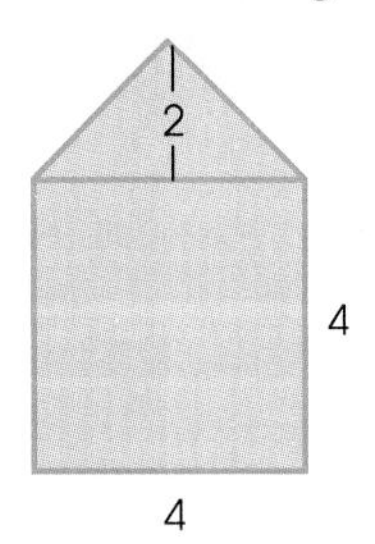


2. 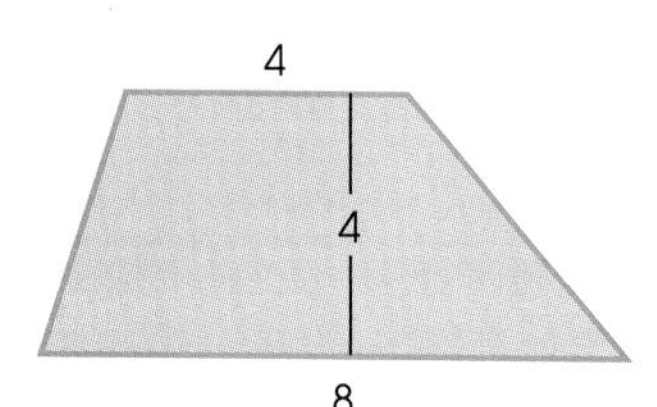


3. 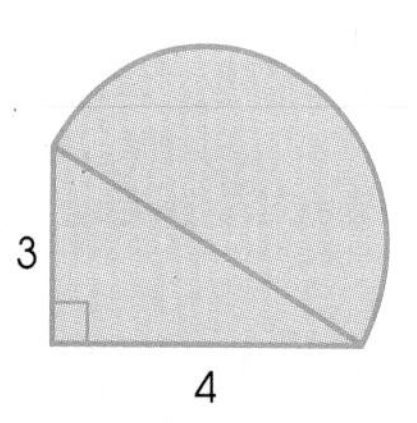


4. 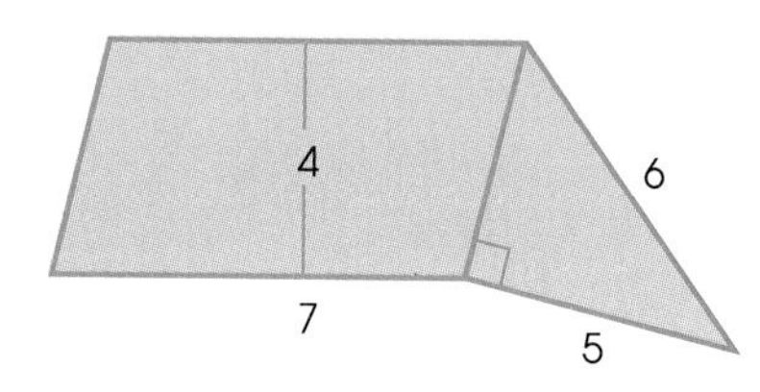


5. 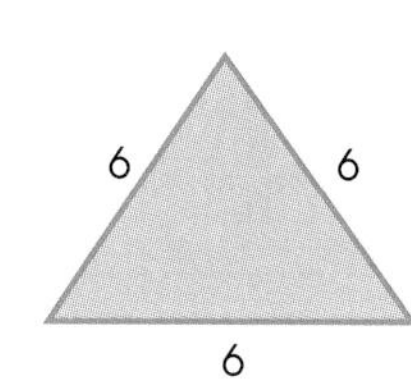


6. 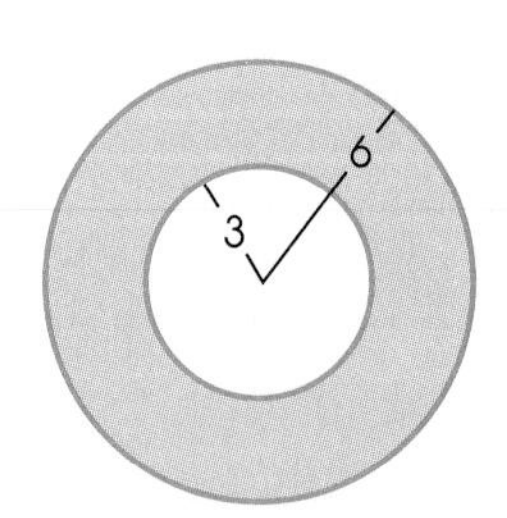

7.

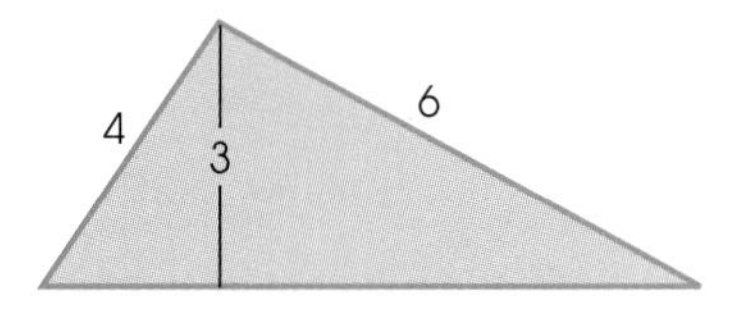


8.

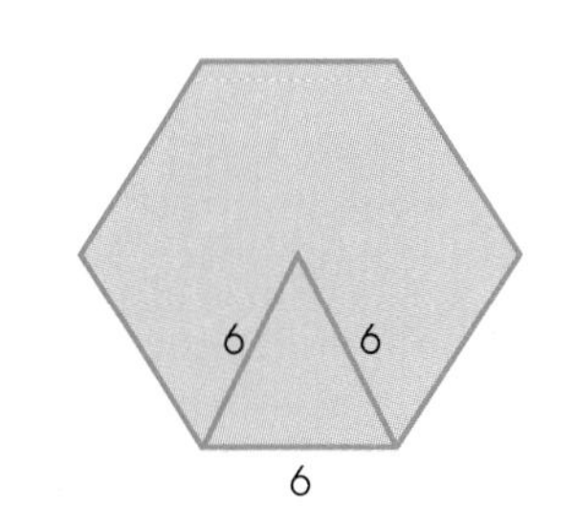


9.

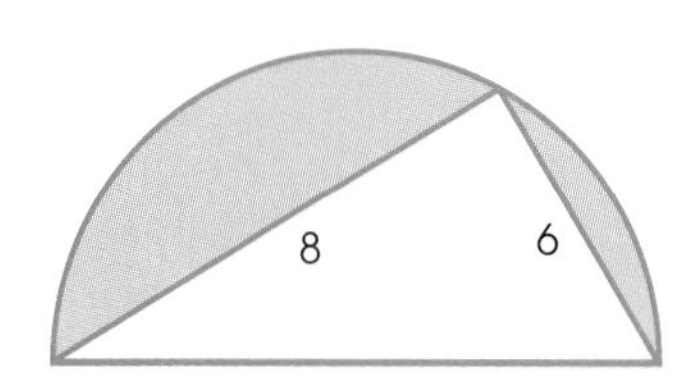


10.

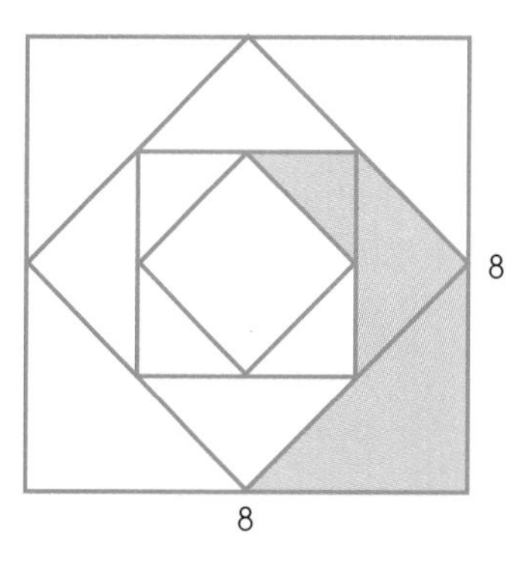


*In Problems 11–22, sketch the graph of f on the interval from a to b and then calculate $\int_a^b f(x)\,dx$. In Problems 17–22, a and b are understood to be the left and right end points of the given domain.*

11. $f(x) = 2x;\ a = -1,\ b = 2$
12. $f(x) = 1.5x;\ a = -4,\ b = 1$
13. $f(x) = -0.5x + 2;\ a = 0,\ b = 6$
14. $f(x) = 2x - 3;\ a = 0,\ b = 4$
15. $f(x) = 3x - 6;\ a = 0,\ b = 2$
16. $f(x) = -3x + 6;\ a = 2,\ b = 6$
17. $f(x) = \begin{cases} 2x & -1 \le x \le 1 \\ 2 & 1 < x \le 3 \end{cases}$
18. $f(x) = \begin{cases} x + 2 & -2 \le x < 0 \\ -x + 2 & 0 \le x \le 4 \end{cases}$
19. $f(x) = \begin{cases} |x| & -2 \le x < 2 \\ 4 - x & 2 \le x \le 5 \end{cases}$
20. $f(x) = \begin{cases} |x - 2| & 0 \le x \le 4 \\ 8 - 2x & 4 < x \le 7 \end{cases}$
21. $f(x) = \begin{cases} x - 1 & 0 \le x \le 2 \\ 1 & 2 < x \le 3 \\ -x + 3 & 3 < x \le 5 \end{cases}$
22. $f(x) = \begin{cases} -x & -2 \le x < 2 \\ x & 2 \le x \le 4 \\ 5 - x & 4 < x \le 5 \end{cases}$
23. Calculate $\int_{-2}^{8} f(x)\,dx$, where $f$ is the function whose graph is shown below.

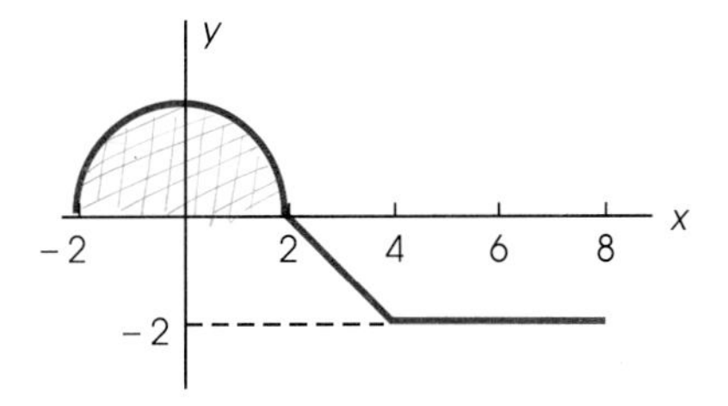


24. Calculate $\int_{-3}^{3} f(x)\,dx$ for the function $f$ whose graph is shown below.

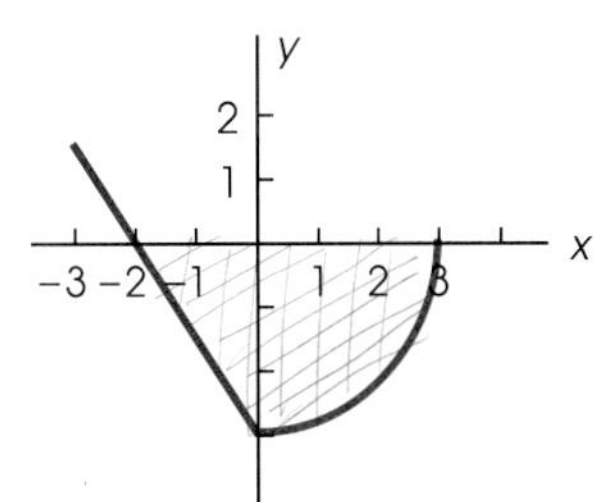


25. Calculate $\int_{-2}^{5} f(x)\,dx$ for the function $f$ whose graph is shown below.

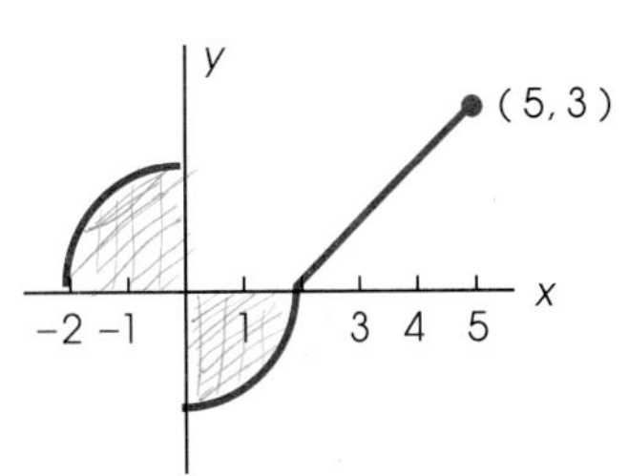

26. Calculate $\int_{-1}^{8} f(x)\,dx$ for the function $f$ whose graph is shown below.

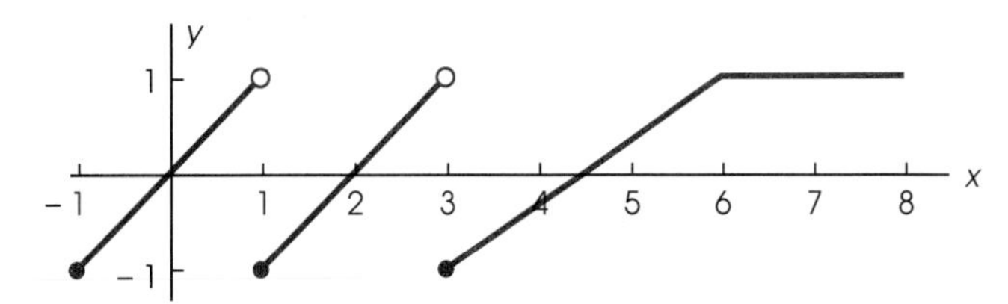


*Use the formula $\int_0^a x^n\,dx = a^{n+1}/(n+1)$ to evaluate each of the definite integrals in Problems 27–32.*

27. $\int_0^3 x^2\,dx$

28. $\int_0^2 x^3\,dx$

29. $\int_0^{\pi} x^4\,dx$

30. $\int_0^{\sqrt{2}} x^3\,dx$

31. $\int_0^4 \sqrt{x}\,dx$

32. $\int_0^8 x^{2/3}\,dx$

33. Decide whether the given function is even, odd, or neither.
(a) $f(x) = 2x^3 - x$ (b) $f(x) = 2x^2 + 5$
(c) $f(x) = x^3 + 2x^2 - 1$
(d) $f(x) = (x^2 + 4)/(x^7 + 4x)$

34. Determine whether the given function is even, odd, or neither.
(a) $f(x) = x^4 - 6$ (b) $f(x) = x^5 - 4x$
(c) $f(x) = |x|$ (d) $f(x) = (x^3 + x)(x^5 - 3x)$

*In Problems 35–42, use symmetry and the formula mentioned before Problem 27 to evaluate each definite integral (see Example E).*

35. $\int_{-2}^{2} 2x^3\,dx$

36. $\int_{-2}^{2} x^2\,dx$

37. $\int_{-1}^{1} x^4\,dx$

38. $\int_{-3}^{3} (x^5 - 4x^3)\,dx$

39. $\int_{-2}^{2} |x|\,dx$

40. $\int_{-4}^{4} \sqrt{|x|}\,dx$

41. $\int_{-5}^{5} 3x/(x^2 + 1)\,dx$

42. $\int_{-\pi}^{\pi} (e^x - e^{-x})\,dx$

---

*In Problems 43–50, sketch the graph of the integrand and then evaluate the definite integral using standard area formulas. Recall that [x] denotes the greatest integer in x.*

43. $\int_0^4 (2x - 3)\,dx$

44. $\int_{-2}^{2} (-3x + 1)\,dx$

45. $\int_0^3 |x - 2|\,dx$

46. $\int_{-1}^{2} |2x - 1|\,dx$

47. $\int_{-3}^{3} \sqrt{9 - x^2}\,dx$

48. $\int_0^4 \sqrt{16 - x^2}\,dx$

49. $\int_0^4 [x]\,dx$

50. $\int_0^4 (x - [x])\,dx$

*If $v(t)$ is the velocity at time $t$ of an object moving along a straight line, then the total distance traveled by the object between time $t = a$ and time $t = b$ is given by $\int_a^b |v(t)|\,dt$. Use this fact to find the total distance traveled by an object having the velocity specified in Problems 51–54. Note that in each case, the velocity is sometimes negative and sometimes positive, which means that the object does some doubling back.*

51. $v(t) = t$; $a = -2$, $b = 3$
52. $v(t) = t^3$; $a = -2$, $b = 2$
53. $v(t) = 2 - t$; $a = 0$, $b = 5$
54. $v(t) = 5 - 2t$; $a = 0$, $b = 4$

## 8.2 THE FUNDAMENTAL THEOREM OF CALCULUS

You know from the title that something very important is coming. It is the theorem that will allow us to evaluate a wide class of definite integrals in an almost trivial manner.

**THE FUNDAMENTAL THEOREM OF CALCULUS**

Let $f(x)$ be continuous on the interval $[a, b]$ and let $F(x)$ be any antiderivative of $f(x)$ there. Then

$$\int_a^b f(x)\,dx = F(b) - F(a)$$

Before offering a proof of this remarkable theorem, we present one example to show you its power.

### EXAMPLE A

Evaluate $\displaystyle\int_1^4 x^2\,dx$.

**Solution** We know from Chapter 7 that $F(x) = x^3/3$ is an antiderivative of $f(x) = x^2$. Thus, by the Fundamental Theorem,

$$\int_1^4 x^2\,dx = F(4) - F(1) = \frac{4^3}{3} - \frac{1^3}{3} = \frac{63}{3} = 21 \qquad \blacksquare$$

## Proof of the Fundamental Theorem

To prepare the way, let us first note that the variable $x$ in the symbol $\int_a^b f(x)\,dx$ is a *dummy variable*. By this we mean that any other variable would serve as well. The result of a definite integration is a number, and this number does not depend on what variable is used in the integrand.

$$\int_a^b f(x)\,dx = \int_a^b f(t)\,dt = \int_a^b f(u)\,du$$

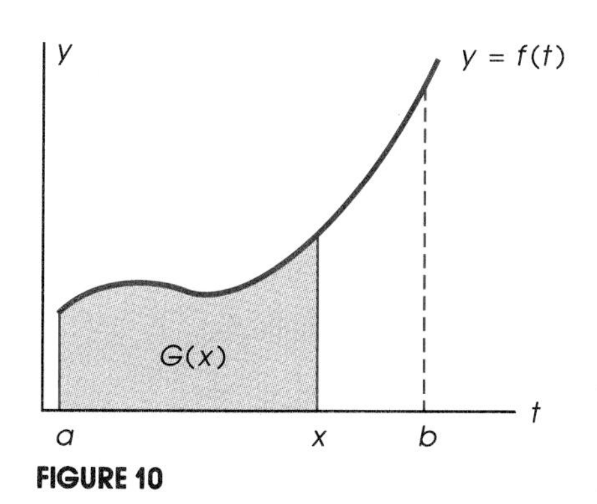


FIGURE 10

We mention this because we now want to let $x$ act as the upper limit on the integral sign, and so we will want to use a different variable in the integrand.

Let $a$ be a fixed number and let $x \geq a$. Consider the function $G$ determined by

$$G(x) = \int_a^x f(t)\,dt$$

Figure 10 gives a geometric interpretation of the function $G$. If $f$ is a nonnegative function, $G(x)$ is the area under the curve $y = f(t)$ between $a$ and $x$. Note that $G(a) = 0$ and $G(b)$ is the definite integral we want to evaluate.

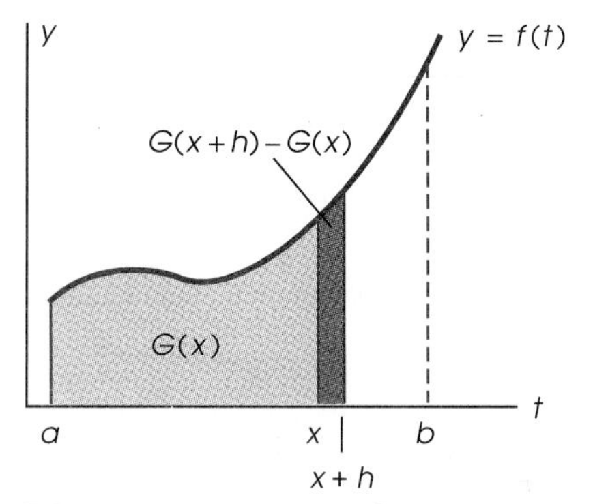


**FIGURE 11**

For reasons that will be clear shortly, we want to find $G'(x)$. From the definition of the derivative, we may write

$$G'(x) = \lim_{h \to 0} \frac{G(x + h) - G(x)}{h}$$

Referring to Figure 11, we see that we can interpret the numerator of this expression (at least if $f$ is nonnegative and $h$ is positive) as the area under the curve between $x$ and $x + h$. But since $f$ is continuous, the value of this area will be very close to $f(x)h$. In fact, it will equal $f(\bar{x})h$, where $\bar{x}$ is some point between $x$ and $x + h$. We conclude that

$$G'(x) = \lim_{h \to 0} \frac{f(\bar{x})h}{h} = \lim_{h \to 0} f(\bar{x}) = f(x)$$

This is just another way of saying that $G(x)$ is an antiderivative of $f(x)$. Now we have two antiderivatives of $f(x)$ in view, namely, $G(x)$ and $F(x)$. Since we know that we can get any antiderivative from another by the addition of a constant, we conclude that there must be a constant $C$ such that $G(x) = F(x) + C$. On the other hand, we noted earlier that $G(a) = 0$, so $F(a) + C = 0$ and consequently $C = -F(a)$. Thus,

$$G(x) = F(x) - F(a)$$

or in integral symbols,

$$\int_a^x f(t)\, dt = F(x) - F(a)$$

When we let $x = b$ in the latter equation, we have the result we wanted.

It should be pointed out that our argument did assume that $f(x)$ was nonnegative and that $h$ was positive. A slight modification will establish the result when these assumptions are dropped.

## The Theorem Illustrated

To increase our confidence in the result, we apply it to the integral of Example B in the previous section where we know the answer should be $-3$.

### EXAMPLE B

Evaluate $\displaystyle\int_{-2}^{4} f(x)\, dx$ where $f(x) = -\dfrac{1}{2}x$.

**Solution** Since $F(x) = -\frac{1}{2}x^2/2 = -\frac{1}{4}x^2$ is an antiderivative of $f(x)$, we obtain

$$\int_{-2}^{4} -\frac{1}{2}x\, dx = F(4) - F(-2) = \left(-\frac{1}{4}\right)4^2 - \left(-\frac{1}{4}\right)(-2)^2$$
$$= -4 + 1 = -3$$

■

### EXAMPLE C

Evaluate $\int_0^2 3e^x\,dx$.

**Solution** Since $F(x) = 3e^x$ is an antiderivative of $f(x) = 3e^x$, we see that

$$\int_0^2 3e^x\,dx = F(2) - F(0) = 3e^2 - 3e^0 \approx 19.167 \qquad \blacksquare$$

Because the expression $F(b) - F(a)$ comes up in every integration problem, we introduce a special notation for it, namely, $[F(x)]_a^b$. Thus,

$$\Big[x^2 - 3\Big]_2^4 = 4^2 - 3 - (2^2 - 3) = 13 - 1 = 12$$

### EXAMPLE D

Evaluate $\int_1^2 (4x^3 + 3x)\,dx$.

**Solution**

$$\int_1^2 (4x^3 + 3x)\,dx = \left[x^4 + \frac{3x^2}{2}\right]_1^2 = 16 + 6 - \left(1 + \frac{3}{2}\right) = \frac{39}{2} \qquad \blacksquare$$

### EXAMPLE E

Find the area of the region under the curve $y = \sqrt{x}$ between $x = 1$ and $x = 4$ (Figure 12).

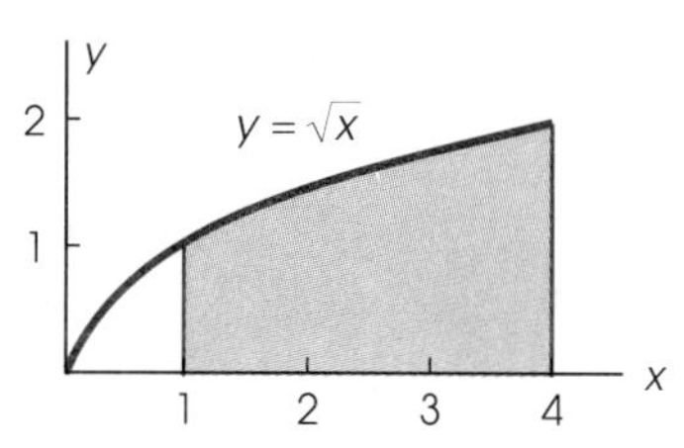


FIGURE 12

**Solution** We calculate this area by means of an integral.

$$\int_1^4 \sqrt{x}\,dx = \int_1^4 x^{1/2}\,dx = \left[\frac{2}{3}x^{3/2}\right]_1^4 = \frac{16}{3} - \frac{2}{3} = \frac{14}{3} \qquad \blacksquare$$

## Some Properties of Definite Integrals

Here are three properties that will often be used.

1. $\int_a^b cf(x)\,dx = c\int_a^b f(x)\,dx$, $c$ a constant.
2. $\int_a^b [f(x) \pm g(x)]\,dx = \int_a^b f(x)\,dx \pm \int_a^b g(x)\,dx$.
3. $\int_a^b f(x)\,dx = \int_a^c f(x)\,dx + \int_c^b f(x)\,dx$.

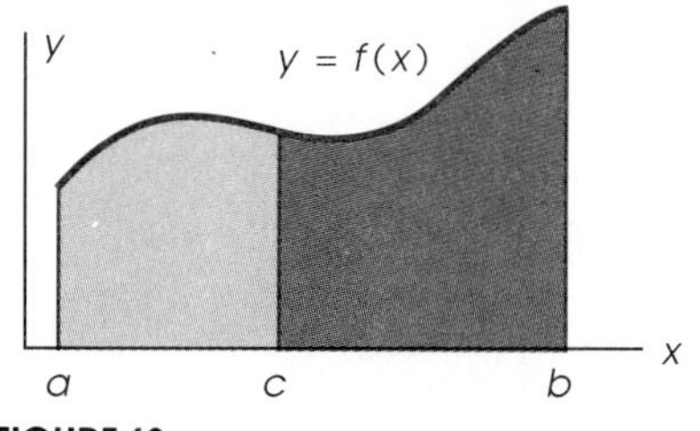


FIGURE 13

The first two properties are inherited from the corresponding properties for antiderivatives. You can see why the third property is true by studying Figure 13, at least for the case where $a < c < b$. Actually, Property 3 is valid no matter how $a$, $b$, and $c$ are arranged provided we make the following agreement.

$$\boxed{\int_b^a f(x)\,dx = -\int_a^b f(x)\,dx}$$

## EXAMPLE F

If $\int_1^3 f(x)\,dx = 4$ and $\int_1^3 g(x)\,dx = -2$, evaluate $\int_1^3 [3f(x) + 5g(x)]\,dx$

**Solution** We use Property 2 and then Property 1 to write

$$\begin{aligned}\int_1^3 [3f(x) + 5g(x)]\,dx &= \int_1^3 3f(x)\,dx + \int_1^3 5g(x)\,dx \\ &= 3\int_1^3 f(x)\,dx + 5\int_1^3 g(x)\,dx \\ &= 3(4) + 5(-2) = 2\end{aligned}$$

■

## EXAMPLE G

Evaluate $\int_{-2}^4 f(x)\,dx$ where

$$f(x) = \begin{cases} -2x & -2 \le x \le 0 \\ x^2 - 2 & 0 < x \le 2 \\ -2x + 8 & 2 < x \le 4 \end{cases}$$

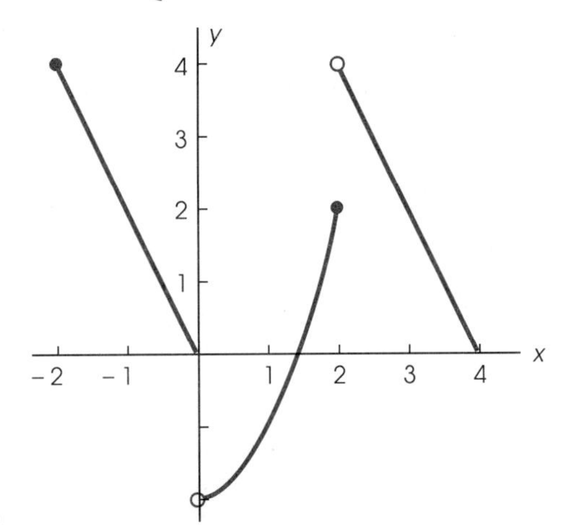


FIGURE 14

**Solution** The graph of $f$ is shown in Figure 14. To evaluate the integral we will need to break it into three parts by using Property 3.

$$\begin{aligned}\int_{-2}^4 f(x)\,dx &= \int_{-2}^0 (-2x)\,dx + \int_0^2 (x^2 - 2)\,dx + \int_2^4 (-2x + 8)\,dx \\ &= \Big[-x^2\Big]_{-2}^0 + \left[\frac{x^3}{3} - 2x\right]_0^2 + \Big[-x^2 + 8x\Big]_2^4 \\ &= [0 - (-4)] + \left[\frac{8}{3} - 4 - (0 - 0)\right] + [-16 + 32 - (-4 + 16)] \\ &= \frac{20}{3}\end{aligned}$$

■

## The Average Value of a Function

We are all familiar with the notion of the *average* of a set of scores. It is simply the sum of the scores divided by the number of scores. Thus, if $S_1$, $S_2$, $S_3$, . . . , $S_n$ are the scores on a test taken by a class of $n$ students, the average score is

$$S_{\text{ave}} = \frac{1}{n}(S_1 + S_2 + S_3 + \cdots + S_n)$$

But what shall we mean by the average value of a function? It won't make sense to add up all its values and divide by the number of values, because a function has infinitely many values.

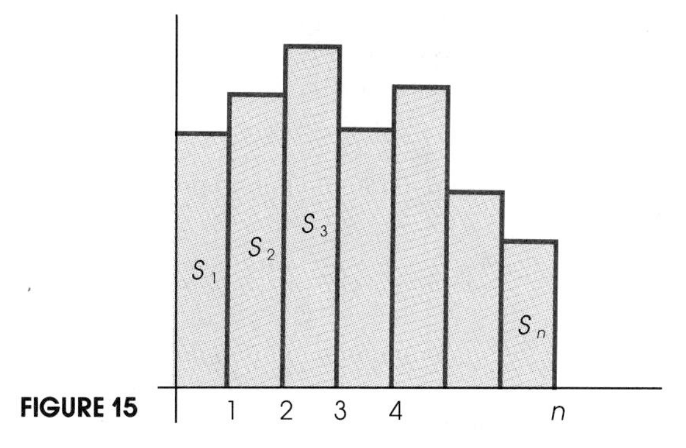

FIGURE 15

We can reinterpret the formula for $S_{\text{ave}}$ displayed above in a geometric way. Refer to Figure 15 and note that $S_1$ is simply the area of the first rectangle, $S_2$ is the area of the second rectangle, and so on. Moreover, $S_{\text{ave}}$ is just the sum of the areas of all the rectangles divided by the length of the interval $[0, n]$. This suggests a definition for the average value of a function. The **average value** of the function $f$ on the interval $[a, b]$ is defined by

$$f_{\text{ave}} = \frac{1}{b-a}\int_a^b f(x)\,dx$$

We will give a more adequate basis for this definition in Section 8.4.

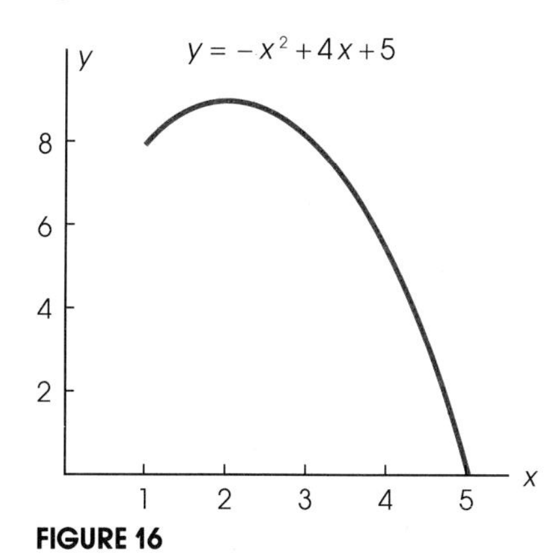

FIGURE 16

### EXAMPLE H

Find the average value of $f(x) = -x^2 + 4x + 5$ on the interval $[1, 5]$.

**Solution** Refer to Figure 16 which should suggest to you that the average value of $f$ is something like 6 or 7. By our definition

$$f_{\text{ave}} = \frac{1}{5-1}\int_1^5 (-x^2 + 4x + 5)\,dx = \frac{1}{4}\left[-\frac{1}{3}x^3 + 2x^2 + 5x\right]_1^5$$

$$= \frac{1}{4}\left(-\frac{125}{3} + 50 + 25\right) - \frac{1}{4}\left(-\frac{1}{3} + 2 + 5\right) \approx 6.66667 \qquad \blacksquare$$

## Problem Set 8.2

*In problems 1–4, evaluate the given definite integral in two ways: (i) by drawing a sketch and then calculating $A(R_u) - A(R_d)$, and (ii) by using the Fundamental Theorem of Calculus.*

1. $\int_{-1}^{3} 2x\,dx$
2. $\int_{-3}^{1} (x + 2)\,dx$
3. $\int_{0}^{3} (-x + 1)\,dx$
4. $\int_{0}^{3} (-2x + 4)\,dx$

*Calculate the definite integrals in Problems 5–20 by using the Fundamental Theorem of Calculus.*

5. $\int_{-1}^{2} -\frac{3}{4}x\,dx$
6. $\int_{-2}^{3} -\frac{2}{3}x\,dx$
7. $\int_{-2}^{3} 6x^2\,dx$
8. $\int_{1}^{2} 4x^3\,dx$
9. $\int_{1}^{3} (3x^2 + 2x - 8)\,dx$
10. $\int_{0}^{2} (x^2 + 4x - 3)\,dx$
11. $\int_{0}^{2} (2x^3 - x - 3)\,dx$
12. $\int_{-1}^{2} (5x^4 - 3x)\,dx$
13. $\int_{-2}^{2} (x - 2)^2\,dx$
14. $\int_{-1}^{1} (x + 1)^3\,dx$
15. $\int_{1}^{8} 4\sqrt[3]{x}\,dx$
16. $\int_{1}^{32} 14x^{2/5}\,dx$
17. $\int_{1}^{e} \frac{1}{x}\,dx$
18. $\int_{1}^{3} -2x^{-2}\,dx$
19. $\int_{0}^{1} (5e^x - 2x)\,dx$
20. $\int_{0}^{1} (4e^{-x} + 1)\,dx$

*In Problems 21–24, sketch the graph of the given function on the interval [a, b] and then find the area of the region under the graph (see Example E).*

21. $f(x) = x^2 + 4;\ a = -1,\ b = 3$
22. $f(x) = x^2 + 4x + 4;\ a = -2,\ b = 1$
23. $f(t) = t^{2/3};\ a = 0,\ b = 8$
24. $f(t) = t^{3/2};\ a = 0,\ b = 4$

*Suppose that $\int_{-1}^{3} f(x)\,dx = -8$, $\int_{3}^{5} f(x)\,dx = 4$, $\int_{-1}^{3} g(x)\,dx = 5$, and $\int_{3}^{5} g(x)\,dx = 2$. Use the properties of the definite integral to calculate the definite integrals in Problems 25–36. See Example F and the discussion preceding it.*

25. $\int_{3}^{-1} f(x)\,dx$
26. $\int_{5}^{3} g(x)\,dx$
27. $\int_{-1}^{3} -2f(x)\,dx$
28. $\int_{-1}^{3} 0.4\,g(x)\,dx$
29. $\int_{3}^{5} [2f(x) + 3g(x)]\,dx$
30. $\int_{3}^{5} [2g(x) - f(x)]\,dx$
31. $\int_{-1}^{5} 3f(x)\,dx$
32. $\int_{-1}^{5} -2g(x)\,dx$
33. $\int_{-1}^{5} [2f(x) + 3]\,dx$
34. $\int_{-1}^{3} [4 - 2g(x)]\,dx$
35. $\int_{-1}^{7} g(x)\,dx + \int_{7}^{5} g(x)\,dx$
36. $\int_{-1}^{7} f(x)\,dx - \int_{5}^{7} f(x)\,dx$

*In Problems 37–40, evaluate $\int_{-2}^{3} f(x)\,dx$ as in Example G.*

37. $f(x) = \begin{cases} x^2 + 4x + 3 & -2 \le x < 0 \\ -x + 3 & 0 \le x \le 3 \end{cases}$

38. $f(x) = \begin{cases} x + 1 & -2 \le x < 1 \\ x^2 + x - 2 & 1 \le x \le 3 \end{cases}$

39. $f(x) = \begin{cases} x^2 & -2 \le x < 0 \\ x & 0 \le x < 2 \\ 4 - x^2 & 2 \le x \le 3 \end{cases}$

40. $f(x) = \begin{cases} -x/2 & -2 \le x < 0 \\ x - 1 & 0 \le x < 1 \\ x^2 - 4 & 1 \le x \le 3 \end{cases}$

*Find the average value of f on the given interval in Problems 41–46. Make a guess first (see Example H).*

41. $f(x) = x^3 + 1$ on $[0, 2]$
42. $f(x) = 3x^2 - 4x$ on $[0, 3]$
43. $f(x) = x^3 + 1$ on $[-1, 2]$
44. $f(x) = 3x^2 - 4x$ on $[0, 2]$
45. $f(x) = 4 - x^2$ on $[-1, 2]$
46. $f(x) = e^x$ on $[-1, 2]$

*In Problems 47–58, perform the indicated calculation.*

47. $\int_{1}^{4} (5x^{3/2} - 3\sqrt{x})\,dx$
48. $\int_{1}^{4} (x^{-2} + 2x^{-1})\,dx$
49. $\int_{0}^{1} (e^x + x^e)\,dx$
50. $\int_{-1}^{1} (x^3 - 2)^2\,dx$

51. $\int_0^3 2|t-2|\,dt$

52. $\int_{-5}^0 4|u+3|\,du$

53. $\int_0^3 f'(x)\,dx$ where $f(x) = 12/(3+x^2)$

54. $\int_0^4 g'(t)\,dt$ where $g(t) = e^{\sqrt{t}}$

55. $G'(0)$ where $G(x) = \int_{-1}^x e^t\,dt$

56. $H'(1)$ where $H(x) = \int_0^x \frac{4}{9+t^2}\,dt$

57. $\int_0^1 2x\,e^{x^2}\,dx$. *Hint:* Make the substitution $u = x^2$.

58. $\int_0^1 3x^2(1+x^3)^4\,dx$

59. The base of a solid is a rectangle 3 inches by 10 inches. The solid is rectangular but with a curved top. The height of the solid $x$ inches from the left end is $e^x$. Find the volume of this solid if the left end is (a) the short end, (b) the long end.

60. Find the area of the region under the curve $y = xe^x$ on the interval $0 \le x \le 2$.

61. The temperature $T$ (in degrees Fahrenheit) in an auditorium during a 12-hour period beginning at 6:00 A.M. was given by $68 + 0.5t + 0.03t^2$, $t$ being the number of hours after 6:00 A.M.
   (a) What was the temperature at 2:00 P.M.?
   (b) What was the average temperature over the 12-hour period?

62. The speed $v$ of a motorist $t$ seconds after she applied the brakes was given by $v = 96 - 2t^2$ in feet per second.
   (a) How long did it take her to come to a full stop?
   (b) How far did she travel during this time interval?
   (c) What was her average speed during this time interval?

63. Suppose that the weight of a steer $t$ months after birth is given by $W = 40 + 50t - 0.6t^2$ pounds for $0 \le t \le 20$. Find the average weight of the steer during this 20-month period.

## 8.3 CALCULATION OF DEFINITE INTEGRALS

We now want to consider the calculation of more complicated definite integrals, integrals in which a substitution will be required. As a first example, consider the problem of evaluating

$$\int_0^1 (2x+1)^5\,dx$$

One way to handle this problem is by a two-step process. First, we find the corresponding indefinite integral; then we apply the Fundamental Theorem of Calculus. This is how it works. From the Substitution Rule (Section 7.2), we may let $u = 2x + 1$ and $du = 2\,dx$ to obtain

$$\int (2x+1)^5\,dx = \frac{1}{2}\int (2x+1)^5\,2\,dx = \frac{1}{2}\int u^5\,du$$

$$= \frac{1}{2}\frac{u^6}{6} + C = \frac{1}{12}(2x+1)^6 + C$$

Then by the Fundamental Theorem of Calculus, we may write

$$\int_0^1 (2x+1)^5\,dx = \left[\frac{1}{12}(2x+1)^6\right]_0^1$$

$$= \frac{3^6}{12} - \frac{1^6}{12} = \frac{728}{12} \approx 60.67$$

Is there a way to short-cut this two-step process? The answer is yes. It is called **substitution in definite integrals** and is useful when we cannot antidifferentiate the integrand in our heads.

## Substitution in Definite Integrals

Substitution in definite integrals is just like that in indefinite integrals, with the exception that when we change from $x$ to $u$, we must make the corresponding change in the limits on the integral sign. We illustrate by redoing the example above.

### EXAMPLE A

Evaluate $\int_0^1 (2x + 1)^5 \, dx$.

**Solution** Let $u = 2x + 1$, so $du = 2\,dx$, but now also note that when $x = 0$, $u = 1$ and when $x = 1$, $u = 3$.Thus,

$$\int_0^1 (2x + 1)^5 \, dx = \frac{1}{2}\int_1^3 u^5 \, du = \left[\frac{u^6}{12}\right]_1^3 = \frac{3^6}{12} - \frac{1^6}{12}$$

$$= \frac{728}{12} \approx 60.67$$

Note that the solution can be written as one chain of equalities; it is simple and elegant. ■

### EXAMPLE B

Evaluate $\int_3^6 [3/(x - 2)] \, dx$.

**Solution** Let $u = x - 2$ so $du = dx$ and note that $u = 1$ when $x = 3$ and $u = 4$ when $x = 6$. Thus,

$$\int_3^6 \frac{3}{x - 2} \, dx = 3\int_1^4 \frac{1}{u} \, du = 3\Big[\ln |u|\Big]_1^4 = 3(\ln 4 - \ln 1) = 3 \ln 4 \approx 4.16$$

■

### EXAMPLE C

Evaluate $\int_0^2 e^{-3t} \, dt$.

**Solution** Let $u = -3t$ so $du = -3dt$ and note that $u = 0$ when $t = 0$ and $u = -6$ when $t = 2$. We conclude that

$$\int_0^2 e^{-3t} \, dt = -\frac{1}{3}\int_0^2 e^{-3t}(-3 \, dt) = -\frac{1}{3}\int_0^{-6} e^u \, du$$

$$= -\frac{1}{3}\Big[e^u\Big]_0^{-6} = -\frac{1}{3}(e^{-6} - e^0) \approx 0.3325$$ ■

**EXAMPLE D**

Evaluate $\int_{\frac{1}{4}}^{1} (\ln 4x)^3 \frac{1}{x}\, dx$.

**Solution** Let $u = \ln 4x$. Then by the Chain Rule, $du = (1/4x)(4)\, dx = (1/x)\, dx$. Also, $u = \ln 1 = 0$ when $x = \frac{1}{4}$ and $u = \ln 4$ when $x = 1$. Thus,

$$\int_{\frac{1}{4}}^{1} (\ln 4x)^3 \frac{1}{x}\, dx = \int_0^{\ln 4} u^3\, du = \left[\frac{u^4}{4}\right]_0^{\ln 4}$$

$$= \frac{(\ln 4)^4}{4} - \frac{0^4}{4} \approx 0.9233$$

■

## Integration by Parts in Definite Integrals

The topic of integration by parts for indefinite integrals was treated in Section 7.3, but that was an optional section so we begin at the beginning. The idea is this. Faced with an integral of the form $\int u(x)v'(x)\, dx$, abbreviated $\int u\, dv$, we may always transform this to an expression involving an integral of the form $\int v\, du$. The latter may be much easier to evaluate than the original integral and, if so, the method gives us a tool for evaluation of integrals. Here is a formal statement of the method as it applies to definite integrals.

**INTEGRATION BY PARTS IN DEFINITE INTEGRALS**

$$\int_a^b u(x)v'(x)\, dx = \Big[u(x)v(x)\Big]_a^b - \int_a^b v(x)u'(x)\, dx$$

or in abbreviated form,

$$\int_a^b u\, dv = \Big[uv\Big]_a^b - \int_a^b v\, du$$

To use integration by parts amounts to making a double substitution, one for $u$ and the other for $dv$, as we now illustrate.

**EXAMPLE E**

Evaluate $\int_1^2 x \ln x\, dx$.

**Solution** We always begin by looking for an ordinary substitution. In this case, none seems to work, so we turn to integration by parts. We first try the double substitution $u = x$ and $dv = \ln x\, dx$ but are stymied because we don't know how to integrate $\ln x$ to obtain $v$. As an alternative, we try letting $u = \ln x$ and $dv = x\,dx$. Then, $du = (1/x)\,dx$ and $v = x^2/2$. The integration by parts formula gives

$$\int_1^2 x \ln x\,dx = \left[\frac{x^2}{2}\ln x\right]_1^2 - \int_1^2 \frac{x^2}{2}\frac{1}{x}\,dx$$

We are no better off unless we can evaluate the latter integral. Fortunately, in this case we can, since the integrand simplifies to $x/2$, which integrates to $x^2/4$. Thus,

$$\begin{aligned}\int_1^2 x \ln x\,dx &= \left[\frac{x^2}{2}\ln x\right]_1^2 - \left[\frac{x^2}{4}\right]_1^2 \\ &= 2\ln 2 - 0 - \left(\frac{4}{4} - \frac{1}{4}\right) \\ &= 2\ln 2 - \frac{3}{4} \approx 0.6363\end{aligned}$$

■

**EXAMPLE F**

Evaluate $\int_1^e \ln x\,dx$.

**Solution** At first glance, this integral would seem to be intractable even to integration by parts. However, the double substitution

$$\begin{aligned} u &= \ln x \qquad & dv &= 1\,dx \\ du &= \frac{1}{x}\,dx \qquad & v &= x \end{aligned}$$

leads to

$$\begin{aligned}\int_1^e \ln x\,dx &= \Big[(\ln x)(x)\Big]_1^e - \int_1^e x\frac{1}{x}\,dx \\ &= (e\ln e - 1\ln 1) - \int_1^e 1\,dx \\ &= e - [x]_1^e = e - (e-1) = 1\end{aligned}$$

■

## Tables of Integrals

We mentioned in Section 7.3 how tables of integrals can help us in finding indefinite integrals. Of course, the same is true for definite integrals. Here is an example which illustrates use of the small table of integrals at the back of the book.

**EXAMPLE G**

Find $\int_0^4 \sqrt{x^2 + 9}\, dx$.

**Solution** Use Integral Formula 44 from the back of the book.

$$\int_0^4 \sqrt{x^2 + 9}\, dx = \left[\frac{x}{2}\sqrt{x^2 + 9} + \frac{9}{2}\ln|x + \sqrt{x^2 + 9}|\right]_0^4$$

$$= 2(5) + \frac{9}{2}\ln(4 + 5) - \frac{9}{2}\ln(0 + 3)$$

$$= 10 + \frac{9}{2}(\ln 9 - \ln 3)$$

$$= 10 + 4.5 \ln\frac{9}{3} = 10 + 4.5 \ln 3 \approx 14.9438$$ ■

## Problem Set 8.3

*Calculate the definite integrals in Problems 1–4 in two ways: (i) by multiplying out the integrand and then integrating, and (ii) by using a substitution.*

1. $\int_0^1 (2x - 1)^2\, 2\, dx$
2. $\int_{-1}^2 (3x + 2)^2\, 3\, dx$
3. $\int_0^2 (x^2 - 3)^2\, x\, dx$
4. $\int_0^1 (x^2 + 4)^2 x\, dx$

*Evaluate the definite integrals in Problems 5–24. You can either antidifferentiate the integrand in your head and apply the Fundamental Theorem of Calculus or use the method illustrated in Examples A–D.*

5. $\int_0^1 (2x - 1)^4\, dx$
6. $\int_0^1 (3x - 2)^3\, dx$
7. $\int_0^4 \sqrt{x^2 + 9}\, x\, dx$
8. $\int_0^3 x/\sqrt{x^2 + 16}\, dx$
9. $\int_5^8 5/(x - 4)\, dx$
10. $\int_{-1}^3 3/(x + 2)\, dx$
11. $\int_0^2 x/(2x^2 + 1)\, dx$
12. $\int_3^5 x/(x^2 - 8)\, dx$
13. $\int_0^1 3e^{2x}\, dx$
14. $\int_0^2 4e^{-2x}\, dx$
15. $\int_0^2 xe^{-x^2}\, dx$
16. $\int_1^4 e^{\sqrt{x}}/\sqrt{x}\, dx$
17. $\int_0^3 (x^2 - 2x)^3(x - 1)\, dx$
18. $\int_0^2 (x^3 - 6x)^4(x^2 - 2)\, dx$
19. $\int_0^1 (\sqrt{x} + 1)^5/\sqrt{x}\, dx$
20. $\int_1^8 (\sqrt[3]{x} - 1)^4 x^{-2/3}\, dx$
21. $\int_0^3 [\ln(1 + 3x)]^2/(1 + 3x)\, dx$
22. $\int_0^1 [\ln(e^x + 1)]^3[e^x/(e^x + 1)]\, dx$
23. $\int_{-2}^1 e^{(x+2)^2}(x + 2)\, dx$
24. $\int_0^2 (x + 2e^{x/2})^3(1 + e^{x/2})\, dx$

*Use integration by parts to evaluate the definite integrals in Problems 25–30.*

25. $\int_0^2 xe^{2x}\, dx$
26. $\int_0^3 xe^{-x}\, dx$
27. $\int_1^e x^{3/2}\ln x\, dx$
28. $\int_e^{e^2} \sqrt{x}\ln x\, dx$

29. $\int_0^1 x(1-x)^4\,dx$

30. $\int_0^2 (x-2)^6\,x\,dx$

*Use the table of integrals at the end of the book to evaluate the definite integrals in Problems 31–36.*

31. $\int_4^5 \sqrt{x^2-16}\,dx$

32. $\int_3^4 \frac{\sqrt{x^2+16}}{x}\,dx$

33. $\int_0^3 \frac{x^2}{\sqrt{x^2+16}}\,dx$

34. $\int_3^4 \frac{1}{x^2\sqrt{25-x^2}}\,dx$

35. $\int_{-2}^1 \frac{x}{2x+5}\,dx$

36. $\int_0^1 x\sqrt{3x+1}\,dx$

*Calculate each integral in Problems 37–42.*

37. $\int_5^{\sqrt{34}} \frac{1}{\sqrt{u^2-9}}\,du$

38. $\int_5^{\sqrt{34}} \frac{u}{\sqrt{u^2-9}}\,du$

39. $\int_5^{\sqrt{34}} \frac{u^2}{\sqrt{u^2-9}}\,du$

40. $\int_2^4 e^{(x-2)^2}(x-2)\,dx$

41. $\int_2^4 (x-2)e^{x-2}\,dx$

42. $\int_0^2 e^{2\ln(x+1)}\,dx$

43. Find the area under the curve $y = x\ln x$ between $x = 1$ and $x = e^3$.

44. Find the area under the curve $y = xe^{-x}$ between $x = 0$ and $x = \ln 4$.

45. Find the average value of $f(x) = (x^3-1)^4x^2$ on the interval $1 \le x \le \sqrt[3]{2}$.

46. How far did an object travel along a line during the time interval $1 \le t \le e$ if its velocity at time $t$ seconds was given by $v = (\ln t)^3/t$ centimeters per second.

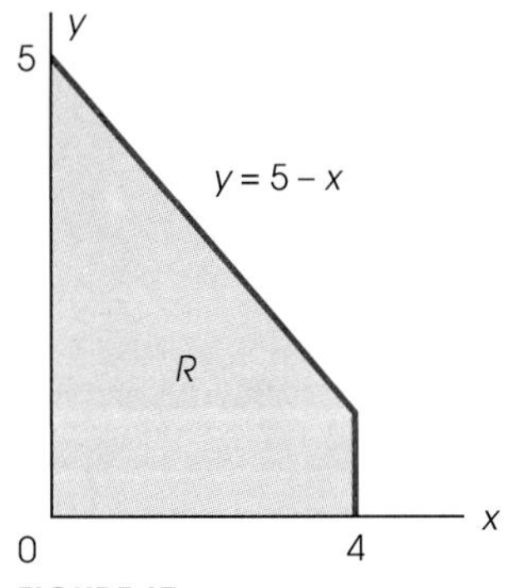

**FIGURE 17**

## 8.4 THE DEFINITE INTEGRAL AS A LIMIT

Our definition of the definite integral was in terms of area. But what is area? Earlier we made little attempt to answer this question and, in fact, it is a deep question. We do intend to give an answer now.

To give the flavor of our approach, consider the region $R$ bounded by the line $y = 5 - x$ and the $x$-axis, between $x = 0$ and $x = 4$ (Figure 17). Partition the interval $[0, 4]$ into 5, then 10, then 20, and in general $n$ equal subintervals and construct the corresponding inscribed rectangular polygons $R_5$, $R_{10}$, $R_{20}$, and $R_n$ as shown in Figure 18. As we increase the number of division points in the partition, the areas of the corresponding rectangular

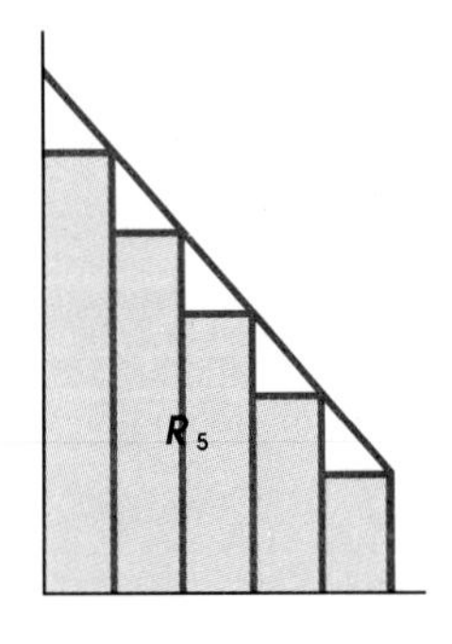

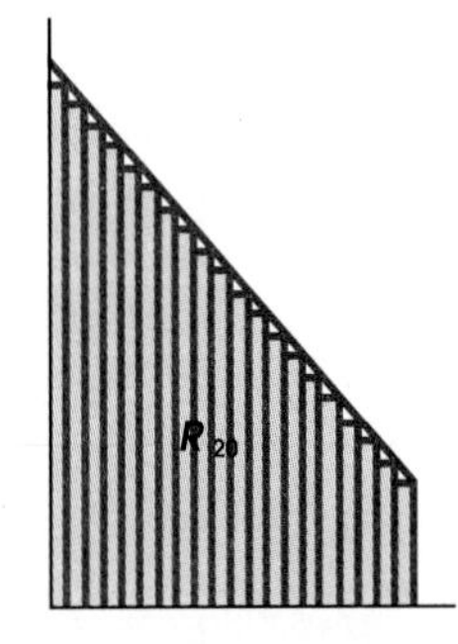
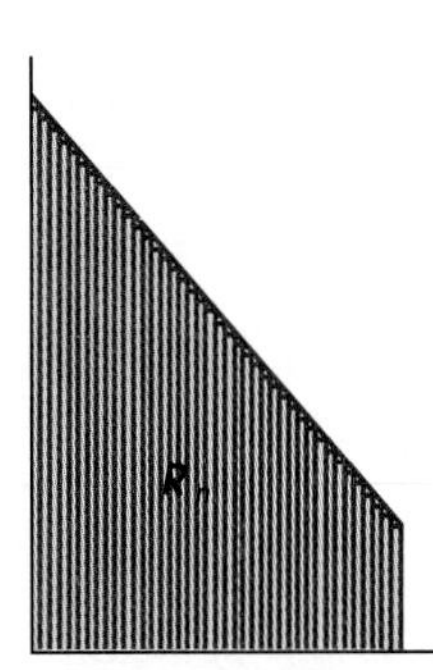

**FIGURE 18**

polygons approach nearer and nearer to what we wish to call the area of $R$, that is,

$$A(R) = \lim_{n\to\infty} A(R_n)$$

Later in this section, we will actually carry out this limit process and show you that the result is the same number you would get by applying the familiar formula for the area of a trapezoid. The significant observation to make now is that the process above can be carried out also for a curved region. In fact, we will use this process in a moment to give the limit definition of the definite integral. It is essentially this definition that is given in all rigorous treatments of calculus. Moreover, though this definition is a bit formidable, it is important because it shows us how to use the definite integral in applications, a subject we take up in Chapter 9.

## A Formal Definition of the Definite Integral

Let $f$ be a continuous function defined on the interval $[a, b]$. Partition this interval into $n$ equal subintervals by means of points.

$$a = x_0 < x_1 < x_2 < x_3 < \cdots < x_{n-1} < x_n = b$$

(see Figure 19) and let $\Delta x = x_i - x_{i-1} = (b - a)/n$. We define the **definite integral of $f(x)$ from $a$ to $b$** as follows.

$$\int_a^b f(x)\,dx = \lim_{n\to\infty} [\,f(x_1)\,\Delta x + f(x_2)\,\Delta x + f(x_3)\,\Delta x + \cdots + f(x_n)\,\Delta x]$$

$$= \lim_{n\to\infty} [\,f(x_1) + f(x_2) + f(x_3) + \cdots + f(x_n)]\,\Delta x$$

Note that each term $f(x_i)\,\Delta x$ is the area of a rectangle or the negative of such an area (depending on whether $f(x_i)$ is positive or negative).

We can write the above definition more compactly after we introduce a special notation for sums, called sigma notation.

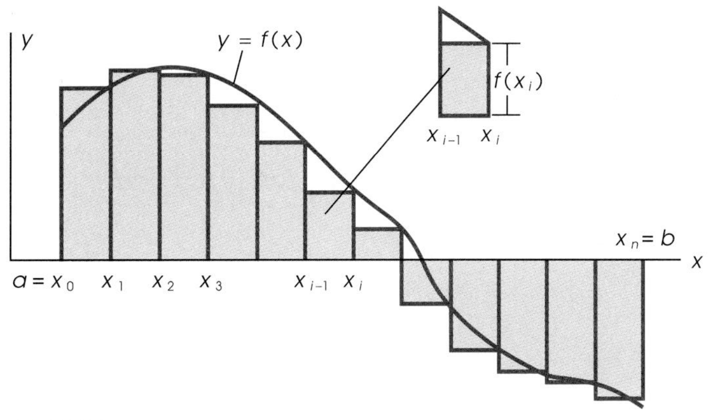


**FIGURE 19**

## Sigma Notation

The Greek letter $\Sigma$ is used throughout mathematics to denote summation. In particular,

$$\sum_{i=1}^{n} x_i = x_1 + x_2 + x_3 + \cdots + x_{n-1} + x_n$$

and

$$\sum_{i=1}^{n} i = 1 + 2 + 3 + \cdots + (n - 1) + n$$

The symbol $\sum_{i=1}^{n}$ is read *sum as i runs from 1 to n* and means that we are to add together all terms of the indicated form as $i$ runs through the integers 1, 2, 3, . . . , $n$.

### EXAMPLE A

Evaluate each of the following sums.

(a) $\sum_{i=1}^{6} i^2$, (b) $\sum_{i=2}^{9} (2i - 3)$

#### Solution

(a) $\sum_{i=1}^{6} i^2 = 1^2 + 2^2 + 3^2 + 4^2 + 5^2 + 6^2$

$= 1 + 4 + 9 + 16 + 25 + 36 = 91$

(b) $\sum_{i=2}^{9} (2i - 3) = 1 + 3 + 5 + 7 + 9 + 11 + 13 + 15 = 64$ ■

### EXAMPLE B

Write each of the following sums in sigma notation.

(a) $1 + \frac{1}{2} + \frac{1}{3} + \frac{1}{4} + \cdots + \frac{1}{99} + \frac{1}{100}$
(b) $b_2 + b_3 + b_4 + \cdots + b_{19} + b_{20}$
(c) $f(x_1)\,\Delta x + f(x_2)\,\Delta x + f(x_3)\,\Delta x + \cdots + f(x_n)\,\Delta x$

#### Solution

(a) $\sum_{i=1}^{100} (1/i)$, (b) $\sum_{i=2}^{20} b_i$, (c) $\sum_{i=1}^{n} f(x_i)\,\Delta x$ ■

We mention two important properties that $\Sigma$ shares with $\int$, the sum property and the constant-multiple property.

1. $\sum_{i=1}^{n} (a_i \pm b_i) = \sum_{i=1}^{n} a_i \pm \sum_{i=1}^{n} b_i$
2. $\sum_{i=1}^{n} ca_i = c \sum_{i=1}^{n} a_i$

Rather than give the simple proofs of these properties, we illustrate them.

**EXAMPLE C**

Given that $\sum_{i=1}^{5} a_i = 20$ and $\sum_{i=1}^{5} b_i = 5$, calculate $\sum_{i=1}^{5} (3a_i - 2b_i)$.

**Solution**

$$\sum_{i=1}^{5} (3a_i - 2b_i) = 3\sum_{i=1}^{5} a_i - 2\sum_{i=1}^{5} b_i = 3(20) - 2(5) = 50$$ ■

Using $\Sigma$ notation, we can write the definition of the definite integral in a particularly compact and suggestive way.

$$\int_a^b f(x)\, dx = \lim_{n\to\infty} \sum_{i=1}^{n} f(x_i)\, \Delta x$$

Thus, we see that the definite integral is the limit of a sum. And if $f$ is a continuous function (or even a function with a finite number of jump discontinuities), the above limit will always exist. For a proof of this latter fact, you will have to go to an advanced calculus book.

## Some Calculations Using the Limit Definition

We are going to need two formulas that you have probably seen at some point in your precalculus training. If not, don't worry too much about where they come from, though you might check that they are correct for a few small values of $n$.

(i) $1 + 2 + 3 + \cdots + n = \sum_{i=1}^{n} i = \dfrac{n(n+1)}{2}$

(ii) $1^2 + 2^2 + 3^2 + \cdots + n^2 = \sum_{i=1}^{n} i^2 = \dfrac{n(n+1)(2n+1)}{6}$

**EXAMPLE D**

Use the limit definition to calculate $\int_0^4 (5 - x)\, dx$. Note that this corresponds to the area of the trapezoid in Figure 17.

**Solution** We partition the interval $[0, 4]$ into $n$ subintervals each of length $\Delta x = (b - a)/n = 4/n$. Also, the division points of the interval are

$x_1 = 4/n$, $x_2 = 2(4/n)$, $x_3 = 3(4/n)$, and so on. Since $f(x) = 5 - x$, we obtain

$$\sum_{i=1}^{n} f(x_i)\,\Delta x$$

$$= [f(x_1) + f(x_2) + f(x_3) + \cdots + f(x_n)]\,\Delta x$$

$$= \left[\left(5 - \frac{4}{n}\right) + \left(5 - 2\frac{4}{n}\right) + \left(5 - 3\frac{4}{n}\right) + \cdots + \left(5 - n\frac{4}{n}\right)\right]\frac{4}{n}$$

$$= \left[5n - \frac{4}{n}(1 + 2 + 3 + \cdots + n)\right]\frac{4}{n}$$

$$= \left[5n - \frac{4}{n}\frac{n(n+1)}{2}\right]\frac{4}{n}$$

$$= 20 - \frac{8}{n}(n + 1) = 20 - 8 - \frac{8}{n} = 12 - \frac{8}{n}$$

We conclude that

$$\int_0^4 (5 - x)\,dx = \lim_{n\to\infty} \sum_{i=1}^{n} f(x_i)\,\Delta x = \lim_{n\to\infty}\left(12 - \frac{8}{n}\right) = 12$$

Naturally, this is the answer you get if you use the formula for the area of a trapezoid. ■

### EXAMPLE E

Use the limit process to evaluate $\int_0^2 x^2\,dx$.

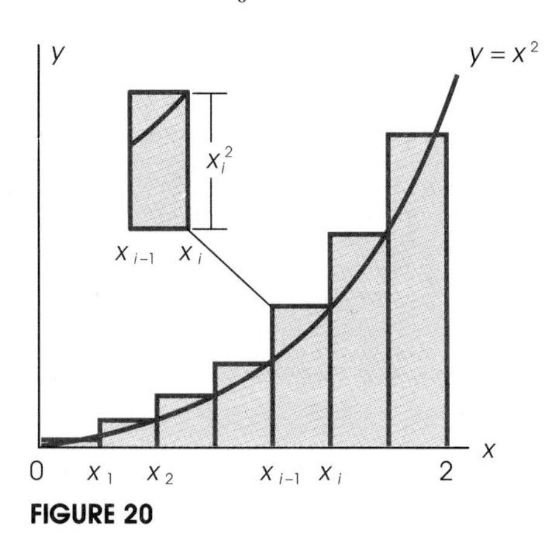


**FIGURE 20**

**Solution** We partition the interval [0, 2] into $n$ equal subintervals each of length $\Delta x = 2/n$ (see Figure 20). The partition points have coordinates $x_1 = 2/n$, $x_2 = 2(2/n)$, $x_3 = 3(2/n)$, and in general $x_i = i(2/n)$. Since $f(x) = x^2$, we see that

$$\sum_{i=1}^{n} f(x_i)\,\Delta x = \sum_{i=1}^{n}\left[i\left(\frac{2}{n}\right)\right]^2\left[\frac{2}{n}\right] = \sum_{i=1}^{n} i^2\left[\frac{8}{n^3}\right]$$

$$= \frac{8}{n^3}\sum_{i=1}^{n} i^2 \qquad \text{(by Property 2 of } \Sigma\text{)}$$

$$= \frac{8}{n^3}\,\frac{n(n+1)(2n+1)}{6} \qquad \text{[by Formula (ii), p. 267]}$$

$$= \frac{4}{3}\,\frac{(n+1)(2n+1)}{n^2}$$

Thus,

$$\int_0^2 x^2\,dx = \lim_{n\to\infty}\frac{4}{3}\,\frac{(n+1)(2n+1)}{n^2} = \frac{4}{3}\lim_{n\to\infty}\frac{2n^2+3n+1}{n^2} = \frac{8}{3}$$

Naturally, this is the same answer we would get by using the Fundamental Theorem of Calculus. ■

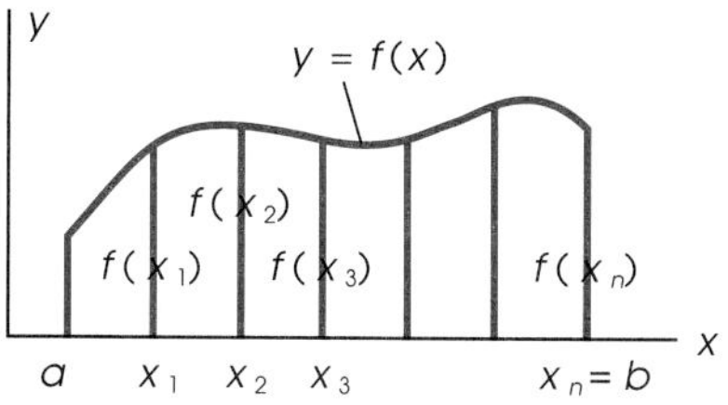

FIGURE 21

## The Average Value of a Function

We can now give the more adequate discussion of the average value of a function that we promised in Section 8.2. Consider a function $f$ defined on the interval $[a, b]$ (see Figure 21). Divide the interval $[a, b]$ into $n$ equal pieces of length $\Delta x = (b - a)/n$ and let $x_i = a + i\,\Delta x$. A reasonable approximation to the average value of $f$ ought to be obtained by averaging the values of $f$ at the sample points $x_1, x_2, \ldots, x_n$, especially if $n$ is large.

$$f_{\text{ave}} \approx \frac{1}{n}[f(x_1) + f(x_2) + \cdots + f(x_n)]$$

$$= \frac{1}{b-a}\,\frac{b-a}{n}[f(x_1) + f(x_2) + \cdots + f(x_n)]$$

$$= \frac{1}{b-a}[f(x_1)\,\Delta x + f(x_2)\,\Delta x + \cdots + f(x_n)\,\Delta x]$$

$$= \frac{1}{b-a}\sum_{i=1}^{n} f(x_i)\,\Delta x$$

This approximation will get better and better as we sample $f$ at more and more points, that is, as $n$ gets larger and larger. Thus, we define the **average value** of $f$ on the interval $[a, b]$ to be the limit of the above expression as $n \to \infty$. This gives

$$f_{\text{ave}} = \frac{1}{b-a}\int_a^b f(x)\,dx$$

**SHARP CURVE**
The basis of our derivation of the formula for the average value of a function was to sample more and more of the infinitely many values of $f$. But we also used a mathematical trick, dividing and multiplying by $b - a$. It is often surprising what a large reward a little trick can bring.

## Problem Set 8.4

*Calculate each of the sums in Problems 1–8 (see Example A).*

1. $\sum_{i=1}^{6} (i/5)$
2. $\sum_{i=1}^{5} (i-1)^2$
3. $\sum_{i=3}^{6} (3i-4)$
4. $\sum_{i=2}^{5} (5i-8)$
5. $\sum_{i=1}^{3} \frac{1}{i(i+1)}$
6. $\sum_{i=4}^{6} \frac{12}{i(i-2)}$
7. $\sum_{i=0}^{4} 2^i$
8. $\sum_{i=1}^{3} (4^i - 5^{i-1})$

*In Problems 9–20, write each sum in sigma notation (see Example B).*

9. $1^3 + 2^3 + 3^3 + \cdots + 10^3$
10. $2 + 4 + 6 + 8 + \cdots + 100$
11. $\frac{3}{2} + \frac{3}{2^2} + \frac{3}{2^3} + \cdots + \frac{3}{2^{100}}$
12. $5 + \frac{5}{2} + \frac{5}{3} + \frac{5}{4} + \cdots + \frac{5}{70}$
13. $x_4 + x_5 + x_6 + \cdots + x_{21}$
14. $t_1 + t_2 + t_3 + \cdots + t_n$
15. $g(t_1)\,\Delta t + g(t_2)\,\Delta t + g(t_3)\,\Delta t + \cdots + g(t_n)\,\Delta t$
16. $f(x_1)\,\Delta x + f(x_2)\,\Delta x + f(x_3)\,\Delta x + \cdots + f(x_m)\,\Delta x$
17. $3x_1\,\Delta x + 3x_2\,\Delta x + 3x_3\,\Delta x + \cdots + 3x_n\,\Delta x$
18. $x_1{}^2\,\Delta x + x_2{}^2\,\Delta x + x_3{}^2\,\Delta x + \cdots x_n{}^2\,\Delta x$
19. $(x_1{}^3 + 4)\,\Delta x + (x_2{}^3 + 4)\,\Delta x + (x_3{}^3 + 4)\,\Delta x + \cdots + (x_n{}^3 + 4)\,\Delta x$
20. $(x_1{}^2 + 5x_1)\,\Delta x + (x_2{}^2 + 5x_2)\,\Delta x + (x_3{}^2 + 5x_3)\,\Delta x + \cdots + (x_n{}^2 + 5x_n)\,\Delta x$

*Calculate each of the sums in Problems 21–28, given that $\sum_{i=1}^{10} x_i = 30$ and $\sum_{i=1}^{10} y_i = 4$ (see Example C). In Problem 25, you will want to use the fact that the constant 2 appears in every term and so contributes 10(2) = 20 to the sum. A similar remark applies in Problems 26–28.*

21. $\sum_{i=1}^{10} 4x_i$
22. $\sum_{i=1}^{10} (-0.5)y_i$
23. $\sum_{i=1}^{10} (3x_i - 20y_i)$
24. $\sum_{i=1}^{10} (2x_i + 11y_i)$
25. $\sum_{i=1}^{10} (x_i + 2)$
26. $\sum_{i=1}^{10} (y_i + 3)$
27. $\sum_{i=1}^{10} (2x_i - 3y_i - 3)$
28. $\sum_{i=1}^{10} (x_i - 3y_i + 1)$

*In Problem 29–38, use the limit process to evaluate each definite integral (see Examples D and E). Check your answer by using the Fundamental Theorem of Calculus.*

29. $\int_0^3 (4 - x)\,dx$
30. $\int_0^4 (6 - x)\,dx$
31. $\int_0^2 (2x + 1)\,dx$
32. $\int_0^3 (3x + 5)\,dx$
33. $\int_0^3 (x^2 + 1)\,dx$
34. $\int_0^4 (x^2 - 2)\,dx$
35. $\int_0^1 (x^2 - 2x)\,dx$
36. $\int_0^1 (2x^2 + x)\,dx$
37. $\int_1^3 (x + 4)\,dx$
    *Hint:* $x_i = 1 + i(2/n)$
38. $\int_2^5 (2x + 1)\,dx$
    *Hint:* $x_i = 2 + i(3/n)$

*Calculate each of the sums in Problems 39–46. For some of them, you will need to use the formulas (i) and (ii) on page 267.*

39. $1 + 2 + 3 + \cdots + 499$
40. $1^2 + 2^2 + 3^2 + \cdots + 20^2$
41. $10 + 11 + 12 + \cdots + 99$
42. $41^2 + 42^2 + 43^2 + \cdots + 100^2$
43. $\sum_{i=1}^{100} (i^2 - 3i)$
44. $\sum_{j=1}^{100} (j^2 + 2j)$
45. $\sum_{k=1}^{100} (2 - 3k + k^2)$
46. $\sum_{i=1}^{100} \left(\frac{1}{i} - \frac{1}{i+1}\right)$ *Hint:* Write out this sum before trying to calculate it.

*Write the expressions in Problems 47 and 48 as definite integrals.*

47. $\lim_{n\to\infty} \sum_{i=1}^{n} \left[ i^2\left(\frac{4}{n}\right)^2 + 3i\frac{4}{n} \right] \frac{4}{n}$

48. $\lim_{n\to\infty} \sum_{i=1}^{n} \left[ 3i^2\left(\frac{2}{n}\right)^2 + i\frac{2}{n} - 2 \right] \frac{2}{n}$

49. Calculate $\sum_{i=1}^{100} \ln\left(\frac{i}{i+1}\right)$

50. Calculate $\sum_{i=2}^{20} \ln\left[1 - \frac{1}{i}\right]$

51. (Statistics) For the set of data $\{x_1, x_2, \cdots, x_n\}$, we define the mean $\bar{x}$ and variance $s^2$ by $\bar{x} = \frac{1}{n}\sum_{i=1}^{n} x_i$ and $s^2 = \frac{1}{n}\sum_{i=1}^{n} (x_i - \bar{x})^2$. Find the mean and variance for the set of data {3, 4, 5, 5, 6, 6, 6, 7, 8, 10}.

## CHAPTER 8 REVIEW PROBLEM SET

1. Calculate $\int_{-3}^{7} f(x)\,dx$, where $f$ is the function whose graph is shown.

(a)

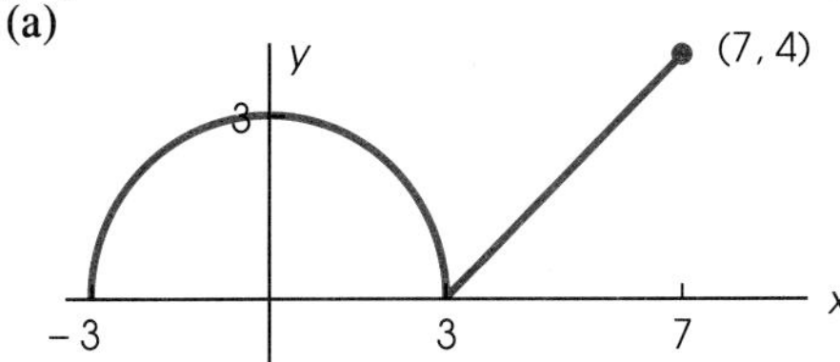


(b)

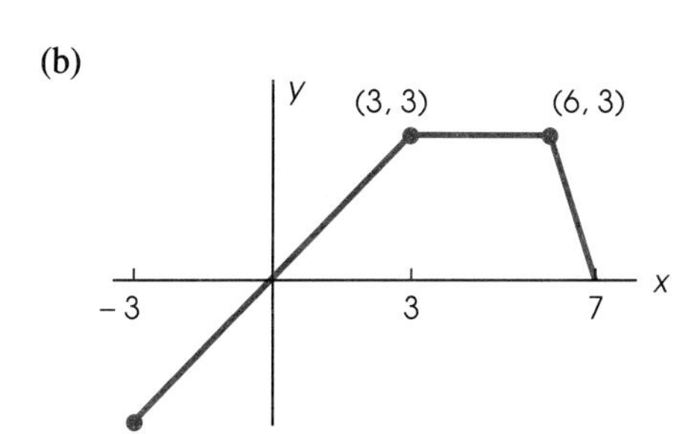


2. Sketch the graph of $f$ on the interval $[-2, 4]$ and then evaluate $\int_{-2}^{4} f(x)\,dx$.

(a) $f(x) = \begin{cases} -x & -2 \le x \le 2 \\ -2 & 2 < x \le 4 \end{cases}$

(b) $f(x) = \begin{cases} -x - 2 & -2 \le x \le 0 \\ (4 - x)/2 & 0 < x \le 4 \end{cases}$

3. Sketch the graph of $f$ on the interval $[a, b]$ and then calculate $\int_a^b f(x)\,dx$.

(a) $f(x) = x - 2;\ a = 0,\ b = 5$

(b) $f(x) = \sqrt{4 - x^2};\ a = -2,\ b = 2$

(c) $f(x) = \begin{cases} |x + 1| & x \le 1 \\ -2x + 4 & 1 < x \le 3 \\ -2 & x > 3 \end{cases}$

$a = -3,\ b = 6$

4. Sketch the graph of $f$ on the interval $[-2, 5]$ and then calculate $\int_{-2}^{5} f(x)\,dx$.

(a) $f(x) = \begin{cases} |x| & -2 \le x \le 2 \\ 2 & 2 \le x \le 5 \end{cases}$

(b) $f(x) = \begin{cases} \sqrt{4 - x^2} & -2 \le x \le 2 \\ (1 - x)/2 & 2 < x \le 5 \end{cases}$

5. Use symmetry and the formula $\int_0^a x^n\,dx = a^{n+1}/n + 1$ to evaluate each integral.

(a) $\int_{-1}^{1} 5x^6\,dx$ (b) $\int_{-8}^{8} \pi x^{101}\,dx$

(c) $\int_{-1}^{1} (x^7 - 2x^4)\,dx$ (d) $\int_{-3}^{3} \frac{2x^7}{\sqrt{x^2 + 4}}\,dx$

6. Suppose that $f$ is an odd function, that $g$ is an even function, and that $\int_0^2 f(x)\,dx = \int_0^2 g(x)\,dx = 3$. Use symmetry to evaluate each of the following.

(a) $\int_{-2}^{2} f(x)\,dx$ (b) $\int_{-2}^{2} |f(x)|\,dx$

(c) $\int_{-2}^{2} g(x)\,dx$ (d) $\int_{-2}^{2} x^3 g(x)\,dx$

7. An object moves along a straight line in such a way that after $t$ seconds its velocity is given by $v(t) = 2t - 3$ feet per second. Find the total distance (backward and forward) traveled by the object between $t = 0$ and $t = 5$.

8. An object moves along a straight line so that after $t$ seconds its velocity is $v = 3t - 2$ feet per second. Find the distance between its position at $t = 0$ and its position at $t = 5$.

*Evaluate the definite integrals in Problems 9–30.*

9. $\int_1^9 3\sqrt{x}\,dx$ 10. $\int_1^8 2\sqrt[3]{x}\,dx$

11. $\int_0^6 \left(\frac{1}{2}x^2 - 4\right) dx$ 12. $\int_0^2 \left(3x^2 - \frac{1}{2}x + 4\right) dx$

13. $\int_1^3 \frac{x + 4}{x^2 + 8x - 3}\,dx$ 14. $\int_0^3 t(t^2 + 16)^{1/2}\,dt$

15. $\int_{1/4}^{1/2} \frac{e^{1/x}}{x^2}\,dx$ 16. $\int_3^5 \frac{t - 2}{3t^2 - 12t + 10}\,dt$

17. $\int_0^3 \frac{2x + 1}{\sqrt{x^2 + x + 4}}\,dx$ 18. $\int_2^5 \frac{4x + 1}{(2x^2 + x - 6)^{3/2}}\,dx$

19. $\int_0^2 [\ln(x^2 + 1)]^3 \frac{x}{x^2 + 1}\,dx$

20. $\int_1^4 \frac{e^{\sqrt{x}}}{\sqrt{x}}\,dx$ 21. $\int_0^5 4|t - 3|\,dt$

22. $\int_0^{\ln 11} \frac{d}{dx}\ln(e^x + 4)\,dx$

23. $\int_0^{\sqrt{e-1}} \frac{d}{dx}[\ln(1 + x^2)]\,dx$

24. $\int_0^{(e-1)/2} \frac{\ln(2x + 1)^4}{2x + 1}\,dx$

25. $\int_{-3}^5 f(x)\,dx$ where $f(x) = \begin{cases} e^{-x} & -3 \le x \le 0 \\ x + 1 & 0 < x \le 2 \\ 3 & 2 < x \le 5 \end{cases}$

26. $\int_0^{e^3} f(x)\,dx$ where

$$f(x) = \begin{cases} 2e^x & 0 \le x \le 1 \\ 2x - 2 & 1 < x < e \\ (\ln x)^2/x & e \le x \le e^3 \end{cases}$$

27. $\int_1^3 (x - 3)^5 x\,dx$ *Hint:* Let $u = x - 3$.
28. $\int_2^4 (x - 4)^5 x\,dx$
29. $\int_0^{\sqrt{5}} (9 - x^2)^{-3/2}\,dx$. *Hint*: Use table of integrals.
30. $\int_0^4 (9 + x^2)^{3/2}\,dx$ *Hint:* Use a table of integrals.
31. Sketch the graph of $y = 2\sqrt[3]{x} + 1$ on the interval $[1, 8]$ and then find the area of the region under this graph.
32. Sketch the graph of $y = 4(x + 1)/\sqrt{x}$ on the interval $[1, 4]$ and then find the area of the region under the graph.
33. Suppose that the height $h$ (in inches) of a tree $t$ months after it was planted was given by

$$h = 52 + 3.2t - 0.02t^2 \qquad 0 \le t \le 30$$

Find the average height of the tree during the 30-month period.

34. After $x$ years, a certain company was making snowblowers at the rate of $140 + 2x + 3x^2$ units per year. On the average, how many units did the company make per year during the first 10 years?
35. Calculate the sum $\sum_{i=1}^{4} (3^i - 2i)$.
36. Calculate $\sum_{k=0}^{4} (k^3 - 2^k)$.
37. Write in sigma notation.
(a) $1 \cdot 2 + 2 \cdot 3 + 3 \cdot 4 + \cdots + 97 \cdot 98$
(b) $(2x_1 + 3)\,\Delta x + (2x_2 + 3)\,\Delta x + \cdots + (2x_n + 3)\,\Delta x$
38. Write the sigma notation.
(a) $1^2 + 2^2 + 3^2 + \cdots + 100^2$
(b) $t_1^3\Delta t + t_2^3\Delta t + t_3^3\Delta t + \cdots + t_n^3\Delta t$
39. Calculate $\sum_{i=1}^{8} (4x_i - 6y_i)$, given that $\sum_{i=1}^{8} x_i = 36$ and $\sum_{i=1}^{8} y_i = 15$.
40. Calculate $\sum_{k=1}^{12} (3x_k - 5y_k + 1)$, given that $\sum_{k=1}^{12} x_k = 40$ and that $\sum_{k=1}^{12} y_k = 20$.
41. Calculate $\sum_{i=1}^{100} (3i^3 + 4i)$ by using the two formulas $\sum_{i=1}^{n} i = \frac{n(n + 1)}{2}$ and $\sum_{i=1}^{n} i^3 = \left[\frac{n(n + 1)}{2}\right]^2$.
42. Calculate $\sum_{i=1}^{10} (i^3 - 2i^2 - 3i)$ given the formulas in Problem 41 and the additional formula $\sum_{i=1}^{n} i^2 = \frac{n(n + 1)(2n + 1)}{6}$.
43. Evaluate $\int_1^e f''(t)\,dt$, given that $f(t) = t \ln t$.
44. Find $\frac{d}{dx}(x \ln x)$. Then evaluate $\int_1^e (1 + \ln x)\,dx$.
45. Use the limit process to evaluate $\int_0^2 (4x + 3)\,dx$. Then think of two ways to check your answer.
46. Use the limit process to find $\int_0^3 (x^2 + 1)\,dx$. Check your answer by using the Fundamental Theorem of Calculus. You will need a formula from Problem 42.
47. Evaluate

$$\lim_{n\to\infty} \frac{4}{n} \sum_{i=1}^{n} \frac{4i/n}{1 + (4i/n)^2}$$

by recognizing this limit as a definite integral and then evaluating the definite integral.

48. Evaluate $\lim_{n\to\infty} \frac{2}{n} \sum_{i=1}^{n} \left[3\left(\frac{2i}{n}\right)^2 + \frac{2i}{n}\right]$.

49. Consider the function $f(x) = 1/x^2$, $x > 0$.
    (a) Sketch the graph of $f$.
    (b) Find the point on the graph where the tangent line has slope $-1/4$.
    (c) If a point moves along the graph of $y = 1/x^2$ so that $dx/dt = 2.5$, find $dy/dt$ at $x = 10$.
    (d) Find the area under the graph of $f$ from $x = 1$ to $x = 1000$.
    (e) Find $\lim_{t\to\infty} \int_1^t \frac{1}{x^2}\,dx$.

50. Consider the function $g(x) = x/\sqrt{x^2 + 16}$.
    (a) Determine the domain of $g$.
    (b) Evaluate $\lim_{x\to\infty} g(x)$.
    (c) Where is $g$ increasing?
    (d) Find the point on the graph of $g$ where the tangent line has slope $1/4$.
    (e) Find the area under the graph between $x = 0$ and $x = 3$.

*Trying to predict weather is a daunting problem involving the collection of masses of data and the construction of complicated probability models. To collect data on temperature, pressure, and humidity everywhere is impossible; rather it must be done at selected points such as the intersection points of the grid shown below.*

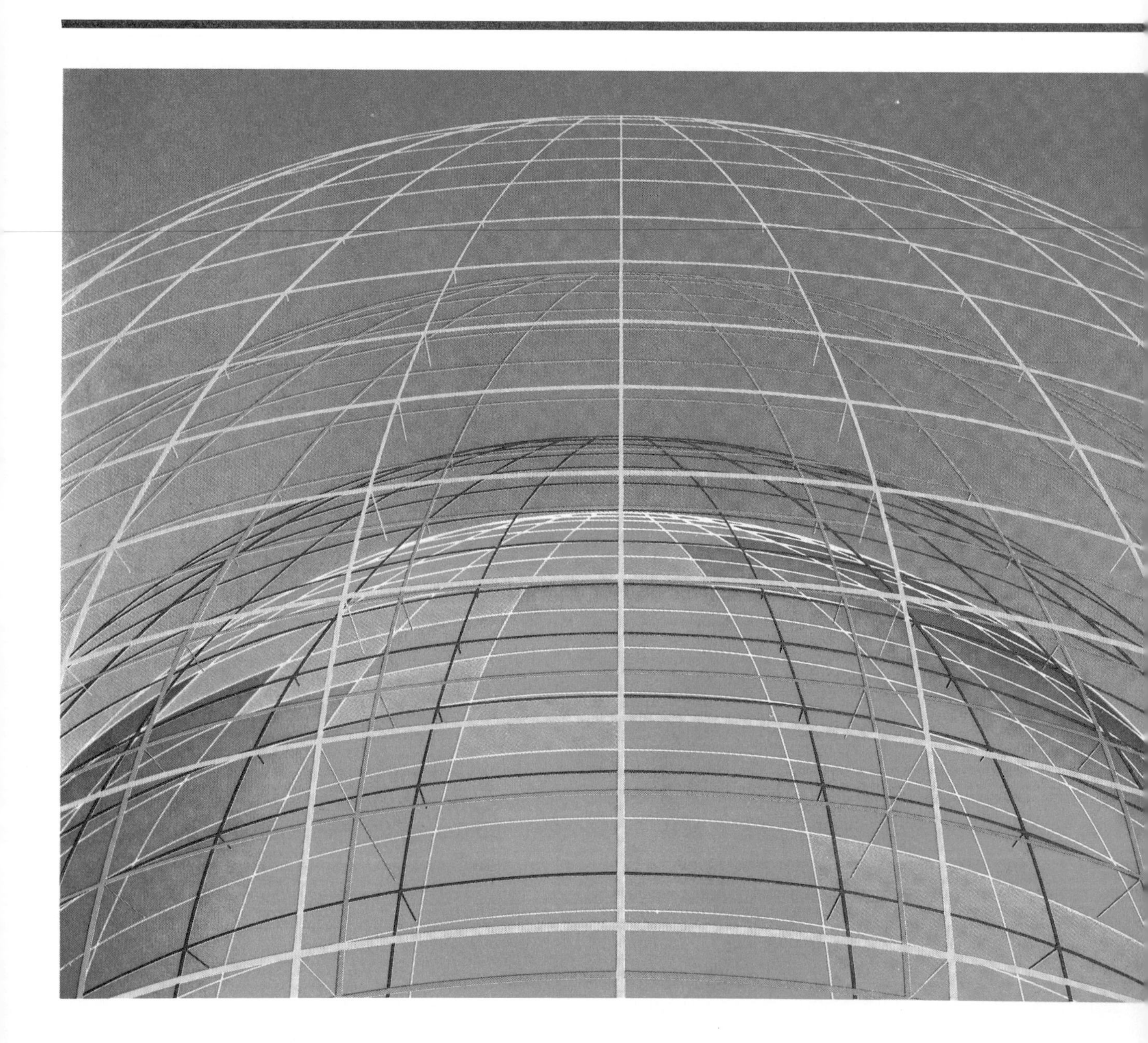

# 9

# Applications of the Definite Integral



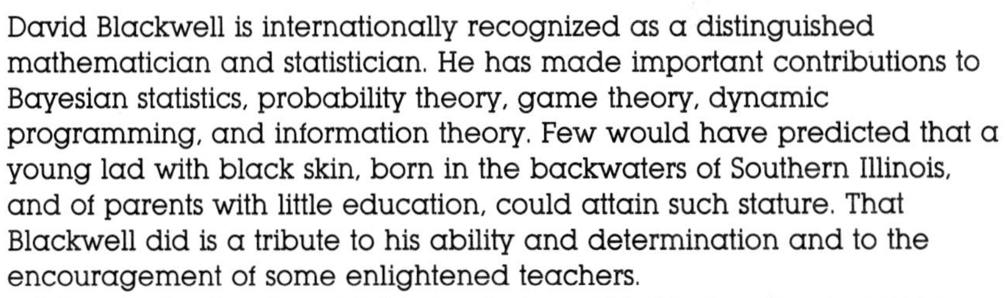

David Blackwell is internationally recognized as a distinguished mathematician and statistician. He has made important contributions to Bayesian statistics, probability theory, game theory, dynamic programming, and information theory. Few would have predicted that a young lad with black skin, born in the backwaters of Southern Illinois, and of parents with little education, could attain such stature. That Blackwell did is a tribute to his ability and determination and to the encouragement of some enlightened teachers.

David Blackwell
(1919– )

After graduation from high school at age 16, Blackwell entered the University of Illinois with the goal of becoming an elementary teacher. When he learned that his father was borrowing money to keep him in college, he insisted on supporting himself by working as a waiter, dishwasher, and lab cleaner. Still, he managed to obtain a bachelor's degree in three years. Mathematics was easy for him, but it was not until he studied G. H. Hardy's *Pure Mathematics* that he began to see the real beauty of the subject. This started a love affair that continues to the present. Soon the young man was setting his sights higher—still teaching but maybe at the college level. This, of course, would require an advanced degree.

Fortunately, Blackwell's talent was recognized with a fellowship. And even more fortunately, one of Illinois's mathematical stars, Joseph Doob, agreed to supervise his Ph.D. work. That degree was awarded in 1941, three short years after his A.B. There followed a year of study at the Institute for Advanced Study at Princeton.

The times did not favor black mathematicians. Blackwell fully expected that he would have to teach at an under-financed and unheralded institution and, in fact, he did teach at several such schools for a number of years. Finally in 1954, he was invited to join the statistics faculty at the University of California at Berkeley, and when the famed Jerzy Neyman gave up chairing this group, it was Blackwell who was asked to take over. There he is part of what is generally recognized as the most prestigious statistics department in the world.

## 9.1 AREA BETWEEN TWO CURVES

The concept of area was used to motivate and actually define the definite integral. In this section, we are going to reverse direction and use the definite integral to calculate areas. We begin by recalling that if the graph of $y = f(x)$ is above the $x$-axis, then the area of the region bounded by this graph and the $x$-axis between $x = a$ and $x = b$ is just $\int_a^b f(x)\,dx$. We refer to this area as *the area under the curve* $y = f(x)$ *between* $a$ *and* $b$.

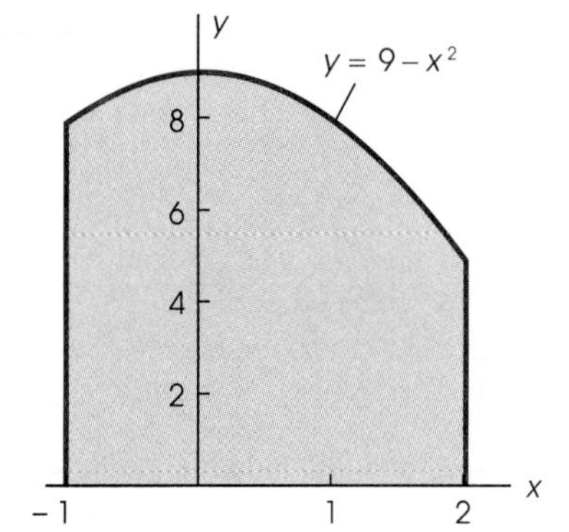

**FIGURE 1**

### EXAMPLE A

Find the area under the curve $y = 9 - x^2$ between $-1$ and 2 (see Figure 1).

**Solution** This is simply a matter of evaluating a definite integral

$$\int_{-1}^{2} (9 - x^2)\,dx = \left[9x - \frac{x^3}{3}\right]_{-1}^{2} = 18 - \frac{8}{3} - \left[-9 - \frac{(-1)^3}{3}\right] = 24 \quad ■$$

### A Helpful Way of Thinking

Consider any area problem of the type illustrated in Example A. If you studied Section 8.4, you know that this area can be approximated by the sum of the areas of a collection of thin rectangles (Figure 2). Concentrate now on a typical thin rectangle (Figure 3) whose right edge has $x$-coordinate $x$, whose height is then $f(x)$, and whose width is $dx$. It will have area $f(x)\,dx$. We want to add up the areas of such rectangles and then take the limit as the number of these rectangles tends to infinity. You may think of $\int_a^b$ as standing for the operation of adding up and taking the limit. This limiting process turns the approximation into an exact value for the desired area. Thus, the required area is $\int_a^b f(x)\,dx$.

From now on, we shall consider the task of finding the area of a region to be a three-step process.

1. Sketch the region and imagine slicing it into thin strips. Show a typical strip as in Figure 3.
2. Approximate the area of this strip by pretending it is a rectangle. Label the right coordinate of the rectangle with $x$, the height with $f(x)$,

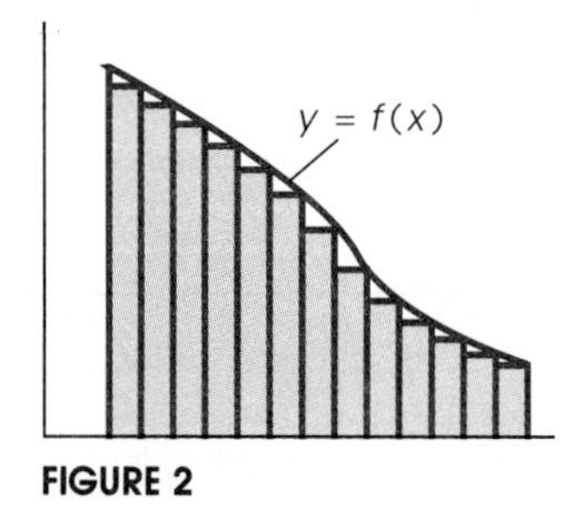

**FIGURE 2**

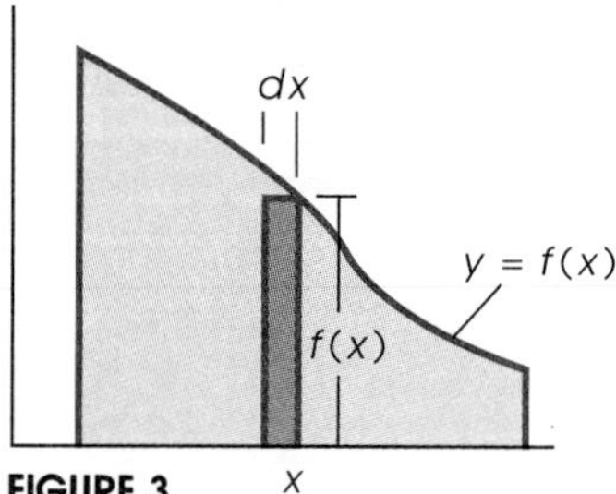

**FIGURE 3**

and the width with $dx$. This gives $f(x)\,dx$ for the area of the rectangle.

3. Add up the areas of these rectangles and take the limit, which is another way of saying "integrate."

We shall describe this three-step process by using the three words *slice, approximate, integrate*. We illustrate with another example.

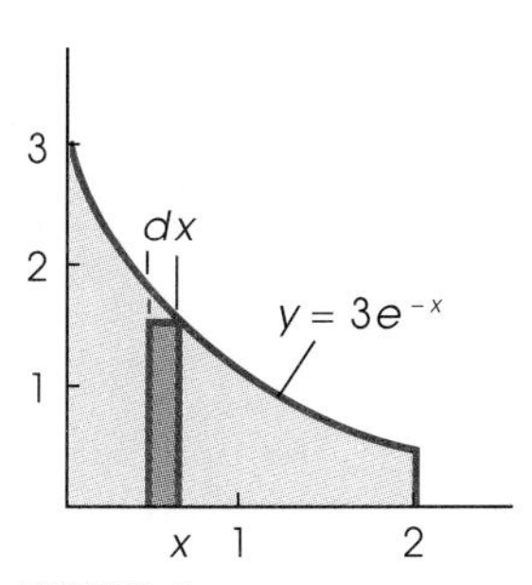

FIGURE 4

### EXAMPLE B

Find the area under the curve $y = 3e^{-x}$ between 0 and 2.

**Solution** Step 1 and 2 (*slice* and *approximate*) are accomplished by drawing a picture and labeling it appropriately (Figure 4). This picture in turn suggests the correct integral to write down, namely, $\int_0^2 3e^{-x}\,dx$. Finally, we carry out the integration (using the substitution $u = -x$ and $du = -dx$).

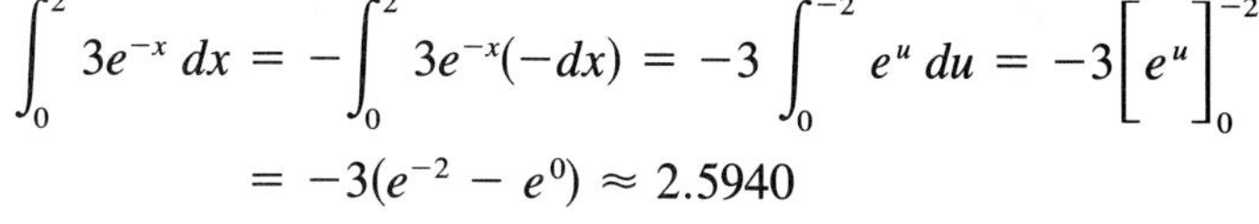

$$\int_0^2 3e^{-x}\,dx = -\int_0^2 3e^{-x}(-dx) = -3\int_0^{-2} e^u\,du = -3\Big[e^u\Big]_0^{-2}$$

$$= -3(e^{-2} - e^0) \approx 2.5940$$

■

## Area Between Two Curves

You will see that the process *slice, approximate, integrate* is the basic notion behind all applications of the definite integral. We next illustrate it for the area of a region between two curves.

### EXAMPLE C

Find the area of the region between the curves $y = f(x) = x^2 + 1$ and $y = g(x) = x - 1$ on the interval $[0, 3]$.

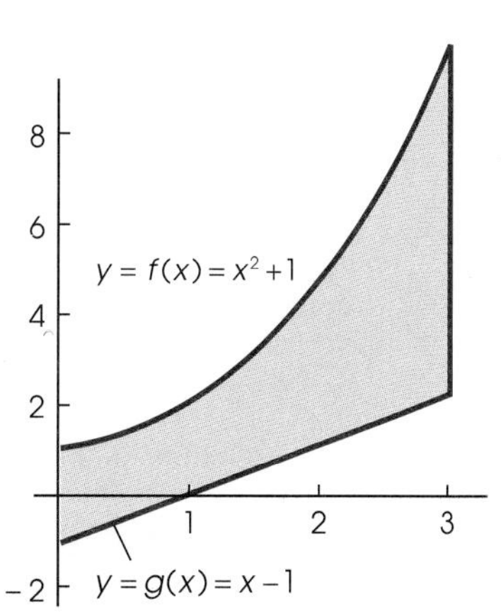

FIGURE 5

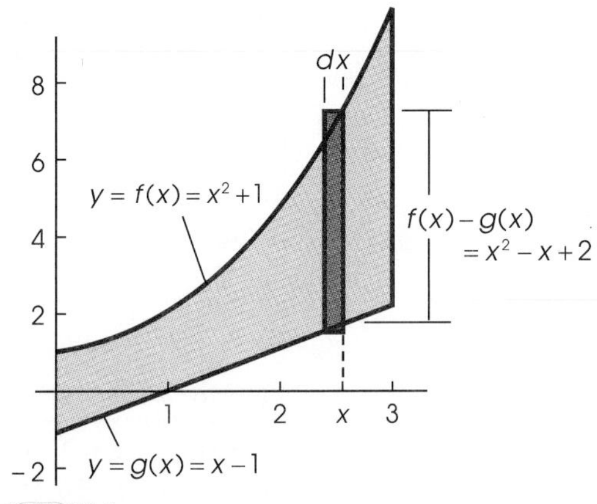


FIGURE 6

**Solution** The region of interest is shown in Figure 5, but it is Figure 6 that represents the steps *slice* and *approximate*. The height of the rectangle at $x$ is $f(x) - g(x)$; that is, it is $x^2 + 1 - (x - 1)$. Note that this expression for the height is correct also in that part of the region to the left of $x = 1$, for there $x - 1$ is negative and subtracting it is really adding just the right positive number. Thus, we see that the area of a typical rectangle is always $(x^2 - x + 2)\,dx$ and the area of the whole region is

$$\int_0^3 (x^2 - x + 2)\,dx = \left[\frac{x^3}{3} - \frac{x^2}{2} + 2x\right]_0^3$$

$$= 9 - \frac{9}{2} + 6 - (0) = \frac{21}{2} \quad \blacksquare$$

## EXAMPLE D

Find the area of the region between the curves $y = x^2 - 4$ and $y = -x + 2$.

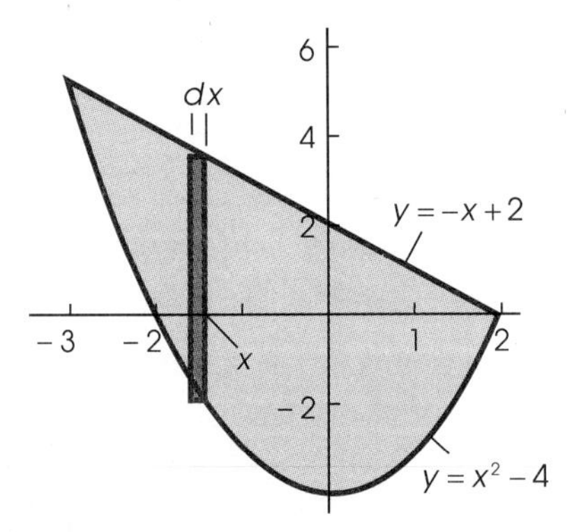


FIGURE 7

**Solution** The appropriate sketch with a typical rectangle is shown in Figure 7. The height of this rectangle is $(-x + 2) - (x^2 - 4)$, and so its area is $(-x^2 - x + 6)\,dx$. But now we face another problem. The limits that go on the integral sign (namely, the smallest and largest $x$-values in the region) are not obvious. They are the $x$-coordinates of the intersection points of the two curves and are obtained by solving the two equations simultaneously. To find them, we equate the two expressions for $y$ and solve as follows.

$$\begin{aligned} x^2 - 4 &= -x + 2 \\ x^2 + x - 6 &= 0 \\ (x + 3)(x - 2) &= 0 \\ x &= -3, 2 \end{aligned}$$

We complete the solution by calculating the required integral.

$$\begin{aligned} \int_{-3}^{2} (-x^2 - x + 6)\,dx &= \left[-\frac{x^3}{3} - \frac{x^2}{2} + 6x\right]_{-3}^{2} \\ &= \left(-\frac{8}{3} - 2 + 12\right) - \left(9 - \frac{9}{2} - 18\right) \\ &\approx 20.8333 \end{aligned}$$

■

## EXAMPLE E

Find the area of the region bounded by the two curves $y = x^2$ and $y = x^3 + 2x^2 - 2x$.

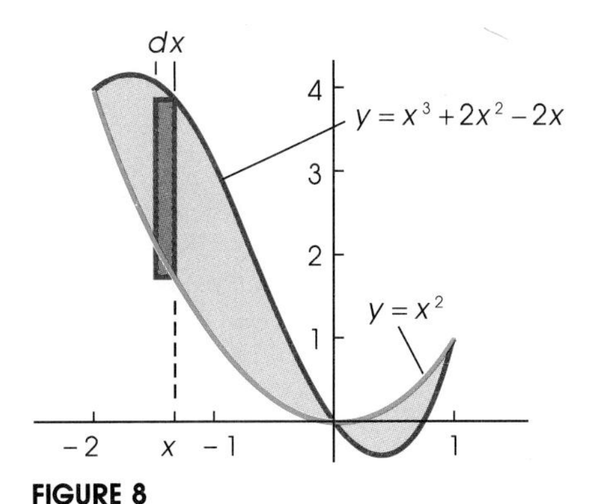


FIGURE 8

**Solution** A sketch of the required region together with a typical rectangle is shown in Figure 8. Note that the region is in two parts and that the curves exchange positions as to which is on top for the two parts. For the left part, the height of the rectangle is $(x^3 + 2x^2 - 2x) - x^2$, whereas for the right part, the height is the negative of this expression. The areas of the corresponding rectangles are $(x^3 + x^2 - 2x)\,dx$ and $(-x^3 - x^2 + 2x)\,dx$. To find the intersection points of the two curves, we solve the equation obtained by setting the two expressions for $y$ equal to each other.

$$x^3 + 2x^2 - 2x = x^2$$
$$x^3 + x^2 - 2x = 0$$
$$x(x + 2)(x - 1) = 0$$
$$x = -2, 0, 1$$

**POINT OF INTEREST**
We can always represent the area $A$ of the region between two curves $y = f(x)$ and $y = g(x)$ by a single integral, namely,

$$A = \int_a^b |f(x) - g(x)|\,dx$$

However, to evaluate this integral may require us to rewrite it as two or more integrals as in Example E.

The required area $A$ is given by

$$\begin{aligned} A &= \int_{-2}^{0} (x^3 + x^2 - 2x)\,dx + \int_0^1 (-x^3 - x^2 + 2x)\,dx \\ &= \left[\frac{x^4}{4} + \frac{x^3}{3} - x^2\right]_{-2}^{0} + \left[-\frac{x^4}{4} - \frac{x^3}{3} + x^2\right]_0^1 \\ &= 0 - \left(4 - \frac{8}{3} - 4\right) + \left(-\frac{1}{4} - \frac{1}{3} + 1\right] - 0 \\ &= \frac{37}{12} \approx 3.0833 \end{aligned}$$

■

## Horizontal Slicing

It will sometimes happen that an area problem is much easier to handle by horizontal slicing than by the vertical slicing that we have so far illustrated. Consider for example the region $R$ shown in Figure 9. Note that vertical slices to the left of $b$ always go from the curve to the line, whereas those to the right of $b$ go from the lower branch of the curve to the upper branch of the curve. Thus to find the area of $R$ by vertical slicing would require that we cut $R$ into two pieces (the parts to the left and right of $b$) and do two separate integrations. A much better choice is to slice horizontally so that a slice always goes from the line on the left to the curve on the right. Then one integration will give the area. We illustrate in an example.

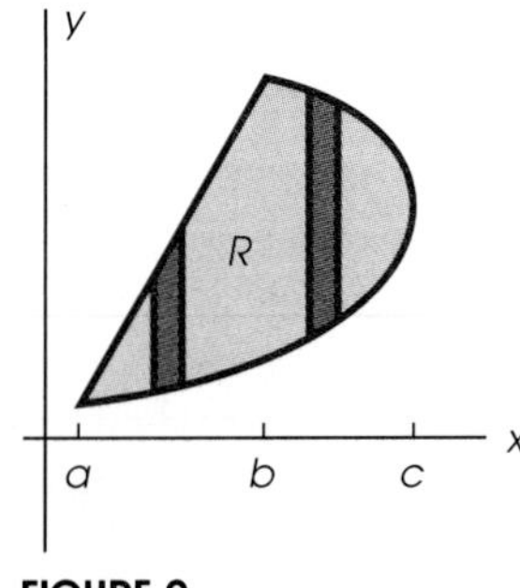


FIGURE 9

### EXAMPLE F

Find the area of the region between the curves $x = y^2$ and $x = y + 2$.

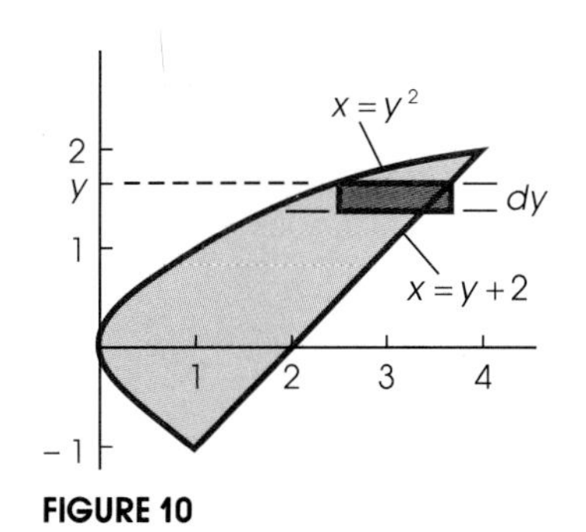


**FIGURE 10**

**Solution** The region with a typical horizontal slice and the corresponding approximating rectangle are shown in Figure 10. The width of this rectangle is denoted by $dy$ (a small change in $y$), and its length at $y$ is the $x$-coordinate on the line minus the $x$-coordinate on the curve; that is, the length is $y + 2 - y^2$. Thus, the area of the rectangle is $(-y^2 + y + 2)\,dy$; this is the expression we must integrate.

Next, we must choose the limits of integration. They are, respectively, the smallest and largest values of $y$ on the region. We obtain them by solving the equations $x = y^2$ and $x = y + 2$ simultaneously, that is, by solving $y^2 = y + 2$. This gives $y = -1$ and $y = 2$.

Finally, we perform the required integration.

$$A(R) = \int_{-1}^{2} (-y^2 + y + 2)\,dy = \left[-\frac{1}{3}y^3 + \frac{1}{2}y^2 + 2y\right]_{-1}^{2} = 4.5 \quad ■$$

## Problem Set 9.1

*In Problems 1–12, find the area of the region under the given curve between a and b (see Example B).*

1. $f(x) = x^2$; $a = 1$, $b = 4$
2. $f(x) = 4 - x^2$; $a = -1$, $b = 2$
3. $f(x) = (x + 1)^2$; $a = 1$, $b = 3$
4. $f(x) = (x - 3)^2$; $a = 0$, $b = 5$
5. $f(x) = x^3 - 4x$; $a = -2$, $b = -1$
6. $f(x) = 9x - x^3$; $a = 1$, $b = 3$
7. $f(x) = 4e^x$; $a = 0$, $b = 2$
8. $f(x) = 3e^{x/2}$; $a = -1$, $b = 0$
9. $f(x) = 1/(x + 3)$; $a = -2$, $b = 0$
10. $f(x) = 12/x^2$; $a = 2$, $b = 4$
11. $f(x) = x/(x^2 + 1)$; $a = 0$, $b = 2$
12. $f(x) = x\sqrt{4 - x^2}$; $a = 0$, $b = 2$

13. In the figure below, we show the graph of a region $R$ with a typical slice, properly labeled. Find the area of $R$.

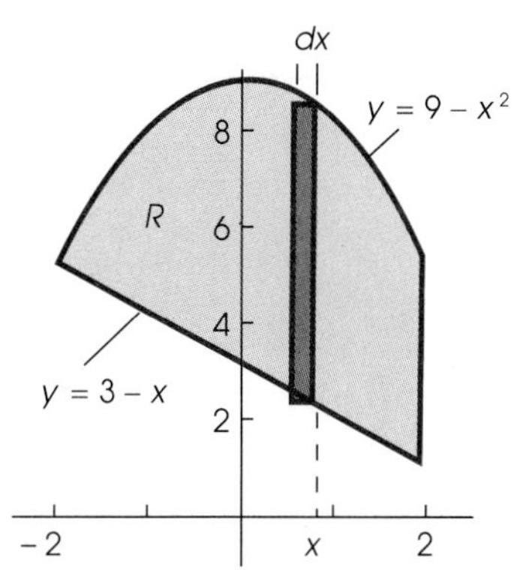

14. Find the area of the region $R$ shown in the figure below.

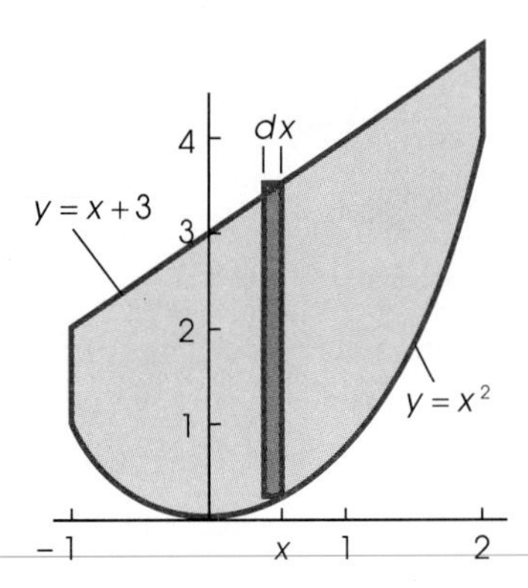


*In Problems 15–22, we give the equations of two curves and an interval. Find the area of the region between the two curves on the given interval. As in Example C, be sure to show an appropriate figure properly labeled (see Figure 6).*

15. $y = x^2 + 2$, $y = -2x^2$; $[0, 2]$
16. $y = x^2 - 1$, $y = 8 - x^2$; $[-1, 2]$
17. $y = x^3 - 1$, $y = x + 2$; $[-2, 1]$
18. $y = 2x^3$, $y = 4 - x^2$; $[-2, 1]$
19. $y = e^x$, $y = 2x + 3$; $[-1, 1]$
20. $y = 2e^{-x}$, $y = (x - 2)^2$; $[3, 5]$
21. $y = 6/(x + 3)$, $y = -x + 1$; $[-2, 2]$
22. $y = (\ln x)/x$, $y = x^2 + x$; $[1, 3]$

*Each of Problems 23–32 gives the equations of two curves, curves that bound a region. Follow the method illustrated in Example D (including a good figure properly labeled) to find the area of this region.*

23. $y = x^2 + 2$, $y = 2x + 5$
24. $y = x^2 - 1$, $y = 2x + 2$
25. $y = 5 - x^2$, $y = -x - 1$
26. $y = 9 - x^2$, $y = -x - 3$
27. $y = x^2 - 3x + 2$, $y = x - 1$
28. $y = x^2 + 6x$, $y = 5x + 2$
29. $y = 2\sqrt{x}$, $y = (2x + 4)/3$
30. $y = \sqrt{x}$, $y = (x + 3)/4$
31. $y = x^2 - 12$, $y = 6 - x^2$
32. $y = x^2 + 3$, $y = 11 - x^2$

*The graphs of each pair of equations in Problems 33–36 bound a region (which may be in two parts). Find the total area of this region as in Example E.*

33. $y = x^3$, $y = 4x$
34. $y = 2x^3$, $y = 2x$
35. $y = 2x^3$, $y = -x^3 + x^2 + 2x$
36. $y = x^2$, $y = x^3 - 2x$

*In Problems 37–42, use horizontal slicing to find the area of the region bounded by the given curves (see Example F).*

37. $x = y^2$, $x = 4$
38. $x = -y^2$, $x = -3$
39. $x = 4y - y^2$, $x = y$
40. $x = y^2 - 6y$, $x = -y$
41. $x = y^2 - 4y + 3$, $x = y - 1$
42. $x = y^2 - 2y - 3$, $x = 2y - 3$

---

*In Problems 43–50, find the area of the region bounded by the given curves.*

43. $y = x^2 - 9$, $y = 0$
44. $x = -y^2 - 2y + 3$, $x = 0$
45. $y = x(x + 2)(x - 3)$, $y = 0$
46. $y = e^x$, $y = e^2$, $x = 0$
47. $x = 1/y$, $x = 0$, $y = 1$, $y = 3$
48. $y = x$, $y = 6 - x$, $y = x/2$
49. $y = x + 6$, $y = x^3$, $2y + x = 0$
50. $y = x^2 - x - 6$, $y = x^3 + x^2 - 5x - 6$
51. $y = x/(x^2 + 1)$, $y = x/2$
52. $y = e^x$, $x = (y - 1) \ln 2$
53. Find the area of the triangle with vertices at $(-1, 0)$, $(4, 5)$, and $(0, 9)$.
54. Find the area of the parallelogram with vertices at $(0, 0)$, $(5, 3)$, $(9, 7)$, and $(4, 4)$.

## 9.2 VOLUMES OF SOLIDS OF REVOLUTION

It is not surprising that the definite integral can be used to calculate areas; it was invented for that purpose. What is somewhat surprising is the number of physical and geometrical quantities that can be calculated by means of definite integrals. In fact, almost any quantity that can be thought of as arising from slicing something into small pieces, approximating each piece, adding up, and taking the limit will finally lead to a definite integral. In particular, this is true of volumes of solids of revolution.

But what is a solid of revolution? Take a plane region that lies entirely on one side of a fixed line called the **axis.** When this region is revolved about the axis, it generates a **solid of revolution** (Figure 11). We wish to calculate the volume of such a solid.

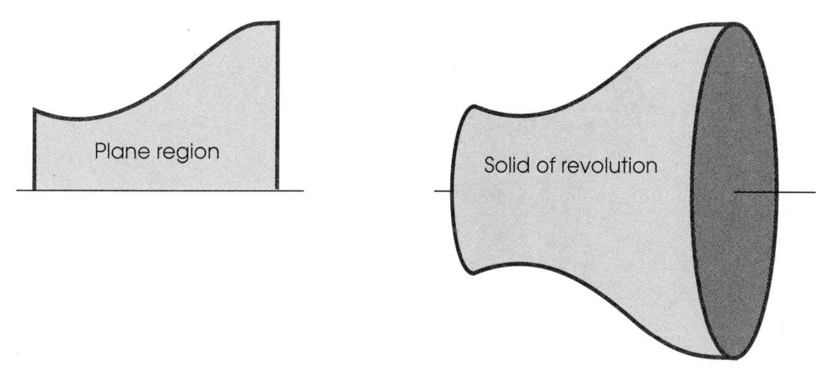


FIGURE 11

Imagine slicing such a solid into thin pieces of equal thickness by means of planes perpendicular to the axis (Figure 12) and then approximating each piece by a disk (Figure 13). If we can find the volume of a disk, add up the volumes of all the disks, and then take the limit as the number of pieces becomes infinite, we should get a definite integral.

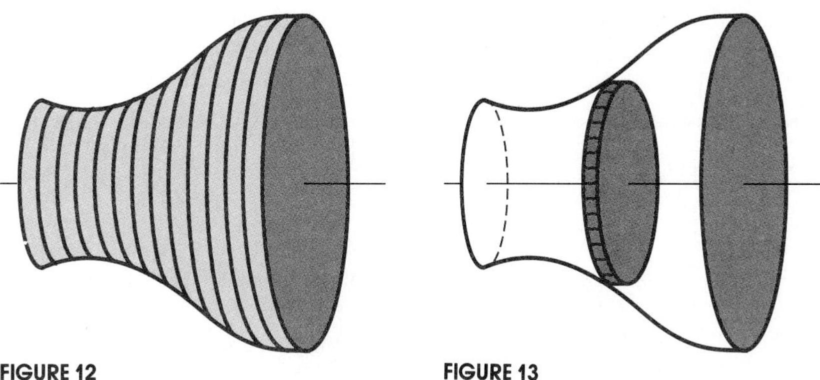

FIGURE 12　　FIGURE 13

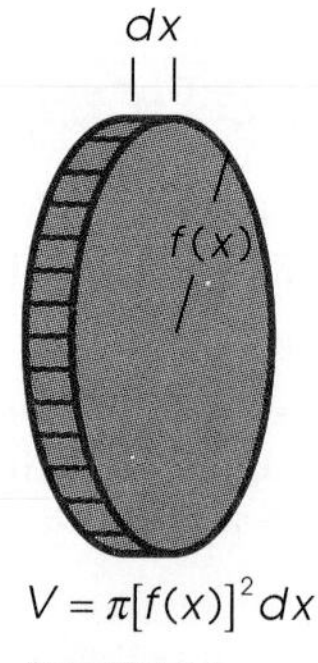


FIGURE 14

The process we have described for calculating volumes of solids of revolution is called the **method of disks.** Let us see how it works in some examples.

## The Method of Disks

Before we can actually do an example, we have to answer one more preliminary question. How do you find the volume of a disk? A disk is just a thin right circular cylinder. We hope you recall that the volume of a right circular cylinder is $\pi r^2 h$; that is, it is the area of the base times the height. For the thin disks we will be considering, this formula will become $\pi[f(x)]^2\,dx$ (Figure 14).

### EXAMPLE A

Find the volume of the solid formed by revolving the region bounded by $y = \sqrt{x}$, $y = 0$, and $x = 4$ about the $x$-axis.

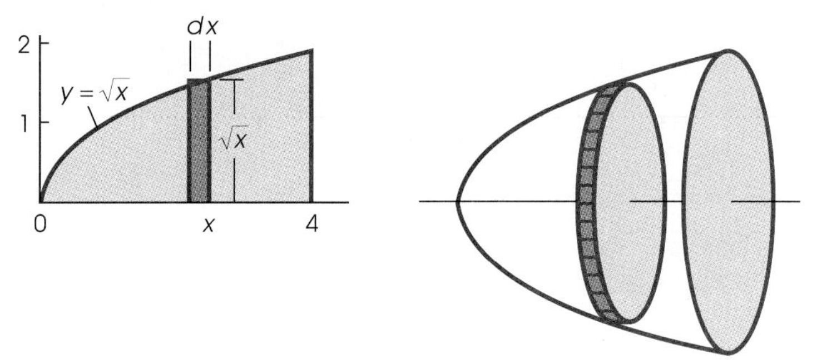


FIGURE 15

**Solution** The region together with a typical rectangle is shown in the left part of Figure 15. The corresponding solid of revolution and the disk generated by the rectangle are shown at the right in Figure 15. Note that the volume of a typical disk is

$$\pi(\sqrt{x})^2\,dx = \pi x\,dx$$

We conclude that the volume of the solid is

$$\int_0^4 \pi x\,dx = \left[\frac{\pi x^2}{2}\right]_0^4 = \pi(8) - 0 = 8\pi$$ ■

Many of us have trouble drawing three-dimensional pictures like the one at the right in Figure 15, but all of us can draw two-dimensional pictures like the one at the left. It is enough to visualize the solid mentally; all the information needed to write down the correct integral appears in the plane figure. We suggest that you concentrate on drawing good, properly labeled two-dimensional figures.

EXAMPLE B

Find the volume of the solid formed by revolving the plane region bounded by $y = \sqrt{9 - x^2}$ and $y = 0$ about the $x$-axis.

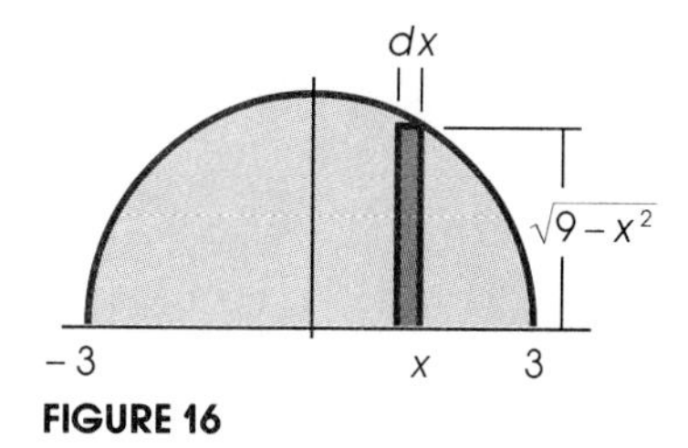


FIGURE 16

**Solution** The equation $y = \sqrt{9 - x^2}$ represents a semicircle (Figure 16), and so the solid generated is a solid sphere. We know that the volume of a sphere is $\frac{4}{3}\pi r^3$, so we will have a good chance to check that our integration method is correct. The rectangle in Figure 16 will generate a disk of volume $\pi(\sqrt{9 - x^2})^2\, dx$, and so the volume of the whole sphere is

$$\begin{aligned}\int_{-3}^{3} \pi(9 - x^2)\, dx &= 2\int_{0}^{3} \pi(9 - x^2)\, dx \\ &= 2\pi\left[9x - \frac{x^3}{3}\right]_0^3 \\ &= 2\pi(27 - 9) = 36\pi\end{aligned}$$

In the first step above, we used the symmetry of the region about the $y$-axis. Note that the answer $36\pi$ is $\frac{4}{3}\pi(3)^3$ as expected. ■

## The Method of Washers

Sometimes slicing a solid of revolution produces disks with holes in the middle; we call them **washers.** A typical washer is shown in Figure 17. Its

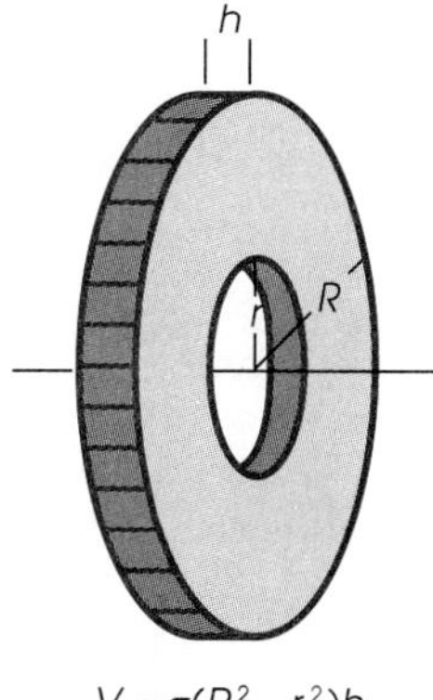


$V = \pi(R^2 - r^2)h$

FIGURE 17

volume is obtained by calculating the volume of a solid disk and then subtracting the volume of the hole. Thus, the volume formula for a washer of thickness $h$ is

$$V = \pi R^2 h - \pi r^2 h = \pi(R^2 - r^2)h$$

where $R$ is the radius of the disk and $r$ is the radius of the hole.

## EXAMPLE C

Find the volume of the solid formed when the region bounded by $y = x^2 + 2$ and $y = 1$ between $x = 0$ and $x = 2$ is revolved about the $x$-axis.

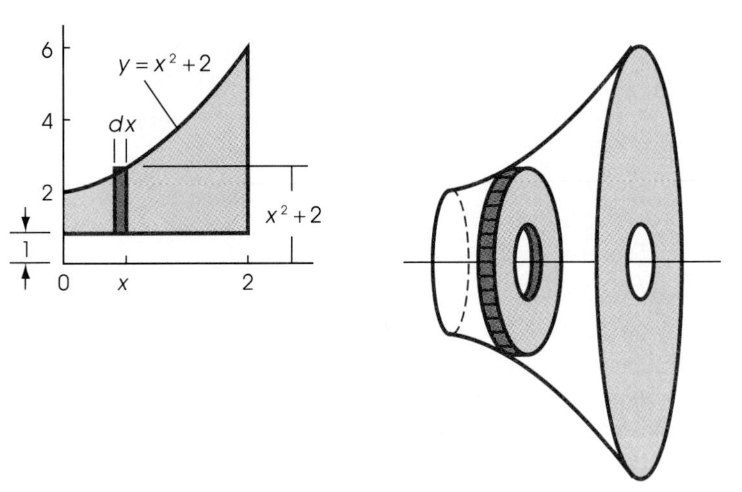


**FIGURE 18**

**Solution** The plane region (with a typical rectangle) and the corresponding solid (with a typical disk) are shown in Figure 18. The volume of this typical disk is

$$\pi[(x^2 + 2)^2 - (1)^2]\, dx$$

and so the volume of the solid is

$$\int_0^2 \pi[(x^2 + 2)^2 - 1]\, dx = \pi \int_0^2 (x^4 + 4x^2 + 3)\, dx$$

$$= \pi\left[\frac{x^5}{5} + \frac{4x^3}{3} + 3x\right]_0^2$$

$$= \pi\left(\frac{32}{5} + \frac{32}{3} + 6\right) \approx 72.4661 \quad ■$$

## Horizontal Slicing

Of course, we can generate a solid of revolution by revolving a region about the $y$-axis or some other vertical line too. This will lead to a $y$-integration, but otherwise all is as before.

### EXAMPLE D

Find the volume of the solid formed by revolving the region bounded by $y = x^3$, $y = 8$, and $x = 0$ about the line $x = -1$.

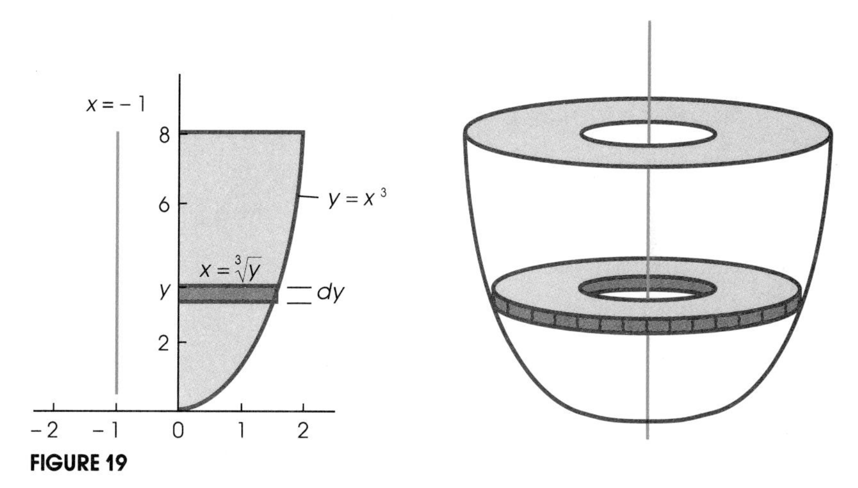


**FIGURE 19**

**Solution** Now we must slice horizontally, and this slicing results in horizontal washers. The plane region (with a typical rectangle) and the corresponding solid (with a typical washer) are shown in Figure 19. Clearly $y = x^3$ means that $x = \sqrt[3]{y}$. Note that $x + 1$ is the outer radius of the washer, whereas 1 is the radius of the hole. Thus, the volume of a typical washer is

$$\pi[(\underbrace{\sqrt[3]{y} + 1}_{\text{Outer radius}})^2 - (\underbrace{1}_{\text{Radius of hole}})^2]\, dy$$

We conclude that the volume of the whole solid is

$$\int_0^8 \pi[(\sqrt[3]{y} + 1)^2 - 1)]\, dy = \pi \int_0^8 (y^{2/3} + 2y^{1/3})\, dy$$

$$= \pi\left[\frac{3}{5}y^{5/3} + \frac{3}{2}y^{4/3}\right]_0^8$$

$$= \pi\left(\frac{3}{5}8^{5/3} + \frac{3}{2}8^{4/3}\right)$$

$$= \pi\left(\frac{96}{5} + 24\right) \approx 135.7168$$

■

### EXAMPLE E

A certain lake has the shape of the solid of revolution generated by revolving the region bounded by

$$y = 30\left[\left(\frac{x}{1000}\right)^2 - 2\right]$$

$y = 0$, and $x = 0$ about the $y$-axis, distances being measured in meters. If a full-grown walleyed pike needs 10,000 cubic meters of water to live comfortably, how many pike can the lake support?

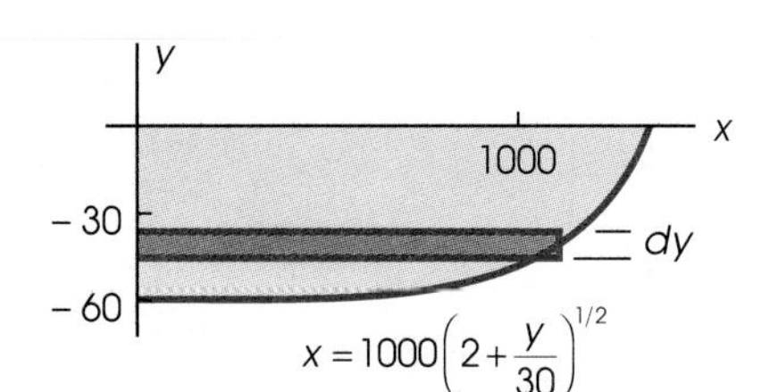


FIGURE 20

**Solution** The problem is simply one of finding the total volume of the lake and dividing the result by 10,000. A picture of the region with a typical rectangle is shown in Figure 20. When we solve the given equation for $x$, we obtain

$$x = 1000\left(2 + \frac{y}{30}\right)^{1/2}$$

Thus, the volume of the disk generated by the displayed rectangle is

$$\pi\left\{\left[1000\left(2 + \frac{y}{30}\right)\right]^{1/2}\right\}^2 dy$$

We conclude that the volume of the lake is

$$\begin{aligned}\int_{-60}^{0} \pi\left[1{,}000{,}000\left(2 + \frac{y}{30}\right)\right] dy &= 1{,}000{,}000\pi \int_{-60}^{0} \left(2 + \frac{y}{30}\right) dy \\ &= 1{,}000{,}000\pi\left[2y + \frac{y^2}{60}\right]_{-60}^{0} \\ &= 1{,}000{,}000\pi[0 - (-120 + 60)] \\ &= 60{,}000{,}000\pi \approx 188{,}500{,}000\end{aligned}$$

The lake will support about 18,850 full-grown pike. ■

## Problem Set 9.2

*In Problems 1–6, find the volume of the solid generated when the indicated region is revolved about the x-axis (see Example A). Begin by showing a typical rectangle properly labeled, as we have already done in Problems 1 and 2.*

1.

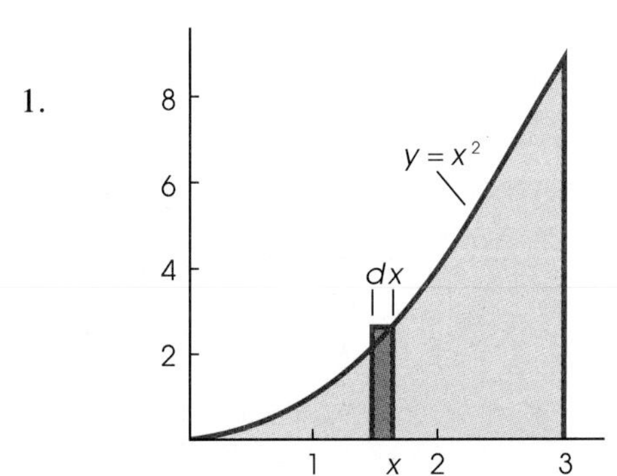

2.

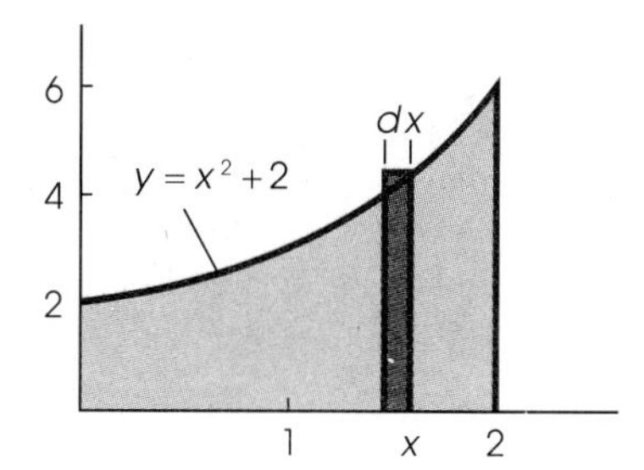


3.

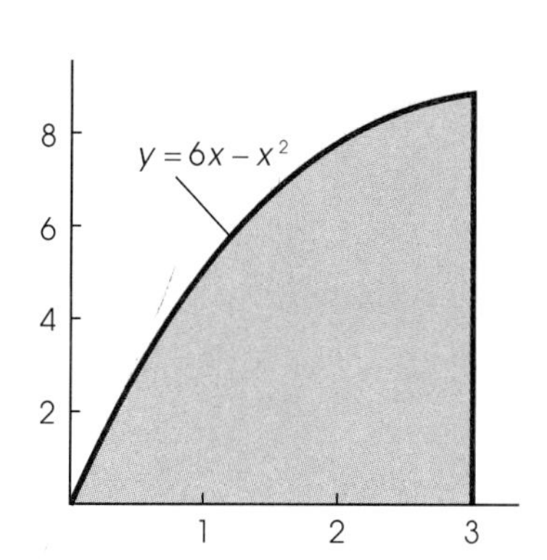


4.

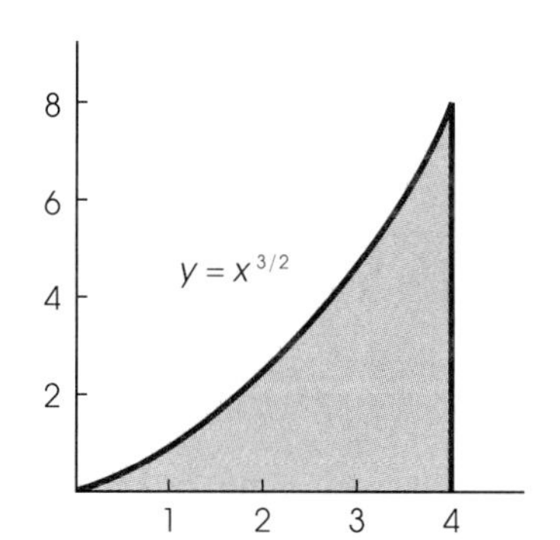


5.

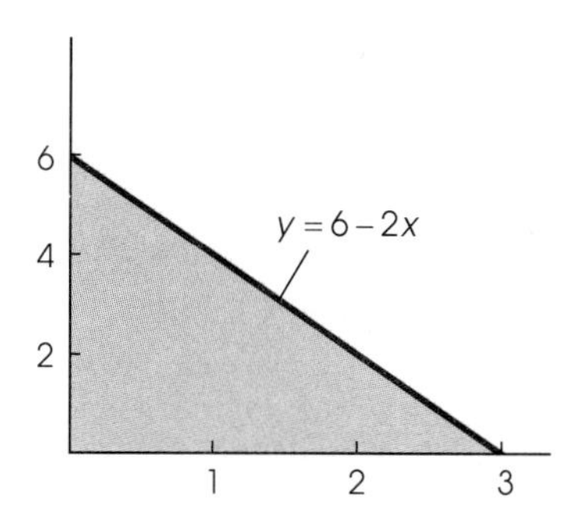


6.

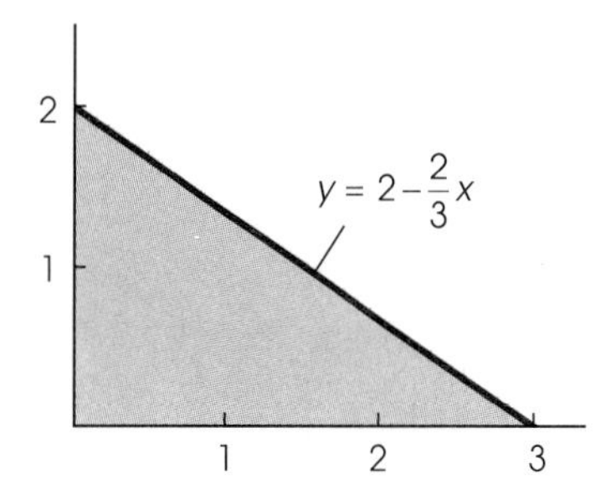


*In Problems 7–12, sketch the region R bounded by the graphs of the given equations, showing a typical vertical rectangle. Then find the volume of the solid generated when R is revolved about the x-axis (see Example B).*

7. $y = 4 - x^2, x = 0, y = 0, x \geq 0$
8. $y = 4x - x^2, y = 0$
9. $y = 4/x, x = 1, x = 4, y = 0$
10. $y = 4/x^2, x = 1, x = 2, y = 0$
11. $y = 2e^{2x}, x = 0, x = 2, y = 0$
12. $y = e^{-x}, x = -2, x = 0, y = 0$

*In Problems 13–18, sketch the region R bounded by the graphs of the given equations. Then use the method of washers to find the volume of the solid generated when R is revolved about the x-axis (see Example C).*

13. $y = 5x^2, y = 5$
14. $y = \sqrt{x}, y = 2, x = 0$
15. $y = x^2, y = x$
16. $y = 2x^2, y = x + 1$
17. $y = 8 - x^2, y = x^2$
18. $y = x^3, y = 4x$

*In Problems 19–26, sketch the region R bounded by the graphs of the given equations. Then use horizontal slicing to find the volume of the solid generated when R is revolved about the y-axis (see Example D).*

19. $x = 4y - y^2, x = 0$
20. $x = y^{1/3}, x = 0, y = 8$
21. $2x + y = 6, x = 0, y = 0$
22. $2x + 3y = 6, x = 0, y = 0$
23. $y = x^3, y = 0, x = 2$
24. $y = x^2, y = 0, x = 3$
25. $y = 2x^2, y = 2x$
26. $y = 4 - x^2, y = -2x + 4$

*In Problems 27–36, find the volume of the solid generated when the given region is revolved about the given line.*

27. Region in Problem 1 about $y = -2$
28. Region in Problem 2 about $y = -1$
29. Region in Problem 3 about $y = 9$
30. Region in Problem 4 about $y = 8$
31. Region in Problem 5 about $x = -3$
32. Region in Problem 6 about $x = -2$
33. Region in Problem 13 about $y = 5$
34. Region in Problem 14 about $y = 2$
35. Region in Problem 25 about $x = 4$
36. Region in Problem 26 about $x = 3$
37. Use the ideas of this section to find the volume of the solid that remains when a hole of radius 3 is drilled through the center of a ball of radius 5.
38. The base of a solid is the region bounded by $y = e^x$, $y = 0$, $x = 0$, and $x = 1$. Cross sections of the solid perpendicular to the $x$-axis are squares. Find the volume of this solid. *Hint:* Slicing through this solid with planes perpendicular to the $x$-axis produces thin slabs (like slices of cheese) whose volumes are easy to approximate.
39. The interior surface of a tank has the shape that is obtained by revolving the curve $y = e^{x^2}$ between $x = 0$ and $x = 2$ about the $y$-axis. Determine the capacity of the tank.

## 9.3 SOCIAL SCIENCE APPLICATIONS

The most impressive applications of the definite integral are in the physical sciences—to such concepts as work, energy, fluid force, center of mass, and moment of inertia. But there are also many applications in the social sciences; it is to the latter that we address our attention in this section.

### Recovering a Quantity from Its Margin

A standard application is to determine an economic quantity $Q(x)$ when we know its marginal rate $Q'(x)$. From the Fundamental Theorem of Calculus,

$$\int_0^b Q'(x)\,dx = Q(b) - Q(0)$$

In other words,

$$Q(b) = Q(0) + \int_0^b Q'(x)\,dx$$

Example A shows how we use this fact.

> **POINT OF INTEREST**
> Since
> $$\int_a^b Q'(x)\,dx = Q(b) - Q(a)$$
> we can always determine the net change in a quantity $Q$ on the interval $a \le x \le b$ by simply integrating its rate of change over that interval.

**EXAMPLE A**

Suppose that the marginal cost (in dollars per item) of manufacturing $x$ units of a product is given by

$$C'(x) = 200e^{-x/100}$$

In simple terms, you can think of $C'(x)$ as representing the cost (approximately) of manufacturing the $x$th unit. For example, it costs about $200e^{-50/100} = \$121.31$ to make the 50th unit. Note that the cost of making the $x$th unit decreases as $x$ increases, presumably as a result of economy of scale. Find the cost of making 300 of these units, assuming that there are \$10,000 of fixed costs (that is, costs that occur even if no units are made). What is the average cost per unit?

**Solution** Note that $C(0)$ corresponds to the fixed costs; that is, $C(0) = 10{,}000$. Thus, the total cost of making 300 units is

$$\begin{aligned} C(300) &= C(0) + \int_0^{300} C'(x)\,dx \\ &= 10{,}000 + \int_0^{300} 200e^{-x/100}\,dx \\ &= 10{,}000 + 200\int_0^{300} e^{-x/100}\,dx \end{aligned}$$

In this integral, we make the substitution $u = -x/100$, and $du = -(\frac{1}{100})\,dx$. We obtain

$$\begin{aligned} C(300) &= 10{,}000 + 200(-100)\int_0^{-3} e^u\,du \\ &= 10{,}000 - 20{,}000\Big[e^u\Big]_0^{-3} \\ &= 10{,}000 - 20{,}000[e^{-3} - e^0] = \$29{,}004.26 \end{aligned}$$

The average cost is $29{,}004.26/300 = \$96.68$. ■

## Consumer and Producer Surplus

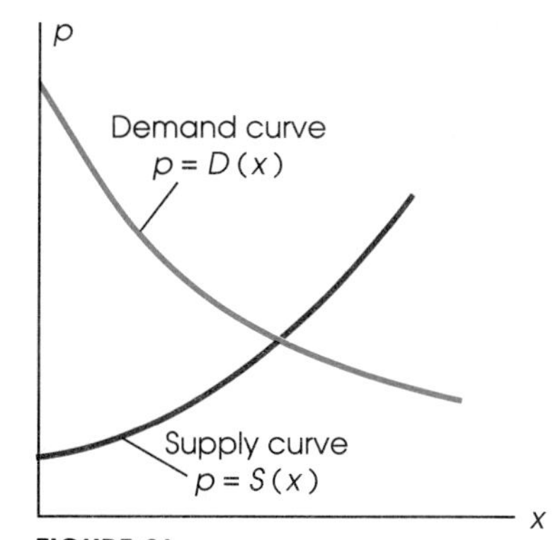


FIGURE 21

In our earlier discussions of demand functions, we usually assumed that $x$, the quantity demanded, is expressed in terms of $p$, the price charged, that is, that $x = f(p)$. Actually, economists are more likely to think of the **demand function** as relating the price $p$ to the quantity $x$ that people will buy, that is, $p = D(x)$. It is really just a matter of perspective; for if $x = -\frac{1}{2}p + 40$, then $p = -2x + 80$, and vice versa. Similarly, the **supply function,** $p = S(x)$, relates the price $p$ to the supply $x$ that is available. Graphs of typical demand and supply functions are shown in Figure 21. Note that the demand function is decreasing (the price must come down for people to buy more) and the supply function is increasing (producers will supply more as prices increase). The **equilibrium price** is the price at which the two curves intersect. This, say economists, is the only price that can persist over the long run.

## EXAMPLE B

Find the equilibrium price if the demand and supply functions (in dollars) for a certain product are given by $p = 500 - 5x$ and $p = \frac{1}{2}x^2 + 200$, respectively (Figure 22a).

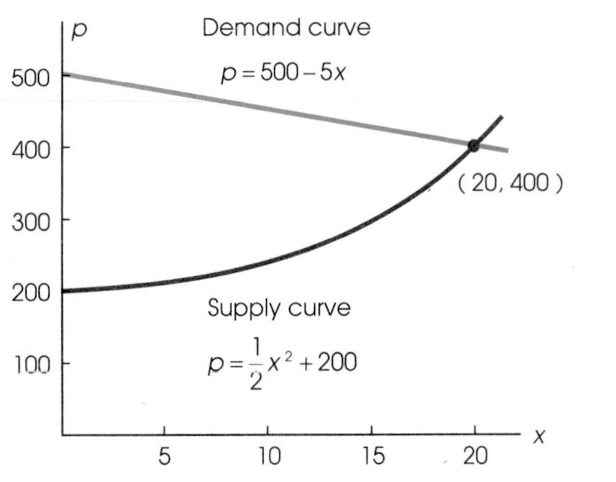


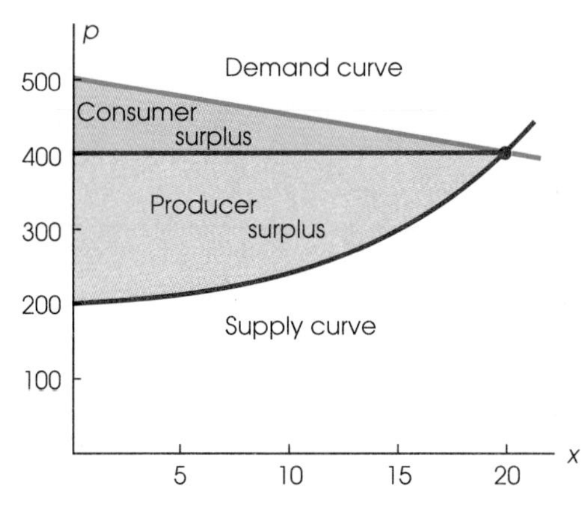


**FIGURE 22**

**Solution** We solve the two equations simultaneously by equating the two expressions for $p$.

$$\frac{1}{2}x^2 + 200 = 500 - 5x$$

$$\frac{1}{2}x^2 + 5x - 300 = 0$$

$$x^2 + 10x - 600 = 0$$

$$(x + 30)(x - 20) = 0$$

$$x = -30 \qquad x = 20$$

Only the positive answer $x = 20$ makes sense. The equilibrium price is $p = 500 - 5(20) = \$400$. ■

In Figure 22b, we have indicated two quantities of economic significance, namely, consumer surplus and producer surplus. The **consumer surplus** is the total amount saved by consumers who are willing to pay a price higher than the equilibrium price and corresponds to the area of the region between the demand curve and the horizontal line through the equilibrium point. Similarly, the **producer surplus** is the increased revenue realized by producers who are willing to sell at a price lower than the equilibrium price. It is the area of the region between the same horizontal line and the supply curve.

## EXAMPLE C

Find the consumer surplus $CS$ and the producer surplus $PS$ for the situation described in Example B.

**Solution**

$$CS = \int_0^{20} [(500 - 5x) - 400]\, dx$$

$$= \left[100x - \frac{5x^2}{2}\right]_0^{20} = \$1000$$

$$PS = \int_0^{20} \left[400 - \left(\frac{x^2}{2} + 200\right)\right] dx$$

$$= \left[200x - \frac{x^3}{6}\right]_0^{20} \approx \$2667$$

■

## Distribution of Income

While no country has been able to achieve equal distribution of income, a desirable goal in the eyes of many political scientists is to reduce the inequalities that exist. The table and corresponding graph (Figure 23) indicate how incomes in the United States were distributed in 1973 (*Source:* Paul Samuelson, *Economics* (10th Ed.), New York: McGraw-Hill, 1976). The left column of the table assumes that people have been ranked according to income.

Note in the table that the lowest paid 40% of the population receives only 18% of the income, whereas if income were absolutely equally distributed, they would get 40% of the income. In fact, if income were equally distributed, the data would lie along the 45-degree line. In actuality, the data always lie along a curve below this line, a curve called a **Lorenz curve.** The further this curve moves from the 45-degree line, the more unequal the in-

| Cumulative percentage of people | Cumulative percentage of income |
|---|---|
| 0 | 0 |
| 20 | 6 |
| 40 | 18 |
| 60 | 35 |
| 80 | 59 |
| 100 | 100 |

Income distribution: 1973

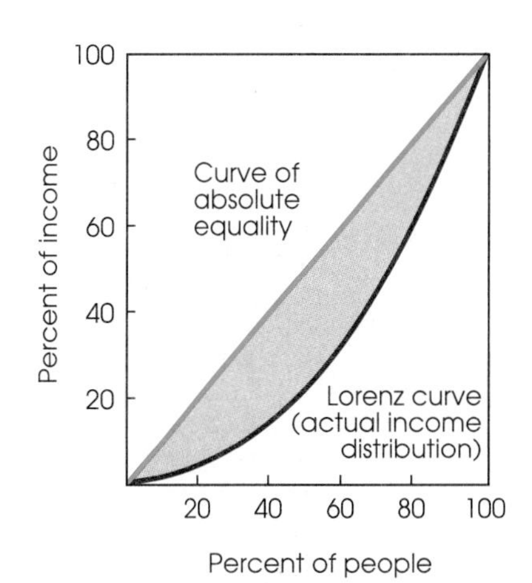


**FIGURE 23**

come distribution. As a measure of inequality, we use the so-called **Gini coefficient of inequality** $g$ defined by

$$g = \frac{\text{area between the Lorenz curve and the line } y = x}{\text{area under the line } y = x}$$

The numerator is a definite integral whose value is between 0 and $\frac{1}{2}$; the denominator has a value of exactly $\frac{1}{2}$. Thus, $g$ is between 0 (absolute equality) and 1 (absolute inequality). If $y = f(x)$ is the equation of the Lorenz curve, the formula for $g$ takes the simple form

$$g = 2\int_0^1 [x - f(x)]\, dx$$

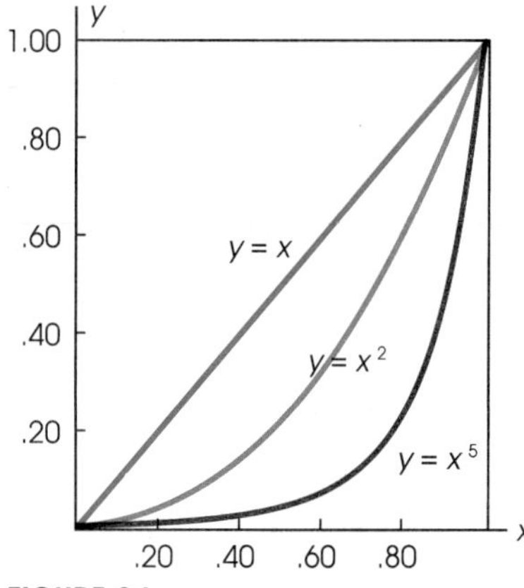

FIGURE 24

### EXAMPLE D

Sketch the Lorenz curves $y = x^2$ and $y = x^5$ on the same graph and determine the corresponding Gini coefficients.

**Solution** The graphs are shown in Figure 24. The required Gini coefficients are

$$g = 2\int_0^1 [x - x^2]\, dx = 2\left[\frac{x^2}{2} - \frac{x^3}{3}\right]_0^1 = \frac{1}{3}$$

and

$$g = 2\int_0^1 [x - x^5]\, dx = 2\left[\frac{x^2}{2} - \frac{x^6}{6}\right]_0^1 = \frac{2}{3}$$ ■

## Problem Set 9.3

*In doing Problems 1–4, keep in mind that $\int_a^b Q'(x)\,dx = Q(b) - Q(a)$. Thus, to find the net change in $Q$ between $a$ and $b$, we simply integrate its derivative between $a$ and $b$.*

1. The population (in hundreds) of a bacteria colony is changing at the rate of $6t^2 + 30t^{1/4} + 100$ per hour at time $t$. How much will the population increase between $t = 1$ and $t = 16$?

C 2. Suppose that the value of a piece of property is growing at a rate of $1000t + 100e^{-t/4}$ dollars per year at time $t$. How much will its value increase between $t = 3$ and $t = 6$?

3. Hermantown has a population of 50,000 now, and it is predicted that $x$ years from now its population will be increasing at the rate of $450 + 35x^{4/3}$ people per year. What is its expected population 8 years from now?

4. A research group has advised a midwestern stock exchange that over the next $t$ months the number of shares traded will change at the rate of $2t + 3\sqrt{t} - 4$ million shares per month. Find the number of shares the stock exchange can expect to trade during the next 16 months.

5. A manufacturer's marginal cost (in dollars per unit) to make $x$ units of a certain product is given by

$C'(x) = 30 - 0.012x$. If there are fixed costs of \$2000, find (a) the total cost of making 150 units, (b) the average cost per unit of making 150 units.

C 6. Suppose that the marginal cost (in dollars per unit) to manufacture $x$ units of a product is given by $C'(x) = 300e^{-x/90}$ and that fixed costs are \$8000. Find (a) the total cost of making 180 units, (b) the average cost per unit of making 180 units.

7. Suppose that the marginal revenue (in dollars per unit) for the manufacturer in Problem 5 from selling $x$ units is given by $R'(x) = 50 + 0.36x - 0.003x^2$.
   (a) Find the total revenue from selling 150 units. [Note that $R(0) = 0$, since you get zero revenue from selling zero units.]
   (b) Find the total profit from making and selling 150 units.

C 8. Suppose that the marginal revenue (in dollars per unit) for the manufacturer in Problem 6 from selling $x$ units is given by $R'(x) = 500e^{-x/80}$.
   (a) Find the total revenue from selling 180 units.
   (b) Find the total profit from making and selling 180 units.

C 9. Based on the information in Problems 5 and 7, what number of units will produce the maximum profit? *Hint:* We can expect the maximum for $P(x)$ to occur at a point where $P'(x) = 0$.

C 10. Based on the information in Problems 6 and 8, what number of units will produce the maximum profit?

*In Problems 11–16, find the equilibrium price corresponding to the given demand and supply functions D(x) and S(x) as in Example B.*

11. $D(x) = 8 - \frac{1}{3}x$, $S(x) = \frac{1}{3}x$
12. $D(x) = 9 - \frac{1}{4}x$, $S(x) = \frac{1}{2}x$
13. $D(x) = 400 - \frac{1}{3}x^2$, $S(x) = \frac{1}{9}x^2$
14. $D(x) = 120 - \frac{1}{5}x^2$, $S(x) = \frac{1}{10}x^2$
15. $D(x) = 240 - 30x - 20x^2$, $S(x) = 20x^2 + 10x$
16. $D(x) = 720 - 180x - 10x^2$, $S(x) = 20x^2 + 120x$

17–22. For each of Problems 11–16, sketch the graphs of the demand and supply functions, showing the regions whose areas correspond to the consumer surplus and producer surplus. Then calculate the consumer surplus *CS* and the producer surplus *PS* (see Example C).

23. Suppose that the demand function for a certain commodity is $p = D(x) = 100 - 6\sqrt{x}$. How many units are demanded when the price is set at \$28? Determine the consumer surplus if the equilibrium price is \$28.

24. If the demand function for a certain product is $D(x) = 45e^{-x/30}$, find the consumer surplus when the equilibrium price is \$15.

25. If the supply function for a certain type of jacket is $S(x) = 10e^{0.02x}$ and the equilibrium price is \$40, determine the producer surplus.

26. The supply function for ABD jogging shoes is $S(x) = 12x^{1/3} + 30$, and the equilibrium price is \$66. Find the producer surplus.

*In Problems 27–32, we give the equation of a Lorenz curve. Sketch this curve on the interval $0 \le x \le 1$ and then calculate the corresponding Gini coefficient.*

27. $y = x^3$
28. $y = x^4$
29. $y = (x^2 + x)/2$
30. $y = (2x^2 + x)/3$
31. $y = (e^x - 1)/(e - 1)$
32. $y = (e^{2x} - 1)/(e^2 - 1)$

---

33. It is estimated that the rate of repair costs in dollars per mile for a city bus after it has been driven $x$ miles is $0.12 + 0.01\sqrt[3]{x}$. Find the total repair costs for driving a bus an additional 61,000 miles if its odometer shows 64,000 miles now.

C 34. A certain country is using oil at the rate $0.05e^{0.015t}$ billion barrels per year, where $t$ is the number of years after January 1, 1985. At this rate, how many billion barrels of oil will it use during the decade of the 1990s?

35. If the total attendance at a racetrack during the first $x$ days of the season was $\int_0^x (10{,}500 + 50t)\,dt$, on what day did the total attendance first surpass 150,000?

36. A factory is expected to discharge pollutants into a lake at the rate of $100t\sqrt{t^2 + 9}$ tons per year, $t$ years after it goes into operation. How much will it discharge during the fifth year of operation?

37. A manufacturer produces small recreational vehicles. The demand function is $D(x) = 20{,}000 - 100x$, and the supply function is $S(x) = 8000 + 500x$, where $x$ is the number of vehicles produced per month. Find (a) the equilibrium price and (b) the consumer surplus.

38. The manager of a paper company estimates that after $x$ years of use a certain machine produces revenue at the rate of $40{,}000 - 60x^2$ dollars per year, but that the cost of running the machine (labor, depreciation, maintenance) is $9675 + 24x^2$ dollars per year.
   (a) How long will it be profitable to use the machine?
   (b) Find the total profit produced by this machine during its useful life.

## 9.4 CONTINUOUS MONEY STREAMS

We introduced the notion of continuous compounding of interest in Section 6.3. Recall that if $P$ dollars is invested at annual rate $r$ compounded continuously, then its value $F$ at the end of $t$ years is given by $F = Pe^{rt}$. Solving this equation for $P$ in terms of $F$ gives the equation

$$P = Fe^{-rt}$$

Written in this form, the equation tells us the present value $P$ of $F$ dollars that will be received $t$ years from now, assuming that interest is compounded continuously at the annual rate $r$.

### EXAMPLE A

Mary knows that she will need to replace her car in 3 years. How much would she have to put in the bank today at 8% interest compounded continuously in order to have the $12,000 she expects to need 3 years from now?

**Solution** We apply the boxed formula, remembering that the interest rate of 8% must be written in its decimal form.

$$P = 12{,}000e^{-0.08(3)} = 12{,}000e^{-0.24} = \$9439.53$$ ■

### Continuous Money Streams

There are many situations in business and industry where it is useful to think of money as flowing continuously into an account. For example, the ABC Company plans to buy an expensive machine which it estimates will increase the company's net income by $10,000 per year. But this income won't come at the end of each year, it will come in dribbles throughout the year. As a model for what will happen, it is convenient to think of the machine as if it will produce income continuously. Our next example raises an important question for the company to answer.

### EXAMPLE B

If the ABC Company expects the machine described above to last for 8 years and if money is considered to be worth 9% compounded continuously, what is the present value $PV$ of the income stream that the machine will produce over its lifetime?

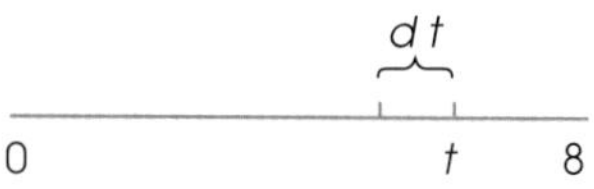

dt

0 t 8

FIGURE 25

**Solution** Recall our basic advice: *slice, approximate, integrate*. Divide the time interval $0 \le t \le 8$ into little pieces with a typical piece at time $t$ being of length $dt$ (Figure 25). The income produced by the machine during this short time interval is approximately \$10,000 $dt$, and the present value of this piece of income is approximately

$$10{,}000\, dt\,(e^{-0.09t})$$

Integrating this amount over the period $0 \le t \le 8$ should give us the present value of the whole income stream; that is,

$$PV = \int_0^8 10{,}000e^{-0.09t}\,dt = -\frac{10{,}000}{0.09}\int_0^8 e^{-0.09t}(-0.09)\,dt$$

$$= -\frac{10{,}000}{0.09}\left[e^{-0.09t}\right]_0^8 = -\frac{10{,}000}{0.09}(e^{-0.72} - 1) \approx \$57{,}027.53$$

The company can afford to pay anything less than this amount for the machine. ■

Let us generalize this result. Suppose that the annual rate of income of an income stream is $R(t)$ at time $t$ [in Example B, $R(t)$ was the constant 10,000]. Suppose further that this income stream will last over the period $0 \le t \le T$. If interest is at rate $r$ compounded continuously, then the present value $PV$ of this income stream is given by

$$PV = \int_0^T R(t)e^{-rt}\,dt$$

## EXAMPLE C

Suppose the ABC Company in Example B takes a more optimistic view and estimates that the new machine will produce income at the rate $R(t) = 10{,}000 + 200t$. What is the present value of this income stream, again assuming interest at 9% compounded continuously and a lifetime of 8 years?

### Solution

$$PV = \int_0^8 (10{,}000 + 200t)e^{-0.09t}\,dt$$

$$= \int_0^8 10{,}000e^{-0.09t}\,dt + \int_0^8 200te^{-0.09t}\,dt$$

We evaluated the first integral in Example B; the second can be evaluated using integration by parts with $u = 200t$ and $dv = e^{-0.09t}\,dt$. We obtain

$$PV = 57{,}027.53 + \left[-\frac{200}{0.09}te^{-0.09t}\right]_0^8 + \frac{200}{0.09}\int_0^8 e^{-0.09t}\,dt$$

$$= 57{,}027.53 - 8653.37 - \frac{200}{0.0081}\left[e^{-0.09t}\right]_0^8 = \$61{,}046.94$$ ■

Let us modify the problem in Example B in another way by assuming that the machine will last indefinitely. More generally, let us ask for the present value $PV$ of a perpetual (meaning infinitely long) income stream which produces income at an annual rate $R(t)$, assuming that interest is at an annual rate $r$ compounded continuously. The result in this case is

$$PV = \int_0^\infty R(t)e^{-rt}\,dt$$

Here, we face a new kind of integral, one with an infinite limit. Such integrals are called **improper integrals** and must be given a clear definition. We do this by defining

$$\int_0^\infty R(t)e^{-rt}\,dt = \lim_{T\to\infty} \int_0^T R(t)e^{-rt}\,dt$$

### EXAMPLE D

Redo Example B assuming the machine will last indefinitely. In other words, find the present value $PV$ of a perpetual income stream with an annual rate of \$10,000 per year, assuming interest is at 9% compounded continuously.

**Solution**

$$\begin{aligned} PV &= \int_0^\infty 10{,}000e^{-0.09t}\,dt = \lim_{T\to\infty} \int_0^T 10{,}000e^{-0.09t}\,dt \\ &= \lim_{T\to\infty} -\frac{10{,}000}{0.09}\left[e^{-0.09t}\right]_0^T = \lim_{T\to\infty} \frac{10{,}000}{0.09}(1 - e^{-0.09T}) \\ &= \frac{10{,}000}{0.09} = \$111{,}111.11 \end{aligned}$$

■

## Improper Integrals

Improper integrals play an important role in mathematics and its applications. For this reason, we now give a general discussion and illustrate with several examples. First, we give the appropriate definitions.

1. $\displaystyle\int_a^\infty f(x)\,dx = \lim_{b\to\infty} \int_a^b f(x)\,dx$
2. $\displaystyle\int_{-\infty}^b f(x)\,dx = \lim_{a\to-\infty} \int_a^b f(x)\,dx$
3. $\displaystyle\int_{-\infty}^\infty f(x)\,dx = \int_{-\infty}^0 f(x)\,dx + \int_0^\infty f(x)\,dx$

If these limits exist and are finite, we say the corresponding improper integrals **converge** and have the indicated values; otherwise, we say the given improper integrals **diverge.**

### EXAMPLE E

Show that $\int_0^\infty e^{-3x}\,dx$ converges and has a value of $\frac{1}{3}$.

**Solution**

$$\int_0^\infty e^{-3x}\,dx = \lim_{b\to\infty}\int_0^b e^{-3x}\,dx = \lim_{b\to\infty}\left[-\frac{1}{3}e^{-3x}\right]_0^b$$

$$= \lim_{b\to\infty}\left(-\frac{1}{3}e^{-3b} + \frac{1}{3}\right) = \frac{1}{3}$$

■

### EXAMPLE F

Show that $\int_1^\infty \frac{1}{x^2}\,dx$ converges, but that $\int_1^\infty \frac{1}{x}\,dx$ diverges.

**Solution**

$$\int_1^\infty \frac{1}{x^2}\,dx = \lim_{b\to\infty}\int_1^b x^{-2}\,dx = \lim_{b\to\infty}\left[-\frac{1}{x}\right]_1^b = \lim_{b\to\infty}\left[-\frac{1}{b} + 1\right] = 1$$

Also,

$$\int_1^\infty \frac{1}{x}\,dx = \lim_{b\to\infty}\int_1^b \frac{1}{x}\,dx = \lim_{b\to\infty}[\ln b] = \infty,$$

which implies that $\int_1^\infty \frac{1}{x}\,dx$ diverges. ■

### EXAMPLE G

Show that the present value of a perpetual money stream at the rate of $P$ dollars per year is $P/r$, assuming that interest is compounded continuously at an annual rate of $r$.

**Solution**

$$PV = \int_0^\infty Pe^{-rt}\,dt = \lim_{b\to\infty}\int_0^b Pe^{-rt}\,dt = \lim_{b\to\infty}\left[-\frac{P}{r}e^{-rt}\right]_0^b$$

$$= \lim_{b\to\infty}\left(-\frac{P}{r}e^{-rb} + \frac{P}{r}\right) = \frac{P}{r}$$

■

### EXAMPLE H

Harry (a business executive) is going through a divorce. His income from investments and salary is estimated at \$100,000 per year and is expected to continue at this rate throughout his lifetime. The court wants to know the present value of this income stream in order to make a fair distribution of assets. A consultant recommended assuming Harry's income to be a perpetual money stream at 8% interest compounded continuously. What present value does this lead to?

**Solution** From Example G, we obtain

$$PV = 100{,}000/0.08 = \$1{,}250{,}000$$

■

## Problem Set 9.4

*In Problems 1–4, find the present value (the value at 0) of each payment (or set of payments) due at the indicated times, assuming that interest is compounded continuously at the given rate $r$. See Example A.*

C 1. $r = 0.075$

| | | | | \$2000 | |
|---|---|---|---|---|---|
| 0 | 1 yr | 2 | 3 | 4 | 5 |

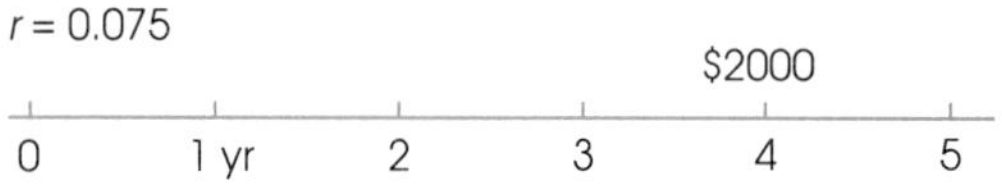

C 2. $r = 0.084$

| | | | | | \$9000 |
|---|---|---|---|---|---|
| 0 | 1 yr | 2 | 3 | 4 | 5 |

C 3. $r = 0.091$

| | | \$2400 | | | \$6200 |
|---|---|---|---|---|---|
| 0 | 1 yr | 2 | 3 | 4 | 5 |

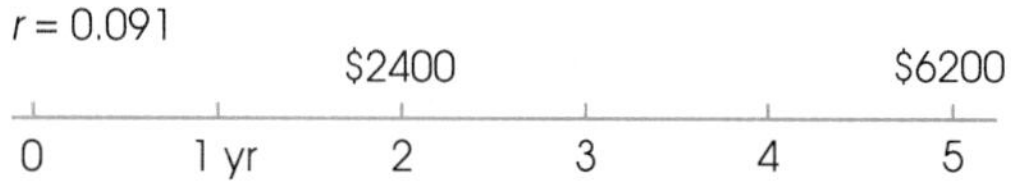

C 4. $r = 0.076$

| | \$3000 | | | \$5400 | |
|---|---|---|---|---|---|
| 0 | 1 yr | 2 | 3 | 4 | 5 |

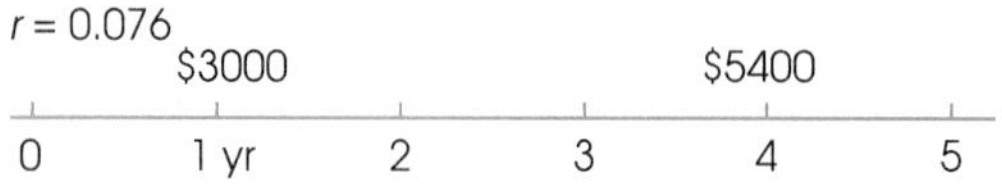

C 5. To settle an estate, Janet has promised to pay her sister \$8500 eight years from now. If she were to pay off this obligation right now, what should she pay, assuming interest is at 8.25% compounded continously?

C 6. Karen is to pay me \$34,000 eight years from now, and Robert is to pay me \$40,000 ten years from now. Which person owes me more and how much more, assuming interest is at 8.5% compounded continuously?

C 7. The XYZ Company plans to buy a new computer which will, according to estimates, produce additional income at the rate of \$25,000 a year for 10 years. Find the present value of the income stream generated by the computer over 10 years, if interest is at 9.5% compounded continuously. See Example B.

C 8. A farmer has decided to erect two new silos at a total cost of \$80,000. He figures that the new silos will produce an income stream at the rate of \$15,000 per year for the next 20 years. Determine the present value of this income stream to see if he has made a wise investment. Assume that interest is at 10.6% compounded continuously.

C 9. Suppose that the XYZ Company in Problem 7 estimates that the new computer will produce additional income at the rate $R(t) = 25{,}000 + 350t$ dollars per year, $t$ being the number of years from now. Assuming again an interest rate of 9.5% compounded continuously, determine the present value of the income stream the computer will generate over the next 10 years. See Example C.

C 10. Suppose that the farmer in Problem 8 predicts more optimistically that the new silos will produce income at the rate $R(t) = 15{,}000 + 300t$ dollars per year, $t$ years after the silos are built. Determine the present value of this income stream over the next 20 years, assuming again an interest rate of 10.6% compounded continuously.

*Evaluate the improper integrals in Problems 11–16 (see Examples D–F). In Problems 15 and 16, you will need the fact that $\lim_{x\to\infty} xe^{-kx} = 0$ provided $k > 0$.*

11. $\displaystyle\int_1^\infty t^{-3}\,dt$

12. $\displaystyle\int_1^\infty 3t^{-4}\,dt$

13. $\int_0^{\infty} e^{-0.09t}\, dt$

14. $\int_0^{\infty} e^{-0.075t}\, dt$

15. $\int_0^{\infty} te^{-0.09t}\, dt$

16. $\int_0^{\infty} te^{-0.075t}\, dt$

*The results for Problems 13–16 should prove helpful in doing problems 17–20.*

17. A restaurant is remodeling its facilities based on the assumption that this remodeling will produce additional income at the rate of \$8000 per year for an indefinitely long period. Determine the present value of this income stream if interest is at 9% compounded continuously. (See Examples D and G.)
18. A small department store is installing a new energy-efficient heating and air conditioning system which will, according to estimates, produce savings at the rate of \$1200 per year for an indefinitely long period of time. If interest is at 7.5% compounded continuously, what is the present value of the savings effected by the new system.
19. [C] Redo Problem 17 assuming that the additional income will be at the rate of $R(t) = 8000 + 100t$ dollars per year, $t$ being the number of years after remodeling.
20. [C] Redo Problem 18, assuming the savings will be at the rate $R(t) = 1200 + 40t$ dollars per year, $t$ being the number of years after installation of the new equipment.

---

21. [C] How much money should Charles Moyer invest today at 10% compounded continuously to have the \$16,500 that he will need for a new car 6 years from now?
22. [C] How much does Sarah actually owe today if she must make payments of \$800 at the end of each of the next 4 years? Assume interest is at 10.8% compounded continuously.
23. [C] Determine the present value of a continuous money stream which will produce income at the rate of \$500 per year for the next 10 years. Assume interest is at 8% compounded continuously.
24. [C] Redo Problem 23 assuming income is produced at the rate of $500 + 20t$ dollars per year for the next 10 years, $t$ being the number of years from now.
25. [C] Redo Problem 23 assuming income will be produced at the rate of \$500 per year forever.
26. [C] Redo Problem 23 assuming income will be produced at the rate of $500 + 20t$ dollars per year forever.

*In Problems 27–32, determine whether the given improper integral converges or diverges and, if convergent, determine its value.*

27. $\int_0^{\infty} e^{0.03x}\, dx$

28. $\int_1^{\infty} \frac{\ln x}{x}\, dx$

29. $\int_0^{\infty} xe^{-x^2}\, dx$

30. $\int_4^{\infty} \frac{4}{(x-3)^5}\, dx$

31. $\int_0^{\infty} \frac{x}{1+x^2}\, dx$

32. $\int_0^{\infty} \frac{x}{(1+x^2)^2}\, dx$

33. Show that $\int_1^{\infty} x^{-p}\, dx$ converges if $p > 1$ and find its value.
34. Show that $\int_0^{\infty} \frac{x}{(1+x^2)^p}\, dx$ converges if $p > 1$ and find its value.
35. J. P. Gorgan has made provision for a perpetual flow of money to Yoohoo University at an annual rate of \$750,000 per year. How much money did he have to set aside to fund this flow if he was able to set up a trust paying 7.4% compounded continuously?

## 9.5*
## CONTINUOUS PROBABILITY

We assume that the reader is familiar with the basic notion of **probability.** In particular, the probability of an event is a number between 0 and 1 which measures the chance of the event occurring in an experiment. Roughly speaking, it gives the percentage of times we expect the event to occur in a large number of trials of the experiment. We are interested in the properties of a continuous **random variable X,** by which we mean a variable that takes values (in a random manner) on an interval of the real line. For example, the weight **X** of people is a random variable which may lie anywhere on

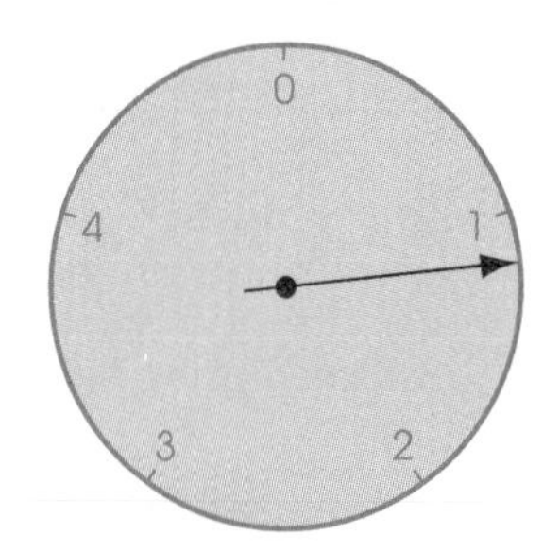


FIGURE 26

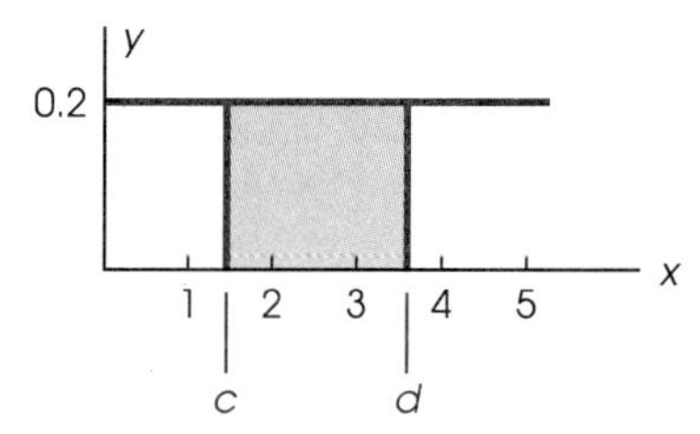


FIGURE 27

the interval, say, from 1 to 1000 pounds. We will be interested in calculating the probabilities of certain events, for example, prob $\{70 \le \mathbf{X} \le 270\}$.

As a particular example, consider the spinner device shown in Figure 26 which has the interval $[0, 5)$ wrapped around its circumference. Our experiment will consist of spinning the spinner, and $\mathbf{X}$ will be the coordinate of the stopping point. The probability of $\mathbf{X}$ taking values in two intervals of equal length should be the same, and the probability of its stopping somewhere in $[0, 5)$ is 1. This leads to the conclusion that if $0 \le c < d < 5$, then prob $\{c \le \mathbf{X} \le d\} = (d - c)/5$. If we let $f(x) = \frac{1}{5}$ for $0 \le x < 5$ and $f(x) = 0$ elsewhere, then (see Figure 27)

$$\text{prob}\ \{c \le \mathbf{X} \le d\} = \int_c^d f(x)\,dx$$

This example suggests some very general ideas.

## Probability Density Functions

Associated with each continuous random variable $\mathbf{X}$ is a function $f$ defined on $(-\infty, \infty)$ satisfying two properties:

1. $f(x)$ is nonnegative; that is, $f(x) \ge 0$ for all $x$.
2. The area under the graph of $y = f(x)$ is 1; that is,

$$\int_{-\infty}^{\infty} f(x)\,dx = 1$$

Moreover,

$$\text{prob}\ \{c \le \mathbf{X} \le d\} = \int_c^d f(x)\,dx$$

The function $f$ is called the **probability density function** associated with the random variable $\mathbf{X}$. If $f(x)$ is a positive constant on an interval (and 0 otherwise) as in the spinner example above, then we say that $\mathbf{X}$ is **uniformly distributed** on that interval.

## EXAMPLE A

Suppose that $\mathbf{X}$ is uniformly distributed on the interval $[1, 9]$. Determine the probability density function $f$ and calculate prob $\{2.3 \le \mathbf{X} \le 3.9\}$.

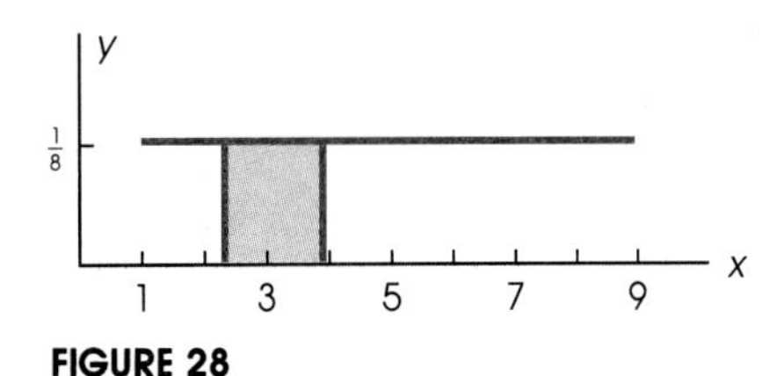


**FIGURE 28**

**Solution** In order for the area under $y = f(x)$ to be 1, it must be that $f(x) = \frac{1}{8}$ on $[1, 9]$ and $f(x) = 0$ elsewhere (Figure 28). Thus,

$$\text{prob}\ \{2.3 \le \mathbf{X} \le 3.9\} = \int_{2.3}^{3.9} \frac{1}{8}\,dx = \frac{1}{8}(3.9 - 2.3) = 0.2 \qquad ■$$

## EXAMPLE B

Let $f(x) = kx$ for $x$ in $[1, 3]$ and $f(x) = 0$ elsewhere. Determine $k$ so that $f$ is a probability density function. Then calculate prob $\{2.1 \le \mathbf{X} \le 2.5\}$.

**Solution** The area under the graph of $y = f(x)$ must be 1, so

$$1 = \int_1^3 kx\,dx = \left[\frac{1}{2}kx^2\right]_1^3 = \frac{9}{2}k - \frac{1}{2}k = 4k$$

Thus, $k = \frac{1}{4}$ and

$$\text{prob}\ \{2.1 \le \mathbf{X} \le 2.5\} = \int_{2.1}^{2.5} \frac{1}{4}x\,dx = \left[\frac{1}{8}x^2\right]_{2.1}^{2.5}$$

$$= \frac{1}{8}[2.5^2 - 2.1^2] = 0.23 \qquad ■$$

## EXAMPLE C

Papa's Ice Cream Shop guarantees that no customer will wait more than 3 minutes for service and that the time **X** one must wait for service has the probability density function $f(x) = \frac{1}{18}(9 - x^2)$ for $x$ on $[0, 3]$ and $f(x) = 0$ elsewhere. Sketch the graph of $f$ and show that it is a probability density function. Then calculate the probability that a customer will wait more than 2 minutes; that is, calculate prob $\{\mathbf{X} \geq 2\}$.

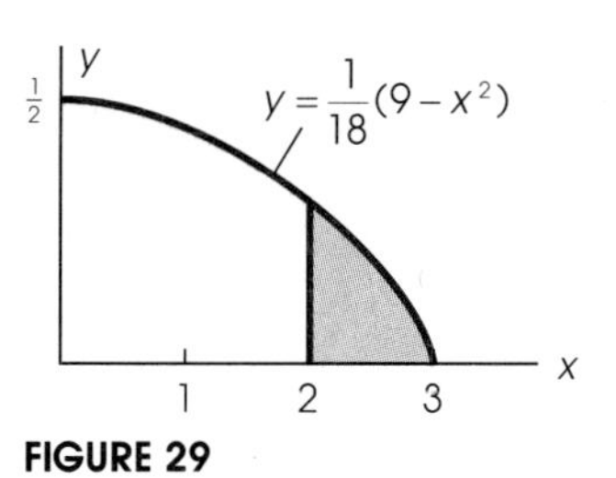


FIGURE 29

**Solution** The graph of $y = f(x)$ is shown in Figure 29. We must demonstrate that the area under this curve is 1.

$$\int_0^3 \frac{1}{18}(9 - x^2)\,dx = \frac{1}{18}\left[9x - \frac{1}{3}x^3\right]_0^3 = \frac{1}{18}(27 - 9) = 1$$

Also,

$$\text{prob}\,\{\mathbf{X} \geq 2\} = \int_2^3 \frac{1}{18}(9 - x^2)\,dx = \frac{1}{18}\left[9x - \frac{1}{3}x^3\right]_2^3$$

$$= \frac{1}{18}\left[27 - 9 - \left(18 - \frac{8}{3}\right)\right] = \frac{4}{27}$$

■

## EXAMPLE D

A transportation engineer claims that the distance **X** (in feet) between cars on a certain freeway during rush hour has probability density function $f(x) = 0.01e^{-0.01x}$ on $(0, \infty)$ and $f(x) = 0$ otherwise. Sketch the graph of $f$, show that it is a density function, and calculate prob $\{\mathbf{X} \leq 200\}$.

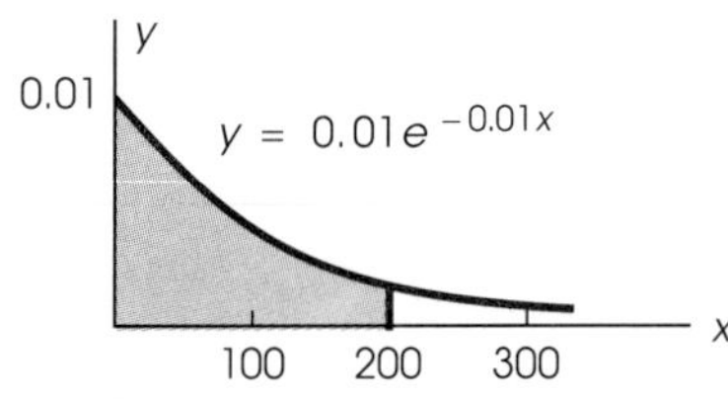


FIGURE 30

**Solution** The graph of $f$ is shown in Figure 30. Note that this model allows for the distance (in feet) between cars to be any positive number but makes it likely that it is a small number. Now

$$\int_0^\infty 0.01e^{-0.01x}\,dx = \lim_{b\to\infty}\int_0^b 0.01e^{-0.01x}\,dx = \lim_{b\to\infty}\left[-e^{-0.01x}\right]_0^b$$
$$= \lim_{b\to\infty}[-e^{-0.01b} + 1] = 1$$

which shows that $f$ really is a density function. Finally,

$$\text{prob}\,\{\mathbf{X} \le 200\} = \int_0^{200} 0.01e^{-0.01x}\,dx = \left[-e^{-0.01x}\right]_0^{200}$$
$$= -e^{-2} + 1 \approx 0.86$$

■

## The Mean of a Random Variable

If we were to choose one number as most representative of the values of a continuous random variable $\mathbf{X}$ with density function $f$, what would it be? The most commonly used such number is the mean $\mu$ defined by

$$\mu = \int_{-\infty}^{\infty} x\,f(x)\,dx$$

Geometrically, the mean is the balance point of the distribution. To be specific, if a rigid wire of total mass 1 is constructed so that its density (mass per unit length) at the point $x$ is $f(x)$, then the wire will balance on a fulcrum at $\mu$ (Figure 31). Thus, the mean of a uniform distribution is the midpoint of this distribution (for example, the mean of the random variable in Example A is 5,as we will soon check). On the other hand, one look at Figure 29 should convince you that the mean in Example C is to the left of 1.5. Our next example gives the exact results for each of the random variables so far considered.

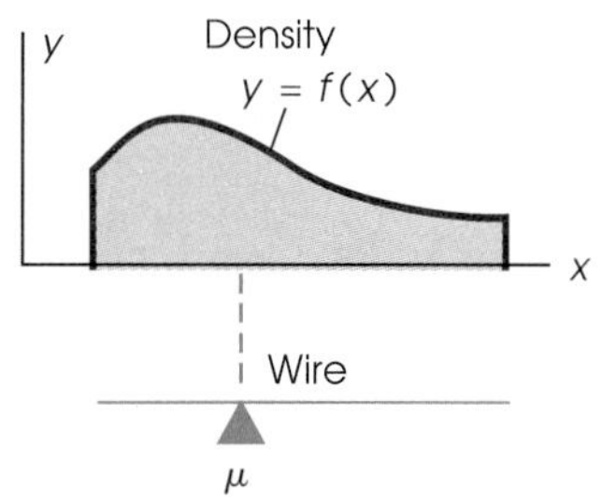


**FIGURE 31**

**SHARP CURVE**

We tried to slip a curve past you in part (d) of Example E in pretending it obvious that

$$\lim_{b\to\infty} be^{-0.01b} = 0$$

Actually, as long as $k > 0$,

$$\lim_{x\to\infty} \frac{x}{e^{kx}} = 0$$

This rests on the fact that any exponential function $e^{kx}$ ultimately grows much faster than the linear function $x$. Do you agree?

### EXAMPLE E

Find the mean of the random variable in (a) Example A, (b) Example B, (c) Example C, and (d) Example D.

**Solution**

(a) $\mu = \int_1^9 x\frac{1}{8}\,dx = \left[\frac{x^2}{16}\right]_1^9 = \frac{81}{16} - \frac{1}{16} = 5$

(b) $\mu = \int_1^3 x\left(\frac{1}{4}x\right)\,dx = \left[\frac{1}{12}x^3\right]_1^3 = \frac{27}{12} - \frac{1}{12} = \frac{13}{6}$

(c) $\mu = \int_0^3 x\left[\frac{1}{18}(9 - x^2)\right]\,dx = \frac{1}{18}\left[\frac{9}{2}x^2 - \frac{1}{4}x^4\right]_0^3 = \frac{9}{8}$

(d) $\mu = \int_0^\infty x(0.01e^{-0.01x})\,dx = \lim_{b\to\infty}\int_0^b x(0.01e^{-0.01x})\,dx$

To evaluate the latter integral, we use integration by parts with $u = x$ and $dv = 0.01e^{-0.01x}\,dx$. We obtain

$$\mu = \lim_{b\to\infty}\left[-xe^{-0.01x}\right]_0^b + \lim_{b\to\infty}\int_0^b e^{-0.01x}\,dx$$

$$= 0 + \lim_{b\to\infty}\left[-\frac{1}{0.01}e^{-0.01x}\right]_0^b = \frac{1}{0.01} = 100$$

■

## Problem Set 9.5

1. Suppose that the random variable **X** is uniformly distributed on [1, 5]. See Example A.
   (a) Determine the probability density function $f$ corresponding to **X**.
   (b) Calculate prob $\{2.9 \le \mathbf{X} \le 4.1\}$.
2. Let **X** denote the result of picking a number at random from the interval [0, 10]. *Note:* The phrase "at random," when it is not qualified, is interpreted to mean that the corresponding random variable is uniformly distributed.
   (a) Determine the probability density function $f$ for **X**.
   (b) Calculate prob $\{|\mathbf{X} - 3| \le 2\}$.
3. A certain traffic light stays red for 30 seconds. You arrive at the light at random and find it red. Calculate the probability that you will have to wait at least 20 seconds before the light turns green.
4. Let **X** be the result of selecting a number at random on the interval $[-a, a]$. Calculate prob $\{|\mathbf{X}| > a/4\}$.
5. Let $f(x) = k(x + 1)$ for $0 \le x \le 4$ and $f(x) = 0$ elsewhere. See Example B.
   (a) Determine $k$ so that $f$ is a probability density function.
   (b) Calculate prob $\{\mathbf{X} > 3\}$.
6. Let $f(x) = k\sqrt{x}$ for $1 \le x \le 4$ and $f(x) = 0$ elsewhere.
   (a) Determine $k$ so that $f$ is a probability density function.
   (b) Calculate prob $\{2 \le \mathbf{X} \le 3\}$.
7. Let $f(x) = k/x$ on $[1, e^2]$ and $f(x) = 0$ otherwise.

(a) Show that $k = \frac{1}{2}$ makes a density function.
(b) Find prob $\{\mathbf{X} \leq 2\}$.
(c) Determine $a$ so that prob $\{\mathbf{X} > a\} = \frac{1}{2}$.

8. Let $f(x) = k/(x + 1)^2$ on $[0, 2]$ and $f(x) = 0$ elsewhere.
(a) Show that $k = \frac{3}{2}$ makes a density function.
(b) Calculate prob $\{\mathbf{X} > 1\}$.
(c) Determine $b$ so that prob $\{\mathbf{X} \leq b\} = \frac{1}{8}$.

9. A chow mein takeout shop claims that the time $\mathbf{X}$ one must wait for service has density function $f(x) = (36 - x^2)/144$ for $0 \leq x \leq 6$ and $f(x) = 0$ otherwise. Note that this implies that no customer will wait more than 6 minutes for service. See Example C.
(a) Show that $f$ is a probability density function.
(b) Sketch the graph of $f$.
(c) Calculate the probability that a customer will wait at least 1 minute for service.
(d) Use your graph to guess at the mean waiting time and then calculate it exactly.

10. Under new management, the chow mein shop in Problem 9 was able to change the probability density function for $\mathbf{X}$ to $f(x) = 3(4 - x^2)/16$ for $0 \leq x \leq 2$ and $f(x) = 0$ otherwise.
(a) Now what is the maximum time any customer has to wait for service?
(b) Calculate the probability that a customer will wait at least 1 minute for service.
(c) Determine the mean waiting time.

11. A restaurant chain has determined that the fraction $\mathbf{X}$ of its waffle shops opened during any month that show a profit during the first 2 years of operation has probability density function $f(x) = 20x(1 - x)^3$ for $0 \leq x \leq 1$ and $f(x) = 0$ otherwise. Calculate the probability that a new shop just opening has at least a 60% chance of showing a profit during the first 2 years.

12. The cells in a certain cell population divide in a maximum of 6 days. The age $\mathbf{X}$ in days of a cell chosen at random from this population has probability density function $f(x) = 2ke^{-kx}$ for $0 \leq x \leq 6$ and $f(x) = 0$ elsewhere.
(a) Show that $k = (\ln 2)/6$.
(b) Calculate the probability that a cell is at least 4 days old.

13. A highway engineer claims that the number of minutes $T$ between automobile accidents on a certain 10-mile stretch of highway has probability density function $f(t) = 0.02e^{-0.02t}$ for $t > 0$ and $f(t) = 0$ otherwise. See Example D.
(a) Calculate the probability that after an accident has occurred there won't be another for at least 40 minutes.
(b) Find the mean length of time between accidents.

14. The number of minutes $T$ between successive telephone calls received at the emergency ward of a large hospital has probability density function $f(t) = 0.08e^{-0.08t}$ for $t > 0$ and $f(t) = 0$ otherwise.
(a) Calculate the probability that the time between calls is at least 1 minute.
(b) Find the mean length of time between calls.

---

15. Let $f(x) = kx(2 - x)$ on $[0, 2]$ and $f(x) = 0$ elsewhere.
(a) Determine $k$ so that $f$ is a probability density function.
(b) Calculate prob $\{\mathbf{X} \leq 0.3\}$.
(c) Find the mean of $\mathbf{X}$.

16. Let $f(x) = k/x^3$ on $[1, \infty)$ and $f(x) = 0$ elsewhere.
(a) Determine $k$ so that $f$ is a density function.
(b) Find the mean.

17. A gasoline station can sell $\mathbf{X}$ thousand gallons of gasoline per week, where $\mathbf{X}$ has probability density function $f(x) = k/(x + 1)^2$ on $[2, \infty)$ and $f(x) = 0$ otherwise.
(a) Determine $k$.
(b) If the manager orders 4000 gallons at the beginning of a week, what is the probability that he will run out before the week is over?

18. Suppose that $\mathbf{X}$ is uniformly distributed on [0, b] and that prob $\{\mathbf{X} > 4\} = \frac{1}{5}$. What is $b$?

19. Let $f(x) = 6x^2$ on $[-a, a]$ and $f(x) = 0$ elsewhere. Determine $a$ so that $f$ is a density function.

20. Determine which of the following improper integrals converge and which diverge and, if convergent, give its value.

(a) $\displaystyle\int_2^{\infty} \frac{1}{\sqrt{x - 1}}\,dx$ (b) $\displaystyle\int_e^{\infty} \frac{1}{x(\ln x)^2}\,dx$

(c) $\displaystyle\int_0^{\infty} t^2 e^{-t^3}\,dt$ (d) $\displaystyle\int_0^{\infty} \frac{t}{2 + t^2}\,dt$

21. Let $f(x) = ke^{-kx}$ for $x > 0$ and $f(x) = 0$ elsewhere. Show that $f$ is a probability density function for every $k > 0$ and determine the corresponding mean.

22. A dart is thrown at random at a circular dart board of radius 1. This means that the probability it will land in a region $R$ is proportional to the area of $R$. Let $\mathbf{X}$ denote the distance from the center to where the dart lands.
(a) For $0 \leq x \leq 1$, determine prob $\{\mathbf{X} \leq x\}$.
(b) Determine the density function for $\mathbf{X}$.

## CHAPTER 9 REVIEW PROBLEM SET

1. Find the area of the region under the curve
$$y = 3x^2 + 2/(x + 1)$$
between $x = 0$ and $x = 2$.
2. Find the area of the region under the curve $y = 4x^3 + 2x + e^{x/2}$ between $x = 0$ and $x = 2$.
3. Find the area of the shaded region in Figure 32.

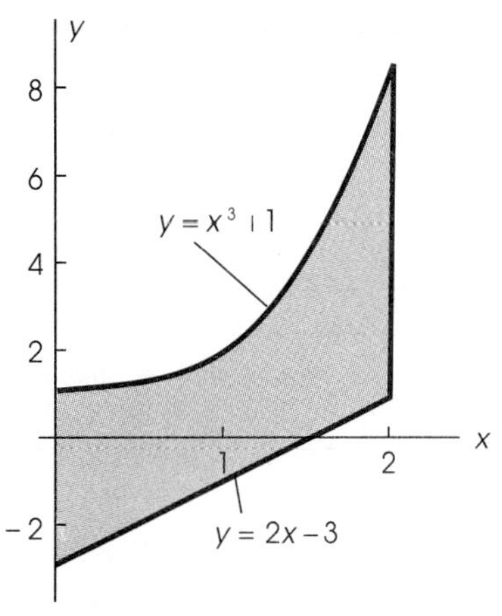


**FIGURE 32**

4. Find the area of the part of the shaded region in Figure 32 that is above the $x$-axis.
5. Find the area of the region bounded by the curve $y = 3 - x^2$ and the lines $y = 4 - x$, $x = -1$, and $x = 2$. Begin by sketching this region.
6. Sketch the region bounded by $y = e^x$ and the lines $y = 4$ and $x = 1$. Then find its area.
7. Find the area of the region bounded by $x = y^2 - y$ and $x = y$.
8. Sketch the region bounded by the parabola $x = 4y - y^2$ and the line $x + 2y = 5$. Then find its area.
9. Sketch the region $R$ bounded by the parabola $y = x^2$ and the line $y = 4$. Then find the volume of the solid generated when $R$ is revolved about (a) the $x$-axis, (b) the $y$-axis.
10. Sketch the region $R$ bounded by the parabola $y = \sqrt{x}$ and the lines $y = 0$ and $x = 4$. Then find the volume of the solid generated when $R$ is revolved about:
    (a) the $x$-axis (b) the $y$-axis
11. Find the volume of the solid generated when the region bounded by the parabolas $x = y^2$ and $x = 4y - y^2$ is revolved about the $y$-axis.
12. Find the volume of the solid generated when the region bounded by the parabola $x = (y - 2)^2$ and the line $x + y = 4$ is revolved about the $y$-axis.
13. Grassway Company makes and sells lawnmowers. Its weekly marginal cost and weekly marginal revenue functions (in making and selling $x$ lawnmowers) are given, respectively, by
$$C'(x) = 270 - 2x + 0.039x^2$$
$$R'(x) = 450$$
both in dollars per mower. If the company has weekly fixed costs of \$640, find:
    (a) the cost of making and selling 100 mowers per week.
    (b) the profit realized in making and selling 100 mowers per week.
14. A manufacturer of bicycles estimates the weekly marginal cost function and weekly marginal revenue function in making and selling $x$ bicycles to be given by
$$C'(x) = 200 - 3x + 0.06x^2$$
$$R'(x) = 400 - 0.03x$$
both in dollars per bicycle. If the weekly fixed costs are \$450, find:
    (a) the cost of making and selling 80 bicycles per week;
    (b) the corresponding weekly profit.
15. Find the equilibrium price if the demand and supply functions (in thousands of dollars) for a certain product are given by $p = 15 + 19/(x + 1)$ and $p = \frac{1}{3}x + 10$, respectively.
16. Suppose that the demand and supply functions (in thousands of dollars) for a certain product are given by $y = 9 - x^2$ and $p = x^2 + 2$, respectively. Find the equilibrium price.
17. For the data in Problem 15, show graphically the regions whose areas correspond to the consumer surplus *CS* and the producer surplus *PS*. Then calculate the values of these quantities.
18. Graph the equations in Problem 16, indicate the regions corresponding to the consumer surplus and the producer surplus, and calculate these latter quantities.
19. Find the Gini coefficient for the Lorenz curve
$$y = \tfrac{1}{2}x^3 + \tfrac{1}{4}x + \tfrac{1}{4}x^{1/5}$$
20. Calculate the Gini coefficient for the Lorenz curve $y = 0.25x^4 + 0.75x^2$.
21. Flowingwell has a population of 10,000 right now,

and it has been predicted that $x$ years from now its population will be increasing at the rate of $100 + 5\sqrt{x}$ people per year.
(a) What is its projected population 9 years from now?
(b) Find the average yearly increase in population over the 9-year period.

22. Let $P(t)$ denote the number of individuals in a bacteria colony $t$ hours from now. Suppose the present population is 12,000 and it is growing at the rate $P'(t) = 9200t + 15{,}000$.
(a) Find the formula for $P(t)$.
(b) Calculate the size of the population 5 hours from now.
(c) Determine the average population during the initial 5-hour period.

C 23. To satisfy his creditors, John has agreed to make a payment of \$3600 three years from now and a second payment of \$2800 six years from now. What is John's present indebtedness, assuming that interest is at 8.8% compounded continuously?

C 24. Helen owes \$8400 to her brother who charges her interest at the rate of 8.5% compounded semiannually. She has agreed to discharge her debt by making a payment of \$4000 3 years from now and a second payment 7 years from now. Find the size of her second payment.

25. How much money must the McMaster Foundation put in trust now if it wishes to provide a perpetual money stream for Morningstar Children's Home at the rate of \$50,000 per year and if interest is at 7.6% compounded continuously?

26. A foundation wishes to provide a perpetual money stream to the Golden Years Nursing Home at the rate of \$75,000 per year. How much money must be put in trust now to provide this money stream if interest is projected at 8.3% compounded continuously?

27. Suppose that $f(x) = k\sqrt[3]{x}$ on $[0, 8]$ and $f(x) = 0$ elsewhere determines the probability density function for a random variable $\mathbf{X}$. Find:
(a) the value of $k$
(b) prob $\{0 \le \mathbf{X} \le 1\}$
(c) the mean of $\mathbf{X}$

28. Let the probability density function for a random variable $\mathbf{X}$ be given by $f(x) = k(5x + 3)\sqrt{x}$ on $[0,4]$ and $f(x) = 0$ elsewhere. Find: (a) $k$, (b) prob $\{1 \le \mathbf{X} \le 4\}$, and (c) the mean of $\mathbf{X}$.

29. The useful life $\mathbf{T}$ in years of a Maxipower vacuum cleaner is a random variable with density function $f(t) = \frac{1}{8}e^{-t/8}$ for $t > 0$ and $f(t) = 0$ otherwise.
(a) What fraction of these cleaners are usable after 8 years?
(b) Determine the mean of $\mathbf{T}$.

30. The time $\mathbf{T}$ in seconds for a rat to learn a certain behavior is a random variable with probability density function $f(t) = 0.03e^{-0.03t}$ for $t > 0$. Calculate the probability that a rat will learn this behavior in 2 minutes or less.

31. Consider the shaded region $R$ shown in Figure 33. Write a definite integral which represents
(a) the area of $R$.
(b) the volume of the solid generated when $R$ is revolved about the $x$-axis.

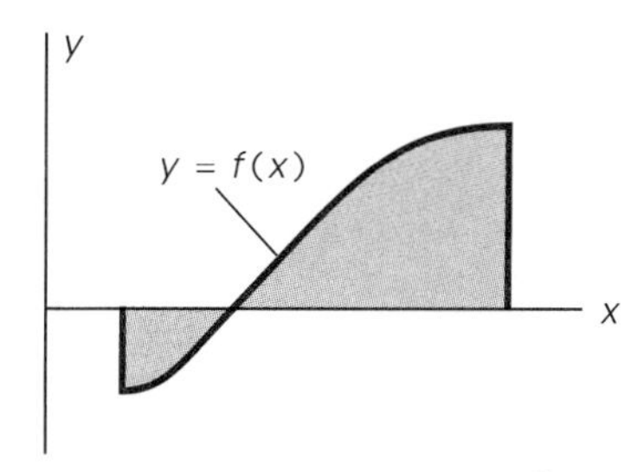


**FIGURE 33**

32. Sketch the region $R$ bounded by the curve $y = 2\sqrt[3]{x} - 3$ and the $x$-axis between $x = 2$ and $x = 11$.
(a) Determine the area of $R$.
(b) Find the volume of the solid generated when $R$ is revolved about the $x$-axis.

33. The function $f(x) = 27x^2/(1 + x^3)^2$ on $[a, \infty)$ and $f(x) = 0$ elsewhere is a probability density function. Determine $a$.

34. Determine $k$ so that $f(x) = kx/(1 + x^2)^2$ is the probability density function for a random variable $\mathbf{X}$ on $[0, \infty)$. Then find the *median value* of $\mathbf{X}$, that is, find $a$ so that prob $\{0 \le \mathbf{X} \le a\} = 0.5$.

35. The work done by a constant force of $F$ pounds in moving an object along a line through a distance $d$ feet is defined to be $Fd$ foot-pounds. Suppose that the variable force required to move an object is $F(x)$ pounds at the point with coordinate $x$. Determine the formula for the work done in moving the object along a coordinate line from $x = a$ to $x = b$. *Hint:* Slice, approximate, integrate.

36. Suppose that the formula for $F(x)$ in Problem 35 is $F(x) = 5x^{2/3}$. Determine the work done in moving the object from $x = 1$ to $x = 8$.

37. $\lim_{n\to\infty} \frac{3}{n} \sum_{i=1}^{n} \left[1 + f\left(\frac{3i}{n}\right)\right]^2$ corresponds to what definite integral?

38. Evaluate $\lim_{n\to\infty} \frac{2}{n} \sum_{i=1}^{n} \sqrt{5 + \left(\frac{2i}{n}\right)^2}\left(\frac{2i}{n}\right)$ by recognizing this limit as a definite integral and then evaluating this integral.

*George Dantzig solved the problem of finding an optimum way of allocating resources by constructing a convex polytope model. His simplex algorithm involves moving from vertex to vertex (ABC . . . M) improving the allocation at each step until an optimum is achieved. The model shown is 3-dimensional; real applications may require thousands of dimensions.*

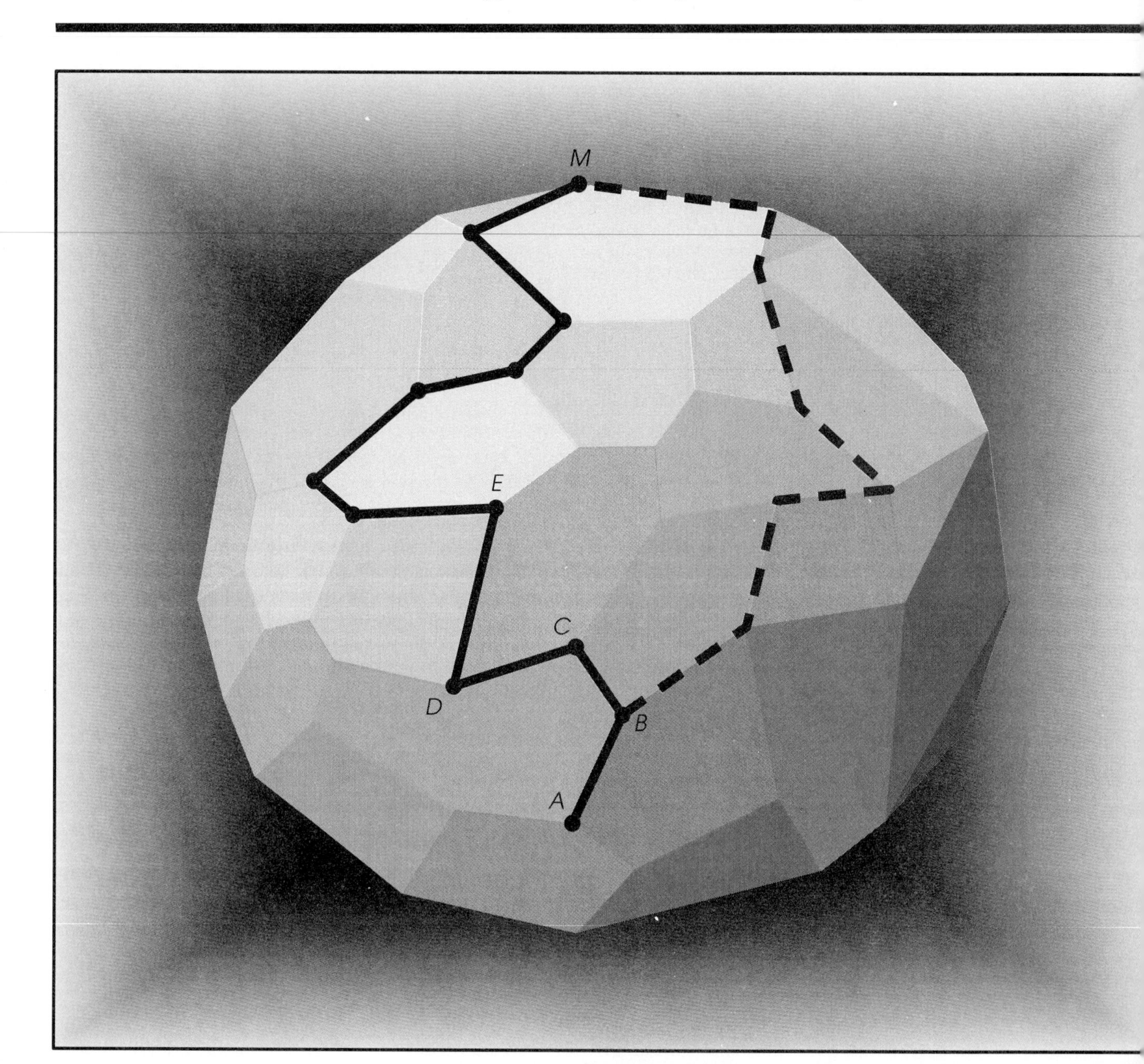

# 10

# Multivariate Calculus

George Bernard Dantzig
(1914–     )

All workers in the fields of management science and operations research know the name George Dantzig. In fact, for all practical purposes, Dantzig founded these fields. The story is one of the most exciting in recent mathematical history.

George, with good mathematical genes inherited from his father, Tobias Dantzig, and a fresh Ph.D. from the University of California at Berkeley, was working for the the planning division of the U.S. Air Force. As anyone who has thought about it realizes, it takes fantastic coordination of supplies, equipment, and personnel to make an operation as big as the Air Force run efficiently. Dantzig formulated the Air Force's planning problem as a *linear program*, that is, as the problem of optimizing a linear function subject to linear constraints. Unfortunately, the problem involved thousands of variables and therefore seemed to be intractable. But Dantzig proceeded to invent an algorithm, the so-called *simplex method*, which offered a clear step-by-step procedure for solving the problem. This method is hardly feasible for hand calculations, but another exciting event occurred almost simultaneously—the invention of the modern high-speed computer. With Dantzig's simplex method and a computer, linear programs with as many as 15,000 variables could be handled with relative ease.

Of course, all the large businesses in the United States face problems similar to the Air Force. Soon they were applying Dantzig's methods to their problems and thereby saving millions of dollars. Perhaps no other mathematical process has so captured the fancy of the corporate world.

But the story is not finished. Many problems of interest are nonlinear in character. Dantzig's early success has attracted a host of workers to these harder problems, and notable advances have been made often with Dantzig himself playing a key role. In 1984, a young Indian-born mathematician, Nerendra Karmarkar, provided the newest breakthrough, a method even better than the simplex method for linear programs with a large number of variables.

## 10.1
# FUNCTIONS OF TWO OR MORE VARIABLES

So far, our interest has focused on functions of one variable, that is, functions whose rule of correspondence has the form $y = f(x)$. We have evaluated such functions, graphed them, differentiated them, and integrated them. The restriction to functions of one variable is very limiting, since most problems that arise in the social sciences involve many variables. In particular, problems in economics often involve hundreds or even thousands of variables. It is time for us to lift our sights. Fortunately, all of calculus can be studied in the context of functions of several variables.

## Examples of Functions of Several Variables

Our first task is to clarify the language and notation. By a function $f$ of two variables $x$ and $y$, we mean a rule that associates with each ordered pair $(x, y)$ of numbers in some set (called the domain) a definite number $f(x, y)$. The rule is usually given by an equation. For example, the equation $f(x, y) = x^2 + y$ associates with the ordered pair $(x, y)$ the number $x^2 + y$. Thus,

$$f(2, 3) = 2^2 + 3 = 7$$

$$f(-4, 5) = (-4)^2 + 5 = 21$$

and

$$f\left(\frac{1}{a}, b^2\right) = \left(\frac{1}{a}\right)^2 + b^2 = \frac{1 + a^2b^2}{a^2}$$

There is no conceptual reason for limiting our discussion to functions $f(x, y)$ of two variables. We can and occasionally will consider a function $f(x, y, z)$ of three variables or, more generally, a function $f(x_1, x_2, \ldots, x_n)$ of $n$ variables. All that we do is valid in the most general context. However, it will simplify life for everyone if we center most of our discussion around cases with two and three variables.

### EXAMPLE A

Let $f(x, y, z) = x^2y/z$. Find and simplify:

(a) $f(3, 2, 12)$, (b) $f(-2, -2, -2)$,
(c) $f(x, x, 1/x)$, (d) $f(x + h, y, z) - f(x, y, z)$

### Solution

(a) $f(3, 2, 12) = (3)^2(2)/12 = 18/12 = 3/2$
(b) $f(-2, -2, -2) = (-2)^2(-2)/(-2) = 4$
(c) $f\left(x, x, \frac{1}{x}\right) = \frac{x^2x}{1/x} = x^4$

$$\text{(d) } f(x+h, y, z) - f(x, y, z) = \frac{(x+h)^2y}{z} - \frac{x^2y}{z}$$
$$= \frac{(x^2 + 2xh + h^2)y - x^2y}{z}$$
$$= \frac{2xhy + h^2y}{z}$$

■

## EXAMPLE B

The material for the top and bottom of a cylindrical can costs 0.05¢ per square inch, while the material for the curved sides costs 0.03¢ per square inch. Let $C(r, h)$ denote the cost of materials for a can of radius $r$ inches and height $h$ inches. Find a formula for $C(r, h)$ and evaluate it when $r = 2$ and $h = 6$.

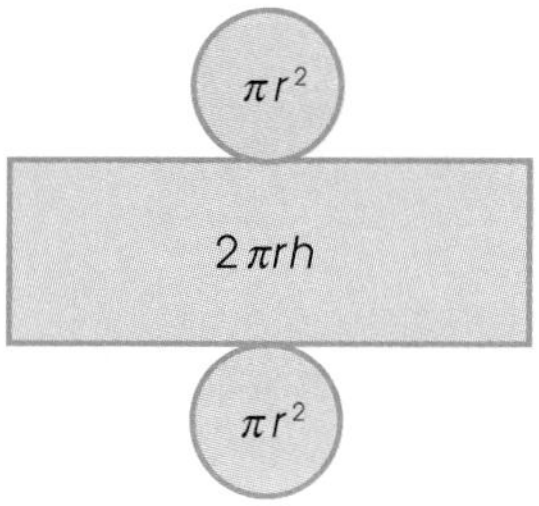

$\pi r^2$

$2\pi rh$

$\pi r^2$

FIGURE 1

**Solution** The top and bottom are circles of area $\pi r^2$, and the curved portion has area $2\pi rh$ (Figure 1). Thus,

$$C(r, h) = 2\pi r^2(0.05) + 2\pi rh(0.03)$$
$$\approx 0.314159r^2 + 0.188496rh$$

and

$$C(2, 6) = 2\pi(4)(0.05) + 2\pi(2)(6)(0.03) \approx 3.52¢$$

■

The costs of a manufacturing process are divided between labor costs $L$ and capital costs $K$ (raw materials, equipment, and so on). In many cases, economists find that a **Cobb-Douglas production function** $F(L, K) = cL^aK^{1-a}$ provides a good model for the number of units that will be produced ($c$ and $a$ are constants, $0 < a < 1$).

## EXAMPLE C

A certain manufacturing process is believed to have the production function

$$F(L, K) = 300L^{1/3}K^{2/3}$$

(a) How many units of output will result from inputs of 64 units of labor and 27 units of capital?
(b) Which would increase output more, doubling the number of units of labor or doubling the number of units of capital?

**Solution**

(a) $F(64, 27) = 300(64)^{1/3}(27)^{2/3} = 300(4)(9) = 10{,}800$
(b) $F(128, 27) = 300(128)^{1/3}(27)^{2/3} = 13{,}607$
$F(64, 54) = 300(64)^{1/3}(54)^{2/3} = 17{,}144$

The calculation shows that output is increased more by doubling the input of capital than by doubling the input of labor. This would be true at any production level, since doubling capital multiplies output by $2^{2/3} \approx 1.5874$, whereas doubling labor multiplies the output by only $2^{1/3} \approx 1.2599$. ■

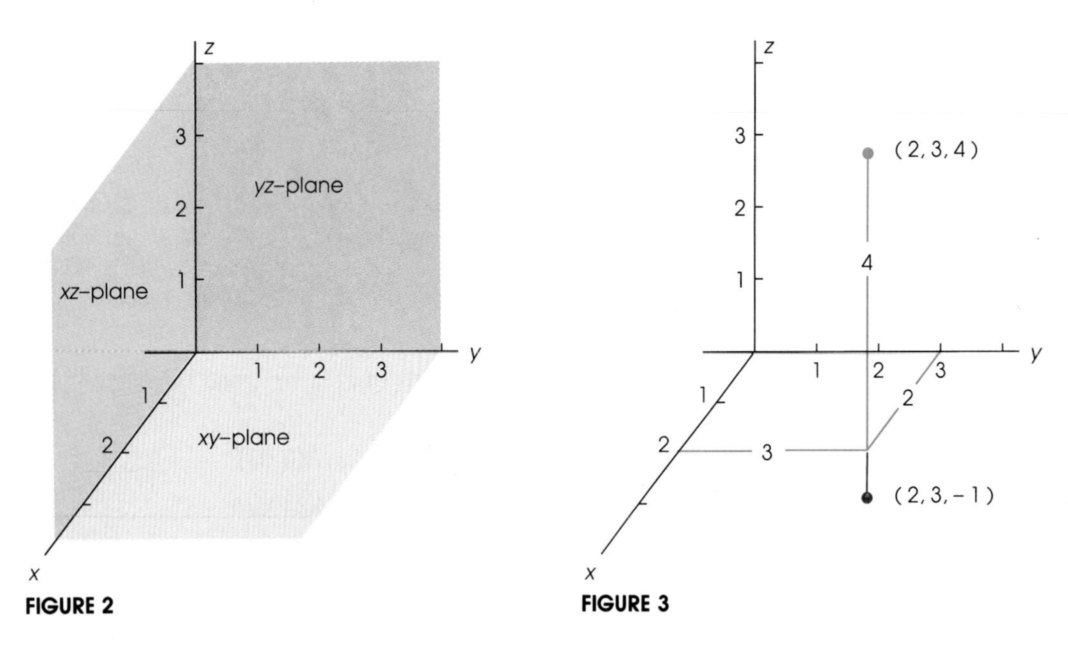


FIGURE 2

FIGURE 3

## Three-Dimensional Space

To graph $y = f(x)$ required the Cartesian plane (two-dimensional space); to graph $z = f(x, y)$ will require Cartesian three-dimensional space. In space, introduce three mutually perpendicular axes, each scaled with the real numbers and intersecting at their common zero point called the origin. We will almost always do this so the $x$- and $y$-axes are in the horizontal plane, with the $x$-axis pointing forward, the $y$-axis pointing to the right, and the $z$-axis pointing straight up (Figure 2). The three axes determine the three coordinate planes (the $xy$-, $xz$- and $yz$-planes). Then every point in space gets a three-number label $(x, y, z)$ called its coordinates and measuring its directed distances from these three planes (Figure 3).

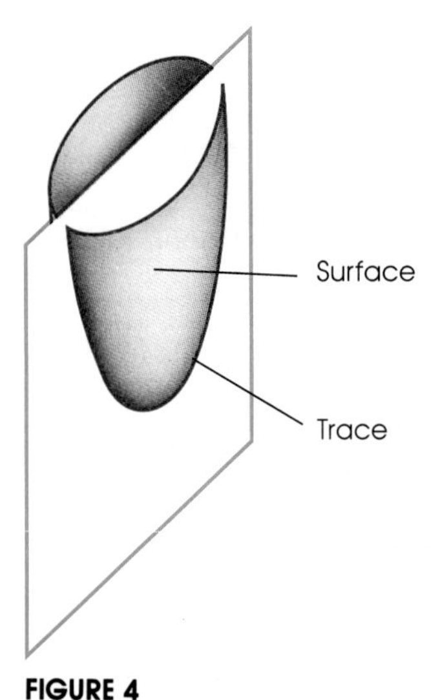


FIGURE 4

## Graphs of Functions of Two Variables

The graph of the function $f(x)$ of one variable consists of all points $(x, y)$ in the plane satisfying $y = f(x)$; in general, it is a curve. The graph of the function $f(x, y)$ of two variables consists of all points $(x, y, z)$ in space satisfying $z = f(x, y)$; it is a surface. To sketch the graph of a surface, we usually begin by finding its **traces** in various planes, especially the three coordinated planes. A trace is just the intersection of the surface with a plane and is therefore a curve (Figure 4).

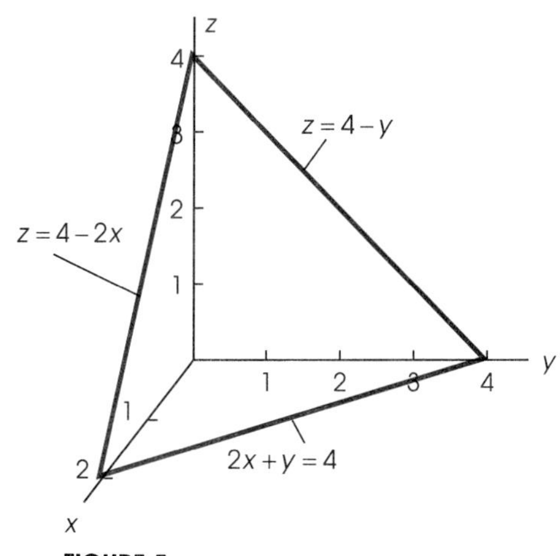


FIGURE 5

## EXAMPLE D

Sketch the graph of $z = f(x, y) = 4 - 2x - y$.

**Solution** Since the variables all occur to the first power, you might guess that the graph will be a plane, the natural generalization of a line. This guess is correct. To sketch the graph, we first sketch the traces in the three coordinate planes. For example, to find the trace in the $xz$-plane, set $y = 0$ in the given equation, obtaining the linear equation $z = 4 - 2x$. The traces in the $yz$-plane and $xy$-plane are obtained in similar fashion by setting $x = 0$ and $z = 0$, respectively. These three traces already suggest the required plane (Figure 5); those with artistic talents can provide a more vivid picture by the use of shading (Figure 6). Of course, we can show only a small part of the plane, since it stretches infinitely far. We choose to show the part of the plane in the first octant (where all coordinates are positive). ∎

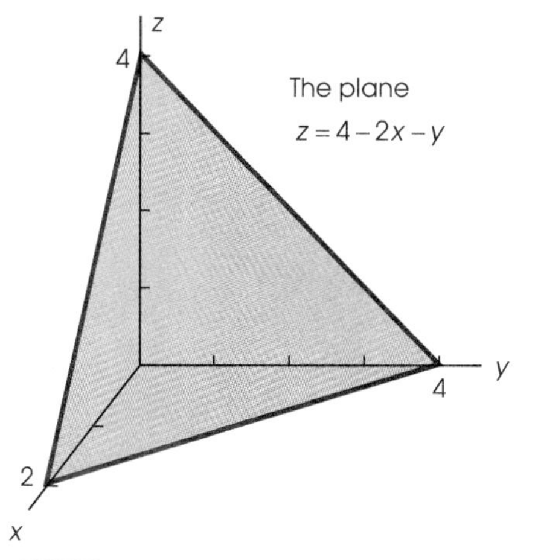


FIGURE 6

## EXAMPLE E

Sketch the graph of $z = f(x, y) = x^2 + y^2$.

**Solution** The traces in the $xz$-plane and $yz$-plane are the graphs of $z = x^2$ and $z = y^2$, respectively. The trace in the $xy$-plane ($z = 0$) reduces to a single point, the origin. The trace in the plane $z = 4$ is the circle $x^2 + y^2 = 4$. All this is shown in Figures 7 and 8. Again, we show only the first octant part of the surface, a surface called a circular paraboloid. The complete surface stretches back behind the $xz$- and $yz$-planes and infinitely high up in the $z$ direction. ∎

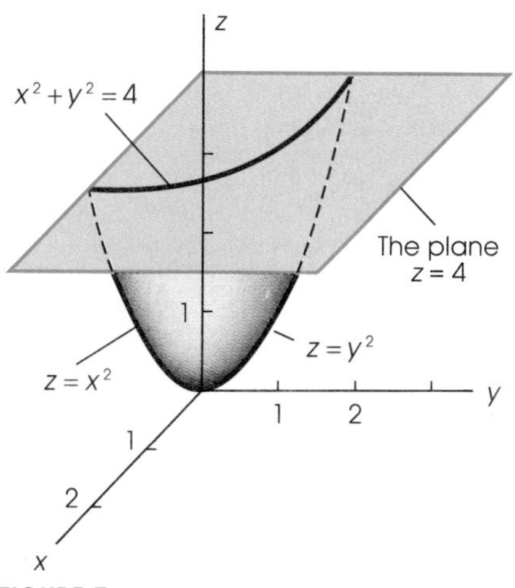


FIGURE 7

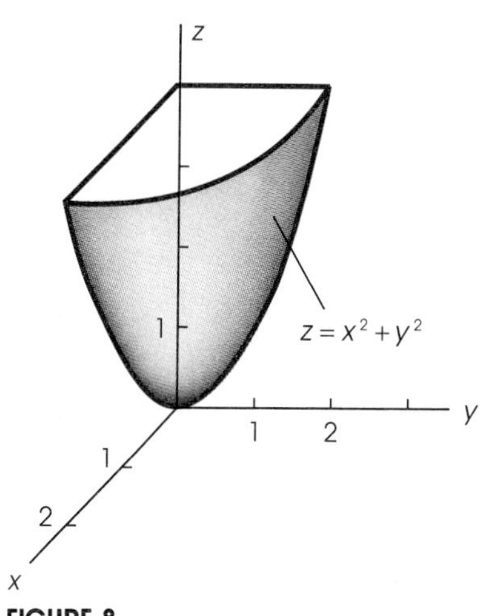


FIGURE 8

## Computer Graphics

With a computer and appropriate software, it is possible to display excellent graphs of some very complicated functions. Figure 9 shows two computer-drawn surfaces. Since we do not assume that all our readers have the required equipment at their fingertips, we will place our emphasis on hand-drawn graphs. It will help to have a little artistic ability.

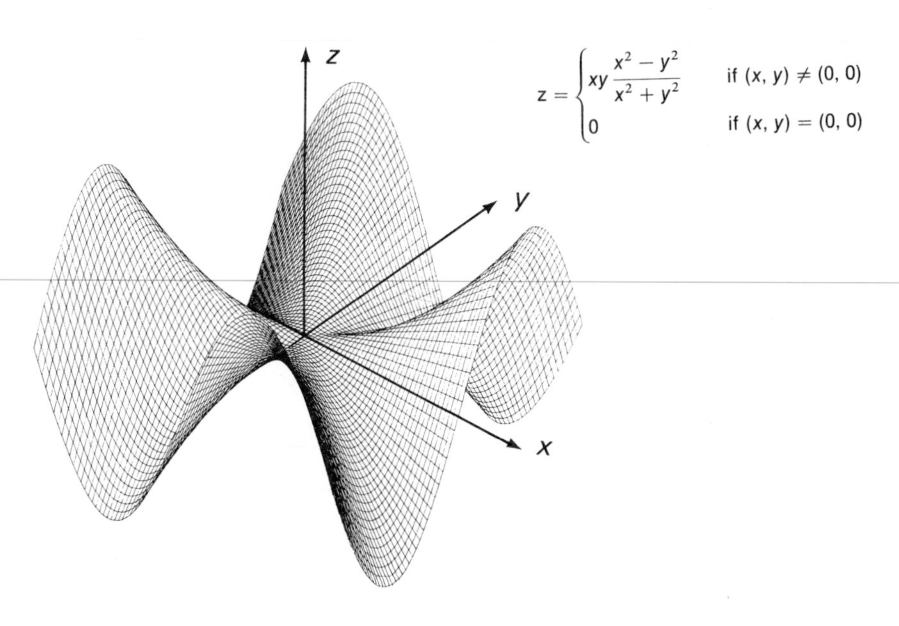


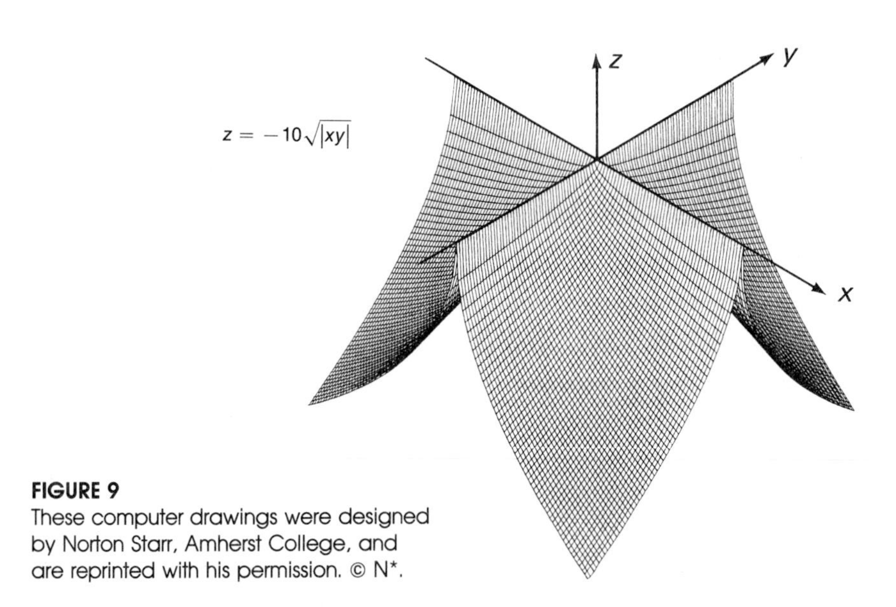


**FIGURE 9**
These computer drawings were designed by Norton Starr, Amherst College, and are reprinted with his permission. © N*.

## Problem Set 10.1

*In Problems 1–6, find and simplify* (a) $f(1, 4)$, (b) $f(2, 2)$, (c) $f(x, 2x)$, (d) $f(x + h, y) - f(x, y)$. *See Example A.*

1. $f(x, y) = x^2 - xy + y^2$
2. $f(x, y) = x^2 + 4\sqrt{y}$
3. $f(x, y) = x/y^4$
4. $f(x, y) = (x + 2y)/(x - 3y)$
5. $f(x, y) = (3x + 1)e^{y-4}$
6. $f(x, y) = x^2 \ln (xy - 3) + y^2 - 5$

*In Problems 7–12, find and simplify* (a) $f(1, 0, 1)$, (b) $f(3, 2, 6)$ (c) $f(-2z, z, z)$, (d) $f(z, y, x)$

7. $f(x, y, z) = 2xy + yz - 3xz$
8. $f(x, y, z) = x^2(y + z) - 2xyz$
9. $f(x, y, z) = (xy - 2z)/(yz - 2x)$
10. $f(x, y, z) = \sqrt{x^2 + y^2 + z^2}$
11. $f(x, y, z) = (x + 2y)(y + 2z)(z + 2x)$
12. $f(x, y, z) = (y^2 + z)e^{x-2y+4z}$
13. For a rectangle of dimensions $x$ and $y$, let $f(x, y)$ denote its perimeter and $g(x, y)$ its area.
    (a) Write explicit expressions for $f(x, y)$ and $g(x, y)$.
    (b) Calculate $f(6, 2)$ and $g(6, 2)$.
    (c) If the perimeter and area of a rectangle have the same numerical value, how is $y$ related to $x$?
14. The picture in Figure 10 shows a garden with dimensions $x$ and $y$, surrounded by a sidewalk of width $z$.
    (a) Write explicit expressions for $f(x, y, z)$, the combined area of the garden and the walk, and $g(x, y, z)$, the area of the walk.
    (b) Calculate $f(12, 8, 2)$ and $g(12, 8, 2)$.
    (c) Suppose that $x = 24$ and $y = 10$. Find $z$ so that the garden and the walk have the same area.

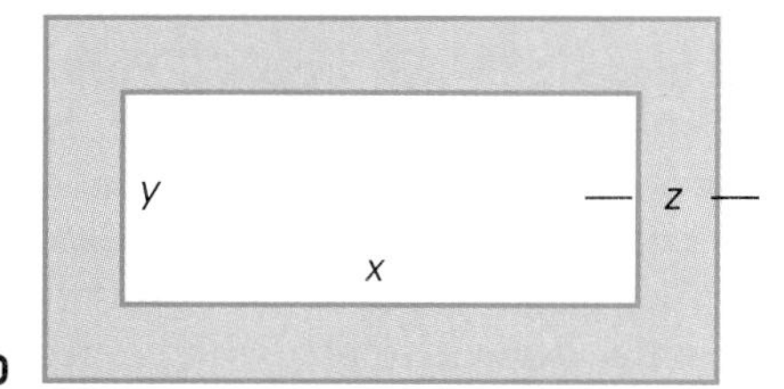


FIGURE 10

15. Let the dimensions of a closed rectangular box with a square base be $x$ by $x$ by $y$ (all in feet).
    (a) Write explicit expressions for its volume $V(x, y)$ and surface area $S(x, y)$.
    (b) What happens to $V(x, y)$ if we double $x$ and halve $y$?
    (c) If the box is made of thin material costing 18¢ per square foot for the sides and 30¢ per square foot for the top and bottom, find an explicit expression for $C(x, y)$, the total cost (in dollars) of material for the box.
16. Consider a rectangular box with base $x$ by $y$ and height $z$ (all in centimeters).
    (a) Write explicit expressions for its volume $V(x, y, z)$ and surface area $S(x, y, z)$.
    (b) Suppose the box is made of thin material costing 30¢ per square centimeter for the top, 25¢ per square centimeter for the sides, and 50¢ per square centimeter for the bottom. Write an explicit expression $C(x, y, z)$ for the cost (in dollars) of materials needed to make the box.
17. Let $U = F(L, K) = 420L^{1/4}K^{3/4}$ be a production function.
    (a) Calculate $F(16, 81)$ and $F(81, 16)$.
    (b) What is the effect on $U$ of doubling $L$? Of doubling $K$? Of doubling both $L$ and $K$?
18. [C] If $F(L, K) = 285L^{0.78}K^{0.22}$, calculate
    (a) $F(200, 200)$ (b) $F(100, 300)$ (c) $F(300, 100)$.
19. Consider the plane whose equation is
    $$z = f(x, y) = 6 - 3x - 2y$$
    (a) Which of the points $(4, 1, -8)$, $(3, 0, 3)$, $(2, -6, 12)$ are in this plane?
    (b) Find the equations of the traces in the three coordinate planes.
    (c) Sketch the graphs of the three traces found in (b) and then the graph of the corresponding plane.
20. Consider the plane with equation $z = 18 - 6x - 2y$.
    (a) Find the equations of the traces in the three coordinate planes.
    (b) Sketch the graphs of these three traces and then the graph of the given plane.

*To find the x-intercept of the graph of the equation $z = f(x, y)$, set $y = z = 0$ and solve for x in the equation. Similiarly, to find the y-intercept, set $x = z = 0$ and solve for y. Thus the x, y, and z-intercepts for the plane $z = 12 + 4x - 3y$ are $x = -3$, $y = 4$, and $z = 12$. The graph of the plane $z = 4x - 8$ has no y-intercept, since setting $x = z = 0$ gives $0 = -8$. Thus this plane never crosses the y-axis but is parallel to it. Finding intercepts can be a great aid in graphing, especially for planes. In Problems 21–26, find the intercepts for the given plane and use this information to help you sketch its graph.*

21. $z = 5 - x - y$
22. $z = 6 - 3x - y$
23. $z = 8 - 2x + y$
24. $z = 6 - 6y + x$
25. $z = 8 - 4x$
26. $z = 10 - 2y$

*In Problems 27–36, sketch the graph of the given equation. First find and sketch the traces (if they exist) in the three coordinate planes. In some cases, it may be wise to find traces in other planes. For example, the graph in Problem 29 has no trace in the xy-plane ($z = 0$), so you might consider its traces in the planes $z = 4$ and $z = 9$ which are circles. See Example E.*

27. $z = x^2 + y^2 - 9$
28. $z = x^2 + y^2 - 16$
29. $z = x^2 + y^2 + 4$
30. $z = x^2 + y^2 + 1$
31. $4z = x^2 + y^2$
32. $z = 4(x^2 + y^2)$
33. $z = 9 - x^2 - y^2$
34. $z = -x^2 - y^2 - 1$
35. $z = 4x^2 + y^2 - 4$
36. $z = x^2 + \frac{1}{9}y^2$

---

37. A shoe manufacturer produces shoes and boots for men. Shoes sell at \$60 a pair and boots at \$75 a pair. Labor and materials cost \$30 per pair for shoes and \$45 per pair for boots, and there are fixed weekly costs of \$2000. Write explicit expressions for the total weekly profit $P(x, y)$ from producing $x$ pairs of shoes and $y$ pairs of boots.
38. A candy store sells milk chocolate at $p$ dollars per pound and fudge at $q$ dollars per pound. The corresponding demand equations for chocolate and fudge in pounds per day are $x = 300 - 2p + 5q$ and $y = 400 + 6p - 2q$, respectively.
    (a) Write an explicit expression for the daily revenue function $R(p, q)$.
    (b) Calculate $R(6, 7.5)$ and $R(5, 8)$.
39. The stopping distance $S = S(w, r)$ for a car after the brakes are applied is directly proportional to its weight $w$ and the square of its speed $r$ (that is, $S = kwr^2$ for some constant $k$). A car weighing 1800 pounds and traveling 40 miles per hour will stop in 86.4 feet. Determine an explicit expression for $S(w, r)$ and use it to find the stopping distance for a car weighing 2400 pounds and traveling 60 miles per hour.
40. A silo has the shape of a right circular cylinder topped with a hemisphere. Suppose that the cylinder has radius $r$ amd height $h$. Write an explicit expression for the volume $V = V(r, h)$ of the silo.
41. Sketch the graphs of the following equations.
    (a) $x = 8 - 2y - 4z$
    (b) $x = y^2 + z^2$
    (c) $z = \sqrt{25 - x^2 - y^2}$
42. Each of the following equations has one of the variables $x$, $y$, or $z$ missing. This means that the graph in three-dimensional space will be parallel to the axis of the missing variable. Sketch the graph of each equation in three-space.
    (a) $3x + y = 9$
    (b) $x + 4z = 8$
    (c) $z = x^2$
    (d) $y = (x - 1)^2$

## 10.2 PARTIAL DERIVATIVES

Let $y = f(x)$ determine a function of one variable. Then the derivative $dy/dx$ measures the rate of change in $y$ with respect to $x$. Let $z = f(x, y)$ determine a function of two variables. Then $\partial z/\partial x$ denotes the rate of change in $z$ with respect to $x$ keeping $y$ fixed, and $\partial z/\partial y$ denotes the rate of change in $z$ with respect to $y$ keeping $x$ fixed. These new derivatives, $\partial z/\partial x$ and $\partial z/\partial y$, are called the **partial derivative of $z$ with respect to $x$** and the **partial derivative of $z$ with respect to $y$,** respectively. Note the use of $\partial$ rather than $d$ to distinguish partial derivatives from ordinary derivatives. Fortunately, calculation of partial derivatives involves no new ideas. Simply treat $y$ as a constant in the calculation of $\partial z/\partial x$, and treat $x$ as a constant in the calculation of $\partial z/\partial y$.

**SHARP CURVE**
You are so used to thinking of $y$ as a variable that it may be hard for you to think of $y$ as a constant. But that is exactly what you must do in calculating $\partial z/\partial x$. Just as

$$\frac{d(3x)}{dx} = 3\frac{dx}{dx} = 3$$

so

$$\frac{\partial(2y^3x)}{\partial x} = 2y^3\frac{dx}{dx} = 2y^3$$

### EXAMPLE A

Let $z = 2xy^3 + 5x^4 - 3y^5$. Find $\partial z/\partial x$ and $\partial z/\partial y$.

**Solution** In calculating $\partial z/\partial x$, we treat $y$ as a constant and obtain (see SHARP CURVE)

$$\frac{\partial z}{\partial x} = 2y^3 + 20x^3 - 0$$

Similarly, in calculating $\partial z/\partial y$, we treat $x$ as a constant and get

$$\frac{\partial z}{\partial y} = 6xy^2 + 0 - 15y^4$$ ■

### EXAMPLE B

For $w = e^{xy} + xyz^2$, find $\partial w/\partial x$, $\partial w/\partial y$, and $\partial w/\partial z$.

**Solution** Note that $w$ is a function of the three variables $x$, $y$, and $z$. Thus, $\partial w/\partial x$ denotes the derivative of $w$ with respect to $x$ holding both $y$ and $z$ fixed, that is, treating $y$ and $z$ as constants. Similar understandings apply to $\partial w/\partial y$ and $\partial w/\partial z$. We obtain (using the Chain Rule on $e^{xy}$)

$$\frac{\partial w}{\partial x} = e^{xy}\frac{\partial}{\partial x}(xy) + yz^2 = ye^{xy} + yz^2$$

$$\frac{\partial w}{\partial y} = e^{xy}\frac{\partial}{\partial y}(xy) + xz^2 = xe^{xy} + xz^2$$

$$\frac{\partial w}{\partial z} = 0 + 2xyz$$ ■

### EXAMPLE C

For $z = x^2 \ln(x/y)$, find $\partial z/\partial x$ and $\partial z/\partial y$.

**Solution** First, we write $z = x^2 \ln x - x^2 \ln y$. Then for $\partial z/\partial x$, we use the Product Rule on the first term.

$$\frac{\partial z}{\partial x} = x^2\frac{1}{x} + 2x\ln x - 2x\ln y$$

$$= x + 2x\ln\frac{x}{y}$$

$$\frac{\partial z}{\partial y} = 0 - x^2\frac{1}{y} = -\frac{x^2}{y}$$ ■

## More Notation

Recall that for a function $f$ of one variable $x$, we have two basic notations for the derivative, namely, $df/dx$ and $f'(x)$. For a function $f$ of two variables $x$ and $y$, the prime notation $f'(x, y)$ would be confusing, since it does not indicate which of $x$ and $y$ is to be treated as a variable and which as a constant. Rather, we use $f_x(x, y)$ for $\partial f/\partial x$ and $f_y(x, y)$ for $\partial f/\partial y$. Thus, we have the following analagous notations.

| Functions of one variable | Functions of two variables |
|---|---|
| $f'(x)$ or $\dfrac{df}{dx}$ | $f_x(x, y)$ or $\dfrac{\partial f}{\partial x}$ |
| $f'(x_0)$ or $\left(\dfrac{df}{dx}\right)_{x=x_0}$ | $f_x(x, y_0)$ or $\left(\dfrac{\partial f}{\partial x}\right)_{y=y_0}$<br>$f_x(x_0, y_0)$ or $\left(\dfrac{\partial f}{\partial x}\right)_{x=x_0,\, y=y_0}$ |

**EXAMPLE D**

For $f(x, y) = xy^2 + xe^{xy} + x^2/2$, find (a) $f_x(x, y)$, (b) $f_x(x, 0)$, (c) $f_x(0, 2)$, (d) $f_y(3, 1)$.

**Solution**

(a) $f_x(x, y) = y^2 + [x(ye^{xy}) + e^{xy}(1)] + x = y^2 + xye^{xy} + e^{xy} + x$
(b) $f_x(x, 0) = 0 + 0 + e^0 + x = 1 + x$
(c) $f_x(0, 2) = 4 + 0 + e^0 + 0 = 5$
(d) $f_y(x, y) = 2xy + x(xe^{xy}) + 0 = 2xy + x^2e^{xy}$, and so
$f_y(3, 1) = 6 + 9e^3$ ■

## Geometric Interpretation

You will recall that $f'(x_0)$ is the slope of the tangent line to the curve $y = f(x)$ at the point $x_0$ (Figure 11). In a similar fashion $f_x(x_0, y_0)$ is the

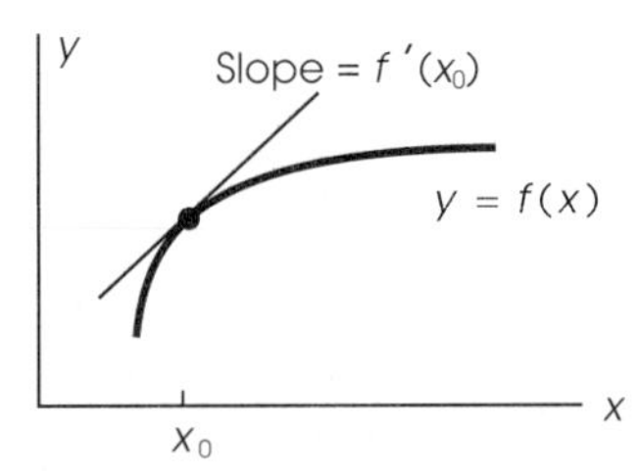


FIGURE 11

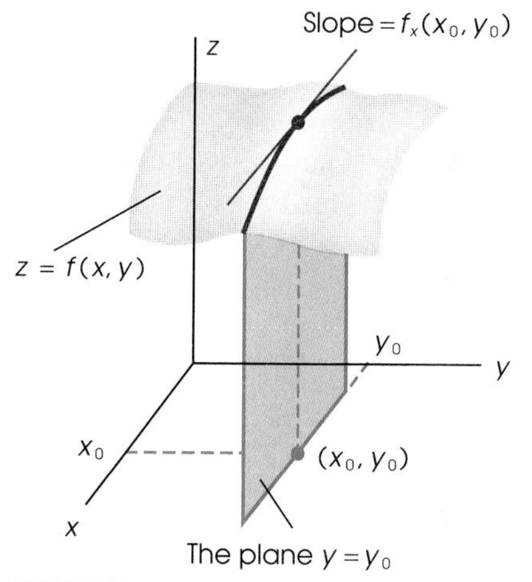

FIGURE 12

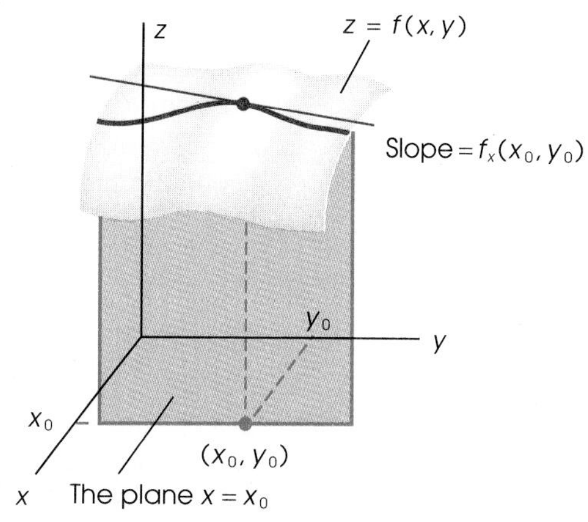


FIGURE 13

slope at $(x_0, y_0)$ of the tangent line to the trace of the surface $z = f(x, y)$ in the plane $y = y_0$ (Figure 12). Likewise, $f_y(x_0, y_0)$ is the slope at $(x_0, y_0)$ of the tangent line to the trace of the surface in the plane $x = x_0$ (Figure 13).

## EXAMPLE E

Jack and Jill ran up the hill $z = f(x, y) = e^{-x^2-y^2}$, running directly above the $x$-axis (Figure 14). What was the grade of their path at the point $(1, 0)$? Distances are in miles.

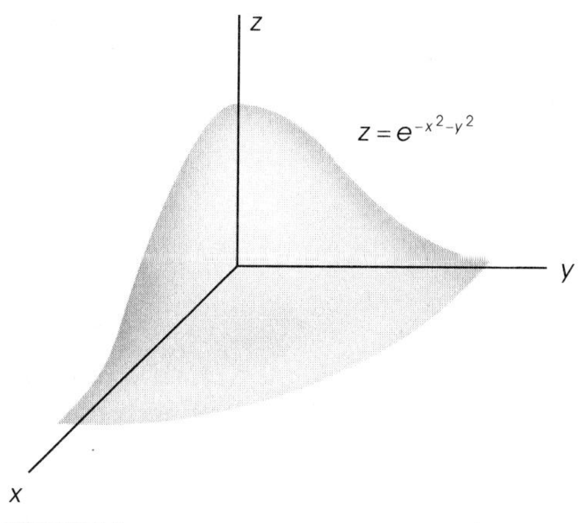


FIGURE 14

**Solution** The grade is just the absolute value of the slope but is given as a percentage (see end of Section 1.3). Since

$$f_x(x, y) = -2xe^{-x^2-y^2}$$

we conclude that the slope $m = f_x(1, 0) = -2e^{-1} \approx -0.73576$. Thus, the grade is about 74%. That is a very steep grade, and it is no wonder that Jack fell down and broke his crown. ■

## Rate of Change Interpretation

Suppose that $w = f(x, y)$ and that $f_x(5, 6) = 10$ and $f_y(5, 6) = 2$. Then at the point (5, 6), $w$ is changing 10 times as fast as $x$ but only 2 times as fast as $y$. Rephrased another way, if $x$ is increased by 1 unit, $w$ will increase by about 10 units; if $y$ is increased by 1 unit, $w$ will increase by about 2 units.

### EXAMPLE F

Let $W = f(L, K) = 90L^{2/3}K^{1/3}$ be the production function for a manufacturing process, with $W$, $L$, and $K$ denoting the number of units of goods produced, labor used, and capital required, respectively. Calculate $f_L(8, 27)$ and $f_K(8, 27)$ and interpret the meaning of these numbers.

**Solution** $f_L(L, K) = 60L^{-1/3}K^{1/3}$ and $f_K(L, K) = 30L^{2/3}K^{-2/3}$, so $f_L(8, 27) = (60)(\frac{1}{2})(3) = 90$ and $f_K(8, 27) = (30)(4)(\frac{1}{9}) \approx 13.3$. Thus, an increase in labor of 1 unit (from 8 to 9 while keeping $K$ fixed at 27) will increase production by about 90 units. On the other hand, an increase in capital of 1 unit (from 27 to 28 while keeping $L$ fixed at 8) will increase production by about 13.3 units. The partial derivatives $f_L$ and $f_K$ are called the *marginal productivity of labor* and the *marginal productivity of capital,* respectively. ■

## Higher-Order Derivatives

The partial derivative $f_x(x, y)$ is itself a function, and so we may take its partial derivative with respect to either $x$ or $y$. We use the following notation.

**Second-Order Partial Derivatives**

| | | | |
|---|---|---|---|
| Second partial of $f$ with respect to $x$ | $\dfrac{\partial^2 f}{\partial x^2} = \dfrac{\partial}{\partial x}\left(\dfrac{\partial f}{\partial x}\right)$ | or | $f_{xx} = (f_x)_x$ |
| Second partial of $f$ with respect to $y$ | $\dfrac{\partial^2 f}{\partial y^2} = \dfrac{\partial}{\partial y}\left(\dfrac{\partial f}{\partial y}\right)$ | or | $f_{yy} = (f_y)_y$ |
| Mixed second partials of $f$ | $\dfrac{\partial^2 f}{\partial y\,\partial x} = \dfrac{\partial}{\partial y}\left(\dfrac{\partial f}{\partial x}\right)$ | or | $f_{xy} = (f_x)_y$ |
| | $\dfrac{\partial^2 f}{\partial x\,\partial y} = \dfrac{\partial}{\partial x}\left(\dfrac{\partial f}{\partial y}\right)$ | or | $f_{yx} = (f_y)_x$ |

Higher-order partial derivatives like $f_{xxy}$ and $f_{xyy}$ are defined in an analagous way, but we shall have little use for them.

### EXAMPLE G

For $f(x, y) = xy^3 + yxe^x$, find (a) $f_{xx}$, (b) $f_{xy}$, (c) $f_{yx}$, (d) $f_{yy}$.

**Solution** First note that

$$f_x(x, y) = y^3 + y(xe^x + e^x) = y^3 + yxe^x + ye^x$$

and

$$f_y(x, y) = 3xy^2 + xe^x$$

(a) $f_{xx}(x, y) = 0 + y(xe^x + e^x) + ye^x = yxe^x + 2ye^x$
(b) $f_{xy}(x, y) = 3y^2 + xe^x + e^x$
(c) $f_{yx}(x, y) = 3y^2 + xe^x + e^x$
(d) $f_{yy}(x, y) = 6xy$ ■

We hope you noted that $f_{xy} = f_{yx}$ in Example G, a circumstance that was not accidental. In fact, these mixed partial derivatives are always equal provided they are continuous.

## An Application to Economics

Economists classify pairs of consumer products by using three words: *competitive, complementary*, and *independent*. Fords and Chevrolets are competitive products. An increase in sales of one is likely to produce a decrease in sales of the other (unless total car sales are expanding rapidly). Cars and gasoline are complementary products. Increases in the sale of one tend to be accompanied by increases in the sale of the other. Cars and rat poison would seem to be independent products. Sales of one are unlikely to have any effect on sales of the other. Let us see what this means in terms of partial derivatives.

Let the two products labeled A and B be selling at prices per unit $p_A$ and $p_B$, respectively. We suppose that the corresponding demands $x_A$ and $x_B$ depend in some way on both $p_A$ and $p_B$, that is,

$$x_A = f(p_A, p_B) \quad \text{and} \quad x_B = g(p_A, p_B)$$

Consider first the competitive situation. Keep $p_A$ fixed but allow $p_B$ to increase. This will cause the demand for product A to increase, that is, $\partial x_A/\partial p_B > 0$, a circumstance shown graphically in Figure 15. Similarly, an increase in $p_A$ with $p_B$ fixed will cause an increase in the demand for product B, that is, $\partial x_B/\partial p_A > 0$. Thus, competition between the two products is characterized by both of these partial derivatives being positive.

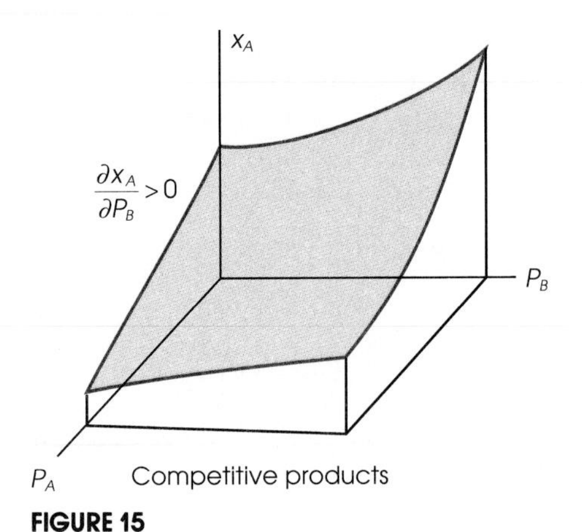

Competitive products

**FIGURE 15**

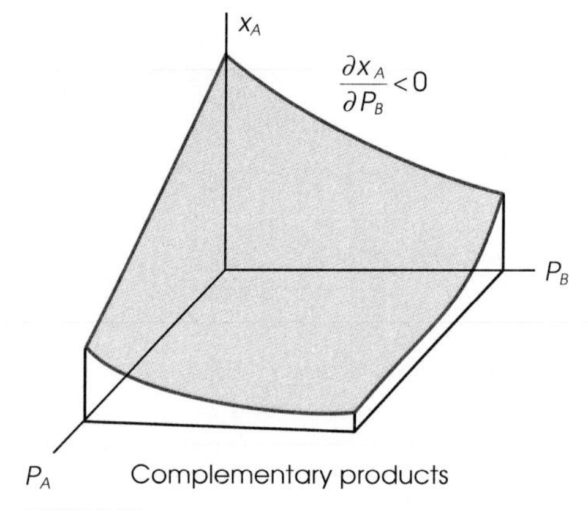

Complementary products

**FIGURE 16**

In the complementary situation, an increase in $p_B$ lowers the demand for product B and therefore also for product A, that is, $\partial x_A/\partial p_B < 0$ (Figure 16). Similarly, $\partial x_B/\partial p_A < 0$. We summarize in the following chart.

| Competitive Products | Complementary Products |
|---|---|
| $\dfrac{\partial x_A}{\partial p_B} > 0$ and $\dfrac{\partial x_B}{\partial p_A} > 0$ | $\dfrac{\partial x_A}{\partial p_B} < 0$ and $\dfrac{\partial x_B}{\partial p_A} < 0$ |

### EXAMPLE H

Suppose that a company makes two products, A and B, with unit prices $p_A$ and $p_B$, respectively. If the corresponding demand equations are

$$x_A = -200p_A + 250p_B$$

$$x_B = 400 + 100p_A - 150p_B$$

decide whether the two products are competitive, complementary, or neither.

**Solution** $\partial x_A/\partial p_B = 250$ and $\partial x_B/\partial p_A = 100$. Both are positive; the products are competitive. ■

## Problem Set 10.2

*In Problems 1–16, find $\partial z/\partial x$ and $\partial z/\partial y$ (see Examples A and C).*

1. $z = x^2y^3$
2. $z = (3x + 2y)^2$
3. $z = xy^2 - 4x^2 + 3y + 5$
4. $z = 3x^2y - 6y^2 + 4x + 2$
5. $z = 2x\sqrt{y} + 4/y^2$
6. $z = x^2/\sqrt{y} - 3x^{2/3}$
7. $z = 4/xy^2$
8. $z = 12/y\sqrt{x}$
9. $z = x^2ye^{2x}$
10. $z = y^2 \ln(2x + 3y)$
11. $z = (x^2 + 2y)^3 \ln y$
12. $z = y^2e^{x^2/y}$
13. $z = \sqrt{2xy + y^2}\, e^{2y}$

14. $z = (y^2/x) \ln(x + 3)$
15. $z = \ln[x^3y^2/(2x + y)^3]$ 16. $z = \ln[e^{xy} \ln(xy^2)]$

*In Problems 17–20, find $\partial w/\partial z$ (see Example B).*

17. $w = 2xyz + 3yz^2$
18. $w = (xyz^3 - z)^4$
19. $w = xye^{z+2} + xz/y$
20. $w = z^2 \ln(x^2y + z)$

*In Problems 21–26, find $f_x(2, 1)$ and $f_y(2, -1)$ as in Example D. To do this, you will first want to find $f_x(x, y)$ and $f_y(x, y)$.*

21. $f(x, y) = x^2 + 2xy^2 - y$
22. $f(x, y) = 3xy - 4x^2 + 5y$
23. $f(x, y) = y^2 \ln(2x + 3y)$
24. $f(x, y) = x^2e^{(x-2)y}$
25. $f(x, y) = (x + 2y)e^{x(y^2-1)}$
26. $f(x, y) = (x + 2y)^2/(y^2 + 1)$

*In Problems 27–30, find the slope of the tangent line at $(x_0, y_0)$ to the trace of $z = f(x, y)$ in the plane $y = y_0$.*

27. $f(x, y) = 9 - x^2 - y^2$; $x_0 = 2$, $y_0 = 1$
28. $f(x, y) = x^2 + y^2 - 4$; $x_0 = 3$, $y_0 = -1$
29. $f(x, y) = \sqrt{25 - x^2 - y^2}$; $x_0 = 3$, $y_0 = 0$
30. $f(x, y) = e^{x^2+2y^2}$; $x_0 = 1$, $y_0 = -1$
31. Suppose that Jack and Jill in Example E ran down the hill $z = f(x, y) = e^{-x^2-y^2}$ directly above the line $y = 2$ (in the $xy$-plane). What was the grade of their path at the point (3, 2)?
32. If Jack and Jill in Example E ran down the given hill directly above the line $x = 3$, what was the grade of their path at the point (3, 2)?
33. Suppose that you traveled down the hill represented by the surface $z = f(x, y) = 4/\ln(2 + 2x^2 + 2y^2)$ on a path directly above the $x$-axis. What was the grade of your path at the point (1, 0)?
34. If in Problem 33 you traveled down the hill directly above the $y$-axis, what was the grade of your path at (0, 2)?
35. Recall that the volume of a right circular cone with radius of base $r$ and height $h$ is given by $V = \pi r^2h/3$. Find the rate of change in volume
    (a) with respect to the radius when $r = 6$ and $h = 8$.
    (b) with respect to the height when $r = 6$ and $h = 8$.
36. The resistance $R$ to blood flow in a human blood vessel depends on the length $L$ and radius $r$ of the vessel according to the formula $R = kL/r^4$, where $k$ is a constant. Find the rate of change in resistance
    (a) with respect to length.
    (b) with respect to radius.

C 37. Empirical evidence suggests that the surface area $A$ (in square centimeters) of an average person's body is related to his or her weight $W$ (in kilograms) and height $H$ (in centimeters) according to the formula $A = 700W^{0.42}H^{0.73}$.
    (a) Determine $\partial A/\partial W$ and $\partial A/\partial H$.
    (b) About how much will $A$ change if a person's weight increases from 70 to 71 kilograms while he or she stays 180 centimeters in height?
    (c) About how much will $A$ change if a person grows from 180 to 181 centimeters in height while maintaining a weight of 70 kilograms?
38. Suppose that the stopping distance in feet for a car weighing $w$ pounds traveling $v$ miles per hour is given by $S = S(w, v) = 0.000012wv^2$.
    (a) Evaluate $S(2000, 70)$, $S_w(2000, 70)$, and $S_v(2000, 70)$.
    (b) About how much will the stopping distance increase for a car of weight 2000 pounds if its velocity is increased from 70 to 71 miles per hour?

*In Problems 39–44, find* (a) $f_{xx}(x, y)$, (b) $f_{xy}(x, y)$, *and* (c) $f_{yx}(x, y)$ *as in Example G.*

39. $f(x, y) = 2xy^3 - 2x^2y$
40. $f(x, y) = y^4 - 2xy^2 + 5x^2$
41. $f(x, y) = e^{xy^2}$
42. $f(x, y) = 2y \ln(x^2 + y)$
43. $f(x, y) = (x^2 + 2y)^3$ 44. $f(x, y) = (y^2 + e^x)^4$

*In Problems 45–48, $p_A$ and $p_B$ represent unit prices of two products A and B, while $x_A$ and $x_B$ represent the corresponding demands. Determine as in Example H whether the products A and B are competitive, complementary, or neither.*

45. $x_A = -300p_A - 250p_B + 300$;
    $x_B = -420p_A - 180p_B + 640$
46. $x_A = -45p_A - 60p_B + 120$;
    $x_B = -50p_A - 35p_B + 80$
47. $x_A = -100p_A + 84p_B + 2000$;
    $x_B = 45p_A - 60p_B + 1200$
48. $x_A = -55p_A + 48p_B + 420$;
    $x_B = -60p_A - 90p_B + 180$

---

49. A corporation estimates that, with annual expenditures of $x$ millions of dollars on research and $y$ millions of dollars on advertising, its revenue in millions of dollars will be given by $R = R(x, y) = 42 + 3x^2 + 2xy$. Evaluate $R_x(20, 30)$ and $R_y(20, 30)$ and interpret the meaning of these answers.

C 50. A ski manufacturer has the weekly production function $W = F(L, K) = 250L^{0.63}K^{0.37}$, where $W$ is the number of pairs of skis produced, $L$ the number of units of labor used, and $K$ the number of units of capital required. Evaluate $F_L(10, 20)$ and $F_K(10, 20)$ and interpret your answers.

51. The temperature $T$ in Celsius degrees in a thin rectangular plate at a point $x$ feet from the left edge and $y$ feet from the bottom edge satisfies $T = T(x, y) = 64 - 0.5x^2 - 1.5y^2$. Find $T_x(4, 2)$ and $T_y(4, 2)$ and interpret your answers.

C 52. In the winter, some weather stations report the temperature $T$ in degrees Celsius, the wind velocity $v$ in miles per hour, and the wind chill $W$ in degrees Celsius using a formula like

$$W = W(v, T) = 30 - 0.05(11 + 9\sqrt{v} - 0.09v)(30 - T)$$

Presumably $W$ measures how cold it feels.
(a) Calculate $W(30, -20)$ and $W_v(30, -20)$.
(b) What significance do you attach to $W_v(30, -20)$?

53. A factory employs $x$ workers at \$6 per hour, $y$ workers at \$9 per hour, and $z$ workers at \$12 per hour.
(a) Write an expression for the average hourly wage $A(x, y, z)$ of the $x + y + z$ workers.
(b) Calculate $A_x(10, 20, 10)$.
(c) Interpret your answer to part (b).

C 54. When $P$ dollars are invested at the rate $r$ compounded continuously, the accumulated amount $A$ at the end of $t$ years is given by $A = A(P, r, t) = Pe^{rt}$. Evaluate $A_r(1000, 0.07, 10)$ and interpret its meaning.

*Recall for functions of one variable $y = f(x)$ that the actual change $\Delta y$ when $x$ goes from $x_0$ to $x_0 + \Delta x$ can be approximated by $dy = f'(x_0)\,\Delta x$. In a similar fashion, the change $\Delta z$ in $z = f(x, y)$ when $(x, y)$ goes from $(x_0, y_0)$ to $(x_0 + \Delta x, y_0 + \Delta y)$ is approximately equal to $dz = f_x(x_0, y_0)\,\Delta x + f_y(x_0, y_0)\,\Delta y$. Use this fact in Problems 55–58.*

55. Find approximately the change in $z = 2x^2\sqrt{y}$ when $x$ changes from 10 to 10.2 and $y$ changes from 16 to 16.1.

56. Approximate the change in $z = 12x/y^2$ when $x$ changes from 20 to 20.5 and $y$ changes from 10 to 10.2.

C 57. About how much did the volume of a right circular cylinder increase when its radius changed from 10 to 10.3 and its height from 30 to 30.4 (all in centimeters)?

C 58. $P$ dollars invested for 1 year at rate $r$ compounded continuously will accumulate to $Pe^r$ at the end of the year. About how much extra will John have at the end of a year if he can increase his investment from \$1000 to \$1050 and get a rate of 0.075 rather than 0.07?

## 10.3 OPTIMIZATION

If optimization was important for functions of a single variable, it is equally important for functions of two or more variables. The behavior of functions of several variables can be quite complicated. Figure 17 illustrates some of the possibilities for a function of two variables.

The language we will use is quite clear from Figure 17, but for completeness, we give a formal definition.

> DEFINITION
>
> Let $D$, the domain of $f(x, y)$, contain the point $(a, b)$. We say that
>
> (i) $f(a, b)$ is the **maximum value** of $f(x, y)$ on $D$ if $f(a, b) \geq f(x, y)$ for all $(x, y)$ in $D$.
>
> (ii) $f(a, b)$ is the **minimum value** of $f(x, y)$ on $D$ if $f(a, b) \leq f(x, y)$ for all $(x, y)$ in $D$.
>
> The same definition applies to **local maximum value** and **local minimum value,** except that we require the inequalities to hold only in a neighborhood of $(a, b)$

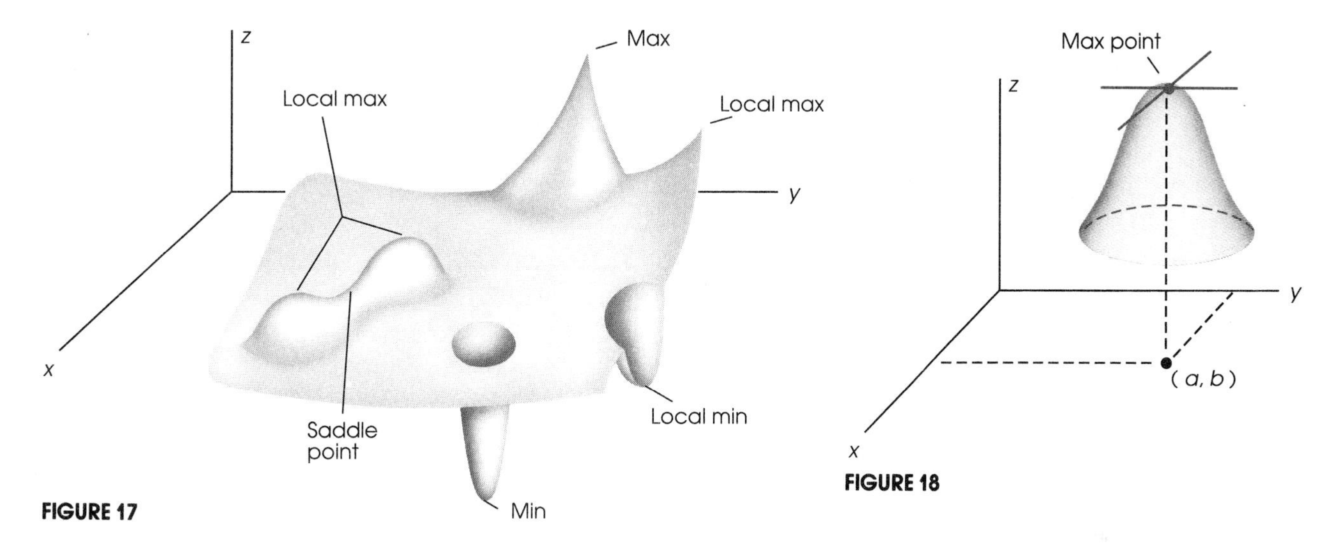


FIGURE 17

FIGURE 18

We use the phrase **extreme value** to refer to either a maximum value or a minimum value. Just as in the one-variable case, an extreme value can occur only at a critical point. A **critical point** is (i) a boundary point of $D$, (ii) a point where the first partial derivatives are 0, or (iii) a point where the function has some type of pathological behavior (its graph has a break, has a corner, or fluctuates wildly). Our interest is mainly in critical points of type (ii). Three types of behavior are typical of this case. They are illustrated in Figures 18, 19, and 20.

Note that in both Figures 18 and 19, the partial derivatives are 0 at $(a, b)$; that is,

$$f_x(a, b) = 0 = f_y(a, b)$$

However, this relation does not guarantee an extremum at $(a,b)$, as the **saddle point** in Figure 20 illustrates.

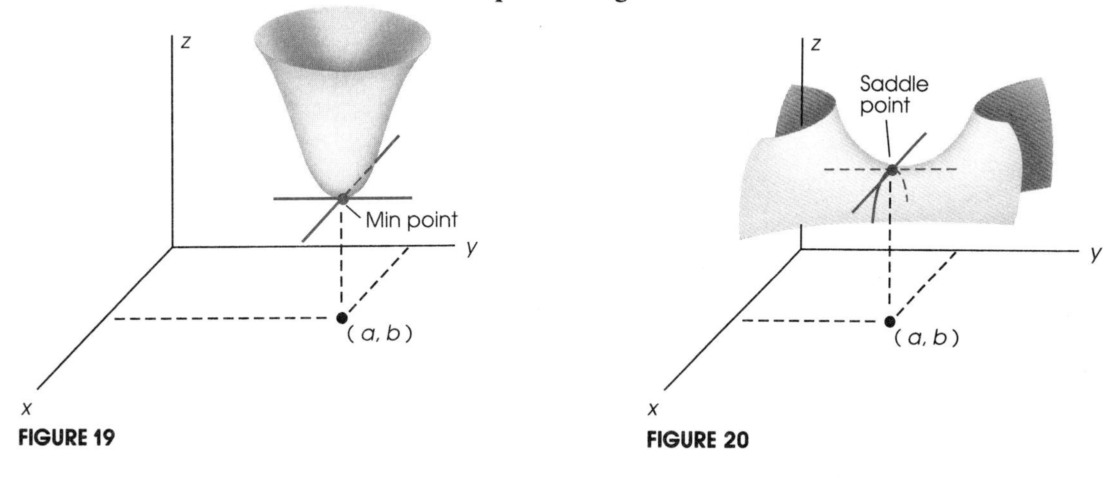


FIGURE 19

FIGURE 20

## Second Partials Test for Extrema

As was true of functions of one variable, we have a test for locating maxima, minima, and saddle points and for distinguishing between them. We state this test as a theorem, a theorem whose proof goes far beyond the level of this course.

**THEOREM (SECOND PARTIALS TEST)**
Suppose that $f(x, y)$ has continuous second partial derivatives in a neighborhood of $(a, b)$ and that

$$f_x(a, b) = 0 = f_y(a, b)$$

Let

$$T(a, b) = f_{xx}(a, b)f_{yy}(a, b) - [f_{xy}(a, b)]^2$$

Then

(i) If $T(a, b) > 0$ and $f_{xx}(a, b) < 0$, $f(a, b)$ is a local maximum value.
(ii) If $T(a, b) > 0$ and $f_{xx}(a, b) > 0$, $f(a, b)$ is a local minimum value.
(iii) If $T(a, b) < 0$, there is a saddle point at $(a, b)$.
(iv) If $T(a, b) = 0$, the test is inconclusive.

### EXAMPLE A

Find the local extrema, if any, of

$$f(x, y) = x^2 + 3y^3 + 4x - 9y + 5$$

**Solution** Since $f_x(x, y) = 2x + 4$ and $f_y(x, y) = 9y^2 - 9$, we find the critical points by solving simultaneously the simple pair of equations

$$2x + 4 = 0$$

$$9y^2 - 9 = 0$$

There are two solutions, namely, $(-2, 1)$ and $(-2, -1)$.
Now $f_{xx}(x, y) = 2$, $f_{xy}(x, y) = 0$, and $f_{yy}(x, y) = 18y$. Thus, at $(-2, 1)$,

$$f_{xx} > 0 \quad \text{and} \quad T = (2)(18) - 0 = 36 > 0$$

We conclude $f(-2, 1) = -5$ is a local minimum value for $f(x, y)$. However, at $(-2, -1)$,

$$T = (2)(-18) - 0 = -36 < 0$$

and therefore there is a saddle point at $(-2, -1)$ (not an extremum). ■

## A Geometric Application

Geometric applications are nice because our intuition guides us along.

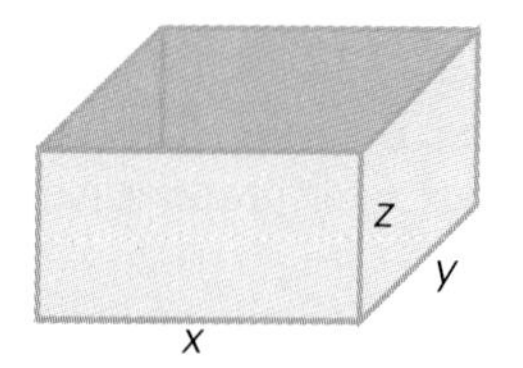


FIGURE 21

### EXAMPLE B

Determine the dimensions of the open rectangular box of volume 32 cubic feet that has the least (outer) surface area.

**Solution** The box, appropriately labeled, is shown in Figure 21. Its surface area is $A = xy + 2xz + 2yz$. Since the volume is 32, we have $xyz = 32$ or, equivalently, $z = 32x^{-1}y^{-1}$. This we substitute in the expression for $A$ to obtain

$$\begin{aligned} A &= xy + 2x(32x^{-1}y^{-1}) + 2y(32x^{-1}y^{-1}) \\ &= xy + 64y^{-1} + 64x^{-1} \end{aligned}$$

Note that $0 < x < \infty$, $0 < y < \infty$ is the appropriate domain, $x$ and $y$ being dimensions of the base. To find the critical values, we calculate the partial derivatives $A_x$ and $A_y$, set them equal to 0, and solve the resulting pair of equations simultaneously.

$$\begin{aligned} A_x &= y - 64x^{-2} = 0 \qquad & A_y &= x - 64y^{-2} = 0 \\ y &= 64x^{-2} & x &= 64y^{-2} \end{aligned}$$

To solve this pair of equations, we substitute the expression for $x$ from the second equation into the first and then solve the resulting equation for $y$.

$$\begin{aligned} y &= 64x^{-2} = 64(64y^{-2})^{-2} = \frac{y^4}{64} \\ 64y &= y^4 \\ y^4 - 64y &= 0 \\ y(y^3 - 64) &= 0 \\ y &= 0, 4 \end{aligned}$$

Only $y = 4$ is in the domain, and substituting this value gives

$$x = 64y^{-2} = 64(4)^{-2} = 4$$

We conclude that (4, 4) is the only critical point for $A$. Now we could use the Second Partials Test to establish that (4, 4) gives a local minimum for $A$ (we leave this to the reader). There is finally the subtle question of whether this local minimum is actually the (global) minimum. Physical considerations suggest that it is. Finally, we calculate $z$ from the equation $z = 32x^{-1}y^{-1}$, obtaining $z = 2$. The dimensions of the box of minimum surface area are 4 by 4 by 2. ■

## A Business Application

We propose a maximum revenue problem.

> **RECALL**
> At the end of Section 10.2 we learned that two products, A and B, are competitive if $\partial x_A/\partial p_B$ and $\partial x_B/\partial p_A$ are both positive. In the present notation, this means that $\partial x/\partial q$ and $\partial y/\partial p$ should be positive. They are, for $\partial x/\partial q = 1$ and $\partial y/\partial p = 1$

### EXAMPLE C

A company sells two competing products A and B at unit prices $p$ and $q$ dollars, respectively. It estimates that the demands $x$ for product A and $y$ for product B satisfy

$$x = 50 - 4p + q$$
$$y = 60 + p - 2q$$

Determine the prices $p$ and $q$ that will maximize the total revenue from these two products.

**Solution** The total revenue $R$ is given by

$$R = px + qy = p(50 - 4p + q) + q(60 + p - 2q)$$
$$= -4p^2 + 2pq - 2q^2 + 50p + 60q$$

Thus, the partial derivatives are

$$R_p = -8p + 2q + 50 \quad \text{and} \quad R_q = -4q + 2p + 60$$

We set these two partial derivatives equal to 0 to obtain the pair of equations

$$8p - 2q = 50$$
$$-2p + 4q = 60$$

This pair of equations can be solved by the method explained in Example D in Section 1.3. The result is $p = \$11.43$, $q = \$20.71$. Since these values for $p$ and $q$ determine the only critical point, we feel sure that they give the maximum revenue. We can, of course, use the Second Partials Test to establish that they at least give a local maximum. Finally, we mention that the total revenue corresponding to these choices of $p$ and $q$ is \$907.14. ■

## Problem Set 10.3

*In Problems 1–14, find all critical points. Indicate which of these gives a local maximum, a local minimum, or a saddle point (see Example A).*

1. $f(x, y) = x^2 + y^2 - 4x + 6y - 4$
2. $f(x, y) = x^2 + y^2 + 8y - 2x + 9$
3. $f(x, y) = (x + 1)^2 + (y - 4)^2 + 4x - 2y$
4. $f(x, y) = (x - 3)^2 + y^2 - 6x + 4y - 10$
5. $f(x, y) = 5 + 6x + 2y - 2x^2 + xy - y^2$
6. $f(x, y) = 3x^2 - xy + y^2 + 5x + y - 10$
7. $f(x, y) = 4y + 2xy - x^2 - 2y^2 + 4$
8. $f(x, y) = 2y - 6x - x^2 - 2y^2 + 3xy$
9. $f(x, y) = x^3 + 2x^2y + 4x - 8y$
10. $f(x, y) = 6xy - x^2 - y^3 + 4$
11. $f(x, y) = x^3 + y^3 - 12xy - 10$
12. $f(x, y) = x^3 + y^3 - 3x - 12y + 24$
13. $f(x, y) = ye^x - 2x - y + 10$
14. $f(x, y) = 2xy + 4 \ln x + y^2 - 4$

15. A rectangular tank open at the top is to hold 256 cubic feet of water. Determine the dimensions of the tank so that it will require the least amount of sheet metal to build. See Example B.
16. A rectangular storage shed will be built with a capacity of 24,000 cubic feet. The cost of the construction is \$8 per square foot for the flat roof, \$6 per square foot for the sides, and \$12 per square foot for the floor. Find the dimensions of the shed that minimize the cost of construction.
17. Mailing instructions for a rectangular box require that the length plus the girth shall not exceed 84 inches (girth is the distance around, as illustrated by the dotted rectangle in Figure 22). Find the dimensions of the box that give the largest possible volume.

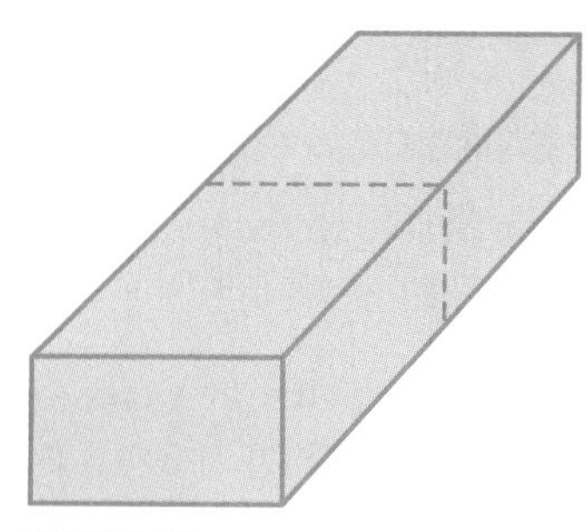

FIGURE 22

18. An open rectangular bin with two partitions is to be built to have a total capacity of 8 cubic feet (Figure 23). Find the dimensions of the bin requiring the least amount of material in its construction.

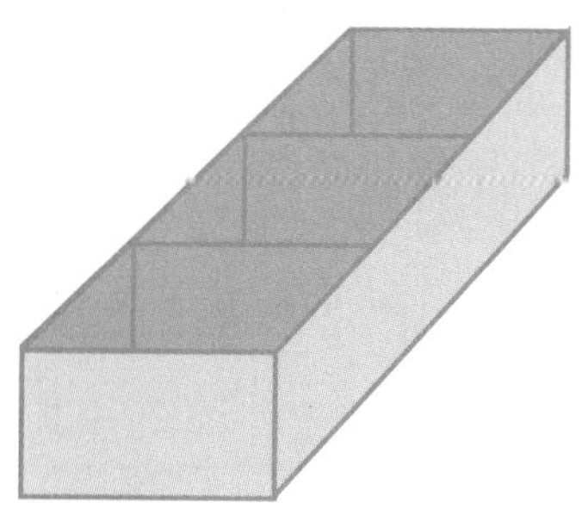

FIGURE 23

19. Suppose that a company sells two products, A and B, at corresponding unit prices $p$ and $q$ and with corresponding demands

$$x = -100p + 84q + 2000$$

$$y = 45p - 60q + 1200$$

respectively. Determine the prices $p$ and $q$ for which the total revenue for the two products is a maximum. See Example C.

20. A department store sells two kinds of women's blazers at prices $p$ and $q$ with corresponding weekly demands

$$x = 40 - 0.3p + 0.2q$$

$$y = 50 + 0.2p - 0.2q$$

respectively. Determine $p$ and $q$ so as to maximize the total revenue.
21. A manufacturer makes two kinds of children's tricycles. Making $x$ units of the first kind and $y$ units of the second kind per day will allow it to set the corresponding prices at $80 - 2.5x$ dollars and $120 - 4y$ dollars, respectively. The total production cost for these tricycles is estimated at $C = 30x + 40y - 4xy$. Determine $x$ and $y$ to maximize profit.
22. The ABC Company can make candy A for 60¢ per pound and candy B for 70¢ per pound. The weekly demands $x$ and $y$ for the two kinds of candy if priced at $p$ cents and $q$ cents per pound, respectively, are

$$x = 300q - 400p$$

$$y = 250(100 + p - q)$$

both measured in pounds. If the ABC Company wants to maximize its profit, how should it set the prices $p$ and $q$?

---

23. Find all critical points for the function $f(x, y) = y^4 - 4xy + 0.5x^2 - 20$. Then identify each as to whether it gives a local maximum, a local minimum, or a saddle point.
24. Find three positive numbers $x$, $y$, and $z$ whose product is 24, such that $2x + 2y + z$ is as small as possible.
25. A large firm estimates that if it spends $x$ million dollars on research and $y$ million dollars on advertising per year, then its annual profit will be $P(x, y) = -3x^2 + 3xy - y^2 + 22x - 8y + 20$ million dollars. Determine $x$ and $y$ so that the annual profit will be a maximum.
26. A campaign committee to reelect the mayor knows that the mayor can count on a certain number of votes with no advertising, that some advertising will increase the vote total, but that too much advertising may actually backfire. It estimates that spending $x$ thousand dollars on TV advertising and $y$ thousand dollars on radio advertising will produce $z$ thousand extra votes for the mayor, where $z = 12xy - 2x^3 - y^2$. Find $x$ and $y$ so as to maximize the mayor's vote total.
27. The reaction (in appropriate units) of a person to $x$ units of a certain drug $t$ hours after it is administered

is given by

$$R(x, t) = 4x(10 - x)te^{-2t}$$

Find the maximum reaction.

28. When $x$ units of drug A and $y$ units of drug B are used in the treatment of one type of cancer, the reaction of the patient (in appropriate units) 12 hours after their administration is given by $R(x, y) = 10xy(20 - x - 2y)$. Find $x$ and $y$ so as to maximize this reaction.

## 10.4*
## CONSTRAINED OPTIMIZATION AND LAGRANGE'S METHOD

Consider the following two minimization problems.

Problem 1. Minimize $z = x^2 + y^2 + 2$.
Problem 2. Minimize $z = x^2 + y^2 + 2$ subject to $x + y = 8$.

The first problem is a free optimization problem; the second is a constrained optimization problem. The geometric interpretation of the two problems is shown in Figures 24 and 25.

We can turn the second problem into one of the first type by solving the constraint for $y$ and substituting in the expression for $z$. Problem 2 then takes the form:

Rephrased Problem 2. Minimize $z = x^2 + (8 - x)^2 + 2$.

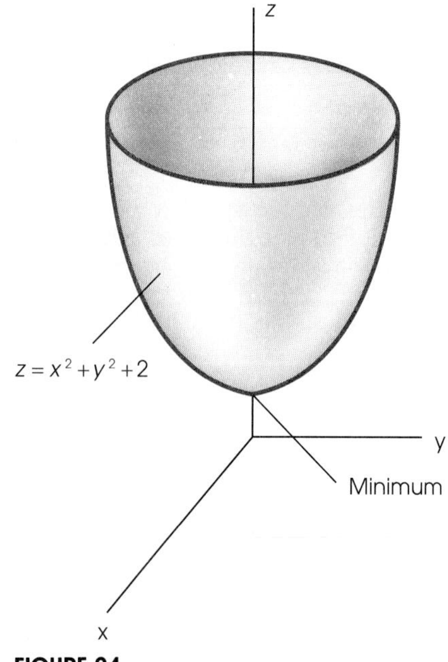


FIGURE 24

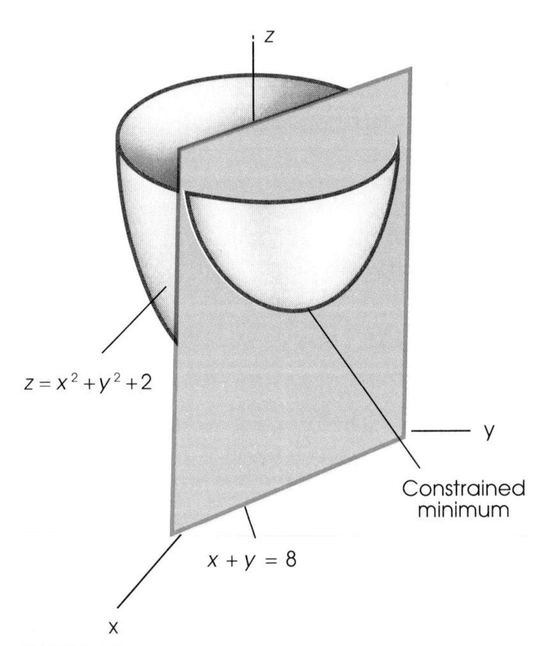


FIGURE 25

**POINT OF INTEREST**
Joseph-Louis Lagrange (1736–1813) called the loss of his father's fortune a lucky disaster, for otherwise "I should probably not have cast my lot with mathematics." In addition to his many theoretical contributions, Lagrange helped perfect the metric system of weights and measures.

In this form the problem is simple enough and can actually be handled by the methods in Section 3.3. The minimum occurs at $x = 4$, which in turn makes $y = 4$ and $z = 34$.

But what if the problem had read:

Minimize $z = x^2 + y^2 + 2$ subject to the constraint $xy^5 + x^5y^2 = 8$.

We could not have solved the constraint for $y$ in terms of $x$ (or vice versa); the standard method breaks down. The eighteenth-century French mathematician, Joseph-Louis Lagrange suggested an alternative. Often it works better than the standard method even when that method works.

## The Method of Lagrange Multipliers

We consider the general problem of optimizing $f(x, y)$ subject to the constraint $g(x, y) = 0$. Here is Lagrange's method.

**LAGRANGE'S METHOD**

For the problem of optimizing $z = f(x, y)$ subject to the constraint $g(x, y) = 0$, let

$$F(x, y, \lambda) = f(x, y) + \lambda g(x, y)$$

The local extreme points $(x, y)$ for this optimization problem will be found among the points $(x, y, \lambda)$ that satisfy the three equations

$$F_x(x, y, \lambda) = 0$$

$$F_y(x, y, \lambda) = 0$$

$$F_\lambda(x, y, \lambda) = 0$$

It would take us too far afield to prove this result, but we will illustrate it with several examples. Our first example was discussed at the beginning of this section.

### EXAMPLE A

Minimize $z = f(x, y) = x^2 + y^2 + 2$ subject to the constraint $g(x, y) = x + y - 8 = 0$.

**Solution** Note that we have expressed the constraint in the form required by Lagrange's method. Let

$$F(x, y, \lambda) = x^2 + y^2 + 2 + \lambda(x + y - 8)$$

Lagrange's three equations take the form

$$F_x = 2x + \lambda = 0$$

$$F_y = 2y + \lambda = 0$$

$$F_\lambda = x + y - 8 = 0$$

To solve them simultaneously, eliminate $\lambda$ between the first two equations, obtaining $2x - 2y = 0$ or $x = y$. Substitute this result in the third equation to get $2x = 8$ or $x = 4$. Then $y = 4$ (and though we are not interested in it, $\lambda = -8$). The only critical point is $(x, y) = (4, 4)$, the point we found earlier by the standard method. The minimum value for $z$ is $f(4, 4) = 34$. ■

## A Three-Variable Problem

Lagrange's method generalizes to three or more variables in the obvious way.

### EXAMPLE B

An open box with two partitions is to be made of cardboard as shown in Figure 26. Maximize the total volume subject to the constraint that only 216 square inches of cardboard can be used.

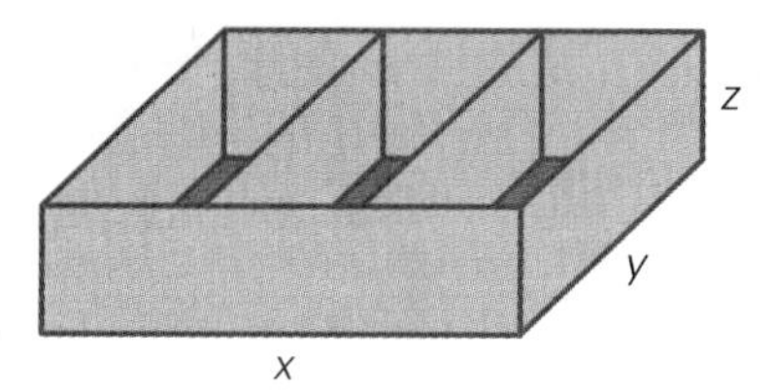


FIGURE 26

**Solution** Using the variables $x$, $y$, and $z$ as indicated in Figure 26, our job is to maximize $V = xyz$ subject to $xy + 2xz + 4yz = 216$, that is, subject to

$$g(x, y, z) = xy + 2xz + 4yz - 216 = 0$$

Let

$$F(x, y, z, \lambda) = xyz + \lambda(xy + 2xz + 4yz - 216)$$

and consider the four equations:

(i) $F_x = yz + \lambda(y + 2z) = 0$
(ii) $F_y = xz + \lambda(x + 4z) = 0$
(iii) $F_z = xy + \lambda(2x + 4y) = 0$
(iv) $F_\lambda = xy + 2xz + 4yz - 216 = 0$

Solving the first two equations for $\lambda$ gives

$$\lambda = \frac{-yz}{y + 2z} \qquad \lambda = \frac{-xz}{x + 4z}$$

$$\frac{-yz}{y + 2z} = \frac{-xz}{x + 4z}$$

$$-xyz - 4yz^2 = -xyz - 2xz^2$$

$$4yz^2 = 2xz^2$$

$$2y = x \qquad \text{(we can assume } z \neq 0\text{)}$$

Similarly, from the second and third equations, we obtain

$$\lambda = \frac{-xz}{x + 4z} \qquad \lambda = \frac{-xy}{2x + 4y}$$

$$\frac{-xz}{x + 4z} = \frac{-xy}{2x + 4y}$$

$$-2x^2z - 4xyz = -x^2y - 4xyz$$

$$2x^2z = x^2y$$

$$z = \frac{y}{2} \qquad \text{(we can assume } x \neq 0\text{)}$$

**CAUTION**
Multiple-variable problems are both messy and difficult, as Examples B and C illustrate. Moreover, we have glossed over important points. How can we be sure in Example B that (12, 6, 3) gives the maximum and not a saddle point? At this stage, we are not going to give a definitive answer. Rather, we ask you to use your physical intuition and note that (12, 6, 3) are reasonable dimensions for the maximum volume box.

Substituting $x = 2y$ and $z = y/2$ in the fourth equation gives

$$(2y)y + 2(2y)(y/2) + 4y(y/2) = 216$$

$$6y^2 = 216$$

$$y^2 = 36$$

We conclude that $y = 6$, $x = 12$, and $z = 3$, and $V = xyz = 216$ cubic inches. ■

## An Economic Application

We introduced the concept of a Cobb-Douglas production function in Section 10.1.

### EXAMPLE C

Suppose the Cobb-Douglas production function for a certain manufacturing process is given by

$$z = f(x, y) = 200x^{3/4}y^{1/4}$$

where $x$ denotes the number of units of labor and $y$ the number of units of capital. If a unit of labor costs \$250 and a unit of capital is \$400 and if total expenditures are limited to \$120,000, find the maximum production level.

**Solution** We want to maximize $f(x, y)$ subject to the constraint $250x + 400y = 120{,}000$, that is, $120{,}000 - 250x - 400y = 0$. Let

$$F(x, y, \lambda) = 200x^{3/4}y^{1/4} + \lambda(120{,}000 - 250x - 400y)$$

and consider the three equations:

$$F_x = 150x^{-1/4}y^{1/4} - 250\lambda = 0$$

$$F_y = 50x^{3/4}y^{-3/4} - 400\lambda = 0$$

$$F_\lambda = 120{,}000 - 250x - 400y = 0$$

If we solve for $\lambda$ in the first two equations, we obtain

$$\lambda = \tfrac{3}{5}x^{-1/4}y^{1/4}$$

and

$$\lambda = \tfrac{1}{8}x^{3/4}y^{-3/4}$$

and therefore

$$\tfrac{3}{5}x^{-1/4}y^{1/4} = \tfrac{1}{8}x^{3/4}y^{-3/4}$$

Multiplying both sides by $x^{1/4}y^{3/4}$ yields

$$\tfrac{3}{5}y = \tfrac{1}{8}x$$

or

$$x = \tfrac{24}{5}y$$

This expression for $x$ can be substituted into the third equation, giving

$$120{,}000 - 250(\tfrac{24}{5})y - 400y = 0$$

$$120{,}000 = 1600y$$

$$75 = y$$

It follows that $x = 360$, and the maximum number of units of production is

$$f(360, 75) = 200(360)^{3/4}(75)^{1/4} \approx 48{,}643$$ ■

In the example just completed,

$$\lambda = (\tfrac{3}{5})(360)^{-1/4}(75)^{1/4} \approx 0.405$$

This value of the Lagrange multiplier $\lambda$ has a nice economic interpretation. It can be shown that the magnitude of $\lambda$ gives the **marginal productivity of money.** One additional dollar spent on production will yield 0.405 additional units of product.

## Problem Set 10.4

*In Problems 1–4, find the minimum value of f(x, y) subject to the constraint g(x, y) = 0 (see Example A).*

1. $f(x, y) = x^2 + 2y^2 + 4$; $g(x, y) = 2x + y - 18$
2. $f(x, y) = 2x^2 + 3y^2 + 12$; $g(x, y) = x + 3y - 14$
3. $f(x, y) = 2x^2 - xy + 4y^2 + 100$; $g(x, y) = 2x + 5y - 76$
4. $f(x, y) = x^2 + 3xy + 4y^2 - 35$; $g(x, y) = 3x + y - 16$
5. Find the maximum and minimum values of $f(x, y) = xy + 10$ subject to the constraint $x^2 + 4y^2 = 32$.
6. Find the maximum and minimum values of $f(x, y) = x^2y$ subject to the constraint $x^2 + y^2 = 20$.
7. A farmer has 300 meters of fence with which to make a rectangular three-pen enclosure as shown in Figure 27. Find the dimensions of the enclosure for which the total area is a maximum.

FIGURE 27

8. Suppose that the long side of the enclosure in Problem 7 is along a building and so does not require fencing. Find the dimensions of the enclosure that give the largest total area.

*In Problems 9–12, find the minimum value of f(x, y, z) subject to the constraint g(x, y, z) = 0 as illustrated in Example B.*

9. $f(x, y, z) = x^2 + y^2 + z^2$; $g(x, y, z) = 2x + y + 3z - 14$
10. $f(x, y, z) = x^2 + 2y^2 + 3z^2 + 20$; $g(x, y, z) = x + y + z - 44$
11. $f(x, y, z) = 2x + y + 4z$; $g(x, y, z) = x^2 + y^2 + z^2 - 84$
12. $f(x, y, z) = xyz$; $g(x, y, z) = x^2 + y^2 + z^2 - 3$
13. A rectangular box with a lid is to be constructed from 36 square feet of material. If the bottom requires a double thickness while the top and sides require only a single thickness, find the dimensions of the box of maximum volume.
14. A rectangular box is to have a volume of 252 cubic feet. The cost of construction is $2 per square foot for the lid, $3 per square foot for the sides, and $5 per square foot for the bottom. Find the dimensions of the box which minimize the cost of construction.
15. A manufacturing process has Cobb-Douglas production function

$$z = f(x, y) = 300x^{2/3}y^{1/3}$$

where $x$ denotes the number of units of labor and $y$ the number of units of capital. A unit of labor costs $400, a unit of capital is $500, and total expenditures for labor and capital are limited to $180,000. Find the maximum production level and the marginal productivity of money. See Example C.

16. Follow the instructions for Problem 15 for the production function

$$z = f(x, y) = 600x^{3/4}y^{1/4}$$

given that a unit of labor costs $600 and a unit of capital is $400 and the total expenditures are limited to $240,000.

---

17. A standard sheet of paper measures 8.5 inches by 11 inches and so has a perimeter of 39 inches. What dimensions, subject to having a perimeter of 39 inches, would give the maximum area?
18. A can is to hold 1 liter (1000 cubic centimeters) of oil. Determine the dimensions of the can so that the surface area is a minimum.
19. Find the least distance between the origin and the plane $2x + 4y - 6z = 14$. *Hint:* Minimize the square of the distance, namely, $x^2 + y^2 + z^2$.
20. Consider a rectangular box with three of its faces in the coordinate planes, one vertex at (0, 0, 0), and the diagonally opposite vertex having positive coordinates and lying in the plane $x/a + y/b + z/c = 1$ ($a$, $b$, and $c$ all positive). Find the maximum volume of the box.
21. A manufacturer makes three types of jackets for which the profits per jacket are $12, $14, and $20, respectively. Suppose that the daily production levels, $x$, $y$, and $z$, for the three jackets must satisfy the constraint $x^2 + y^2 + z^2 = 740$. Find $x$, $y$, and $z$ for maximum daily profit and calculate this profit.
22. A firm with $92,000 available for advertising each month will apportion $x$ thousand dollars to television, $y$ thousand dollars to radio, and $z$ thousand dollars to

newspapers. It estimates the corresponding monthly profit $P$ thousand dollars to be given by

$$P = P(x, y, z)$$
$$= x^2 + y^2 + z^2 - 2x - 4y - 4z + 20$$

Find $x$, $y$, and $z$ that will maximize profit.

23. If the temperature $T$ in degrees Celsius at any point $(x, y, z)$ in space is given by the formula $T = xy + xz$, find the hottest point on the sphere $x^2 + y^2 + z^2 = 25$.

24. Let the Cobb-Douglas production function for a manufacturing process be given by $z = f(x, y) = kx^a y^b$, where $a + b = 1$ ($a > 0$, $b > 0$), with the associated cost constraint $cx + dy = M$. Show that $z$ is maximized when $x = aM/c$ and $y = bM/d$.

## 10.5 * APPLICATION: THE LEAST SQUARES LINE

Surely you have wondered how social scientists obtain the mathematical models they use. How, for example, can an economist claim that the relation between the demand $x$ for a certain product and its unit price $p$ is given by

$$x = -0.345p + 400$$

A serious economist will have looked at the available data (or perhaps gathered new data) on the sales of the product at various prices. When plotted in the $px$-plane, the data typically look something like Figure 28. Often there is a tendency for them to lie along a line (as there is in Figure 28). The economist's problem reduces to finding the line that best fits the data (Figure 29). The problem of finding a line (or, more generally, a curve of some special type) that best fits a set of data is one of the most important problems in the sciences.

But what do we mean by *best*? What is the criterion for saying that a line best fits a set of data? There is general agreement that we should choose the line that minimizes the sum of the squares of the vertical deviations of the data from the line, hence the name **least squares line.**

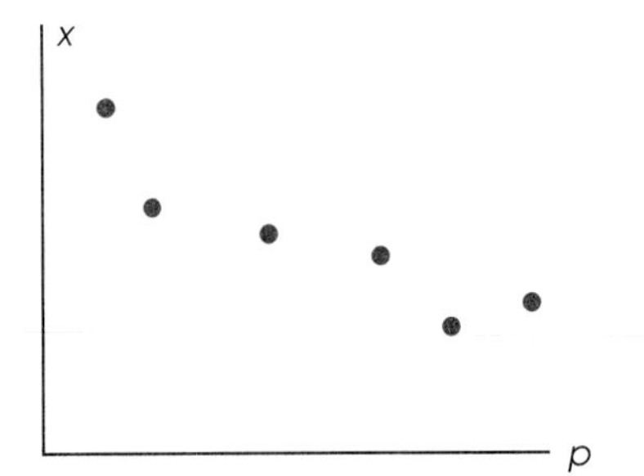


FIGURE 28

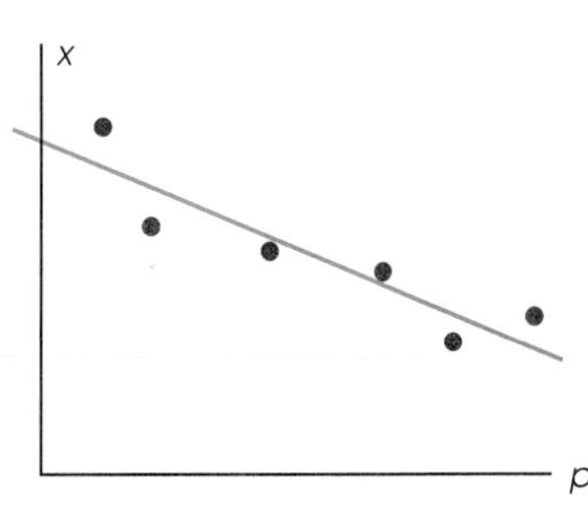


FIGURE 29

## A Simple Example

To get a handle on the mathematics involved in least squares, we begin with an example that has only three data points: (1, 2), (2, 5), and (3, 4). Our task is to choose $m$ and $b$ so as to minimize $S = d_1^2 + d_2^2 + d_3^2$, where $d_1$, $d_2$, and $d_3$ are the vertical deviations of the data from the line $y = mx + b$ (Figure 30).

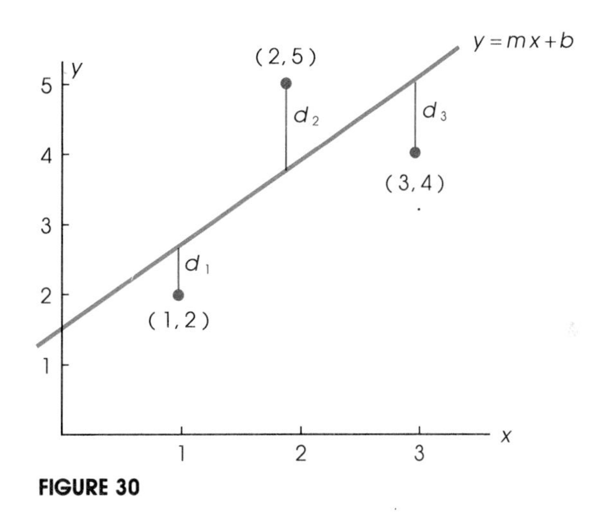


**FIGURE 30**

The point on the line with $x$-coordinate 1 has $y$-coordinate $m(1) + b$, and so

$$d_1^2 = (m + b - 2)^2$$

Similarly,

$$d_2^2 = (2m + b - 5)^2$$

and

$$d_3^2 = (3m + b - 4)^2$$

Thus,

$$S = (m + b - 2)^2 + (2m + b - 5)^2 + (3m + b - 4)^2$$

We want to minimize $S$, a function of the two variables $m$ and $b$. The theory in Section 10.3 applies: Take partial derivatives and set them equal to 0.

$$S_m = 2(m + b - 2) + 4(2m + b - 5) + 6(3m + b - 4) = 0$$

$$S_b = 2(m + b - 2) + 2(2m + b - 5) + 2(3m + b - 4) = 0$$

This pair of equations simplifies to

$$28m + 12b = 48$$

$$12m + 6b = 22$$

These equations are easily solved, giving $m = 1$ and $b = \frac{5}{3}$. Since there is only one critical point, we are confident that it produces the minimum $S$. The least squares line is $y = x + \frac{5}{3}$.

## The General Case

Consider now a set of $n$ data points $(x_1, y_1), (x_2, y_2), (x_3, y_3), \ldots, (x_n, y_n)$. The corresponding sum of squares $S$ to be minimized is

$$S = d_1^2 + d_2^2 + d_3^2 + \cdots + d_n^2$$

where

$$d_i = mx_i + b - y_i$$

To simplify our work, we are going to use the summation notation introduced in Section 8.4, namely, $\Sigma$. And since all of our sums will have $i$ running from 1 to $n$, we write simply $\Sigma$ rather than the more complete $\Sigma_{i=1}^{n}$. We will also make use of the properties of $\Sigma$ discussed in Section 8.4. Get ready for a bit of fancy mathematical footwork.

$$S = \sum d_i^2 = \sum (mx_i + b - y_i)^2$$

Thus,

$$\begin{aligned} S_m &= \sum 2x_i(mx_i + b - y_i) \\ &= \sum (2mx_i^2 + 2bx_i - 2x_i y_i) \\ &= 2m \sum x_i^2 + 2b \sum x_i - 2 \sum x_i y_i \end{aligned}$$

Also,

$$\begin{aligned} S_b &= \sum 2(mx_i + b - y_i) \\ &= \sum (2mx_i + 2b - 2y_i) \\ &= 2m \sum x_i + 2b(n) - 2 \sum y_i \end{aligned}$$

Setting $S_m$ and $S_b$ equal to 0 and dividing by 2 gives the pair of equations

$$\left(\sum x_i^2\right)m + \left(\sum x_i\right)b = \sum x_i y_i$$

$$\left(\sum x_i\right)m + nb = \sum y_i$$

When these equations are solved for $m$ and $b$, we obtain the following boxed formulas. To use them, $m$ must be calculated first, since the formula for $b$ involves $m$.

**POINT OF INTEREST**

The important task of finding least squares lines is routinely done with computer software packages such as Minitab and SPSS. Note that the formulas we give for $m$ and $b$ are easy to implement on a computer and a computer doesn't much care whether there are 10 or 10,000 pairs of data.

$$m = \frac{(\Sigma x_i)(\Sigma y_i) - n(\Sigma x_i y_i)}{(\Sigma x_i)^2 - n(\Sigma x_i^2)} \qquad b = \frac{\Sigma y_i - m(\Sigma x_i)}{n}$$

We summarize. The least squares line for a set of data $(x_1, y_1)$, $(x_2, y_2)$, . . . , $(x_n, y_n)$ is $y = mx + b$, where $m$ and $b$ are given by the formulas in the box. We mention that the least squares line is also called the **regression line of $y$ on $x$.**

## Predicting

A common use of the least squares line is in predicting the future based on data collected in the past. To cut down on the arithmetic, we illustrate with an example that uses only 10 pairs of data, but you can imagine that $n$ is 1000 rather than $n = 10$.

### EXAMPLE A

Suppose that we have collected the data shown in the left two columns in Figure 31. The value of $x_i$ indicates high school percentile rank in graduating class. The corresponding $y_i$ is the grade point average in the first term of college (A = 4, B = 3, C = 2, D = 1, and F = 0). Use these data to find the least squares line, and use it to make your best prediction for the college calculus grade of a student who had a high school percentile rank of 76.

| | $x_i$ | $y_i$ | $x_i^2$ | $x_i y_i$ |
|---|---|---|---|---|
| | 80 | 3.0 | 6400 | 240.0 |
| | 72 | 2.6 | 5184 | 187.2 |
| | 51 | 2.2 | 2601 | 112.2 |
| | 90 | 3.3 | 8100 | 297.0 |
| | 95 | 3.5 | 9025 | 332.5 |
| | 99 | 3.1 | 9801 | 306.9 |
| | 82 | 2.9 | 6724 | 237.8 |
| | 85 | 3.0 | 7225 | 255.0 |
| | 77 | 2.5 | 5929 | 192.5 |
| | 64 | 2.0 | 4096 | 128.0 |
| $\Sigma$ | 795 | 28.1 | 65,085 | 2289.1 |

**FIGURE 31**

**Solution** In the table in Figure 31, we have arranged our data in a form convenient for using the boxed formulas for $m$ and $b$. Note the calculations of $\Sigma x_i$, $\Sigma y_i$, $\Sigma x_i^2$, and $\Sigma x_i y_i$ obtained by adding the corresponding columns.

We conclude that

$$m = \frac{(795)(28.1) - (10)(2289.1)}{(795)^2 - (10)(65085)} \approx 0.029$$

and

$$b = \frac{(28.1) - (0.029)(795)}{10} \approx 0.505$$

Thus, the least squares line is $y = 0.029x + 0.505$, and a student with a high school percentile rank of 76 can expect a grade point average of

$$y = (0.029)(76) + 0.505 \approx 2.7$$

■

You can see from Example A that finding $m$ and $b$ involves a large amount of calculation. Fortunately, computers can easily be programmed to do these calculations, and even some hand-held calculators have buttons that give $m$ and $b$ automatically as soon as the $xy$-data are entered.

## Problem Set 10.5

*In Problems 1–4, find the least squares line for the given data by going through the following steps.*

*(i) Determine the sum $S$ of the squares of the deviations of the given data points from the line $y = mx + b$. For example, in Problem 1,*

$$S = (m + b - 3)^2 + (4m + b - 2)^2 + (5m + b - 2)^2$$

*(ii) Find the partial derivatives $S_m$ and $S_b$ and set them equal to 0.*

*(iii) Solve the resulting pair of equations simultaneously.*

1. (1, 3), (4, 2), (5, 2)
2. (1, 3), (4, 5), (6, 8)
3. (−1, 1) (1, 2), (2, 2), (5, 3)
4. (−1, 2), (1, 2), (2, 1), (5, 0)

*In Problems 5–8, use the formulas for m and b given in the text to find the equation of the least squares line for the given set of data.*

5. (1, 1), (3, 2), (4, 5), (5, 5)
6. (1, 3), (3, 2), (4, 1), (6, 2)
7. (2, 3), (3, 4), (4, 3), (5, 4), (6, 7)
8. (2, 5), (4, 4), (4, 2), (5, 1), (7, 0)
9. Table 1 gives the per capita yearly income $y$ in thousands of dollars for a certain country in the years 1960, 1965, and so on. Note that $x$ represents the number of years after 1960.
   (a) Find the least squares line for this set of data.
   (b) Use this equation to estimate the per capita income in 1978.
   (c) Predict the per capita income in the year 2000.

**TABLE 1**

| Year | $x$ | $y$ |
|---|---|---|
| 1960 | 0 | 1.8 |
| 1965 | 5 | 2.1 |
| 1970 | 10 | 2.6 |
| 1975 | 15 | 3.0 |
| 1980 | 20 | 4.2 |
| 1985 | 25 | 5.6 |

10. Table 2 gives the amount $y$ in billions of dollars that a certain country spent on new factories and equipment during the years 1970, 1975, and so on. Note that $x = 0$ corresponds to 1970, and $x = 1$ to 1975 (what value of $x$ corresponds to 1971?). Find the least squares line for these data and use it to predict the expenditures for new factories and equipment during 2000 and 2002.

**TABLE 2**

| *Year* | $x$ | $y$ |
|---|---|---|
| 1970 | 0 | 1.8 |
| 1975 | 1 | 2.1 |
| 1980 | 2 | 2.4 |
| 1985 | 3 | 3.2 |
| 1990 | 4 | 3.8 |

11. Table 3 shows the appraised value in thousands of dollars of a house in various years. Find a least squares line for these data and use it to predict the value of the house in the year 2000. *Hint:* See Problems 9 and 10 for two possible ways of introducing variables $x$ and $y$.

**TABLE 3**

| *Year* | *Value* |
|---|---|
| 1975 | 180 |
| 1980 | 195 |
| 1985 | 215 |
| 1990 | 225 |

12. Table 4 reports the number of new cases of flu in a certain city. Use these data and a least squares line to predict the number of new cases of flu on day 9.

**TABLE 4**

| $x$ *(day number)* | 1 | 2 | 3 | 4 | 5 | 6 | 7 |
|---|---|---|---|---|---|---|---|
| $y$ *(new cases)* | 10 | 15 | 18 | 21 | 24 | 33 | 35 |

---

C 13. A company wishes to determine the demand equation for one of its products. At four outlets, A, B, C, and D, located in comparable communities, the number $x$ of units sold in 1 month with the product priced at $p$ dollars was shown in Table 5. Find the linear demand equation which best fits these data in the least squares sense and use the result to predict the demand if the price is set at $12.

**TABLE 5**

| *Price* $p$ | *Demand* $x$ |
|---|---|
| 10.00 | 1200 |
| 10.50 | 1100 |
| 11.50 | 880 |
| 13.50 | 720 |

C 14. A stage show has run for 4 weeks with the attendance as shown in Table 6. Find the equation of the least squares line $y = mx + b$ for these data, with $x$ being the week number and $y$ the attendance for the week. Then predict the week in which the attendance will fall below 2500.

**TABLE 6**

| *Week no.* | *Attendance* |
|---|---|
| 1 | 6400 |
| 2 | 5400 |
| 3 | 4600 |
| 4 | 4100 |

C 15. Find the curve $y = ae^{mx}$ that fits the data in Table 7 best in the least squares sense. *Hint:* The equation $y = ae^{mx}$ is equivalent to $\ln y = \ln a + mx$. The substitutions $Y = \ln y$ and $b = \ln a$ transform this equation to the linear equation $Y = mx + b$. Use the method of least squares to find the values of $m$ and $b$ that give the best fit to the data pairs $(x, Y) = (x, \ln y)$. Then determine $a$ and write the corresponding equation $y = ae^{mx}$.

**TABLE 7**

| $x$ | $y$ |
|---|---|
| 1 | 3.1 |
| 2 | 6.3 |
| 3 | 11.9 |
| 4 | 24.1 |
| 5 | 47.8 |

C 16. A certain city spent for services (police, fire, administration) the amount $y$ (in millions of dollars) shown in Table 8 for the years 1970, 1975, and so on. Find the equation of the form $y = ae^{mx}$ that best fits these data and use it to predict expenditures in the year 2000.

**TABLE 8**

| *Year* | $x$ *(year no.)* | $y$ *(expenditure)* |
|---|---|---|
| 1970 | 0 | 40 |
| 1975 | 1 | 44 |
| 1980 | 2 | 49 |
| 1985 | 3 | 54 |
| 1990 | 4 | 60 |

# 10.6*
# DOUBLE INTEGRALS

Our goal for this section is to extend integration theory to functions $f(x, y)$ of two variables. It would be well to review Chapter 8, for it is that material we wish to generalize. Recall that for $f(x) \geq 0$, $\int_a^b f(x)\,dx$ is defined to be the area of the region under the curve $y = f(x)$ between $a$ and $b$ (Figure 32).

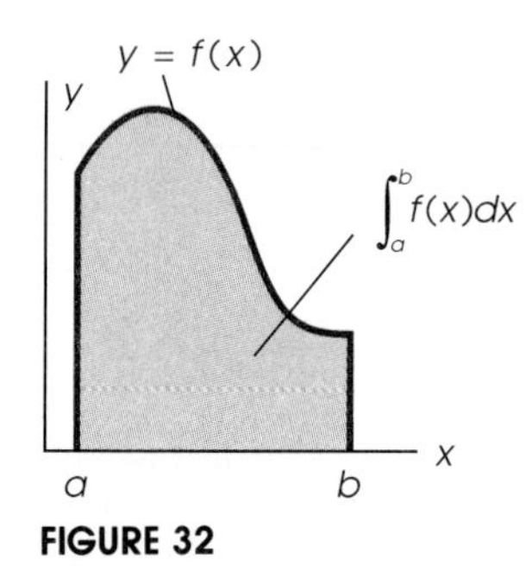


FIGURE 32

Let $f(x, y) \geq 0$ on a domain $D$. Then the graph of $z = f(x, y)$ is a surface and we define the **double integral of $f(x, y)$ over $D$** to be the volume of the solid region under this surface and above the domain $D$ (Figure 33). Our symbol for this double integral is

$$\iint_D f(x, y)\,dy\,dx$$

The basic tool for evaluating integrals is the Fundamental Theorem of Calculus. But for double integrals, we also need the concepts of partial integrals and iterated integrals. We digress to discuss these ideas.

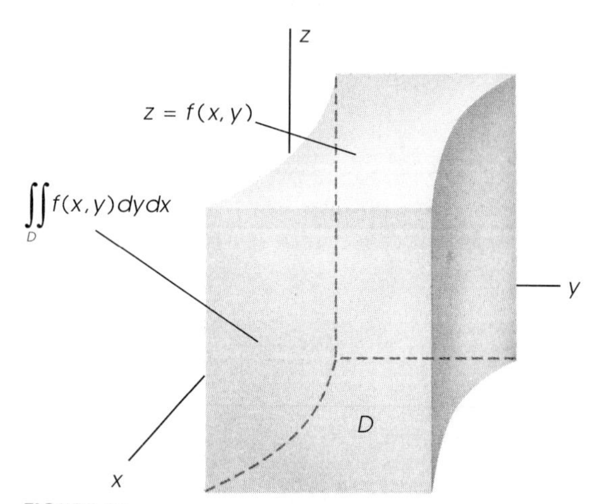


FIGURE 33

## Partial Integrals and Iterated Integrals

Consider the symbol $\int_a^b f(x, y)\, dy$. We call it a **partial integral** because, though the integrand is a function of two variables, only one integration is indicated. By analogy with partial derivatives, we are to treat $x$ as a constant and integrate with respect to $y$ (the presence of $dy$ tells us that $y$ is the variable of integration).

### EXAMPLE A

Evaluate the partial integral $\displaystyle\int_1^2 3xy^2\, dy$.

**Solution** We treat $x$ as a constant, so $3x$ can be passed across the integral sign.

$$\int_1^2 3xy^2\, dy = 3x \int_1^2 y^2\, dy = 3x \left[\frac{y^3}{3}\right]_1^2 = 3x\left(\frac{8}{3} - \frac{1}{3}\right) = 7x$$

We say that we integrated out the variable $y$, leaving an answer that involves $x$. ■

Our first example had the constant limits 1 and 2 on the integral sign. However, since $x$ is being treated as a constant, these limits could just as well be functions of $x$. We illustrate.

### EXAMPLE B

Evaluate $\displaystyle\int_{x^2}^{2x} 3xy^2\, dy$.

**Solution**

$$\begin{aligned}\int_{x^2}^{2x} 3xy^2\, dy &= 3x \int_{x^2}^{2x} y^2\, dy = 3x\left[\frac{y^3}{3}\right]_{x^2}^{2x} \\ &= 3x\left[\frac{(2x)^3}{3} - \frac{(x^2)^3}{3}\right] \\ &= 8x^4 - x^7\end{aligned}$$

■

Now we are ready for the concept of an **iterated integral.** The symbol we use is

$$\int_a^b \left[\int_{g(x)}^{b(x)} f(x, y)\, dy\right] dx$$

We perform the partial integration with respect to $y$ first, which results in a function of $x$. Then we integrate this function of $x$ between the constants $a$ and $b$, which results in a numerical answer.

### EXAMPLE C

Evaluate $\int_0^2 \left[\int_{x^2}^{2x} 3xy^2\, dy\right] dx$.

**Solution** We evaluated the inner partial integral in Example B. Thus,

$$\int_0^2 \left[\int_{x^2}^{2x} 3xy^2\, dy\right] dx = \int_0^2 (8x^4 - x^7)\, dx$$
$$= \left[\frac{8x^5}{5} - \frac{x^8}{8}\right]_0^2$$
$$= \frac{256}{5} - \frac{256}{8} = \frac{96}{5}$$ ■

There is no reason why $y$ has to be the first variable of integration; it can just as well be $x$, as we now illustrate.

### EXAMPLE D

Evaluate $\int_1^2 \left[\int_0^y (xy + 4x^3)\, dx\right] dy$.

**Solution** We work on the inner partial integral first.

$$\int_0^y (xy + 4x^3)\, dx = \left[\frac{yx^2}{2} + x^4\right]_0^y = \frac{y^3}{2} + y^4$$

Thus,

$$\int_1^2 \left[\int_0^y (xy + 4x^3)\, dx\right] dy = \int_1^2 \left(\frac{y^3}{2} + y^4\right) dy$$
$$= \left[\frac{y^4}{8} + \frac{y^5}{5}\right]_1^2$$
$$= 2 + \frac{32}{5} - \left(\frac{1}{8} + \frac{1}{5}\right) = \frac{323}{40}$$ ■

## Evaluation of Double Integrals

We return to the main problem—how to evaluate the double integral

$$\iint_D f(x, y)\, dy\, dx$$

As we said earlier, this integral represents the volume of the solid region under the surface $z = f(x, y)$ and above the region $D$ in the $xy$-plane [see Figure 33 and remember that we are assuming $f(x, y) \geq 0$]. We now point out that this volume can also be represented as an iterated integral. We suppose that the left boundary of $D$ is the graph of $y = g(x)$ and the right

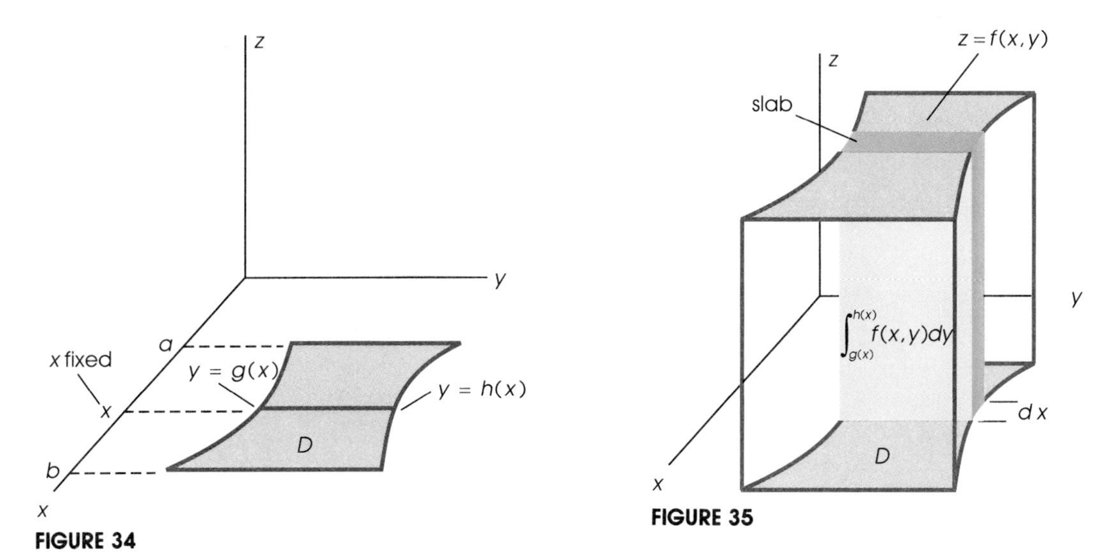


FIGURE 34

FIGURE 35

boundary is the graph of $y = h(x)$ (Figure 34). Then, if $x$ is fixed, $\int_{g(x)}^{h(x)} f(x, y)\, dy$ is the area of a cross section of the solid determined by the plane at $x$ perpendicular to the $x$-axis. If we multiply this area by $dx$, we get (approximately) the volume of a slab (Figure 35), and if we add the volumes of these slabs (integrate), we get the total volume of the solid. Thus, the double integral can be expressed as an iterated integral; that is,

$$\iint_D f(x, y)\, dy\, dx = \int_a^b \left[\int_{g(x)}^{h(x)} f(x, y)\, dy\right] dx$$

## EXAMPLE E

Find the volume of the solid under the surface $z = 3xy^2$ and above the region $D$ in the $xy$-plane bounded by $y = x^2$ and $y = 2x$ (Figure 36).

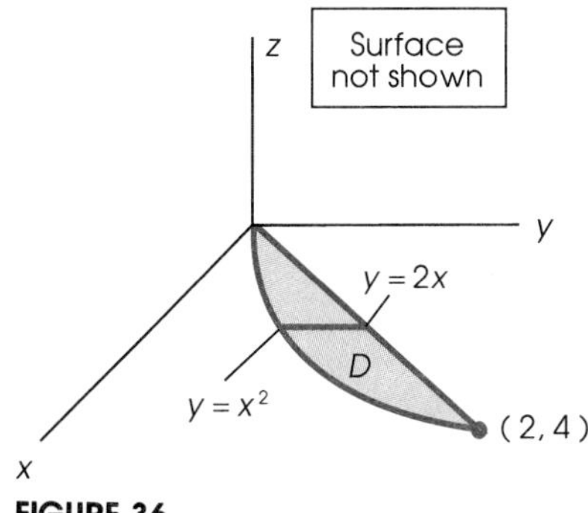


FIGURE 36

**Solution** This volume is $\iint_D 3xy^2\,dy\,dx$. Moreover, this double integral can be written as an iterated integral, in fact, the iterated integral in Example C. The answer is $\frac{96}{5}$. ■

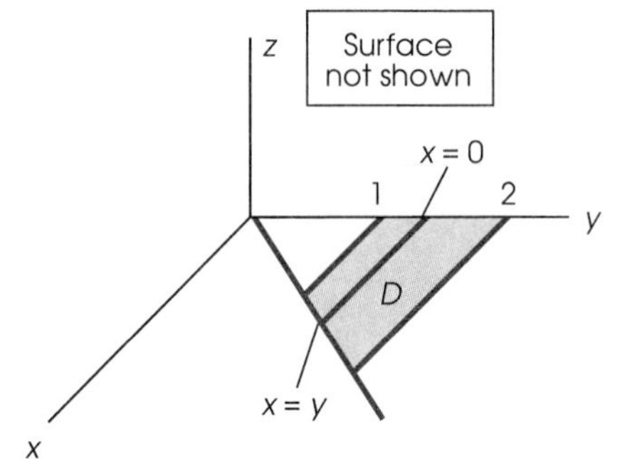


FIGURE 37

### EXAMPLE F

Evaluate $\iint_D (xy + 4x^3)\,dy\,dx$, where $D$ is the region in the $xy$-plane bounded by $x = y$ and $x = 0$ between $y = 1$ and $y = 2$ (Figure 37).

**Solution** This corresponds exactly to the iterated integral in Example D. The answer is $\frac{323}{40}$. ■

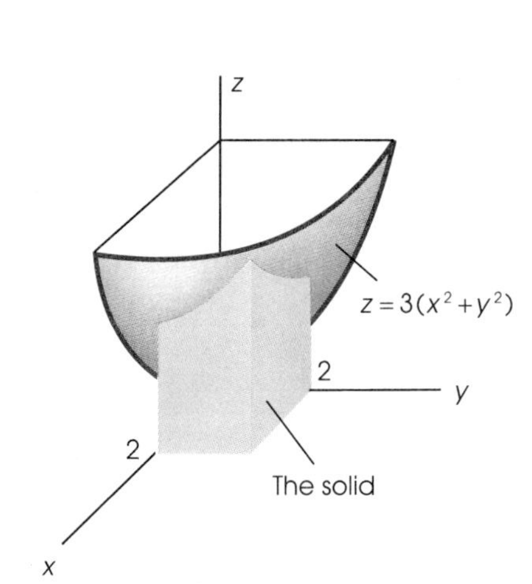


FIGURE 38

### EXAMPLE G

Find the volume of the solid (Figure 38) under the surface $z = 3(x^2 + y^2)$ and above the first quadrant square bounded by $x = 0$, $x = 2$, $y = 0$, and $y = 2$.

**Solution** We choose to keep $x$ fixed in the first integration, in which case $y$ ranges between $y = 0$ and $y = 2$. The appropriate iterated integral is therefore

$$\int_0^2 \left[\int_0^2 (3x^2 + 3y^2)\,dy\right] dx$$

The inner integral has the value

$$\left[3x^2y + y^3\right]_0^2 = 6x^2 + 8$$

Consequently, the value of the iterated integral reduces to

$$\int_0^2 (6x^2 + 8)\,dx = \left[2x^3 + 8x\right]_0^2 = 16 + 16 = 32$$ ■

## Problem Set 10.6

*Evaluate the partial integrals in Problems 1–14 as in Examples A and B.*

1. $\int_2^4 2x^3y\,dy$
2. $\int_0^3 3x^2y^2\,dy$
3. $\int_2^6 (3x + 2y)\,dx$
4. $\int_1^4 (5y - 2x)\,dx$
5. $\int_1^2 \frac{2x^2}{y + 1}\,dy$
6. $\int_0^2 (y^2 + x)^3 2y\,dy$
7. $\int_0^7 y^2e^{2x}\,dx$
8. $\int_0^1 (xy + e^{-x})\,dx$

9. $\int_0^{1/x} e^{xy}\,dy$

10. $\int_1^x \frac{x}{xy-1}\,dy$

11. $\int_{\sqrt{x}}^x 2x^{3/2}y\,dy$

12. $\int_y^{y^2} (\ln y)(2x+3)\,dx$

13. $\int_{2/x}^{4/x^2} \frac{x}{y^2}\,dy$

14. $\int_{2y}^{4y} \frac{y+2}{x}\,dx$

*Evaluate the iterated integrals in Problems 15–24 as in Examples C and D. Here the order dy dx or dx dy indicates the order of integration.*

15. $\int_1^3\int_0^2 12x^2y\,dy\,dx$

16. $\int_0^1\int_{-1}^2 6x^2y^3\,dy\,dx$

17. $\int_{-1}^0\int_0^3 2x(x+y)\,dx\,dy$

18. $\int_{-2}^1\int_0^2 (3x^2y+6xy^2)\,dx\,dy$

19. $\int_1^3\int_0^{\sqrt{x}} 4xy\,dy\,dx$

20. $\int_1^2\int_0^{x^2} 3x^2\sqrt{y}\,dy\,dx$

21. $\int_0^1\int_y^{y^2} (3x^2+y^2)\,dx\,dy$

22. $\int_0^2\int_{1-2y}^{-y} 20(2y+x)^3\,dx\,dy$

23. $\int_1^3\int_0^{x+1} \frac{y}{x+1}\,dy\,dx$

24. $\int_0^1\int_{4/x^2}^{1/x} x^2 e^{x^2y}\,dy\,dx$

*The area of a region D in the xy-plane is equal to the volume of the solid under the surface z = 1 and above the region D. Thus the area of D is given by the double integral*

$$A = \iint_D 1\,dy\,dx$$

*Write an iterated integral that represents the area of each region D in Problems 25–28. Then evaluate A.*

25.

y

y = x

D

$y = x^2$

x

26.

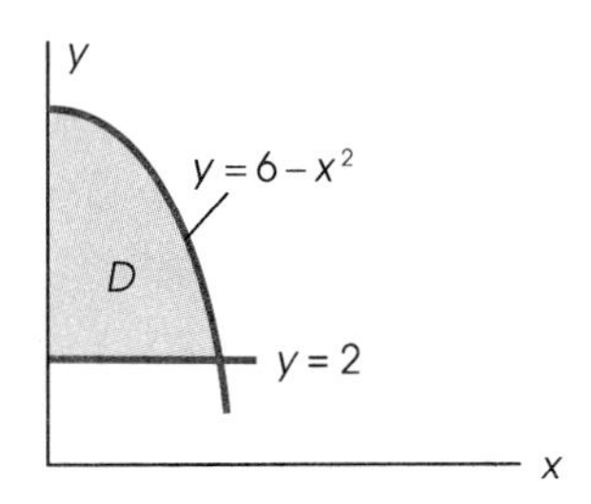


27.

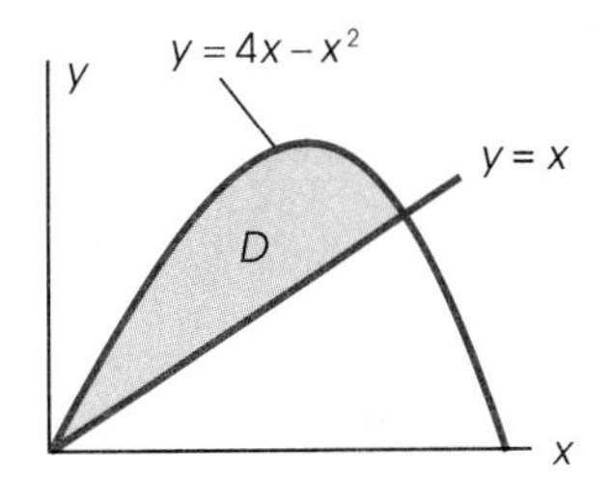


28.

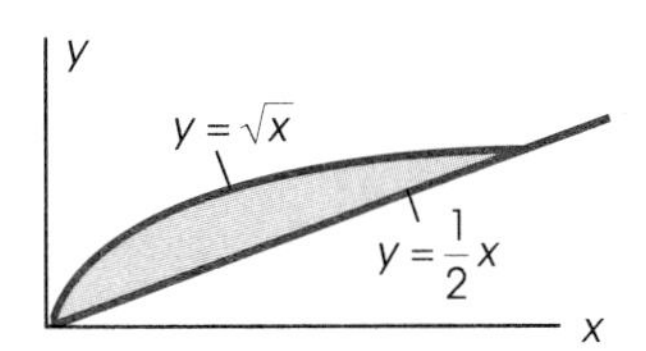


*In Problems 29–34, sketch the region D bounded by the given curves and then write an iterated integral that represents its area.*

29. $y = x$, $y = 2x$, and $x = 3$.
30. $y = x^3$, $y = 8$, and $x = 0$.
31. $y = x^2$ and $y = 8 - x^2$.
32. $y = x + 6$, $y = (x-2)^2$, $x = 1$, and $x = 4$.
33. $y = x^2 + x + 4$ and $y = 5x + 1$.
34. $y = 2x^2$ and $y = x^2 + 4$.

*In Problems 35–44, find the volume of the solid under the given surface z = f(x, y) and above the given region D in the xy-plane.*

35. $z = x^2 + y^2$; $D$ is the region in Problem 25.
36. $z = xy$; $D$ is the region in Problem 26.
37. $z = x^2 + 4y$; $D$ is the region in Problem 27.
38. $z = 2x^2 + e$; $D$ is the region in Problem 28.
39. $z = 3x\sqrt{y}$; $D$ is the region bounded by $x = 0$, $x = 2$, $y = 1$, and $y = 4$.

40. $z = x/y^2$; $D$ is the region bounded by $x = 0$, $x = 2$, $y = 1$, and $y = 3$.
41. $z = xe^y$; $D$ is the region bounded by $x = 0$, $y = 0$, $x = 2$ and $y = x + 1$.
42. $z = 6 - 2x - 3y$; $D$ is the region bounded by $2x + 3y = 6$, $x = 0$, and $y = 0$.
43. $z = x^2 + y^2$; $D$ is the region bounded by $y = x^2$ and $y = \sqrt{x}$.
44. $z = 2x$; $D$ is the region bounded by $x = 2 - y$, $x = \frac{1}{2}(2 - y)$, and $y = 0$.

---

*In Problems 45–48, find the volume of the first octant solid shown in the figure.*

45.

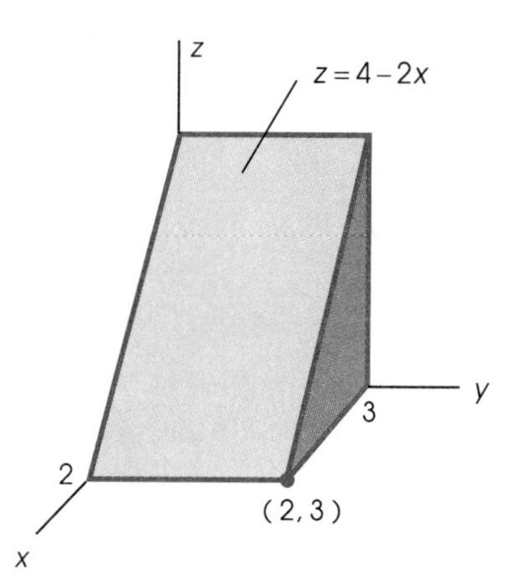


46.

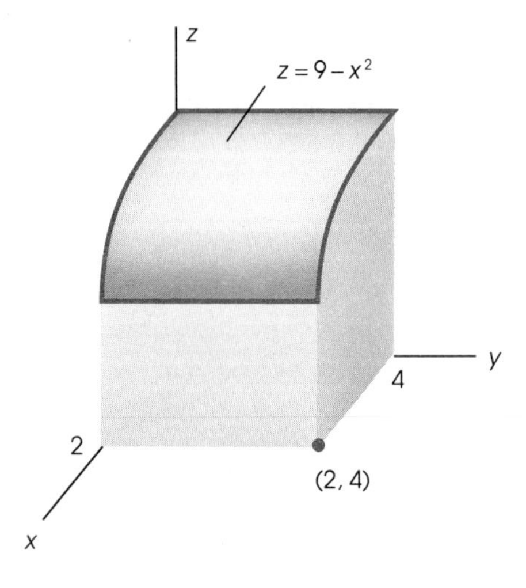


47.

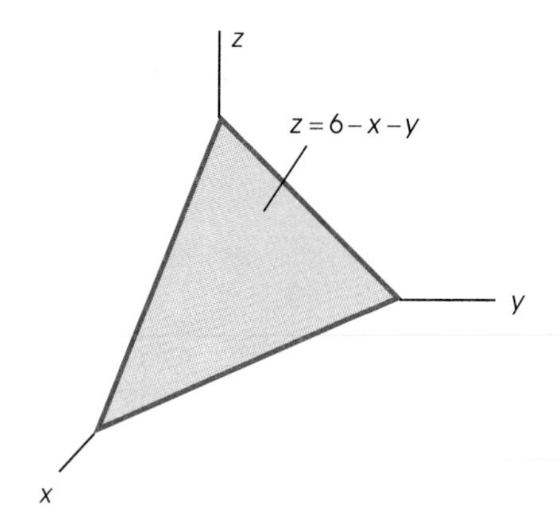


48.

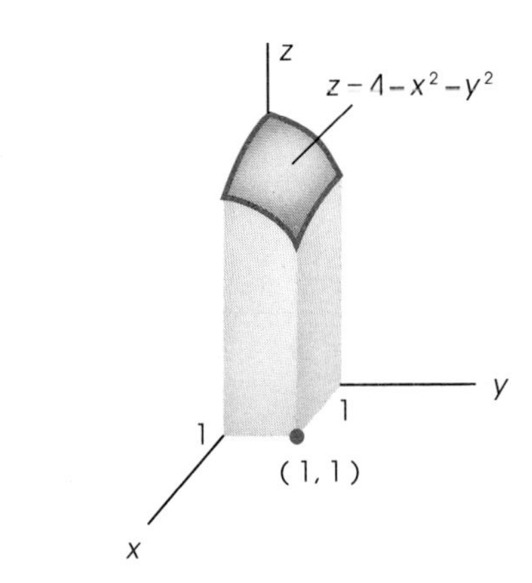


*The average value A of a function f(x, y) on a region D is defined by*

$$A = \iint_D f(x, y)\, dy\, dx/\text{area of } D$$

*Find the average value of the functions f(x, y) on region D in Problems 49–52.*

49. $f(x, y) = 3x^2 - 4y$; $D$ is the region in Problem 25.
50. $f(x, y) = 4x - 2y$; $D$ is the region in Problem 26.
51. $f(x, y) = -2x + 3$; $D$ is the region in Problem 27.
52. $f(x, y) = 4y - 5$; $D$ is the region in Problem 28.
53. A manufacturing company has two plants. The weekly cost function for the first plant at production level $x$ is $C_1(x) = 0.2x^2 + 40x + 3600$, and at the second plant at production level $y$ it is $C_2(y) = 0.4y^2 + 24y + 6000$. Find the average cost [that is, the average value of $C_1(x) + C_2(y)$] over the domain $300 \le x \le 400$, $350 \le y \le 550$.

54. A shed has a roof of parabolic shape $z = 30 - 10(x/15)^2$. The four sides of the shed have equations $x = -20$, $x = 20$, $y = 0$, and $y = 80$ (all dimensions in feet). Sketch a picture of the shed and compute its volume.

## CHAPTER 10 REVIEW PROBLEM SET

1. If $f(x, y, z) = x^2y^3z^4$, find and simplify:
   (a) $f(3, -1, -2)$ (b) $f(x, 2x, x/2)$
   (c) $f(2a, 2b, c/2)/f(a, b, c)$
2. If $f(x, y, z) = x^3yz^{1/3}$, find and simplify:
   (a) $f(2, 4, \frac{1}{8})$ (b) $f(y^2, y, y^6)$
   (c) $f(2a, 2b, 2c)/f(a, b, c)$
3. Let $f(r, h)$ denote the total surface area (inner plus outer) of a spherical shell of inner radius $r$ and thickness $h$. Write a formula for $f(r, h)$ and calculate $f(3, 0.5)$. *Hint:* The formula for the surface area of a sphere is $S = 4\pi r^2$.
4. Let $g(r, h)$ denote the volume of the spherical shell of Problem 3. Write a formula for $g(r, h)$ and calculate $g(4, 0.3)$. *Hint:* The formula for the volume of a solid sphere of radius $r$ is $V = \frac{4}{3}\pi r^3$.
5. Write the formula for the volume $g(a, b)$ of a prism of length $a$ if each cross section is an equilateral triangle of side $b$.
6. Write the formula for the volume of a prism of height $h$ if each cross section is a trapezoid whose parallel sides have lengths $a$ and $b$ and whose altitude is $d$.
7. Sketch the graphs of each of the following equations, showing the intercepts and appropriate traces.
   (a) $4x + y - 2z = 8$ (b) $x^2 + y^2 + z^2 = 16$
   (c) $x + z = 4$ (d) $z = 9 - x^2 - y^2$
8. Sketch the graph of each equation, showing the intercepts and appropriate traces.
   (a) $3x - 2y + 4z = 12$
   (b) $x^2 + y^2 + (z - 2)^2 = 4$
   (c) $4x + 5z = 20$ (d) $z = x^2 + y^2 - 9$
9. For each of the following, find $f_x(x, y)$ and $f_y(1, 0)$.
   (a) $f(x, y) = 2x^2 - 3xy + y^3$
   (b) $f(x, y) = x^2 \ln(2x + 3y)$
   (c) $f(x, y) = e^{x^2y}$
   (d) $f(x, y) = \left(3x - \dfrac{2y}{x}\right)^4$
10. For each of the following, find $f_y(x, y)$ and $f_x(0, 2)$.
    (a) $f(x, y) = (x - 2y)^3$ (b) $f(x, y) = y^2e^{2xy}$
    (c) $f(x, y) = y + 3\ln(x^2 + y^2 - x - 1)$
    (d) $f(x, y) = \left(xy + \dfrac{8}{y}\right)^4$
11. Find the slope of the tangent line at (2, 3) to the trace of $z = 25 - 4x^2 - y^2$ in the plane $x = 2$.
12. Find the slope of the tangent line at (3, 1) to the trace of the surface $z = x^2 - 3xy + 4y^2 + 12$ in the plane $y = 1$.
13. Suppose you ran down the hill represented by the surface $z = 12/(x^2 + y^2 + 1)^2$ directly above the line $y = 1$. What was the slope of your path at (2, 1)?
14. The plane $x = 2$ intersects the surface $z = 24e^{9-x^2-y^2}$ in a curve. What is the slope of this curve at (2, 2)?
15. [C] If $P$ dollars is invested for 20 years at the rate $r$ compounded annually, the accumulated amount at the end of that time will be $f(P, r) = P(1 + r)^{20}$. Calculate $f_P(2000, 0.08)$ and $f_r(2000, 0.08)$ and give interpretations of these numbers.
16. Suppose that the maximum safe load $S(x, y, z)$ for a horizontal beam supported at both ends is given by $S(x, y, z) = 1200xy^2/z$ where $x$ and $y$ are the width and depth in inches, $z$ is the length in feet, and $S$ is in pounds. Calculate $S_x(2, 4, 10)$ and $S_y(2, 4, 10)$ and interpret these numbers.
17. Find all critical points of $f(x, y) = x^2 + y^3 - 6xy$ and indicate which of them gives a local maximum, a local minimum, or a saddle point.
18. Find all critical points of $f(x, y) = x^3 + y^3 - 9xy$ and indicate which of them gives a local maximum, a local minimum, or a saddle point.
19. Find three positive numbers, $x$, $y$, and $z$, satisfying the condition $2x + y + z = 2$ for which the product $xyz$ is a maximum.
20. Find the dimensions of the box with maximum volume for which the length plus twice the width plus twice the height equals 24.
21. A manufacturing company sells two kinds of riding mowers at prices $p$ dollars and $q$ dollars with corresponding weekly demands

$$x = 300 - 0.5p + 0.3q$$

$$y = 400 + 0.2p - 0.4q$$

Determine $p$ and $q$ so as to maximize weekly revenue.
22. Find the coordinates of the point on the surface $z = 4 + x^2 + y^2$ that is nearest to the point (1, 2, 0). *Hint:* Minimize $d^2 = (x - 1)^2 + (y - 2)^2 + z^2$ with $(x, y, z)$ on the surface.

23. Use Lagrange's method to find the maximum value of
$$f(x, y) = xy$$
subject to the constraints $4x^2 + 9y^2 = 36$, $x > 0$, $y > 0$.
24. Use Lagrange's method to find the maximum value of $f(x, y) = x^2y$ subject to the constant $x^2 + y^2 = 27$.
25. Jane Kelly recently opened an ice cream parlor. The weekly revenue (in hundreds of dollars) for the first 6 weeks was as shown in the accompanying table.

| *x (week number)* | 1 | 2 | 3 | 4 | 5 | 6 |
|---|---|---|---|---|---|---|
| *y (revenue)* | 32 | 36 | 35 | 44 | 46 | 50 |

(a) Find the equation of the least squares line for these data.
(b) Use this equation to predict revenue for the tenth week.

26. For the first six months after Jim MacMillan opened an automotive center, the monthly profit (in hundreds of dollars) was as shown.

| *x (month number)* | 1 | 2 | 3 | 4 | 5 | 6 |
|---|---|---|---|---|---|---|
| *y (profit)* | 42 | 48 | 50 | 54 | 57 | 62 |

(a) Find the equation of the least squares line.
(b) Use this equation to predict the profit for the 12th month.

27. Evaluate each of the following iterated integrals.
(a) $\int_1^4 \left[\int_0^3 \frac{2y}{\sqrt{x}}\, dy\right] dx$
(b) $\int_0^1 \left[\int_{x^2}^{x} (3x^2 + 4xy)\, dy\right] dx$
28. Evaluate each iterated integral.
(a) $\int_0^3 \left[\int_1^{e^2} \frac{2x^2}{y}\, dy\right] dx$
(b) $\int_0^1 \left[\int_{4y}^{y^2} (6x - 3\sqrt{xy})\, dx\right] dy$
29. Sketch the region $D$ bounded by the curves $y = x^2 - 2$ and $y = 6 - x^2$. Then write an iterated integral that represents its area. *Hint:* The area of $D$ is the same as the volume of a cylinder with base $D$ and height 1.
30. Sketch the region $D$ bounded by the curve $y = 4x - x^2$ and the line $y = x$. Then write an iterated integral that represents its area.
31. Find the volume of the solid under the surface $z = (\ln x)/y$ and above the region in the $xy$-plane bounded by $y = 1$ and $y = e^x$ between $x = 1$ and $x = 3$.
32. Find the volume of the solid under the surface $4z = 12 - 2x - 3y$ and above the region in the $xy$-plane bounded by $x = 0$, $y = 0$, and $2x + 3y = 8$.
33. How would you define $f_x(a, b)$ as a limit?
34. Evaluate the partial integral $\int_2^3 f_y(2, y)\, dy$.
35. The maximum safe load $L$ that can be supported by a beam of width $w$, depth $d$, and length $s$ is given by $L = k(wd^2/s)$, where $k$ is a constant (Figure 39).

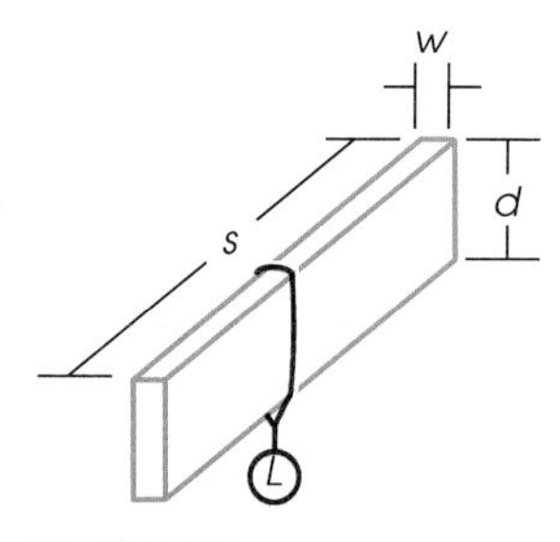


**FIGURE 39**

(a) Determine $k$ if a beam 2 inches wide, 1 foot deep, and 20 feet long, can safely support a load of 1400 pounds.
(b) Use a partial derivative to determine approximately what changing the dimensions from 2, 1, 20 to 2, 1, 21 does to the safe load.

36. In analogy with the limit definition of $\int_0^1 f(x)\, dx$, indicate how one might give a limit definition for the double integral $\iint_D f(x, y)\, dy\, dx$ where $D = \{(x, y)$: $0 \le x \le 1, 0 \le y \le 1\}$.
37. By changing the order of integration, evaluate
$$\int_0^4 \left[\int_{x/2}^{2} e^{y^2}\, dy\right] dx$$
*Hint:* Begin by sketching the domain of integration.
38. Write $\int_0^1 \left[\int_{x^2}^{x} f(x, y)\, dy\right] dx$ as an iterated integral with the opposite order of integration.

# Appendix

# Basic Algebra Review

## A1 NUMBERS

The fundamental characters in this book are the **real numbers.** Among the real numbers are the *integers:* . . . , −3, −2, −1, 0, 1, 2, 3, . . . ; the *rational numbers* which consist of ratios of integers such as $\frac{2}{3}$, $\frac{21}{13}$, $\frac{-7}{2}$, and $\frac{5}{1}$; and the *irrational numbers* such as $\sqrt{2}$, $\sqrt{3}$, $\pi$, and $\sqrt{5}/3$. The real numbers can be made to correspond to points on a line as indicated in Figure 1. Each point on the line has a real number label, and each real number determines a point on the line.

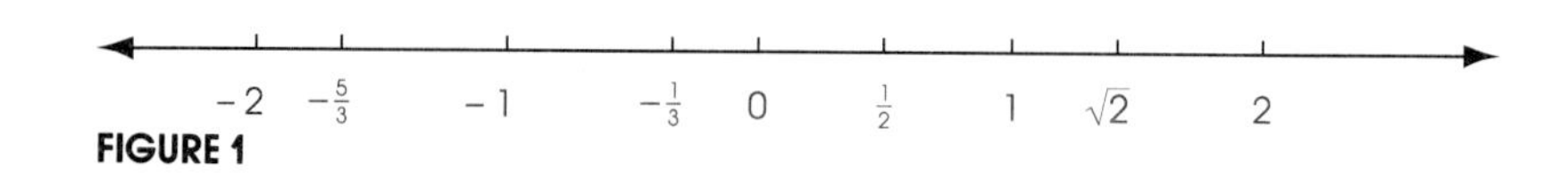


FIGURE 1

Another way to view the real numbers is as unending decimals. For example,

$$
\begin{aligned}
3 &= 3.0000\ldots &&= 3.\overline{0} \\
\tfrac{2}{3} &= 0.6666\ldots &&= 0.\overline{6} \\
\tfrac{122}{99} &= 1.232323\ldots &&= 1.\overline{23} \\
\sqrt{2} &= 1.4142135623\ldots \\
\pi &= 3.1415926535\ldots
\end{aligned}
$$

The bar over a group of digits indicates that this group of digits repeats indefinitely. The rational numbers correspond to real numbers whose decimal expansions ultimately repeat; the irrational numbers have decimal expansions with no repeating pattern. The real numbers consist of the rational numbers and the irrational numbers.

While the real numbers consist of unending decimals, we normally work with rounded versions of them. For example, we often use 3.14159 as an approximation for $\pi$. Our rule for rounding is to round up if the first neglected digit is 5 or more; otherwise, we round down. Thus, $\pi$ rounded to 5 digits is 3.1416, to 4 digits is 3.142, and to 3 digits is 3.14. Electronic calculators typically round to either 8 or 10 digits.

We operate on numbers using the five operations of addition, subtraction, multiplication, division, and exponentiation. It is important that we understand these operations and follow the established agreements as to the order in which they are performed. In particular, we have the following hierarchy of operations.

**HIERARCHY OF OPERATIONS**

In an expression without parentheses

(i) exponentiations are performed first.
(ii) multiplications and divisions are performed next going from left to right.
(iii) additions and subtractions are performed last, again going from left to right.

An expression with parentheses is evaluated beginning inside the innermost parentheses and moving outward.

These rules are the same rules followed by most scientific calculators, though you must learn the special features of your own calculator to use it with confidence.

## EXAMPLE A

Perform the following calculations both by hand and using your calculator.

(a) $3 \cdot 6 + 2 \cdot 5 - \frac{6}{3}$
(b) $3(6 + 2)(5 - 2^3)/3$
(c) $3 \cdot 6 + 2 \cdot 5^2 - 24/2^3$
(d) $3(6 - 2)^3 - (\frac{6}{3})^4$
(e) $3[2^3 - 2(4 - 3^2) + 2]$
(f) $-2[-3 - (-2)^3]^2$

**Solution** (a) 26, (b) −24, (c) 65, (d) 176, (e) 60, (f) −50 ■

The ability to handle numerical fractions is critical in understanding how to handle the more abstract algebraic fractions that will play an important role in calculus. Let us review the basic principles.

**RULES FOR FRACTIONS**

(i) $\dfrac{a}{b} = c$ if and only if $a = bc$.

(ii) $\dfrac{a}{1} = a$, $\dfrac{0}{a} = 0$ provided $a \neq 0$, and $\dfrac{a}{0}$ is undefined.

(iii) $\dfrac{ac}{bc} = \dfrac{a}{b}$ provided $c \neq 0$.

(iv) $\dfrac{-a}{b} = \dfrac{a}{-b} = -\dfrac{a}{b}$ and $\dfrac{-a}{-b} = \dfrac{a}{b}$.

(v) $\dfrac{a}{c} + \dfrac{b}{c} = \dfrac{a+b}{c}$ and $\dfrac{a}{c} - \dfrac{b}{c} = \dfrac{a-b}{c}$

(vi) $\dfrac{a}{c} \cdot \dfrac{b}{d} = \dfrac{ab}{cd}$ and $\dfrac{a}{c} \Big/ \dfrac{b}{d} = \dfrac{ad}{bc}$

We note that all three parts of Rule (ii) follow from Rule (i). Note in particular that $a/0$ cannot be defined, for if $a/0 = c$, Rule (i) would say that $a = 0 \cdot c = 0$, which is nonsense. To put it bluntly, *we must never divide by zero*.

### EXAMPLE B

Simplify the following numerical expressions.

(a) $2(\frac{1}{6} - \frac{7}{6} + \frac{2}{6})$ (b) $\frac{2}{3} - \frac{3}{8} + \frac{5}{6}$

(c) $2 - \dfrac{\frac{2}{3}}{4}$ (d) $\dfrac{1 - \frac{1}{8}}{\frac{3}{4} - \frac{2}{3}}$

**Solution**

(a) $2(\frac{1}{6} - \frac{7}{6} + \frac{2}{6}) = 2(\frac{-4}{6}) = 2(\frac{-2}{3}) = -\frac{4}{3} = -1.333\ldots = -1.\overline{3}$

(b) $\frac{2}{3} - \frac{3}{8} + \frac{5}{6} = \frac{16}{24} - \frac{9}{24} + \frac{20}{24} = \frac{27}{24} = \frac{9}{8} = 1.125$

(c) $2 - \dfrac{\frac{2}{3}}{4} = 2 - \frac{2}{3}\frac{1}{4} = 2 - \frac{1}{6} = \frac{12}{6} - \frac{1}{6} = \frac{11}{6} = 1.8333\ldots = 1.8\overline{3}$

(d) $\dfrac{1 - \frac{1}{8}}{\frac{3}{4} - \frac{2}{3}} = \dfrac{\frac{8}{8} - \frac{1}{8}}{\frac{9}{12} - \frac{8}{12}} = \dfrac{\frac{7}{8}}{\frac{1}{12}} = \frac{7}{8}\frac{12}{1} = \frac{21}{2} = 10.5$ ■

We use the symbols $<$ and $\leq$ to abbreviate "is less than" and "is less than or equal to." Geometrically, $a < b$ means that $a$ is to the left of $b$ on the real line (Figure 2). It is correct to say that $3 < 5$, $3 \leq 5$, and $3 \leq 3$. Equivalently, we may write these three facts as $5 > 3$, $5 \geq 3$, and $3 \geq 3$. The rules for $\leq$ are summarized below. There are corresponding rules for $<$, $>$, and $\geq$.

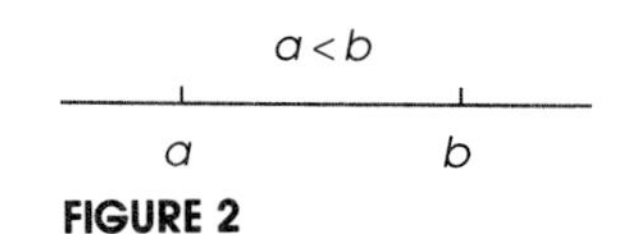


FIGURE 2

**RULES FOR $\leq$**

(i) If $a \leq b$ and $b \leq c$, then $a \leq c$.

(ii) If $a \leq b$, then $a + c \leq b + c$.

(iii) If $a \leq b$ and $c \geq 0$, then $a \cdot c \leq b \cdot c$.

(iv) If $a \leq b$ and $c < 0$, then $a \cdot c \geq b \cdot c$.

(v) If $0 < a \leq b$, then $\dfrac{1}{a} \geq \dfrac{1}{b}$.

(vi) $\dfrac{a}{b} \leq \dfrac{c}{d}$ provided $bd > 0$ and $ad \leq bc$.

### EXAMPLE C

Arrange the numbers $3.1415\overline{8}$, $\frac{22}{7}$, $\frac{60}{19}$, and $\pi$ in increasing order.

**Solution** Let us first note from Rule (vi) that $\frac{22}{7} < \frac{60}{19}$, since $(22)(19) < (7)(60)$. Next use long division (or a calculator) to write $\frac{22}{7}$ as a decimal ($\frac{22}{7} = 3.\overline{142857}$) and compare with $3.1415\overline{8}$ and with $\pi = 3.141592635$ . . . . We conclude that

$$3.1415\overline{8} < \pi < \frac{22}{7} < \frac{60}{19}$$ ■

Occasionally, we will use the symbol $|a|$, read the *absolute value of a*. The absolute value of $a$ is the distance from $a$ to 0 on the real line (Figure 3). Algebraically, $|a| = a$ if $a \geq 0$, and $|a| = -a$ if $a < 0$. Here are the main facts about absolute value.

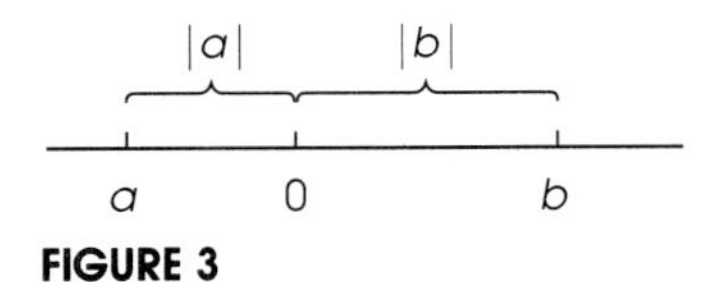


FIGURE 3

**PROPERTIES OF ABSOLUTE VALUE**

(i) $|a \cdot b| = |a| \cdot |b|$

(ii) $\left|\dfrac{a}{b}\right| = \dfrac{|a|}{|b|}$

(iii) $|a + b| \leq |a| + |b|$

## A2 EXPONENTS

We used positive integral exponents in Section A1 when we wrote $2^3$ in place of $2 \cdot 2 \cdot 2$ and $3^2$ in place of $3 \cdot 3$. In general, if $n$ is a positive integer,

$$a^n = \underbrace{a \cdot a \cdot a \ . \ . \ . \ a}_{n \text{ factors}}$$

This definition leads immediately to five rules for exponents.

**RULES FOR EXPONENTS**

(i) $a^m a^n = a^{m+n}$ (ii) $a^m/a^n = a^{m-n}$

(iii) $(a^m)^n = a^{mn}$ (iv) $(ab)^n = a^n b^n$

(v) $(a/b)^n = a^n/b^n$

In stating these rules, we temporarily assume that $m$ and $n$ are positive integers with $m > n$. But it is precisely because we want these rules to be universally true that we are led to extend the notion of exponent. First, if Rule (ii) is to be true in general, $1 = a^2/a^2 = a^{2-2} = a^0$, which forces $a^0 = 1$. Also by Rule (ii), $1/a^n = a^0/a^n = a^{0-n} = a^{-n}$, which forces $a^{-n} = 1/a^n$. Then by Rule (iii), $(a^{1/n})^n = a^1$, which means that $a^{1/n}$ should mean the same thing as $\sqrt[n]{a}$. Finally, $a^{m/n} = (a^{1/n})^m = (\sqrt[n]{a})^m$. We summarize in the following definitions in which we suppose $m$ and $n$ to be positive integers.

**ZERO, NEGATIVE, AND RATIONAL EXPONENTS**

(vi) $a^0 = 1$ (vii) $a^{-n} = 1/a^n$

(viii) $a^{1/n} = \sqrt[n]{a}$ (ix) $a^{m/n} = (\sqrt[n]{a})^m$

With these understandings, Rules (i) to (v) are valid in general. There is one small caveat. In this course, we do not attach meaning to $a^{m/n}$ if $a < 0$ and $n$ is an even integer, since this would lead us into a discussion of imaginary numbers, numbers not needed in this course.

### EXAMPLE D

Simplify each of the following, writing your answer without any negative exponents.

(a) $2^{-3}$ (b) $8^{2/3}$ (c) $(2^{1/2}3^{3/2})^2$

(d) $a^{-2}b^3a^4b^{-5}$ (e) $\dfrac{a^2b^{-3}c^3}{a^4b^5c}$ (f) $\left(\dfrac{a^{1/3}b^2a^{4/3}}{b^4}\right)^3$

**Solution**

(a) $2^{-3} = 1/2^3 = \frac{1}{8}$

(b) $8^{2/3} = (\sqrt[3]{8})^2 = 2^2 = 4$

(c) $(2^{1/2}3^{3/2})^2 = (2^{1/2})^2(3^{3/2})^2 = 2 \cdot 3^3 = 2 \cdot 27 = 54$

(d) $a^{-2}b^3a^4b^{-5} = a^{-2}a^4b^3b^{-5} = a^2b^{-2} = a^2/b^2$

(e) $\dfrac{a^2b^{-3}c^3}{a^4b^5c} = \dfrac{a^2}{a^4} \cdot \dfrac{b^{-3}}{b^5} \cdot \dfrac{c^3}{c} = a^{2-4}b^{-3-5}c^{3-1} = a^{-2}b^{-8}c^2 = \dfrac{c^2}{a^2b^8}$

(f) $\left(\dfrac{a^{1/3}b^2a^{4/3}}{b^4}\right)^3 = \left(\dfrac{a^{5/3}}{b^2}\right)^3 = \dfrac{a^5}{b^6}$ ■

## A3 ALGEBRAIC EXPRESSIONS

We illustrate the problem of multiplying two binomials (two-term expressions) with a simple example.

$$\begin{aligned}(2x + y)(3x + 4y) &= (2x + y)(3x) + (2x + y)(4y)\\ &= (2x)(3x) + (y)(3x) + (2x)(4y) + (y)(4y)\\ &= 6x^2 + 3xy + 8xy + 4y^2\\ &= 6x^2 + 11xy + 4y^2\end{aligned}$$

There is a short cut called the FOIL method. Multiply the Firsts, the Outers, the Inners, and the Lasts; add the results. Let us apply this in another example.

$$\begin{aligned}(3a - 2b)(4a + b) &= \underset{F}{(3a)(4a)} + \underset{O}{(3a)(b)} + \underset{I}{(-2b)(4a)} + \underset{L}{(-2b)(b)}\\ &= 12a^2 - 5ab - 2b^2\end{aligned}$$

### EXAMPLE E

Expand each of the following.

(a) $(2x - 3y)^2$
(b) $(4m - 3n)(4m + 3n)$
(c) $(2xy - 3z^2)(5xy - 2z^2)$

**Solution**

(a) $(2x - 3y)^2 = (2x - 3y)(2x - 3y) = 4x^2 - 12xy + 9y^2$
(b) $(4m - 3n)(4m + 3n) = 16m^2 + 12mn - 12mn - 9n^2$
$= 16m^2 - 9n^2$
(c) $(2xy - 3z^2)(5xy - 2z^2) = 10x^2y^2 - 19xyz^2 + 6z^4$ ■

Parts (a) and (b) in Example E are important enough to be emphasized.

$$(a + b)^2 = a^2 + 2ab + b^2$$
$$(a - b)(a + b) = a^2 - b^2$$

Factoring is the reverse of expanding (multiplying) and is equally important. Becoming proficient at it is mainly a matter of practice.

### EXAMPLE F

Factor each of the following.

(a) $9x^2 - 16$
(b) $4x^2 + 4xy + y^2$
(c) $x^2 - x - 6$
(d) $3x^2 + 4xy - 4y^2$

### Solution

(a) $9x^2 - 16 = (3x - 4)(3x + 4)$
(b) $4x^2 + 4xy + y^2 = (2x + y)^2$
(c) $x^2 - x - 6 = (x - 3)(x + 2)$
(d) $3x^2 + 4xy - 4y^2 = (3x - 2y)(x + 2y)$ ■

We handle algebraic fractions just like numerical fractions.

### EXAMPLE G

Perform the indicated operations and simplify.

(a) $\dfrac{3}{x+y} + \dfrac{a}{x+y} - \dfrac{5}{x+y}$

(b) $\dfrac{3}{x^2y} + \dfrac{2}{xy^2}$

(c) $\dfrac{2}{x+y} - \dfrac{4y}{x^2-y^2} + \dfrac{2}{x-y}$

(d) $\dfrac{x^2-x-6}{6x}\ \dfrac{2x^2}{x^2+5x+6}$

(e) $\dfrac{x^2+6x+8}{x-3} \bigg/ \dfrac{x+2}{x^2-9}$

(f) $\dfrac{1 - \dfrac{x}{x+y}}{y - \dfrac{y^2}{x+y}}$

### Solution

(a) $$\frac{3}{x+y} + \frac{a}{x+y} - \frac{5}{x+y} = \frac{3+a-5}{x+y} = \frac{a-2}{x+y}$$

(b) $$\frac{3}{x^2y} + \frac{2}{xy^2} = \frac{3y}{x^2y^2} + \frac{2x}{x^2y^2} = \frac{3y+2x}{x^2y^2}$$

(c) $$\begin{aligned}\frac{2}{x+y} - \frac{4y}{x^2-y^2} + \frac{2}{x-y} &= \frac{2(x-y)}{(x-y)(x+y)} - \frac{4y}{(x-y)(x+y)} \\ &\quad + \frac{2(x+y)}{(x-y)(x+y)} \\ &= \frac{2x-2y-4y+2x+2y}{(x-y)(x+y)} \\ &= \frac{4x-4y}{(x-y)(x+y)} = \frac{4(x-y)}{(x-y)(x+y)} \\ &= \frac{4}{x+y}\end{aligned}$$

(d) $$\begin{aligned}\frac{x^2-x-6}{6x}\ \frac{2x^2}{x^2+5x+6} &= \frac{(x-3)(x+2)}{6x}\ \frac{2x^2}{(x+2)(x+3)} \\ &= \frac{x(x-3)}{3(x+3)}\end{aligned}$$

(e) $$\begin{aligned}\frac{x^2+6x+8}{x-3} \bigg/ \frac{x+2}{x^2-9} &= \frac{(x+4)(x+2)}{x-3}\ \frac{(x-3)(x+3)}{x+2} \\ &= (x+4)(x+3) = x^2+7x+12\end{aligned}$$

(f) $$\frac{1-\dfrac{x}{x+y}}{y-\dfrac{y^2}{x+y}} = \frac{\dfrac{x+y}{x+y}-\dfrac{x}{x+y}}{\dfrac{xy+y^2}{x+y}-\dfrac{y^2}{x+y}} = \frac{x+y-x}{x+y}\cdot\frac{x+y}{xy+y^2-y^2}$$
$$= \frac{y}{xy} = \frac{1}{x}$$ ■

Occasionally, we will need to work with radical expressions.

**EXAMPLE H**

Simplify the following.

(a) $(\sqrt{x}-\sqrt{2y})(\sqrt{x}+\sqrt{2y})$ (b) $\sqrt{x+y}-\dfrac{x}{\sqrt{x+y}}$

**Solution**

(a) $(\sqrt{x}-\sqrt{2y})(\sqrt{x}+\sqrt{2y}) = (\sqrt{x})^2-(\sqrt{2y})^2 = x-2y$

(b) $$\sqrt{x+y}-\frac{x}{\sqrt{x+y}} = \frac{\sqrt{x+y}\sqrt{x+y}}{\sqrt{x+y}}-\frac{x}{\sqrt{x+y}} = \frac{x+y-x}{\sqrt{x+y}}$$
$$= \frac{y}{\sqrt{x+y}}$$ ■

## A4 EQUATIONS

Solving equations is one of the principal tasks of mathematics. For a linear (first-degree) equation, this is a simple task. We illustrate by solving $3x - 4 = 7$.

$$3x - 4 = 7$$
$$3x = 11 \quad \text{(adding 4 to each side)}$$
$$x = \tfrac{11}{3} \quad \text{(dividing both sides by 3)}$$

Our procedure was to perform the same operations on both sides of the equation a step at a time until we isolated $x$ on one side and a number (the solution) on the other. But what operations are legal? That is, what operations on both sides of an equation result in an equation with the same solutions as the original equation?

**OPERATIONS THAT PRESERVE SOLUTIONS**

(i) Adding the same quantity to (or subtracting the same quantity from) both sides of an equation.

(ii) Multiplying or dividing both sides of an equation by the same nonzero quantity.

Solving quadratic (second-degree) equations is harder but always possible in principle. The simplest method, the method of factoring, depends on the following property of numbers.

$$ab = 0 \text{ if and only if } a = 0 \text{ or } b = 0 \text{ (or both)}$$

Thus, for example, to solve the quadratic equation $x^2 - x - 6 = 0$, write it in factored form and then set each factor equal to 0.

$$x^2 - x - 6 = 0$$
$$(x + 2)(x - 3) = 0$$
$$x + 2 = 0 \quad \text{or} \quad x - 3 = 0$$
$$x = -2 \qquad\qquad x = 3$$

We conclude that the equation has two solutions, namely, $-2$ and $3$.

### EXAMPLE 1

Solve each of the following equations.

(a) $x^2 - 2x = 15$ (b) $6x^2 + 13x - 5 = 0$
(c) $4y^2 - 4y + 1 = 0$ (d) $3x^2 - 4 = 0$

**Solution**

(a)
$$\begin{aligned} x^2 - 2x &= 15 \\ x^2 - 2x - 15 &= 0 \\ (x + 3)(x - 5) &= 0 \\ x &= -3, 5 \end{aligned}$$

(b)
$$\begin{aligned} 6x^2 + 13x - 5 &= 0 \\ (2x + 5)(3x - 1) &= 0 \\ x &= -\tfrac{5}{2}, \tfrac{1}{3} \end{aligned}$$

(c)
$$\begin{aligned} 4y^2 - 4y + 1 &= 0 \\ (2y - 1)(2y - 1) &= 0 \\ y &= \tfrac{1}{2} \end{aligned}$$

Note that while a quadratic equation normally has two solutions, part (c) illustrates a case where there is only one solution.

(d)
$$\begin{aligned} 3x^2 - 4 &= 0 \\ (\sqrt{3}x + 2)(\sqrt{3}x - 2) &= 0 \\ x &= -2/\sqrt{3}, 2/\sqrt{3} \end{aligned}$$

An alternative way to solve this last equation would be to first write

$$3x^2 = 4$$
$$x^2 = \tfrac{4}{3}$$

This means that $x$ is a square root of $\frac{4}{3}$. There are two such square roots, namely, $-2/\sqrt{3}$ and $2/\sqrt{3}$. ■

Many quadratic equations are too complicated to factor by inspection. For them we can always resort to the *Quadratic Formula*.

> **THE QUADRATIC FORMULA**
> The solutions of $ax^2 + bx + c = 0$ are given by the formula
> $$x = \frac{-b \pm \sqrt{b^2 - 4ac}}{2a}$$

### EXAMPLE J

Solve by the Quadratic Formula.

(a) $2x^2 - 3x - 1 = 0$ (b) $6x^2 + 13x - 5 = 0$

### Solution

(a) $a = 2$, $b = -3$, and $c = -1$. Thus,

$$x = \frac{3 \pm \sqrt{9 + 8}}{4} = \frac{3 \pm \sqrt{17}}{4}$$

(b) $a = 6$, $b = 13$, and $c = -5$. Thus

$$x = \frac{-13 \pm \sqrt{169 + 120}}{12} = \frac{-13 \pm \sqrt{289}}{12} = \frac{-13 \pm 17}{12}$$

Using the plus sign gives $\frac{4}{12} = \frac{1}{3}$, whereas using the minus sign gives $\frac{-30}{12} = -\frac{5}{2}$. Note that these are the same answers we got by factoring in part (b) of Example I. ■

## A5 INEQUALITIES

Solving a linear inequality is very similar to solving a linear equation. About all we need to remember is that multiplying (or dividing) both sides of an inequality by a negative number reverses the direction of the inequality.

### EXAMPLE K

Solve $-2x + 3 \leq 14$.

### Solution

$$-2x + 3 \leq 14$$
$$-2x \leq 11$$
$$x \geq -\tfrac{11}{2}$$

$-\frac{11}{2}$

FIGURE 4

Note that the solution is a whole set of numbers: all numbers greater than or equal to $-\frac{11}{2}$. The solution set is shown in Figure 4. ■

As in solving quadratic equations, so in solving quadratic inequalities, it is best to start by factoring.

**EXAMPLE L**

Solve the following inequalities.

(a) $x^2 - x - 6 \leq 0$ (b) $x^2 - 6x + 8 > 0$

**Solution**

(a)

$$x^2 - x - 6 \leq 0$$

$$(x + 2)(x - 3) \leq 0$$

−2 3

FIGURE 5

The expression on the left can change sign only at $-2$ and 3. These points, the so-called *split points* for the given inequality, divide the real line into three intervals as shown in Figure 5. By checking with one value from each of these intervals (say $-3$, 0, and 4), we discover that only the points on the middle interval satisfy the inequality. The solution set consists of the interval $-2 \leq x \leq 3$.

(b)

$$x^2 - 6x + 8 > 0$$

$$(x - 2)(x - 4) > 0$$

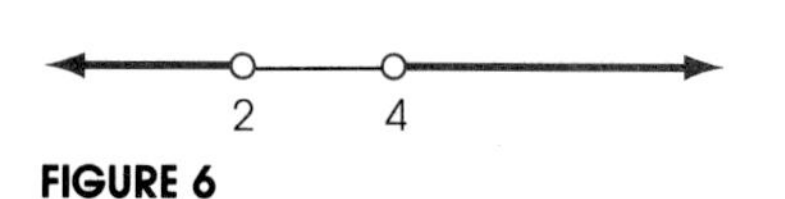


FIGURE 6

The split points are 2 and 4, dividing the real line into the three intervals shown in Figure 6. A check at 0, 3, and 5 shows that the first and last of the three intervals satisfy the inequality. The solution set consists of the two intervals $x < 2$ and $x > 4$. ■

## Basic Algebra Review Problem Set

*Evaluate the following expressions without using a calculator.*

1. $2 \cdot 5 - 36/3^2 + 4^2$
2. $4(-2)^2 - (2 - 4)^2 - 2^2$
3. $5^2/(3^2 + 4^2) + 2^3$
4. $3[22 - (5 - 2)^3]^2 + \sqrt{3^2 + 4^2}$
5. $[(21^2 - 8^2)/(21 + 8)]^2$
6. $\{2[31 - (15 - 9)^2]\}^2$

*Write each rational number as a repeating decimal (using the bar notation). Then round this number correctly to three decimal places.*

7. $\frac{5}{12}$
8. $\frac{11}{24}$
9. $\frac{5}{7}$
10. $\frac{5}{13}$

*Simplify each of the following, first writing the exact fractional answer and then its correct four-decimal approximation.*

11. $\frac{2}{5} - \frac{5}{12} + \frac{11}{20}$
12. $\frac{8}{3}(\frac{5}{12} - \frac{3}{4} + \frac{7}{24})$
13. $(\frac{2}{3} - \frac{4}{5})/\frac{7}{15}$
14. $(\frac{3}{4} - \frac{5}{6})/(\frac{3}{4} + \frac{5}{6})$
15. $\frac{3}{2}[\frac{3}{4} - (\frac{2}{3} - \frac{1}{2})]$
16. $\frac{5}{8} + \frac{2}{3}/(\frac{4}{9} - \frac{7}{18})$
17. $-\frac{3}{4} - \dfrac{\frac{5}{6}}{2}$
18. $9 + 9/(1 - \frac{5}{2})^2 + \frac{2}{3}$

*In each case, arrange the numbers in increasing order.*

19. $\frac{13}{40}, \frac{19}{59}, 0.334, 0.3\overline{3}$
20. $0.5\overline{3}, 0.\overline{52}, \frac{13}{25}, \frac{41}{79}$

*Use your calculator to evaluate each expression, rounding your answer to four decimal places.*

21. $\dfrac{(3.134)^2 + \sqrt{11.321}}{5.432 + 21.351 - 4.652}$
22. $\dfrac{2.5(3.123 - 41.534)^2}{5.123 - (6.541)^3}$
23. $\dfrac{52.12 - (5.432/0.521)^2}{6.312}$
24. $\dfrac{5 \times 10^{-4} + 6 \times 10^{-3}}{4.1 \times 10^{-3} - 2.2 \times 10^{-4}}$
25. $\dfrac{|41.24 - (7.325)^2|}{4.412 \times 10^2}$
26. $2.32 + \sqrt{54.12}/|7.31 - 10.75|$

*Simplify, writing your answer with no negative exponents.*

27. $\sqrt{3}(3^{3/2} - 3^{1/2})$
28. $(27)^{2/3}(25)^{3/2}$
29. $\dfrac{x^{-2}y^3x^{2/3}y^{3/4}}{x^{-3}y^4}$
30. $\left(\dfrac{a^{2/3}b^{5/6}}{c^{-1}}\right)^6$
31. $\dfrac{(a^{-5/3}b^{-3/4})^2}{(a^{-2}b^{-2/3})^3}$
32. $\dfrac{(\sqrt{a}\,\sqrt[3]{a}\,\sqrt[4]{a})^{12}}{a^{12}}$

*Expand each of the following.*

33. $(3xy - 2)^2$
34. $(1 - 4xy)^2$
35. $(3xy - 2)(3xy + 2)$
36. $(1 - 4xy)(1 + 4xy)$
37. $(2x - 3y)(x + 4y)$
38. $(5x - 1)(4x - 2)$
39. $(a^{1/2} - b^{1/2})^2$
40. $(a^{1/3} + a^{2/3})^2$
41. $(a^{1/2} - a^{-1/2})^2$
42. $(x - y)(x + y)(x^2 + y^2)$

*Factor each expression.*

43. $25a^2 - 9$
44. $25a^2 - 10a + 1$
45. $5a^2 - 2a - 3$
46. $a^4 - 1$
47. $3x^2y^2 + 5xyz - 2z^2$
48. $x^2 + 4x + 4 - y^2$

*Perform the indicated operations and simplify.*

49. $\dfrac{2}{x - y} + \dfrac{3}{y - x} - \dfrac{4}{x - y}$
50. $\dfrac{3}{a^2b^2} - \dfrac{2}{ab^3} + \dfrac{1}{a^2b}$
51. $\dfrac{2}{a - 3} + \dfrac{2a}{a^2 - 9} - \dfrac{4}{a + 3}$
52. $\dfrac{4}{x + 3} - \dfrac{x}{x^2 + x - 6} + \dfrac{2}{x - 2}$
53. $\dfrac{3x}{x^2 + 7x + 12} \cdot \dfrac{x^2 + x - 6}{9x^2}$
54. $\dfrac{2xy}{x^2 - 2xy - 8y^2} \Big/ \dfrac{x^2}{x^2 - 16y^2}$
55. $\dfrac{1 - x^2/y^2}{x - y}$
56. $\dfrac{(x/y)^2 - (y/x)^2}{\dfrac{x^2 - y^2}{x^2y^2}}$

*Simplify.*

57. $(\sqrt{a} + \sqrt{b^2})(\sqrt{a} - \sqrt{b^2})$
58. $(\sqrt{x} - 2\sqrt{y})^2$
59. $2\sqrt{x + h} - \dfrac{4x}{2\sqrt{x + h}}$
60. $\left(\sqrt{x + 2 + h} - \dfrac{x}{\sqrt{x + 2 + h}}\right) \Big/ \sqrt{x + 2 + h}$
61. $\sqrt{4x^4 + 9}$
62. $\sqrt{4x^4 + 12x^2 + 9}$

*Solve the following equations.*

63. $3x - 2 = 8x + 4$
64. $5 = 11x - 3$
65. $x^2 + 5x = 0$
66. $x^2 + 5x - 36 = 0$
67. $a^2 + 2a = 8$
68. $4x^2 - 12x = -9$
69. $y^4 - 1 = 0$
70. $x^3 - x^2 - 6x = 0$
71. $6y^2 - y - 2 = 0$
72. $4y^2 - 19y + 12 = 0$
73. $5x^2 = 15$
74. $(x + 1)^2 = 4$
75. $x^2 - 2x - 2 = 0$
76. $x^2 + 4x - 4 = 0$
77. $2a^2 - 5a + 1 = 0$
78. $3a^2 + 7a + 3 = 0$

*Solve the following inequalities. Then show the solution set on the real line.*

79. $3x - 4 \geq x + 3$
80. $5x - 7 < 8x + 1$
81. $x^2 + x - 6 < 0$
82. $x^2 + 3x - 18 \geq 0$
83. $2x^2 - x - 3 > 0$
84. $3x^2 - 4x - 4 \leq 0$
85. $(x + 2)(x + 1)(x - 2) \leq 0$
86. $x^3 - 2x^2 - 8x > 0$

## Table A

# THE NATURAL EXPONENTIAL FUNCTION

| $x$ | $e^x$ | $e^{-x}$ |
|---|---|---|
| 0.00 | 1.0000 | 1.0000 |
| 0.01 | 1.0101 | 0.9900 |
| 0.02 | 1.0202 | 0.9802 |
| 0.03 | 1.0305 | 0.9704 |
| 0.04 | 1.0408 | 0.9608 |
| 0.05 | 1.0513 | 0.9512 |
| 0.06 | 1.0618 | 0.9418 |
| 0.07 | 1.0725 | 0.9324 |
| 0.08 | 1.0833 | 0.9231 |
| 0.09 | 1.0942 | 0.9139 |
| 0.10 | 1.1052 | 0.9048 |
| 0.11 | 1.1163 | 0.8958 |
| 0.12 | 1.1275 | 0.8869 |
| 0.13 | 1.1388 | 0.8781 |
| 0.14 | 1.1503 | 0.9694 |
| 0.15 | 1.1618 | 0.8607 |
| 0.16 | 1.1735 | 0.8521 |
| 0.17 | 1.1853 | 0.8437 |
| 0.18 | 1.1972 | 0.8353 |
| 0.19 | 1.2092 | 0.8270 |
| 0.20 | 1.2214 | 0.8187 |
| 0.21 | 1.2337 | 0.8106 |
| 0.22 | 1.2461 | 0.8025 |
| 0.23 | 1.2586 | 0.7945 |
| 0.24 | 1.2712 | 0.7866 |
| 0.25 | 1.2840 | 0.7788 |
| 0.26 | 1.2969 | 0.7711 |
| 0.27 | 1.3100 | 0.7634 |
| 0.28 | 1.3231 | 0.7558 |
| 0.29 | 1.3364 | 0.7483 |
| 0.30 | 1.3499 | 0.7408 |
| 0.31 | 1.3634 | 0.7334 |
| 0.32 | 1.3771 | 0.7261 |
| 0.33 | 1.3910 | 0.7189 |
| 0.34 | 1.4049 | 0.7118 |
| 0.35 | 1.4191 | 0.7047 |
| 0.36 | 1.4333 | 0.6977 |
| 0.37 | 1.4477 | 0.6907 |
| 0.38 | 1.4623 | 0.6839 |
| 0.39 | 1.4770 | 0.6771 |
| 0.40 | 1.4918 | 0.6703 |
| 0.41 | 1.5068 | 0.6637 |
| 0.42 | 1.5220 | 0.6570 |
| 0.43 | 1.5373 | 0.6505 |
| 0.44 | 1.5527 | 0.6440 |
| 0.45 | 1.5683 | 0.6376 |
| 0.46 | 1.5841 | 0.6313 |
| 0.47 | 1.6000 | 0.6250 |
| 0.48 | 1.6161 | 0.6188 |
| 0.49 | 1.6323 | 0.6126 |
| 0.50 | 1.6487 | 0.6065 |
| 0.51 | 1.6653 | 0.6005 |
| 0.52 | 1.6820 | 0.5945 |
| 0.53 | 1.6989 | 0.5886 |
| 0.54 | 1.7160 | 0.5827 |
| 0.55 | 1.7333 | 0.5769 |
| 0.56 | 1.7507 | 0.5712 |
| 0.57 | 1.7683 | 0.5655 |
| 0.58 | 1.7860 | 0.5599 |
| 0.59 | 1.8040 | 0.5543 |
| 0.60 | 1.8221 | 0.5488 |
| 0.61 | 1.8044 | 0.5434 |
| 0.62 | 1.8589 | 0.5379 |
| 0.63 | 1.8776 | 0.5326 |
| 0.64 | 1.8965 | 0.5273 |
| 0.65 | 1.9155 | 0.5220 |
| 0.66 | 1.9348 | 0.5169 |
| 0.67 | 1.9542 | 0.5117 |
| 0.68 | 1.9739 | 0.5066 |
| 0.69 | 1.9937 | 0.5016 |
| 0.70 | 2.0138 | 0.4966 |
| 0.71 | 2.0340 | 0.4916 |
| 0.72 | 2.0544 | 0.4868 |
| 0.73 | 2.0751 | 0.4819 |
| 0.74 | 2.0959 | 0.4771 |
| 0.75 | 2.1170 | 0.4724 |
| 0.76 | 2.1383 | 0.4677 |
| 0.77 | 2.1598 | 0.4630 |
| 0.78 | 2.1815 | 0.4584 |
| 0.79 | 2.2034 | 0.4538 |
| 0.80 | 2.2255 | 0.4493 |
| 0.81 | 2.2479 | 0.4449 |
| 0.82 | 2.2705 | 0.4404 |
| 0.83 | 2.2933 | 0.4360 |
| 0.84 | 2.3164 | 0.4317 |
| 0.85 | 2.3396 | 0.4274 |
| 0.86 | 2.3632 | 0.4232 |
| 0.87 | 2.3869 | 0.4190 |
| 0.88 | 2.4109 | 0.4148 |
| 0.89 | 2.4351 | 0.4107 |

| $x$ | $e^x$ | $e^{-x}$ |
|---|---|---|
| 0.90 | 2.4596 | 0.4066 |
| 0.91 | 2.4843 | 0.4025 |
| 0.92 | 2.5093 | 0.3985 |
| 0.93 | 2.5345 | 0.3946 |
| 0.94 | 2.5600 | 0.3906 |
| 0.95 | 2.5857 | 0.3867 |
| 0.96 | 2.6117 | 0.3829 |
| 0.97 | 2.6379 | 0.3791 |
| 0.98 | 2.6645 | 0.3753 |
| 0.99 | 2.6912 | 0.3716 |
| 1.00 | 2.7183 | 0.3679 |
| 1.05 | 2.8577 | 0.3499 |
| 1.10 | 3.0042 | 0.3329 |
| 1.15 | 3.1582 | 0.3166 |
| 1.20 | 3.3201 | 0.3012 |
| 1.25 | 3.4903 | 0.2865 |
| 1.30 | 3.6693 | 0.2725 |
| 1.35 | 3.8574 | 0.2592 |
| 1.40 | 4.0552 | 0.2466 |
| 1.45 | 4.2631 | 0.2346 |
| 1.50 | 4.4817 | 0.2231 |
| 1.55 | 4.7115 | 0.2122 |
| 1.60 | 4.9530 | 0.2019 |
| 1.65 | 5.2070 | 0.1920 |
| 1.70 | 5.4739 | 0.1827 |
| 1.75 | 5.7546 | 0.1738 |
| 1.80 | 6.0496 | 0.1653 |
| 1.85 | 6.3598 | 0.1572 |
| 1.90 | 6.6859 | 0.1496 |
| 1.95 | 7.0287 | 0.1423 |
| 2.00 | 7.3891 | 0.1353 |
| 2.05 | 7.7679 | 0.1287 |
| 2.10 | 8.1662 | 0.1225 |
| 2.15 | 8.5849 | 0.1165 |
| 2.20 | 9.0250 | 0.1108 |
| 2.25 | 9.4877 | 0.1054 |
| 2.30 | 9.9742 | 0.1003 |
| 2.35 | 10.486 | 0.0954 |
| 2.40 | 11.023 | 0.0907 |
| 2.45 | 11.588 | 0.0863 |
| 2.50 | 12.182 | 0.0821 |
| 2.55 | 12.807 | 0.0781 |
| 2.60 | 13.464 | 0.0743 |
| 2.65 | 14.154 | 0.0707 |
| 2.70 | 14.880 | 0.0672 |

| $x$ | $e^x$ | $e^{-x}$ |
|---|---|---|
| 2.75 | 15.643 | 0.0639 |
| 2.80 | 16.445 | 0.0608 |
| 2.85 | 17.288 | 0.0578 |
| 2.90 | 18.174 | 0.0550 |
| 2.95 | 19.106 | 0.0523 |
| 3.00 | 20.086 | 0.0498 |
| 3.05 | 21.115 | 0.0474 |
| 3.10 | 22.198 | 0.0450 |
| 3.15 | 23.336 | 0.0429 |
| 3.20 | 24.533 | 0.0408 |
| 3.25 | 25.790 | 0.0388 |
| 3.30 | 27.113 | 0.0369 |
| 3.35 | 28.503 | 0.0351 |
| 3.40 | 29.964 | 0.0334 |
| 3.45 | 31.500 | 0.0317 |
| 3.50 | 33.115 | 0.0302 |
| 3.55 | 34.813 | 0.0287 |
| 3.60 | 36.598 | 0.0273 |
| 3.65 | 38.475 | 0.0260 |
| 3.70 | 40.447 | 0.0247 |
| 3.75 | 42.521 | 0.0235 |
| 3.80 | 44.701 | 0.0224 |
| 3.85 | 46.993 | 0.0213 |
| 3.90 | 49.402 | 0.0202 |
| 3.95 | 51.935 | 0.0193 |
| 4.00 | 54.598 | 0.0183 |
| 4.10 | 60.340 | 0.0166 |
| 4.20 | 66.686 | 0.0150 |
| 4.30 | 73.700 | 0.0136 |
| 4.40 | 81.451 | 0.0123 |
| 4.50 | 90.017 | 0.0111 |
| 4.60 | 99.484 | 0.0101 |
| 4.70 | 109.95 | 0.0091 |
| 4.80 | 121.51 | 0.0082 |
| 4.90 | 134.29 | 0.0074 |
| 5.00 | 148.41 | 0.0067 |
| 5.20 | 181.27 | 0.0055 |
| 5.40 | 221.41 | 0.0045 |
| 5.60 | 270.43 | 0.0037 |
| 5.80 | 330.30 | 0.0030 |
| 6.00 | 403.43 | 0.0025 |
| 7.00 | 1096.6 | 0.0009 |
| 8.00 | 2981.0 | 0.0003 |
| 9.00 | 8103.1 | 0.0001 |
| 10.00 | 22026. | 0.00005 |

## Table B

# THE NATURAL LOGARITHM FUNCTION

| | 0.00 | 0.01 | 0.02 | 0.03 | 0.04 | 0.05 | 0.06 | 0.07 | 0.08 | 0.09 |
|---|---|---|---|---|---|---|---|---|---|---|
| 1.0 | 0.0000 | 0.0100 | 0.0198 | 0.0296 | 0.0392 | 0.0488 | 0.0583 | 0.0677 | 0.0770 | 0.0862 |
| 1.1 | 0.0953 | 0.1044 | 0.1133 | 0.1222 | 0.1310 | 0.1398 | 0.1484 | 0.1570 | 0.1655 | 0.1740 |
| 1.2 | 0.1823 | 0.1906 | 0.1989 | 0.2070 | 0.2151 | 0.2231 | 0.2311 | 0.2390 | 0.2469 | 0.2546 |
| 1.3 | 0.2624 | 0.2700 | 0.2776 | 0.2852 | 0.2927 | 0.3001 | 0.3075 | 0.3148 | 0.3221 | 0.3293 |
| 1.4 | 0.3365 | 0.3436 | 0.3507 | 0.3577 | 0.3646 | 0.3716 | 0.3784 | 0.3853 | 0.3920 | 0.3988 |
| 1.5 | 0.4055 | 0.4121 | 0.4187 | 0.4253 | 0.4318 | 0.4383 | 0.4447 | 0.4511 | 0.4574 | 0.4637 |
| 1.6 | 0.4700 | 0.4762 | 0.4824 | 0.4886 | 0.4947 | 0.5008 | 0.5068 | 0.5128 | 0.5188 | 0.5247 |
| 1.7 | 0.5306 | 0.5365 | 0.5423 | 0.5481 | 0.5539 | 0.5596 | 0.5653 | 0.5710 | 0.5766 | 0.5822 |
| 1.8 | 0.5878 | 0.5933 | 0.5988 | 0.6043 | 0.6098 | 0.6152 | 0.6206 | 0.6259 | 0.6313 | 0.6366 |
| 1.9 | 0.6419 | 0.6471 | 0.6523 | 0.6575 | 0.6627 | 0.6678 | 0.6729 | 0.6780 | 0.6831 | 0.6881 |
| 2.0 | 0.6931 | 0.6981 | 0.7031 | 0.7080 | 0.7130 | 0.7178 | 0.7227 | 0.7275 | 0.7324 | 0.7372 |
| 2.1 | 0.7419 | 0.7467 | 0.7514 | 0.7561 | 0.7608 | 0.7655 | 0.7701 | 0.7747 | 0.7793 | 0.7839 |
| 2.2 | 0.7885 | 0.7930 | 0.7975 | 0.8020 | 0.8065 | 0.8109 | 0.8154 | 0.8198 | 0.8242 | 0.8286 |
| 2.3 | 0.8329 | 0.8372 | 0.8416 | 0.8459 | 0.8502 | 0.8544 | 0.8587 | 0.8629 | 0.8671 | 0.8713 |
| 2.4 | 0.8755 | 0.8796 | 0.8838 | 0.8879 | 0.8920 | 0.8961 | 0.9002 | 0.9042 | 0.9083 | 0.9123 |
| 2.5 | 0.9163 | 0.9203 | 0.9243 | 0.9282 | 0.9322 | 0.9361 | 0.9400 | 0.9439 | 0.9478 | 0.9517 |
| 2.6 | 0.9555 | 0.9594 | 0.9632 | 0.9670 | 0.9708 | 0.9746 | 0.9783 | 0.9821 | 0.9858 | 0.9895 |
| 2.7 | 0.9933 | 0.9969 | 1.0006 | 1.0043 | 1.0080 | 1.0116 | 1.0152 | 1.0188 | 1.0225 | 1.0260 |
| 2.8 | 1.0296 | 1.0332 | 1.0367 | 1.0403 | 1.0438 | 1.0473 | 1.0508 | 1.0543 | 1.0578 | 1.0613 |
| 2.9 | 1.0647 | 1.0682 | 1.0716 | 1.0750 | 1.0784 | 1.0818 | 1.0852 | 1.0886 | 1.0919 | 1.0953 |
| 3.0 | 1.0986 | 1.1019 | 1.1053 | 1.1086 | 1.1119 | 1.1151 | 1.1184 | 1.1217 | 1.1249 | 1.1282 |
| 3.1 | 1.1314 | 1.1346 | 1.1378 | 1.1410 | 1.1442 | 1.1474 | 1.1506 | 1.1537 | 1.1569 | 1.1600 |
| 3.2 | 1.1632 | 1.1663 | 1.1694 | 1.1725 | 1.1756 | 1.1787 | 1.1817 | 1.1848 | 1.1878 | 1.1909 |
| 3.3 | 1.1939 | 1.1970 | 1.2000 | 1.2030 | 1.2060 | 1.2090 | 1.2119 | 1.2149 | 1.2179 | 1.2208 |
| 3.4 | 1.2238 | 1.2267 | 1.2296 | 1.2326 | 1.2355 | 1.2384 | 1.2413 | 1.2442 | 1.2470 | 1.2499 |
| 3.5 | 1.2528 | 1.2556 | 1.2585 | 1.2613 | 1.2641 | 1.2669 | 1.2698 | 1.2726 | 1.2754 | 1.2782 |
| 3.6 | 1.2809 | 1.2837 | 1.2865 | 1.2892 | 1.2920 | 1.2947 | 1.2975 | 1.3002 | 1.3029 | 1.3056 |
| 3.7 | 1.3083 | 1.3110 | 1.3137 | 1.3164 | 1.3191 | 1.3218 | 1.3244 | 1.3271 | 1.3297 | 1.3324 |
| 3.8 | 1.3350 | 1.3376 | 1.3403 | 1.3429 | 1.3455 | 1.3481 | 1.3507 | 1.3533 | 1.3558 | 1.3584 |
| 3.9 | 1.3610 | 1.3635 | 1.3661 | 1.3686 | 1.3712 | 1.3737 | 1.3762 | 1.3788 | 1.3813 | 1.3838 |
| 4.0 | 1.3863 | 1.3888 | 1.3913 | 1.3938 | 1.3962 | 1.3987 | 1.4012 | 1.4036 | 1.4061 | 1.4085 |
| 4.1 | 1.4110 | 1.4134 | 1.4159 | 1.4183 | 1.4207 | 1.4231 | 1.4255 | 1.4279 | 1.4303 | 1.4327 |
| 4.2 | 1.4351 | 1.4375 | 1.4398 | 1.4422 | 1.4446 | 1.4469 | 1.4493 | 1.4516 | 1.4540 | 1.4563 |
| 4.3 | 1.4586 | 1.4609 | 1.4633 | 1.4656 | 1.4679 | 1.4702 | 1.4725 | 1.4748 | 1.4770 | 1.4793 |
| 4.4 | 1.4816 | 1.4839 | 1.4861 | 1.4884 | 1.4907 | 1.4929 | 1.4952 | 1.4974 | 1.4996 | 1.5019 |
| 4.5 | 1.5041 | 1.5063 | 1.5085 | 1.5107 | 1.5129 | 1.5151 | 1.5173 | 1.5195 | 1.5217 | 1.5239 |
| 4.6 | 1.5261 | 1.5282 | 1.5304 | 1.5326 | 1.5347 | 1.5369 | 1.5390 | 1.5412 | 1.5433 | 1.5454 |
| 4.7 | 1.5476 | 1.5497 | 1.5518 | 1.5539 | 1.5560 | 1.5581 | 1.5602 | 1.5623 | 1.5644 | 1.5665 |
| 4.8 | 1.5686 | 1.5707 | 1.5728 | 1.5748 | 1.5769 | 1.5790 | 1.5810 | 1.5831 | 1.5851 | 1.5872 |
| 4.9 | 1.5892 | 1.5913 | 1.5933 | 1.5953 | 1.5974 | 1.5994 | 1.6014 | 1.6034 | 1.6054 | 1.6074 |
| 5.0 | 1.6094 | 1.6114 | 1.6134 | 1.6154 | 1.6174 | 1.6194 | 1.6214 | 1.6233 | 1.6253 | 1.6273 |
| 5.1 | 1.6292 | 1.6312 | 1.6332 | 1.6351 | 1.6371 | 1.6390 | 1.6409 | 1.6429 | 1.6448 | 1.6467 |
| 5.2 | 1.6487 | 1.6506 | 1.6525 | 1.6544 | 1.6563 | 1.6582 | 1.6601 | 1.6620 | 1.6639 | 1.6658 |
| 5.3 | 1.6677 | 1.6696 | 1.6715 | 1.6734 | 1.6752 | 1.6771 | 1.6790 | 1.6808 | 1.6827 | 1.6845 |
| 5.4 | 1.6864 | 1.6882 | 1.6901 | 1.6919 | 1.6938 | 1.6956 | 1.6974 | 1.6993 | 1.7011 | 1.7029 |

$\ln(N \cdot 10^m) = \ln N + m \ln 10, \qquad \ln 10 = 2.3026$

| | 0.00 | 0.01 | 0.02 | 0.03 | 0.04 | 0.05 | 0.06 | 0.07 | 0.08 | 0.09 |
|---|---|---|---|---|---|---|---|---|---|---|
| 5.5 | 1.7047 | 1.7066 | 1.7084 | 1.7102 | 1.7120 | 1.7138 | 1.7156 | 1.7174 | 1.7192 | 1.7210 |
| 5.6 | 1.7228 | 1.7246 | 1.7263 | 1.7281 | 1.7299 | 1.7317 | 1.7334 | 1.7352 | 1.7370 | 1.7387 |
| 5.7 | 1.7405 | 1.7422 | 1.7440 | 1.7457 | 1.7475 | 1.7492 | 1.7509 | 1.7527 | 1.7544 | 1.7561 |
| 5.8 | 1.7579 | 1.7596 | 1.7613 | 1.7630 | 1.7647 | 1.7664 | 1.7682 | 1.7699 | 1.7716 | 1.7733 |
| 5.9 | 1.7750 | 1.7766 | 1.7783 | 1.7800 | 1.7817 | 1.7834 | 1.7851 | 1.7867 | 1.7884 | 1.7901 |
| 6.0 | 1.7918 | 1.7934 | 1.7951 | 1.7967 | 1.7984 | 1.8001 | 1.8017 | 1.8034 | 1.8050 | 1.8066 |
| 6.1 | 1.8083 | 1.8099 | 1.8116 | 1.8132 | 1.8148 | 1.8165 | 1.8181 | 1.8197 | 1.8213 | 1.8229 |
| 6.2 | 1.8245 | 1.8262 | 1.8278 | 1.8294 | 1.8310 | 1.8326 | 1.8342 | 1.8358 | 1.8374 | 1.8390 |
| 6.3 | 1.8406 | 1.8421 | 1.8437 | 1.8453 | 1.8469 | 1.8485 | 1.8500 | 1.8516 | 1.8532 | 1.8547 |
| 6.4 | 1.8563 | 1.8579 | 1.8594 | 1.8610 | 1.8625 | 1.8641 | 1.8656 | 1.8672 | 1.8687 | 1.8703 |
| 6.5 | 1.8718 | 1.8733 | 1.8749 | 1.8764 | 1.8779 | 1.8795 | 1.8810 | 1.8825 | 1.8840 | 1.8856 |
| 6.6 | 1.8871 | 1.8886 | 1.8901 | 1.8916 | 1.8931 | 1.8946 | 1.8961 | 1.8976 | 1.8991 | 1.9006 |
| 6.7 | 1.9021 | 1.9036 | 1.9051 | 1.9066 | 1.9081 | 1.9095 | 1.9110 | 1.9125 | 1.9140 | 1.9155 |
| 6.8 | 1.9169 | 1.9184 | 1.9199 | 1.9213 | 1.9228 | 1.9242 | 1.9257 | 1.9272 | 1.9286 | 1.9301 |
| 6.9 | 1.9315 | 1.9330 | 1.9344 | 1.9359 | 1.9373 | 1.9387 | 1.9402 | 1.9416 | 1.9430 | 1.9445 |
| 7.0 | 1.9459 | 1.9473 | 1.9488 | 1.9502 | 1.9516 | 1.9530 | 1.9544 | 1.9559 | 1.9573 | 1.9587 |
| 7.1 | 1.9601 | 1.9615 | 1.9629 | 1.9643 | 1.9657 | 1.9671 | 1.9685 | 1.9699 | 1.9713 | 1.9727 |
| 7.2 | 1.9741 | 1.9755 | 1.9769 | 1.9782 | 1.9796 | 1.9810 | 1.9824 | 1.9838 | 1.9851 | 1.9865 |
| 7.3 | 1.9879 | 1.9892 | 1.9906 | 1.9920 | 1.9933 | 1.9947 | 1.9961 | 1.9974 | 1.9988 | 2.0001 |
| 7.4 | 2.0015 | 2.0028 | 2.0042 | 2.0055 | 2.0069 | 2.0082 | 2.0096 | 2.0109 | 2.0122 | 2.0136 |
| 7.5 | 2.0149 | 2.0162 | 2.0176 | 2.0189 | 2.0202 | 2.0215 | 2.0229 | 2.0242 | 2.0255 | 2.0268 |
| 7.6 | 2.0282 | 2.0295 | 2.0308 | 2.0321 | 2.0334 | 2.0347 | 2.0360 | 2.0373 | 2.0386 | 2.0399 |
| 7.7 | 2.0412 | 2.0425 | 2.0438 | 2.0451 | 2.0464 | 2.0477 | 2.0490 | 2.0503 | 2.0516 | 2.0528 |
| 7.8 | 2.0541 | 2.0554 | 2.0567 | 2.0580 | 2.0592 | 2.0605 | 2.0618 | 2.0631 | 2.0643 | 2.0656 |
| 7.9 | 2.0669 | 2.0681 | 2.0694 | 2.0707 | 2.0719 | 2.0732 | 2.0744 | 2.0757 | 2.0769 | 2.0782 |
| 8.0 | 2.0794 | 2.0807 | 2.0819 | 2.0832 | 2.0844 | 2.0857 | 2.0869 | 2.0882 | 2.0894 | 2.0906 |
| 8.1 | 2.0919 | 2.0931 | 2.0943 | 2.0956 | 2.0968 | 2.0980 | 2.0992 | 2.1005 | 2.1017 | 2.1029 |
| 8.2 | 2.1041 | 2.1054 | 2.1066 | 2.1078 | 2.1090 | 2.1102 | 2.1114 | 2.1126 | 2.1138 | 2.1150 |
| 8.3 | 2.1163 | 2.1175 | 2.1187 | 2.1190 | 2.1211 | 2.1223 | 2.1235 | 2.1247 | 2.1258 | 2.1270 |
| 8.4 | 2.1282 | 2.1294 | 2.1306 | 2.1318 | 2.1330 | 2.1342 | 2.1353 | 2.1365 | 2.1377 | 2.1389 |
| 8.5 | 2.1401 | 2.1412 | 2.1424 | 2.1436 | 2.1448 | 2.1459 | 2.1471 | 2.1483 | 2.1494 | 2.1506 |
| 8.6 | 2.1518 | 2.1529 | 2.1541 | 2.1552 | 2.1564 | 2.1576 | 2.1587 | 2.1599 | 2.1610 | 2.1622 |
| 8.7 | 2.1633 | 2.1645 | 2.1656 | 2.1668 | 2.1679 | 2.1691 | 2.1702 | 2.1713 | 2.1725 | 2.1736 |
| 8.8 | 2.1748 | 2.1759 | 2.1770 | 2.1782 | 2.1793 | 2.1804 | 2.1815 | 2.1827 | 2.1838 | 2.1849 |
| 8.9 | 2.1861 | 2.1872 | 2.1883 | 2.1894 | 2.1905 | 2.1917 | 2.1928 | 2.1939 | 2.1950 | 2.1961 |
| 9.0 | 2.1972 | 2.1983 | 2.1994 | 2.2006 | 2.2017 | 2.2028 | 2.2039 | 2.2050 | 2.2061 | 2.2072 |
| 9.1 | 2.2083 | 2.2094 | 2.2105 | 2.2116 | 2.2127 | 2.2138 | 2.2148 | 2.2159 | 2.2170 | 2.2181 |
| 9.2 | 2.2192 | 2.2203 | 2.2214 | 2.2225 | 2.2235 | 2.2246 | 2.2257 | 2.2268 | 2.2279 | 2.2289 |
| 9.3 | 2.2300 | 2.2311 | 2.2322 | 2.2332 | 2.2343 | 2.2354 | 2.2364 | 2.2375 | 2.2386 | 2.2396 |
| 9.4 | 2.2407 | 2.2418 | 2.2428 | 2.2439 | 2.2450 | 2.2460 | 2.2471 | 2.2481 | 2.2492 | 2.2502 |
| 9.5 | 2.2513 | 2.2523 | 2.2534 | 2.2544 | 2.2555 | 2.2565 | 2.2576 | 2.2586 | 2.2597 | 2.2607 |
| 9.6 | 2.2618 | 2.2628 | 2.2638 | 2.2649 | 2.2659 | 2.2670 | 2.2680 | 2.2690 | 2.2701 | 2.2711 |
| 9.7 | 2.2721 | 2.2732 | 2.2742 | 2.2752 | 2.2762 | 2.2773 | 2.2783 | 2.2793 | 2.2803 | 2.2814 |
| 9.8 | 2.2824 | 2.2834 | 2.2844 | 2.2854 | 2.2865 | 2.2875 | 2.2885 | 2.2895 | 2.2905 | 2.2915 |
| 9.9 | 2.2925 | 2.2935 | 2.2946 | 2.2956 | 2.2966 | 2.2976 | 2.2986 | 2.2996 | 2.3006 | 2.3016 |

# Answers to Odd-Numbered Problems

## Problem Set 1.1 (p. 6)

1. $y = 2x^2$ 3. $V = 2x^3$
5. (a) 11; (b) $-17$; (c) $-3$; (d) 5; (e) $-\frac{3}{2}$
7. (a) 1; (b) $-5$; (c) 233; (d) $\frac{5}{4}$; (e) $\frac{9}{4}$
9. (a) $-2$; (b) $-2 - 5h - 2h^2$; (c) $-5 - 2h$; (d) $3 - 4x - 2d$
11. $-4/(a^2 + ah)$ 13. $-2 + 6x + 3h$
15. (a) $\frac{1}{2}$; (b) 2; (c) undefined; (d) $1/(2x^2 - 8x + 8)$; (e) $x^4/2$
17. (a) $\sqrt{14}$; (b) 0; (c) $\frac{1}{2}$; (d) $\sqrt{x^2 - 2}$
19. $2x^2 - 4x + 3, -2x^2$ 21. $x^6 + 2x^3 + 1, x^6 + 1$
23. $4/x^2, (7 - x^2)/(x^2 + 1)$
25. (a) $-0.4313416$; (b) 0.6296419; (c) $-0.5075616$
27. $f(x) = \ln x$, $g(x) = \cos x$, $h(x) = \sqrt{x}$
29. (a) All real numbers; (b) all real numbers except 2; (c) all real numbers except $\pm 3$; (d) all real numbers except $\pm 1$; (e) all real numbers greater than 1.
31. $4x^3 + 6hx^2 + 4h^2x + h^3$
33. $f(x) = 700 + 60x$ 35. $f(x) = 0.015x - 180$
37. $g(x) = \sqrt{4x^2 + 12x + 5}$
39. (a) 0, 54.48, 96.30; (b) 37

5.

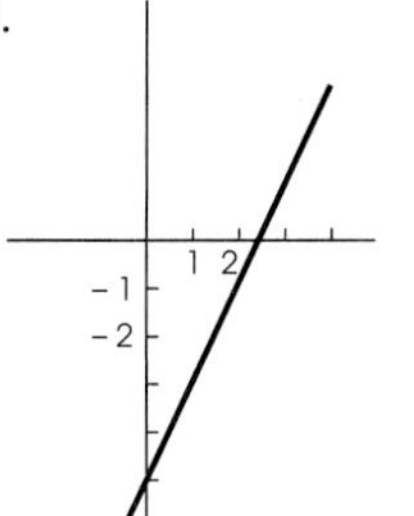


7.

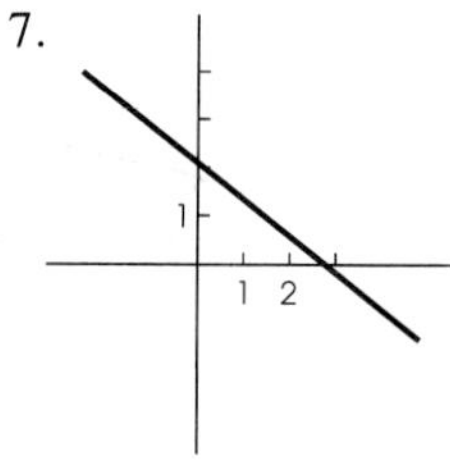


9.

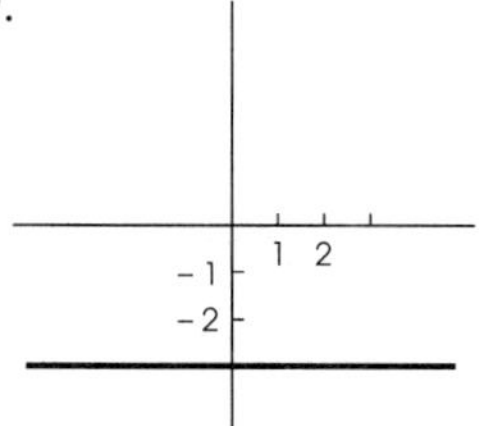


11.

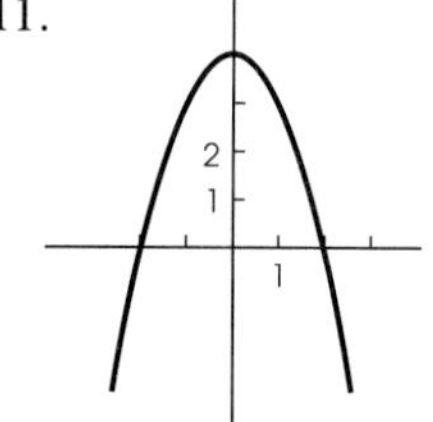


13.

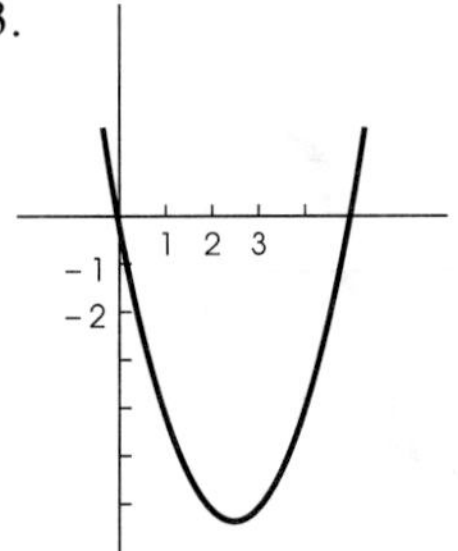


15.

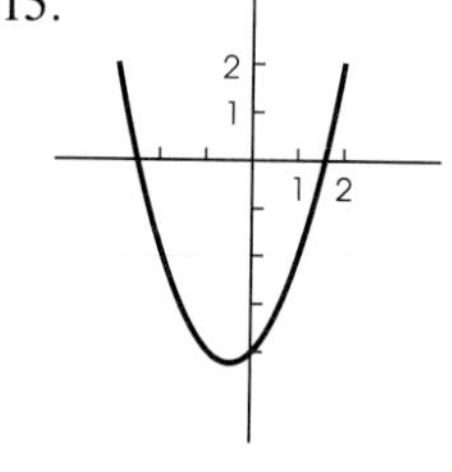


## Problem Set 1.2 (p. 13)

1. $f, h, G, H$
3. (a) Odd; (b) even; (c) neither; (d) neither; (e) even; (f) odd.

17.

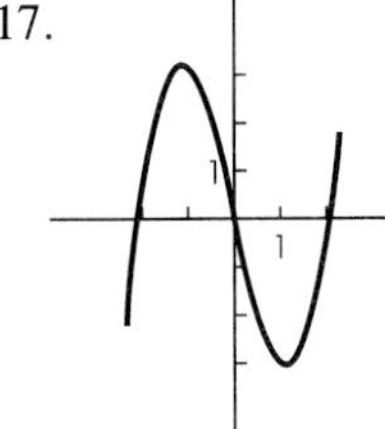

19.

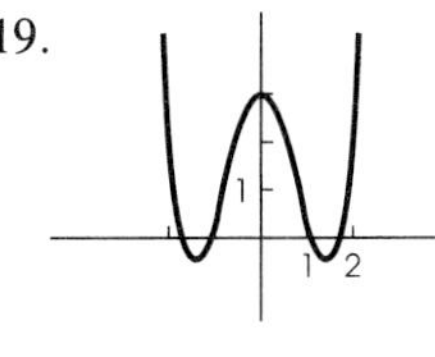

21.

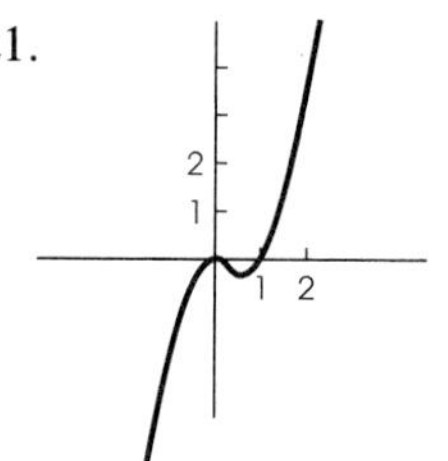

23.

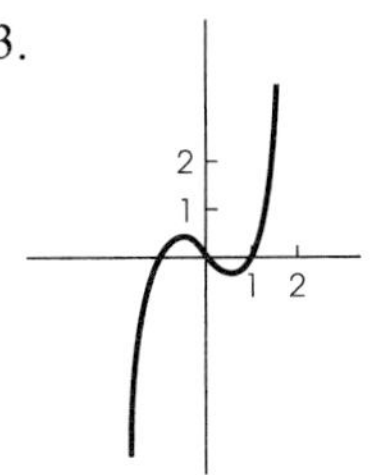

25.

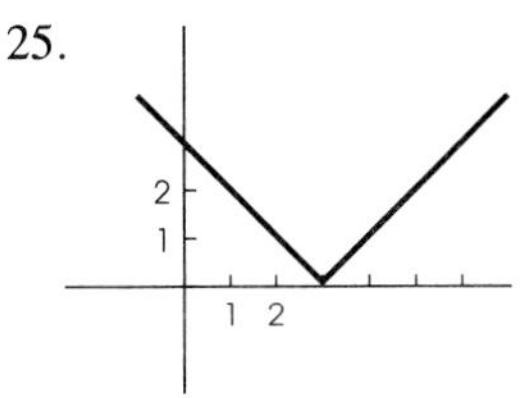

27.

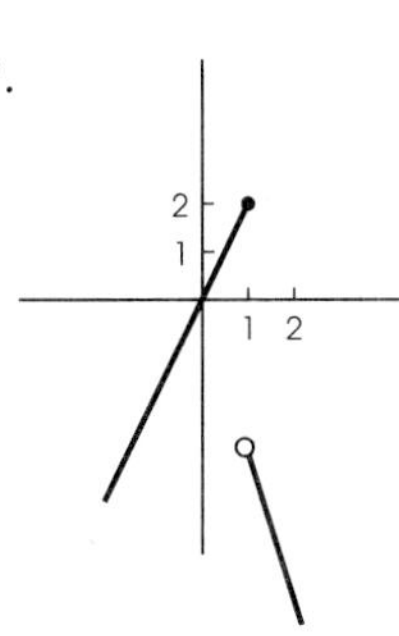

29.

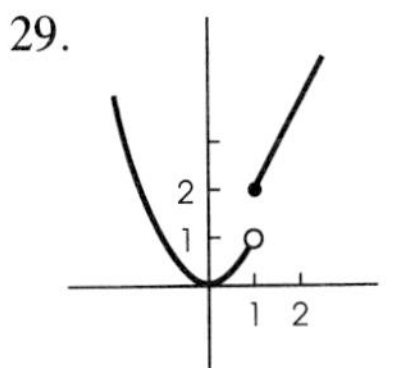

31.

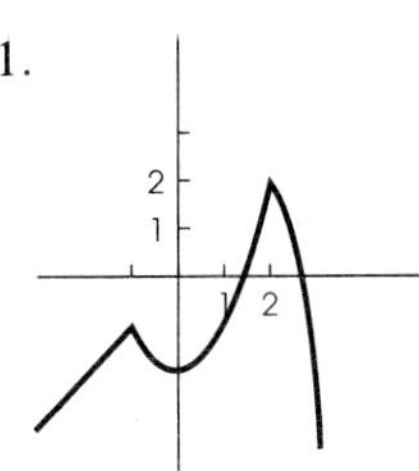

33.

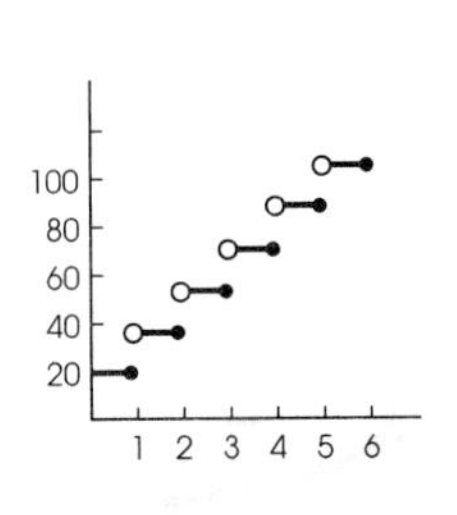


35. (a) Odd; (b) odd; (c) even; (d) even; (e) even; (f) even; (g) neither; (h) even.

37.

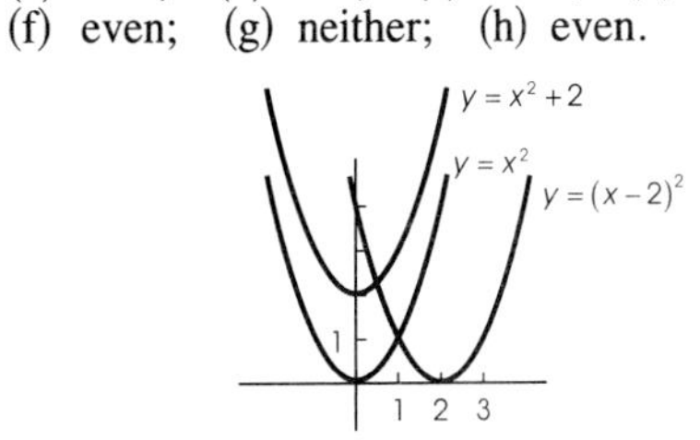
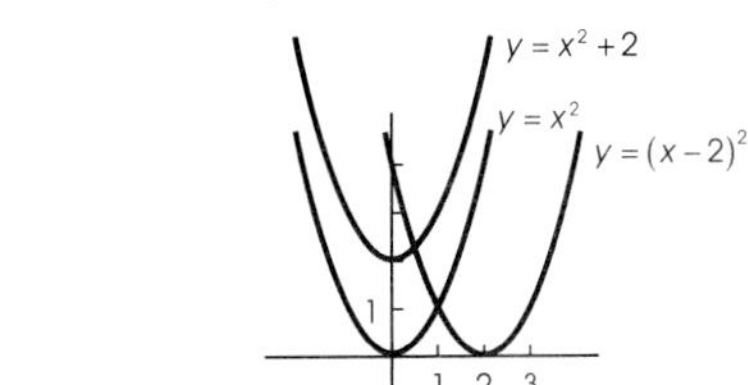


39. $R(x) = \begin{cases} 160x & \text{if } x \leq 30 \\ 220x - 2x^2 & \text{if } x > 30 \end{cases}$
$R(35) = 5250$, $R(70) = 5600$

41. (a) $908.67; (b) $1559.10; (c) $8305.40

43. (a) $1312.65; (b) $1858.50: (c) $7474.18

## Problem Set 1.3 (p. 21)

1. 4
3. $-\frac{2}{7}$
5. $-22.56$
7. $y - 1 = 3(x - 2)$
9. $y - 1 = 4(x - 3)$
11. $y + 2 = \frac{9}{2}(x - 1)$
13. $4x - y = 3$
15. $5x - y = 11$
17. $2x - y = 5$
19. $4x + y = 12$
21. $3x - 4y = 17$
23. $3x - 2y = -9$
25. $f(x) = 3x, f(5) = 15$
27. $f(x) = -\frac{2}{5}x + \frac{16}{5}, f(5) = \frac{6}{5}$
29. (a) $-\frac{2}{3}, 2$; (b) $-\frac{4}{3}, -12$; (c) $\frac{3}{4}, -\frac{11}{4}$; (d) $\frac{3}{2}, -\frac{8}{3}$
31. (a) $3x + 2y = -4$; (b) $2x + y = -1$; (c) $y = -5$ (d) $5x + 3y = -5$
33. $f(x) = -1800x + 20{,}000$
35. (a) $F = 1.8C + 32$; (b) 89.6; (c) 27.22
37. 8393 ft

## Problem Set 1.4 (p. 26)

1. 196.4 cm
3. (a) $S(x) = 1200 + 0.04x$; (b) $2640
5. (a) $F(x) = 150 + 0.12x$; (b) $1110
7. (a) $V(n) = 75{,}000 - 6700n$;
(b)

60,000
40,000
20,000
2 4 6 8 10

(c) $m = -6700$; annual decrease in value.

9. $C(x) = 42{,}000 + 250x$
11. $R(x) = 375x$, 336
13. (a) $C(x) = 20{,}000 + 40x$; (b) $R(x) = 100x$; (c) $P(x) = -20{,}000 + 60x$; (d) $P(1000) = 40{,}000$
15. $s = 480 - 60t$
17. (a) Sarah: $s = 55t$, John: $s = 70(t - 2)$ for $t \geq 2$; (b) after $9\frac{1}{3}$ hours or at 9:20 P.M.
19. (a) $L = 64 + 0.00228T$; (b) 64.228 ft; (c) 0.00228 ft/degree
21. (a) $\frac{9}{5}$; (b) $C = \frac{5}{9}(F - 32)$, $\frac{5}{9}$
23. (a) $f(n) = 1{,}200{,}000 + 48{,}000n$; (b) 48,000 cases/yr
25. (a) $C(x) = 12{,}500 + 35x$; (b) \$12,500; (c) \$35 per suit; (d) \$35

## Problem Set 1.5 (p. 32)

1. $-2$
3. $-\frac{1}{3}$
5. 2
7. 0
9. (a) $A, E$; (b) $C$; (c) $B, D$
11.

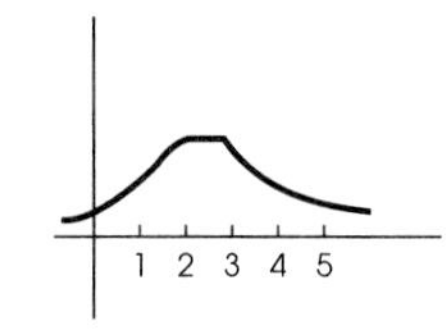


13. $y - 9 = 6(x - 3)$
15. 48
17. 5
19. $30x^5 + 12x^2 - 12$
21. $8x^3 - 9x^2 + 10x$
23. $-15x^2 + x$
25. $y + 64 = 160(x + 2)$
27. $y + 30 = 53(x - 2)$
29. $(-2, 20)$, $(2, -8)$
31. (a) $6x^2 + 6x(\Delta x) + 2(\Delta x)^2$; (b) $6x^2$
33. $y - 20 = 3(x + 2)$, $y = 3(x - 2)$
35. All values of $x$.
37. $x = 3$
39. 19.384

## Problem Set 1.6 (p. 39)

1. (a) $16\pi$ in.$^3$/in.; (b) $1.6\pi$ in.$^3$
3. 1100 gal/min
5. (a) $-1$; (b) 3
7. (a)

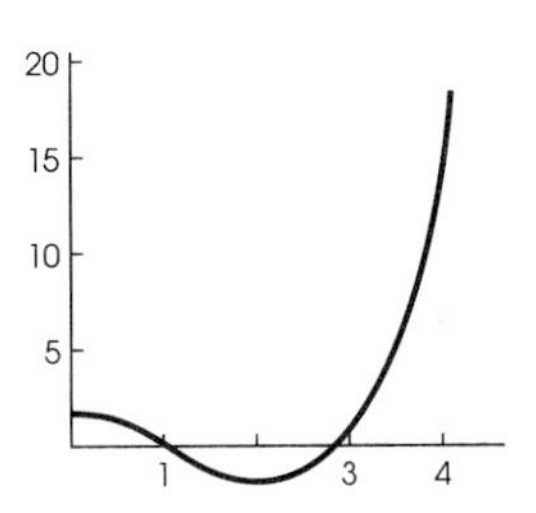


(b) 0, $-3$, 0, 24; (c) 0, 2
9. (a) 6.1 ft/s; (b) 6.01 ft/s; (c) 6.001 ft/s; (d) 6 ft/s
11. (a) 64 ft/s, $-64$ ft/s; (b) 4 s; (c) 256 ft
13. $-80$ ft/s
15. (a) 1500 people/yr; (b) 1965; (c) 3000 people/yr
17. (a) \$136; (b) \$0.90; (b) \$0.90; (d) 125
19. (a) $3.5 - \frac{1}{150}x$, $7 - \frac{1}{100}x$, $3.5 - \frac{1}{300}x$; (b) \$0.17 per wagon; (c) \$2.00 per wagon; (d) \$1.83 per wagon, \$0.17 per wagon
21. (a) About \$8000; (b) about 38%
23. (a) $-3$ cm/s, 12 cm/s; (b) 10 s; (c) 2 s; (d) 17 cm/s
25. (a) $t = 0$, $t = 1.5$; (b) $t = 0.75$; (c) $t = 0.5$, $t = 0.75$
27. (a) 25,000; (b) 900 bacteria/h; (c) 1.4 h
29. $f'(t) > 0, f''(t) < 0; f'(t) < 0$

## Chapter 1 Review Problem Set (p. 42)

1. (a) $x \geq \frac{2}{3}$; (b) 3, $\sqrt{7}$; (c) $2x + h - 3$
3. (a) Neither; (b) odd; (c) even
5. (a)

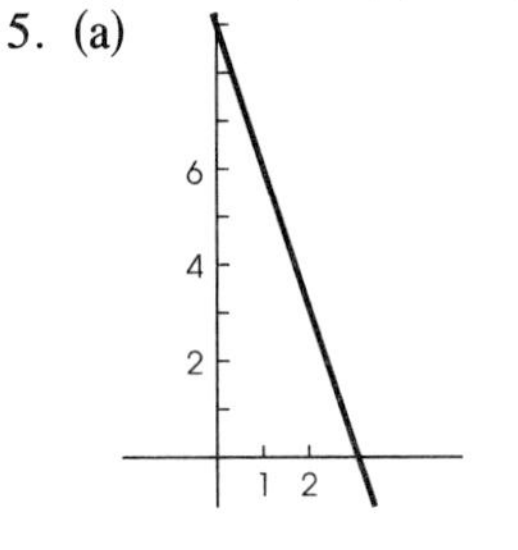


(b)

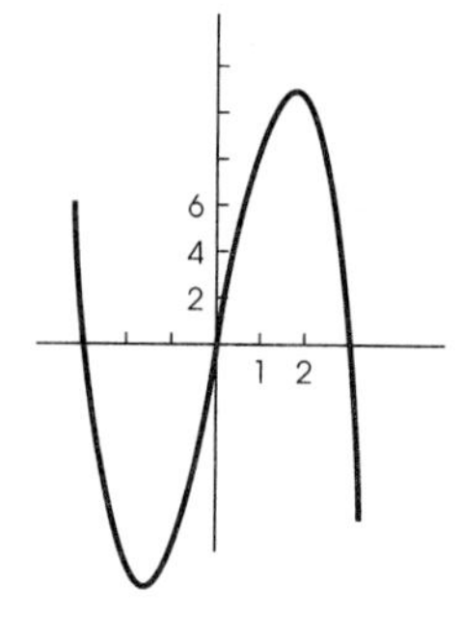


(c)

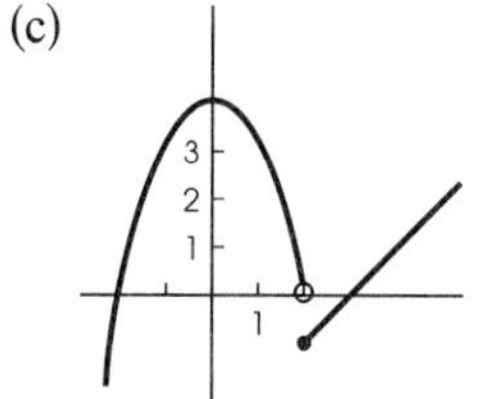

7. $f(x) = 2x + 1,\ g(x) = x^2 + 3x,\ h(x) = \sqrt{x}$
9. (a) $\frac{3}{2}, -6$; (b) 3, 5; (c) $-\frac{4}{3}, 4$
11. $y = \frac{2}{5}x + \frac{9}{5}$
13. (a) 11; (b) 0; (c) $-\frac{3}{2}$
15. $y + 7 = -14(x - 1)$
17. (a) $F(x) = 180 + 0.15x$; (b) \$1080
19. (a) $S(x) = 1800 + 0.05x$; (b) \$10,000
21. $f(t) = 90{,}000 - 11{,}000t$
23. (a) $1 < x < 3$; (b) 0.5 and 3.5; (c) $2 < x < 4$; (d) 2 and $4 < x < 5$; (e) 1
25. (a) 96 ft/s, −128 ft/s; (b) 6 s; (c) 576 ft; (d) 12 s; (e) 192 ft/s
27. (a) 6%/yr; (b) 10%/yr
29. (a) \$27, \$31, \$4; (b) \$27; (c) 1500
31. (a) $A(x) = x^2 + \frac{\pi}{8}x^2$; $P(x) = 3x + \frac{\pi}{2}x$; (b) $20 + \frac{5\pi}{2}$, $3 + \frac{\pi}{2}$

## Problem Set 2.1 (p. 52)

1. (a) 11; (b) 22; (c) 9; (d) 17
3. 2
5. 2
7. Does not exist
9. (a)

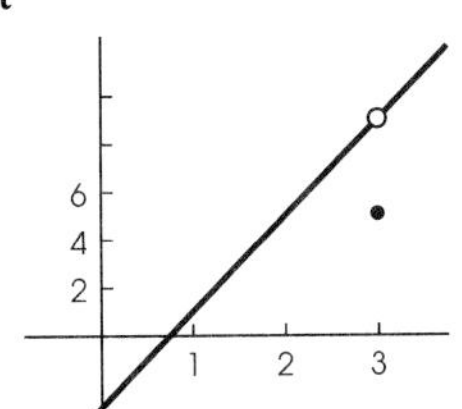


(b) 9

11. (a) 3; (b) does not exist; (c) 4; (d) 2; (e) 3; (f) 1

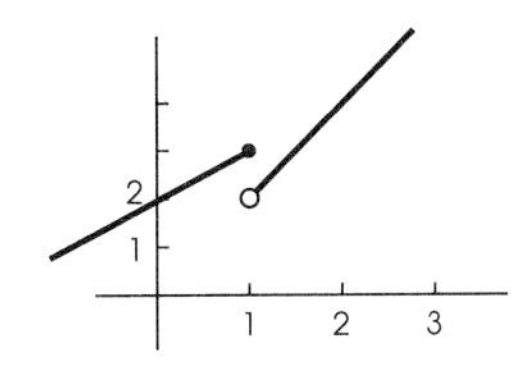


13. 4
15. −5
17. 11
19. 2
21. $\frac{5}{8}$
23. Does not exist.
25. $-\frac{1}{7}$
27. $\frac{3}{2}$
29.
$$\lim_{x\to 2}(5x^3 - 4x^2 + 3x + 5) \overset{④}{=} \lim_{x\to 2}(5x^3 - 4x^2) + \lim_{x\to 2}(3x + 5)$$
$$\overset{④}{=} \lim_{x\to 2} 5x^3 - \lim_{x\to 2} 4x^2 + \lim_{x\to 2} 3x + \lim_{x\to 2} 5$$
$$\overset{③,①}{=} 5\lim_{x\to 2} x^3 - 4\lim_{x\to 2} x^2 + 3\lim_{x\to 2} x + 5$$
$$\overset{⑦}{=} 5(\lim_{x\to 2} x)^3 - 4(\lim_{x\to 2} x)^2 + 3\lim_{x\to 2} x + 5$$
$$\overset{②}{=} 5(2)^3 - 4(2)^2 + 3(2) + 5 = 35$$

31. −1
33. 5
35. 64
37. −2
39. 9
41. $-\frac{4}{3}$
43.
$$\lim_{x\to 2}\frac{3f(x) + 2x - 1}{2g(x)} \overset{⑥}{=} \frac{\lim_{x\to 2}[3f(x) + 2x - 1]}{\lim_{x\to 2} 2g(x)}$$
$$\overset{④}{=} \frac{\lim_{x\to 2} 3f(x) + \lim_{x\to 2} 2x - \lim_{x\to 2} 1}{\lim_{x\to 2} 2g(x)}$$
$$\overset{③,①}{=} \frac{3\lim_{x\to 2} f(x) + 2\lim_{x\to 2} x - 1}{2\lim_{x\to 2} g(x)}$$
$$\overset{②}{=} \frac{3(-3) + 2(2) - 1}{2(4)} = -\frac{3}{4}$$

45. (a) 16; (b) $\frac{1}{16}$; (c) 3; (d) 125
47. 4
49. $-\frac{1}{49}$
51. 2
53. Does not exist.
55. Does not exist.
57. 3
59. (a) 1; (b) 0.5
61. (a) 0; (b) does not exist; (c) 36; (d) $\frac{1}{8}$

## Problem Set 2.2 (p. 60)

1.
$$f'(x) = \lim_{h\to 0}\frac{(x + h)^2 + 2(x + h) - (x^2 + 2x)}{h}$$
$$= \lim_{h\to 0}\frac{2xh + h^2 + 2h}{h}$$
$$= \lim_{h\to 0}(2x + h + 2) = 2x + 2$$

3.
$$f'(x) = \lim_{h\to 0}\frac{(x + h)^3 - x^3}{h} = \lim_{h\to 0}\frac{3x^2h + 3xh^2 + h^3}{h}$$
$$= \lim_{h\to 0}(3x^2 + 3xh + h^2) = 3x^2$$

5. $f'(x) = \lim_{h\to 0} \dfrac{\dfrac{4}{x+h} - \dfrac{4}{x}}{h} = \lim_{h\to 0} \dfrac{4x - 4(x+h)}{(x+h)xh}$
$= \lim_{h\to 0} \dfrac{-4h}{(x+h)xh} = \lim_{h\to 0} \dfrac{-4}{(x+h)x} = -\dfrac{4}{x^2}$

7. $f'(x) = \lim_{h\to 0} \dfrac{\dfrac{x+h}{x+h+1} - \dfrac{x}{x+1}}{h}$
$= \lim_{h\to 0} \dfrac{(x+1)(x+h) - x(x+h+1)}{(x+h+1)(x+1)h}$
$= \lim_{h\to 0} \dfrac{h}{(x+h+1)(x+1)h}$
$= \lim_{h\to 0} \dfrac{1}{(x+h+1)(x+1)} = \dfrac{1}{(x+1)^2}$

9. $f'(x) = \lim_{h\to 0} \dfrac{\sqrt{x+h+2} - \sqrt{x+2}}{h}$
$= \lim_{h\to 0} \dfrac{(\sqrt{x+h+2} - \sqrt{x+2})(\sqrt{x+h+2} + \sqrt{x+2})}{h(\sqrt{x+h+2} + \sqrt{x+2})}$
$= \lim_{h\to 0} \dfrac{h}{h(\sqrt{x+h+2} + \sqrt{x+2})}$
$= \lim_{h\to 0} \dfrac{1}{\sqrt{x+h+2} + \sqrt{x+2}} = \dfrac{1}{2\sqrt{x+2}}$

11. $f(x) = x^{4/3}, f'(x) = \frac{4}{3}x^{1/3}$
13. $f(x) = x^{1.9}, f'(x) = 1.9x^{0.9}$
15. $2x^{-2/3}$
17. $-\frac{5}{2}x^{1/4}$
19. $3.27x^{0.09}$
21. $-11\sqrt{3}\,x^{\sqrt{3}-1}$
23. (a) $2x^2$; (b) $x^{7/2}$; (c) $x^{1/2}$; (d) $x^{7/4}$; (e) $x^3/8$; (f) $x^3 - 2x^{7/2} + x^4$
25. (a) $12x^3$; (b) $9x^{7/2}$; (c) $3x^{1/2}$; (d) $4x^3$
27. (a) 0; (b) $28x^6$; (c) $31x^{2.1}$; (d) $14x^{1/6}$; (e) $-(6\pi + 3)x^{2\pi}$; (f) 0

29. $f'(x) = \lim_{h\to 0} \dfrac{\dfrac{1}{(x+h)^2} - \dfrac{1}{x^2}}{h} = \lim_{h\to 0} \dfrac{x^2 - (x+h)^2}{(x+h)^2x^2h}$
$= \lim_{h\to 0} \dfrac{-2xh - h^2}{(x+h)^2x^2h} = \lim_{h\to 0} \dfrac{-2x - h}{(x+h)^2x^2}$
$= -\dfrac{2x}{x^4} = -\dfrac{2}{x^3}$

31. $50\pi$ cm³/cm
33. $-0.8$ dyn/cm
35. (27, 243)
37. (a) 0.24998, 0.25; (b) $-0.12498$, $-0.125$

## Problem Set 2.3 (p. 67)

1. $6x^2 - 2x^{-1/3} - 10x^{-3}$
3. $6x^2 + \frac{5}{2}x^{-1/2} + 20x^{-6}$
5. $x^{-2/3} + 2x^{-3/2}$
7. $\frac{7}{2}x^{5/2} - 2 + \frac{5}{2}x^{-1/2}$
9. $y + 3 = 17(x + 1)$
11. $y - 2 = \frac{3}{16}(x - 16)$
13. $4x^3 - 10x$
15. $10x^4 + 21x^2 + 18x - 15$
17. $2x + 6x^{1/2} + 2 + 5x^{-1/2}$
19. $\frac{10}{3}$
21. $-4$
23. $(2x^2 + 3)/x^2$
25. $(4x^2 - 4x - 10)/(2x - 1)^2$
27. $6/[\sqrt{x}(3 + \sqrt{x})^2]$
29. $2 - 2x^{-1/2} - 15x^{-4}$
31. $8 - 6x^{-4} - 12x^{-5}$
33. $4x + 3 + 4(x - 4)^{-2}$
35. $y = \frac{13}{3}(x - 1)$
37. (a) $6(3x + 1)$; (b) $2(x^2 - x + 4)(2x - 1)$; (c) $16 + 20x^{-1/2}$; (d) $2(4x + 3x^{-2})(4 - 6x^{-3})$
39. 2.01 cm/s
41. (a) $P(x) = -0.02x^2 + 220x - 3000$; (b) $-0.04x + 220$, \$180 per mower; (c) 5500; this production level produces maximum profit.

## Problem Set 2.4 (p. 73)

1. $36(3x - 4)^{11}$
3. $8(3x^2 + x + 1)^7(6x + 1)$
5. $5(x + 2x^{-2})^4(1 - 4x^{-3})$
7. $10(x^2 - 2x^{1/2})^9(2x - x^{-1/2})$
9. $\frac{1}{2}(2 + 3x - x^2)^{-1/2}(3 - 2x)$
11. $\frac{2}{3}(2x^3 - 7x + 6)^{-1/3}(6x^2 - 7)$
13. $\frac{3}{4}(2x^3 + x)^{-1/4}(6x^2 + 1)$
15. $-60(3x - 1)^{-6}$
17. $\frac{8}{3}(2x - 1)^{-5/3}$
19. $-35/4374$
21. $3\left(\dfrac{2x - 1}{3x + 2}\right)^2 \dfrac{7}{(3x + 2)^2}$
23. $5\left(\dfrac{x^2 + 4}{2x - 5}\right)^4 \dfrac{2x^2 - 10x - 8}{(2x - 5)^2}$
25. $(8x + 5)(5)(2x - 1)^4 + (2x - 1)^5(8) = 2(2x - 1)^4(48x + 21)$
27. $(3x^2)(\frac{5}{3})(2x^2 + 3)^{2/3}(4x) + (2x^2 + 3)^{5/3}(6x) = 2x(2x^2 + 3)^{2/3}(16x^2 + 9)$
29. $(x^2 + 1)^3(2)(x^2 + 4)(2x) + (x^2 + 4)^2(3)(x^2 + 1)^2(2x) = 2x(x^2 + 1)^2(x^2 + 4)(5x^2 + 14)$
31. $\dfrac{(3x - 1)^4(2x) - (x^2)(12)(3x - 1)^3}{(3x - 1)^8} = -\dfrac{2x(3x + 1)}{(3x - 1)^5}$
33. 114
35. $y - \frac{1}{6} = -\frac{7}{18}(x - 1)$

37. (a) \$41.00;
  (b) increasing production from 100 to 101 increases daily profit by about \$41.00;
  (c) \$148.33

39. 13.5
41. 322.5
43. 2

45. (a) $20{,}000(1 + r/2)^7$; (b) 7969.24; 30,072.61

47. (a) $Y = 1{,}000{,}000r/(8.5r + 40{,}000)$;
  (b) $dy/dr = -8{,}500{,}000/(8.5r + 40{,}000)^2 < 0$
  $dY/dr = 4 \times 10^{10}/(8.5r + 40{,}000)^2 > 0$;
  (c) Yield per plant decreases as density increases, but yield per hectare increases as density increases. Within limits, this does make sense.

## Problem Set 2.5 (p. 81)

1. $60x^2 + 18$
3. $-24x^{-4}$
5. $-\frac{3}{4}x^{-3/2} - 24x$
7. $-9/256$
9. $-45/1024$
11. $6(x^2 - 3x + 4)(5x^2 - 15x + 13)$
13. $-1/64$
15. (a) $s = -16t^2 + 96t + 256$; (b) 384 ft; (c) 32 ft/s; (d) $t = 3$ s; (e) $t = 8$ s; (f) $-160$ ft/s; (g) $-32$ ft/s² both times.
17. (a) $-1$ cm, 24 cm/s, 30 cm/s²; (b) $-13.5$ cm/s; (c) 18 cm/s²; (d) $-1 < t < 2$
19. (a) 1/1/72 to 12/31/77; (b) after 6/30/74; (c) 7/1/77; (d) 7/1/74;
  (e)

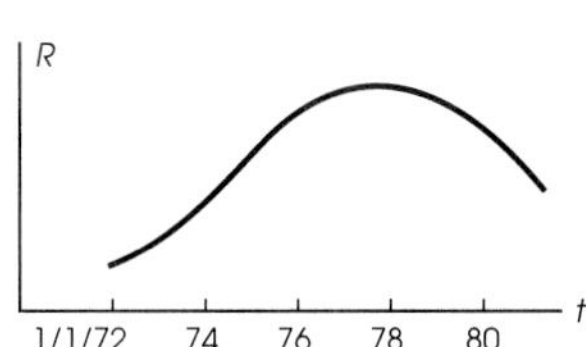


21. (a) $dI/dt = 0$ last year; $dI/dt > 0$ and $d^2I/dt^2 > 0$ in future;
  (b) $dP/dt$ is quite negative now; $d^2P/dt^2 > 0$ in future; $dP/dt > 0$ after 2 years;
  (c) $dT/dt > 0$ but $d^2T/dt^2 < 0$;
  (d) $d^2s/dt^2 > 0$ for $0 < t < 2$, but $d^2s/dt^2 = 0$ for $t > 2$.
23. $-90$
25. 192
27. 30
29. (a) 64 ft/s; (b) 160 ft/s

## Chapter 2 Review Problem Set (p. 82)

1. (a) 0; (b) 2; (c) 12; (d) 4; (e) 0 (f) does not exist.

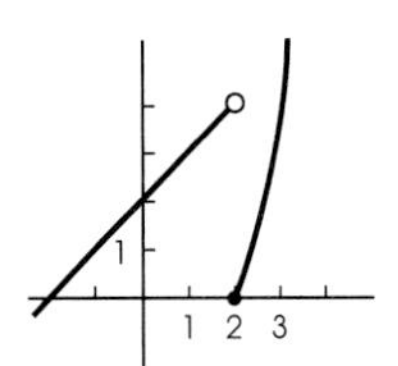


3. 2
5. 10
7. 3
9. $\frac{2}{5}$
11. $\frac{2}{3}$
13. 0
15. $-\frac{5}{3}$
17. $\frac{3}{2}$
19. $\frac{19}{9}$
21. $\frac{8}{3}$
23. 3
25. $-\frac{5}{3}$
27. $\frac{1}{6}$
29. $-1$
31. $2x - 3$
33. $-x^{-3/2}$
35. $x^{-2/3}$
37. $14x^{4/3}$
39. $3x^{1/2}$
41. $\frac{3}{4}x^{-1/4}$
43. $12(2x - 1)^5$
45. $(x^2 - 2x + 5)^{-1/2}(x - 1)$
47. $3(2x - 4x^{-1/2})^2(2 + 2x^{-3/2})$
49. $\frac{4}{3}(x^3 - 3x)^{1/3}(3x^2 - 3) = 4(x^2 - 1)(x^3 - 3x)^{1/3}$
51. $3\left(\frac{2x^2 + 1}{x^3 - 2x + 4}\right)^2 \frac{-2x^4 - 7x^2 + 16x + 2}{(x^3 - 2x + 4)^2}$
53. $(2x^2 + 5)(2x + 5)^{-1/2} + \sqrt{2x + 5}(4x) = (10x^2 + 20x + 5)/\sqrt{2x + 5}$
55. $\frac{3}{2}(2x^2 + 5)^{1/2}(4x) = 6x\sqrt{2x^2 + 5}$
57. $5[(x^2 + 2x)^4 + 3]^4(4)(x^2 + 2x)^3(2x + 2) = 40[(x^2 + 2x)^4 + 3]^4(x^2 + 2x)^3(x + 1)$
59. $\frac{3}{4}$
61. $y + 1 = 13(x - 1)$
63. (16, 16)
65. (a) $1000(1 + \frac{x}{1200})^{119}$; (b) \$24,513.57; \$2433.11
67. (a) 14 ft, 6 ft/s, $-3$ ft/s²; (b) $-\frac{3}{2}$ ft/s²; (c) $0 \le t < 4$
69. (a) $0 \le t \le 17$; (b) $0 \le t < 10$; (c) $t = 10$

## Problem Set 3.1 (p. 91)

1. Continuous.
3. Not continuous.
5. Continuous.
7. Continuous.
9. $\lim_{x\to 1} f(x) = \lim_{x\to 1} (x^2 - 4x)/(x - 4) = 1 = f(1)$; $f$ is not defined at 4.
11. $\lim_{x\to 4} f(x) = \lim_{x\to 4} (x^2 - 4)/(3x + 6) = \frac{12}{18} = f(4)$; $f$ is not defined at $-2$.
13. $f(4) = 2$
15. $f(1) = -1$

17. $-1$ and 6; $f(6) = \frac{8}{7}$
19. 1 and 4
21. $-2$ and 2; $f(-2) = f(2) = 3$
23. None.
25. None.
27. Continuous on its domain, $x > 0$.
29. None.
33. None.
37. $x = 0$
41. $x = 2$
31. $x = 1$
35. $x = -1$
39. $x = -\frac{1}{2}$
43. $x = 2$
45. (a) $p, r, t$; (b) $p, q, t$; (c) $p, t$; (d) $t$
47. Continuous, not differentiable.
49. Not continuous.
51. Not continuous.
53. Continuous, not differentiable.
55. Discontinuous at every integer.

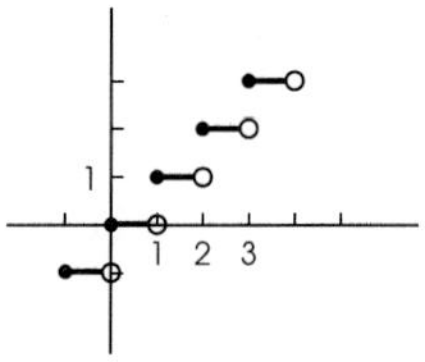


57.

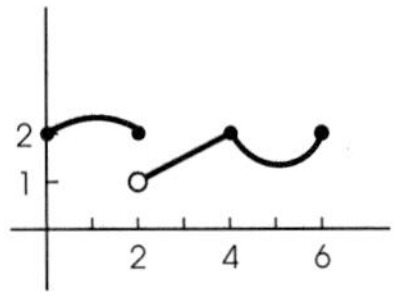


## Problem Set 3.2 (p. 98)

1.

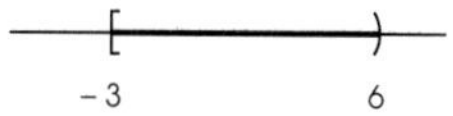


3.

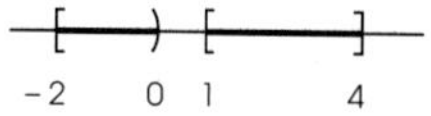


5.

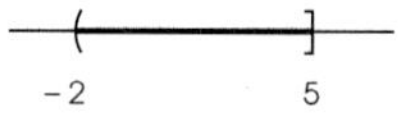


7. $(-6, \infty)$
9. $(-\infty, -\frac{7}{3}]$
11. $(-\infty, 6)$
13. $[-7, 3]$

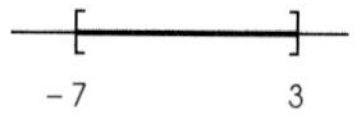


15. $(-8, 2)$

17. $(-\infty, -2] \cup [3, \infty)$

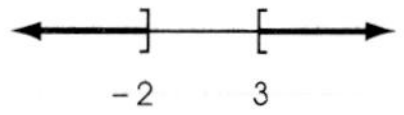


19. $(-4, \frac{1}{2})$

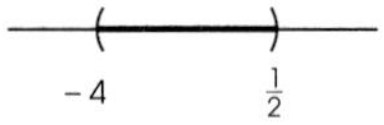


21. Inc $(1, \infty)$, dec $(-\infty, 1)$
23. Inc $(-\infty, 2)$, dec $(2, \infty)$
25. Inc $(-\infty, -2) \cup (2, \infty)$, dec $(-2, 2)$
27. Inc $(-\infty, 1 - \sqrt{5}) \cup (1 + \sqrt{5}, \infty)$, dec $(1 - \sqrt{5}, 1 + \sqrt{5})$
29. Up $(0, \infty)$, down $(-\infty, 0)$
31. Up $(1, \infty)$, down $(-\infty, 1)$
33. Up $(-\infty, -2) \cup (2, \infty)$, down $(-2, 2)$
35.

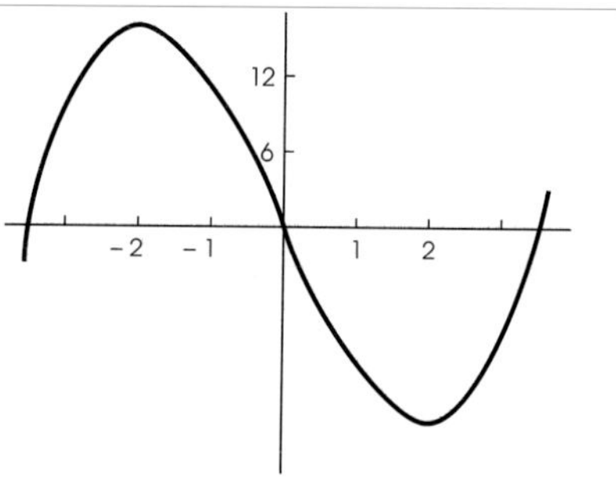


37.

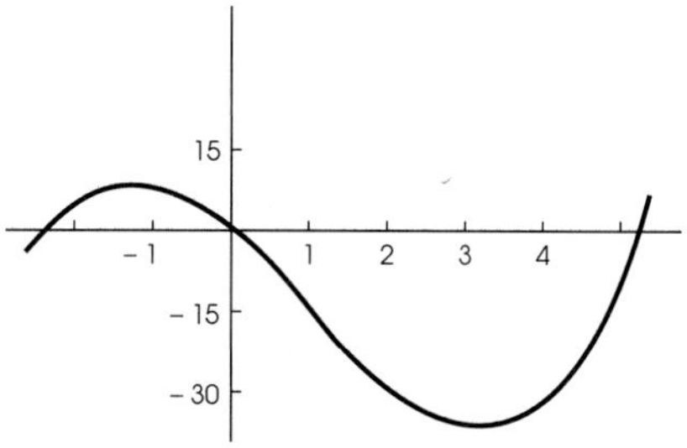


39. (a) $-2, 1, 3.5$; (b) $(1, 3.5)$; (c) $(-2, 0) \cup (2, 4)$
41. (a) $(0, 2)$; (b) $(3, 4)$; (c) $(2, 3)$
43.

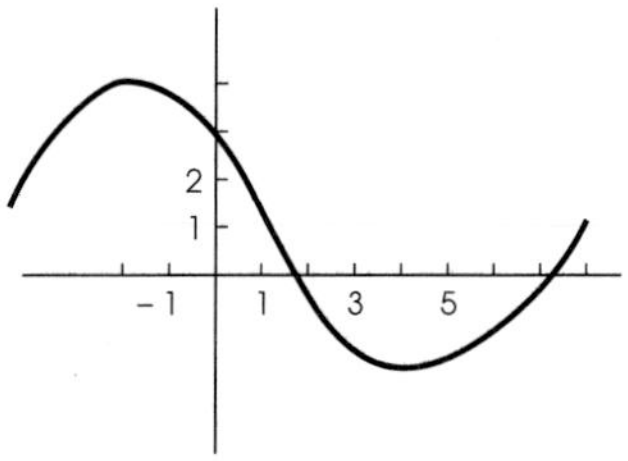

45.

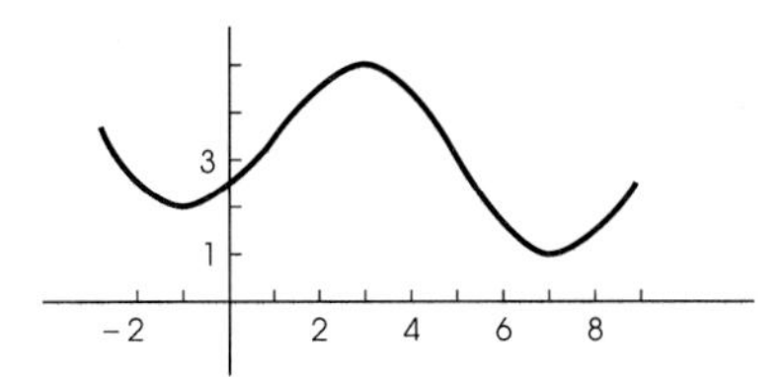


47.

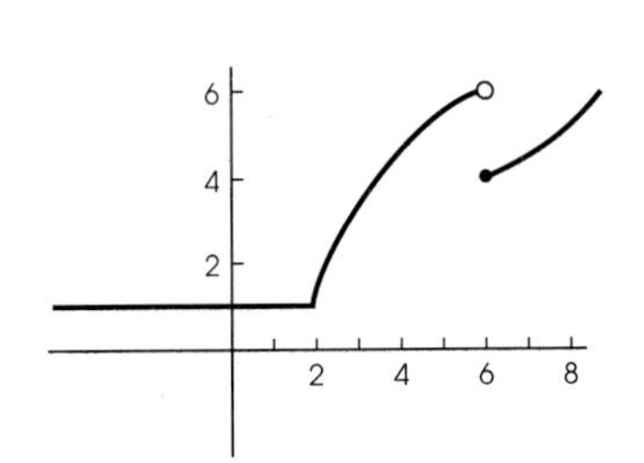


49.

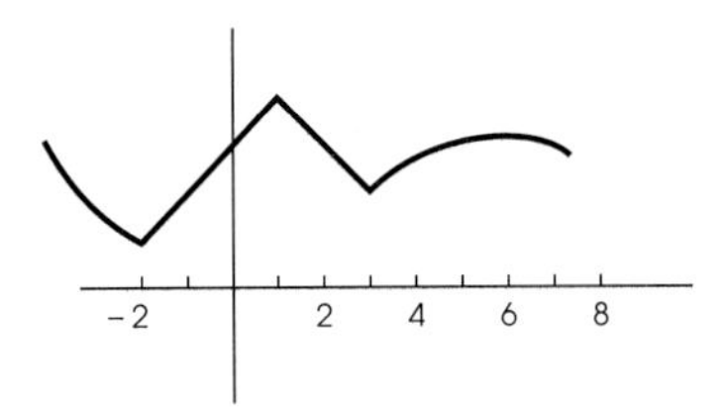


51. $(-\frac{29}{12}, \infty)$

53. $(-\infty, 2] \cup [8, \infty)$

55. $(-3, 2) \cup (2, 4)$

57. $(-\infty, -3) \cup (2, \infty)$

59. $f'(t) > 0$ and $f''(t) < 0$ for $1970 < t < 1990$

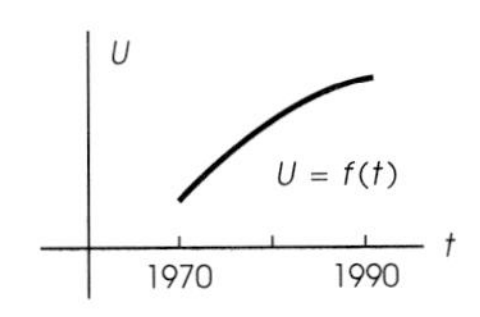


61. $f'(t) < 0$ and $f''(t) > 0$ for $0 < t < 10$, $f'(t) = 0$ for $t > 10$

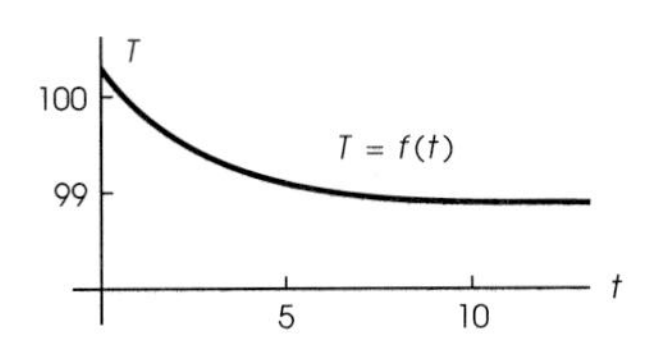


63.

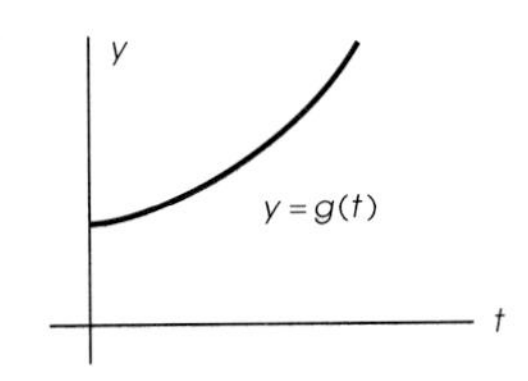


65.

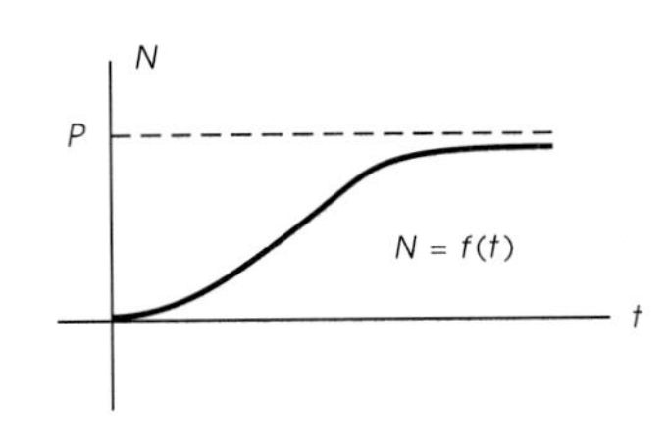

## Problem Set 3.3 (p. 107)

1. Global max at $t$, global min at $u$, local max at $r$, local min at $p$ and $s$.
3. Global max at x $= 2, f(2) = 2$; no global min.
5. Global min at $x = 0, f(0) = 0$; no global max.
7. Global max at $x = 10, f(10) = 2$; global min at $x = 0$, $f(0) = 0$
9. Max $f(0) = 0$; min $f(2) = -8$
11. Max $f(0) = 3$; min $f(3) = -6$
13. Max $f(3) = 19$; min $f(-2) = f(1) = -1$
15. Max $f(-1) = 9$; min $f(-3) = -43$
17. Max $f(1) = 2$; min $f(-8) = -4$
19. Max $f(3) = 3$; min $f(-6) = -6$
21. Max $f(-6) = f(3) = 0$; min $f(1) = -4$
23. Max $f(1) = \frac{3}{2}$; min $f(-1) = -\frac{3}{2}$
25. Max $f(3) = \frac{9}{10}$; min $f(0) = 0$
27. 0, 1; local max at 0, local min at 1
29. $-2$, 2; local max at 2, local min at $-2$
31. 1, 4; local max at 1, local min at 4
33. $-2$, 2; local max at $-2$, local min at 2
35. 1, 3; local max at 1, local min at 3
37. $-2$, 0, 2; local max at 0, local min at $\pm 2$
39. $-2$; local max at $-2$
41. (0, 0)
43. (2, $-1$)
45. ($-1$, $-3$), (3, $-179$)
47. ($-1$, 2), (1, 2)

49.

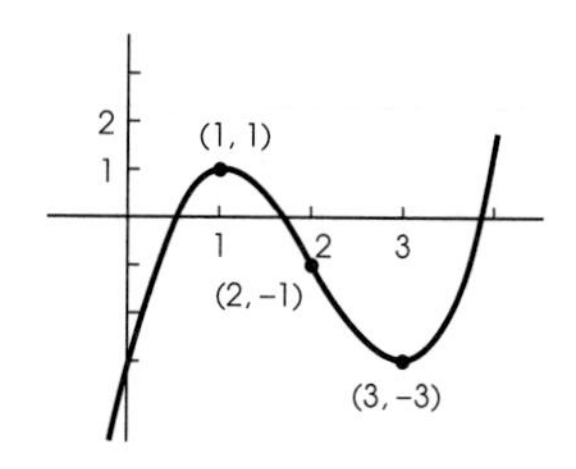


51.

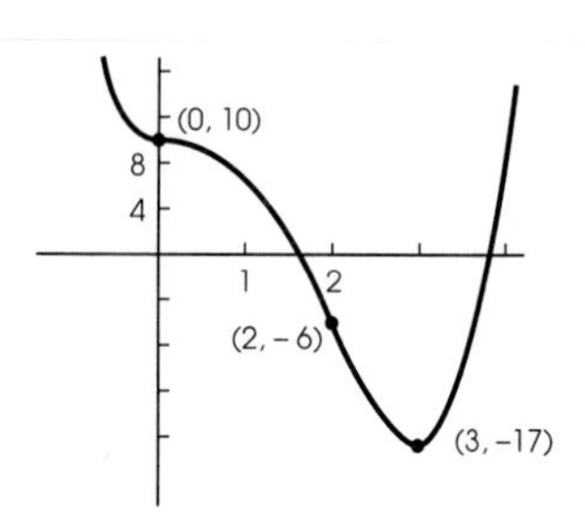


53. Min $f(2) = 8$; max $f(1) = 10$

55. Min $f(2) = 2$; max $f(0) = 6$

57.

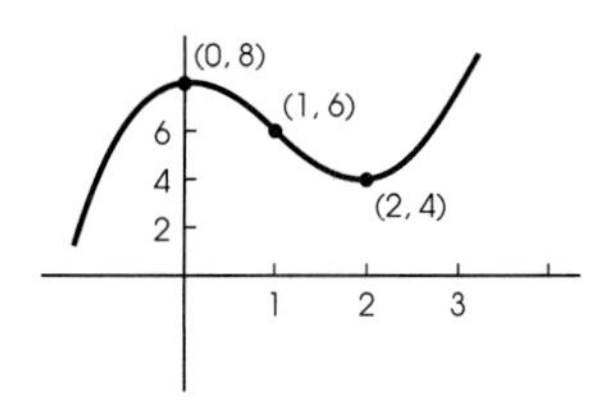


59.

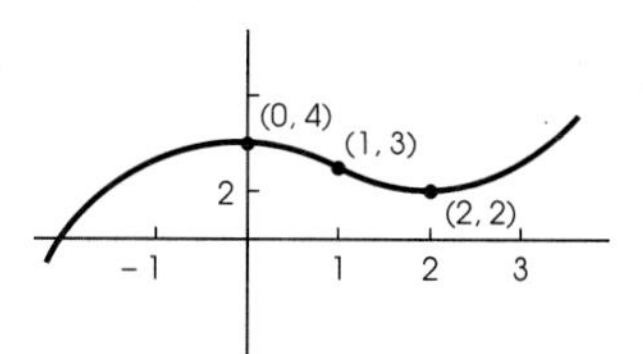


61.

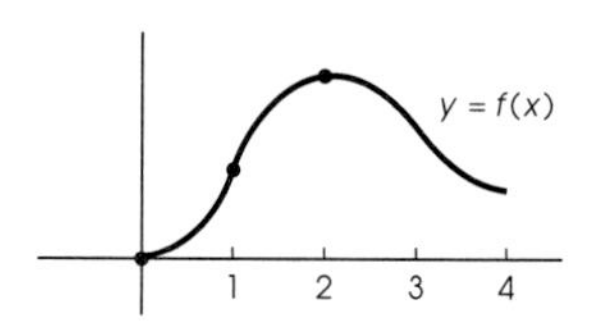


63.

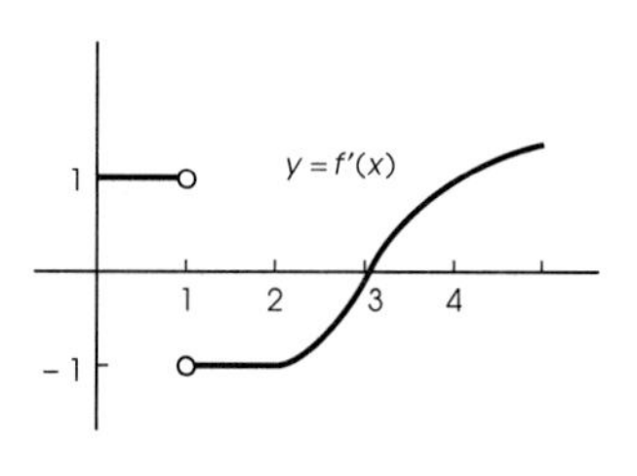


65.

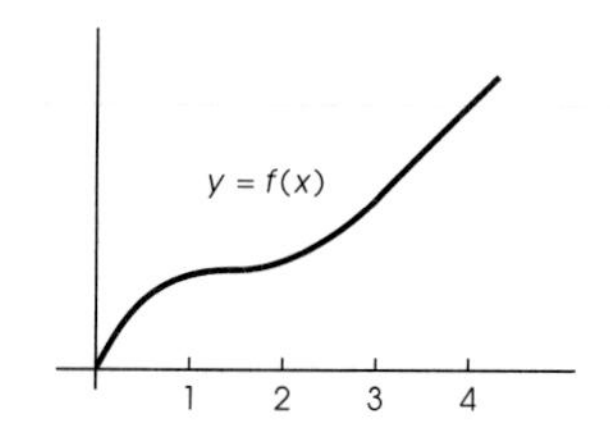


67.

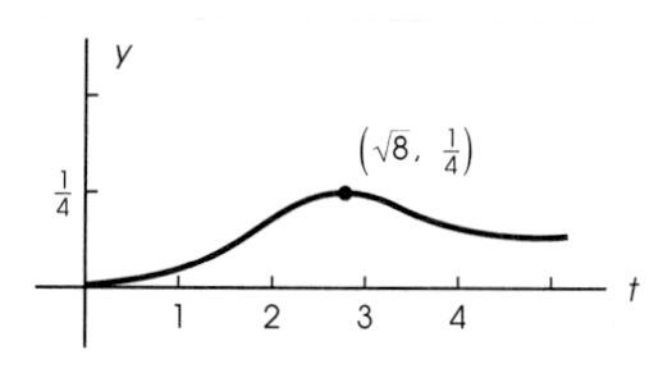


## Problem Set 3.4 (p. 116)

1. $f(x)$ gets closer and closer to 10 as $x$ becomes more and more negative (that is, as $x$ gets larger and larger in the negative direction).
3. $f(x)$ gets larger and larger as $x$ gets larger and larger.
5. $f(x)$ becomes more and more negative as $x$ gets closer and closer to 2 from the right.

7. 0
9. $\infty$
11. $\infty$
13. $-\infty$
15. $\infty$
17. 4
19. $\frac{5}{4}$
21. 0
23. 0
25. $-\frac{4}{3}$
27. 2
29. $\frac{1}{2}$
31. 0
33. 5
35. $-1$
37. $-\frac{1}{2}$
39. $y = 0$; $x = 0$
41. $y = -1$; $x = -2$
43. $y = -2$; $x = 0$, $x = 5$
45. $y = 1$; $x = -1$, $x = 3$
47. None; none.
49. $y = 1$; $x = -1$
51.

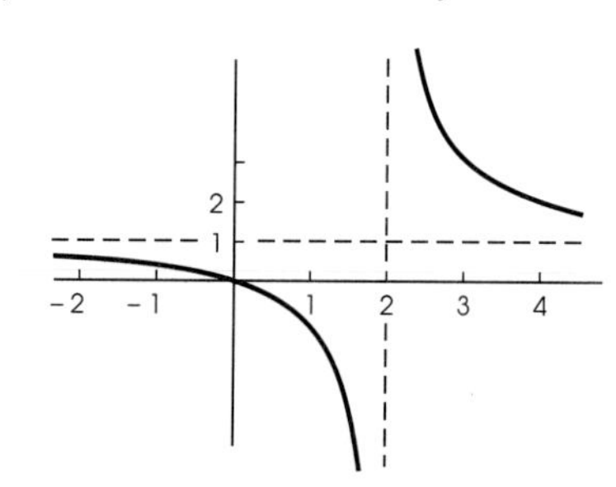

53.

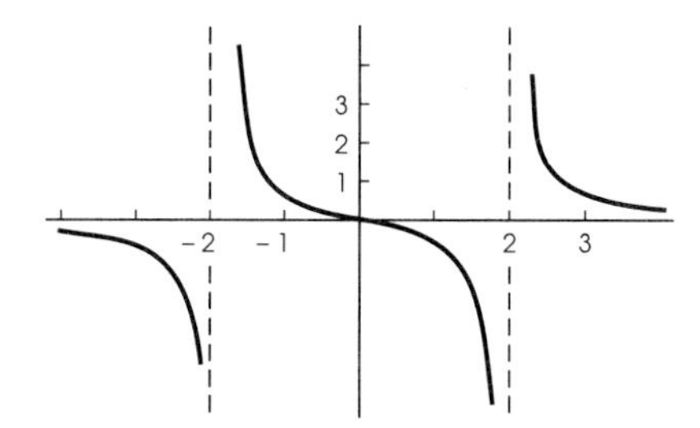


55.

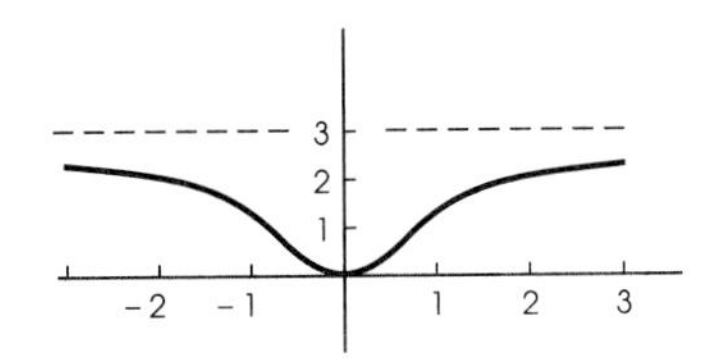


57.

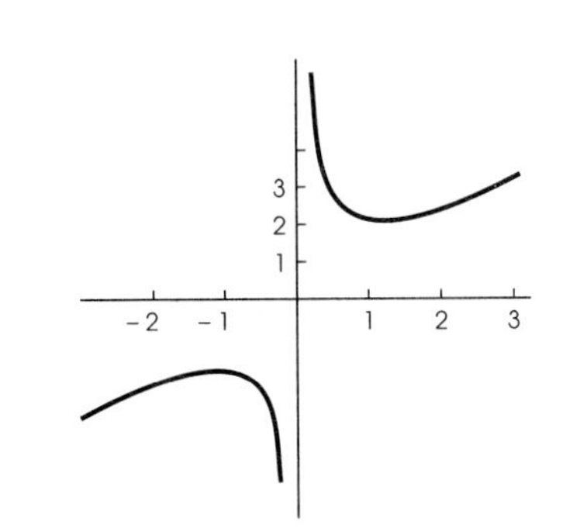


59.

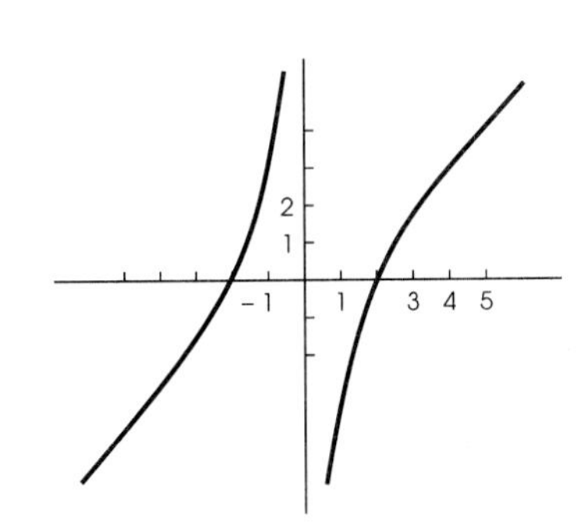


61. $y = 0$; $x = -3$, $x = 0$, $x = 3$

63. $f(x) = \dfrac{4x^2}{x^2 - x - 6}$

65. (a) 0; (b) 2

67. \$250

69. 12 min

71. $\infty$

## Chapter 3 Review Problem Set (p. 118)

1. (a) $\lim_{x \to 1} f(x) = \lim_{x \to 1} \dfrac{x^3 - 8}{x^2 - 4} = \dfrac{7}{3} = f(1)$; (b) no; (c) 3
3. (a) $[-\frac{7}{5}, \infty)$; (b) $(-3, \frac{1}{2})$
5. Continuous, differentiable.
7. Not continuous.
9. Continuous, not differentiable.
11. Continuous, differentiable.
13. Max value: $f(1) = \frac{32}{3}$; Min value: $f(27) = 4$
15. Local max at $x = -4$; local min at $x = 2$
17. $(-2, -162)$ and $(2, -18)$
19. (a) Inc $(3, \infty)$; dec $(-\infty, 3)$;
(b) up $(-\infty, 0) \cup (2, \infty)$; down $(0, 2)$;
(c)

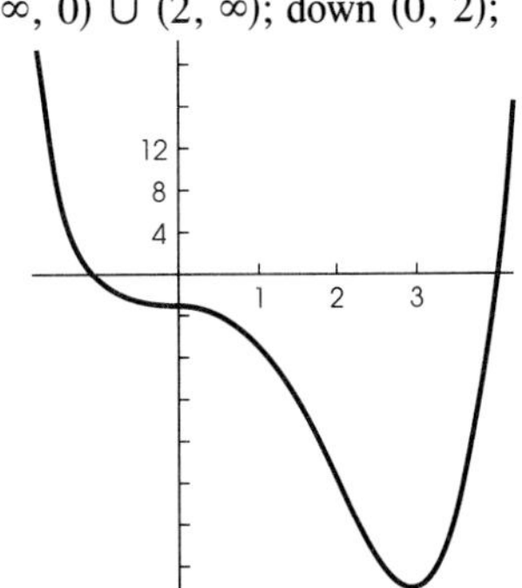


21. $\infty$

23. $-\infty$

25. 0

27. $-1$

29. (a) Inc $(5, \infty)$; dec $(0, 5)$; (b) up $(0, \infty)$;
(c) local min: $f(5) = 10$;
(d)

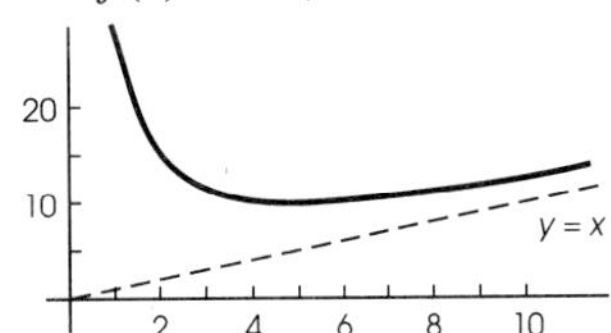


31.

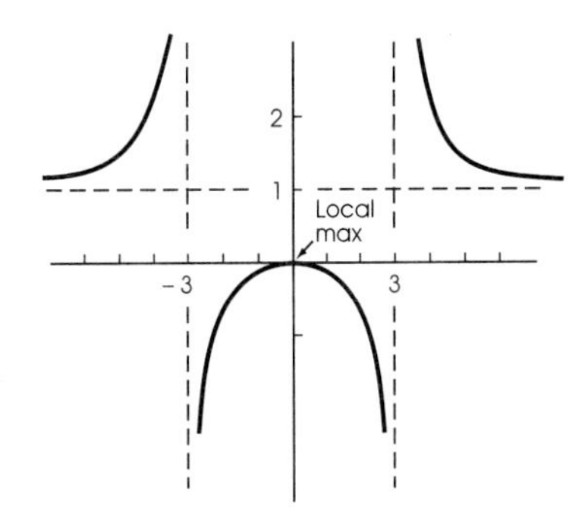

33. $f(x) = \dfrac{2x^3 + 1}{x(x+1)(x-2)}$

35. (a) $\{x: x \neq -2\}$; (b) $x = -2$; (c) 0;
(d) $x = 0$ and $x = -2$; (e) $-\frac{7}{600}$

37. This is $f'(8)$ where $f(x) = \sqrt[3]{x}$ and so has value $\frac{1}{12}$.

## Problem Set 4.1 (p. 129)

1. 5000
3. $\frac{9}{8}$
5. 32
7. $250\pi$
9. $x = 4$, $y = 2$, $A = 8$
11. $x = 75$ m, $y = 50$ m, $A = 3750$ m$^2$
13. $x = 75$ m, $y = 37.5$ m, $A = 2812.5$ m$^2$
15. $x = 20\sqrt{2}$ ft, $y = 30\sqrt{2}$ ft
17. 2 in. by 5 in. by 20 in.
19. 20 ft by 20 ft by 40 ft
21. $2000/\sqrt{3} \approx 1155$ ft
23. 14 in. wide and 21 in. high
25. $w = 2\sqrt{3}/3$ ft, $d = 2\sqrt{6}/3$ ft
27. (a) 17.596 cm (use it to make circle);
(b) 0 (use all 40 cm to make circle).
29. $h = 8\sqrt{3}$ in., $r = 4\sqrt{6}$ in.
31. $(1 \pm \sqrt{2}, 2)$
33. $x = 4 + \sqrt[3]{16} \approx 6.52$, $y \approx 5.17$

## Problem Set 4.2 (p. 136)

1. (a) $-1000 + 8x - 0.008x^2$; (b) $8 - 0.016x$;
(c) 500
3. (a) $-1000 + 30x - 0.05x^2$; (b) $30 - 0.10x$;
(c) 300
5. $28
7. 16
9. (a) 31; (b) $36.96
11. (a) 1000, $2; (b) 625
13. (a) $10{,}000 + 120x$; (b) $\frac{2}{3}(500x - x^2)$; (c) 200
15. (a) $1200x - x^2$; (b) 550; (c) $272,500;
(d) $650; (e) $800/unit; (f) $800
17. 145
19. 63,250. This might suggest 2 runs per year of 50,000 each.
21. 8.4%
23. 9.66667%
25. $600
27. 5
29. Sell immediately.
31. (c) 9.53518%; (d) $108,794

## Problem Set 4.3 (p. 142)

1. 35
3. (a) 2652 lb; (b) 55 trees/acre
5. Charges of $3.70 and $3.80 both produce a maximum revenue of $421.80.
7. (a) 63.25 mi/h; (b) $286.49
9. (a) 44.72 mi/h; (b) $69.77
11. (a) $\frac{8}{9}c$; (b) $\frac{4}{9}c$
13. 37,500; 28,125
15. (a) $\frac{20}{3}$; (b) $\frac{10}{3}$
17. (a) 10; (b) 1156 lb
19. $90.40 resulting in 107 rented rooms.
21. (a) 12 days; (b) 20 days
23. 5.18 mi

## Problem Set 4.4 (p. 149)

1. $y = (3 - x^2)/2$
3. $y = (2 - x)/(4 + 3x)$
5. $y = (-x^2 + x + 3)/(x - 1)$
7. $y = (4x^2 + 7x + 2)/(x^2 + x)$
9. $y = \pm\sqrt{x^2 - 2x + 4}$
11. $y = 4 \pm \sqrt{x^2 + 5}$
13. $(-x^2 + 2x - 4)/(x - 1)^2$; $(1 - y - 2x)/(x - 1)$
15. $-2y/x$
17. $2x(1 - y)/(x^2 + 4)$
19. $-y/(x + y^2)$
21. $-(x + y)/(x + 3y)$
23. $-2x/y$
25. $y - 5 = \frac{12}{5}(x + 12)$
27. $y - 1 = -\frac{1}{2}(x - 2)$
29. $y - 1 = -3(x - 4)$
31. 12 units/s
33. 1.5 ft/s
35. (a) $8\sqrt{5}/5 \approx 3.58$ units/s; (b) 6 units$^2$/s
37. $-\$30{,}000$/wk
39. (a) $\frac{4}{3}$ lb/wk; (b) $0.96/wk
41. $y - 1 = -\frac{5}{16}(x - 2)$
43. (a) 6.31 cm/min; (b) 14.31 cm/min;
(c) 27 cm$^2$/min
45. $51.2\pi$ cm$^3$/day
47. $-\$5\frac{1}{3}$ hundred thousand

## Problem Set 4.5 (p. 156)

1. $dy = 3x^{1/2}\,dx$
3. $dy = (2x^{-1/2} - 5x^{-2})\,dx$
5. $dy = -15(4 - 3x)^4\,dx$
7. $\Delta y = 1.13$; $dy = 1.1$

9. $\Delta y = 0.0049938$; $dy = 0.005$
11. $\Delta y = -0.1184275$; $dy = -0.12$
13. $N \approx 3.0067$
15. $N \approx 2.991$
17. $N \approx 29.525$
19. (a) $-937.5$ in.$^3$; (b) $-150$ in.$^2$
21. $108.72 \pm 0.47$
23. (a) 10.71%; (b) 1.12%
25. 3%
27. (a) Inelastic; (b) increased; (c) \$46.40
29. (a) $15{,}000 - 1000p$; (b) $p/(15 - p)$; (c) $7.5 < p < 15$
31. (a) $(-3x^{-2} + 2x)\,dx$; (b) $x(x^2 + 9)^{-1/2}\,dx$; (c) $12(3t + 1)^3\,dt$; (d) $[-8/(s - 2)^2]\,ds$
33. 42.8008
35. (a) $\sqrt{p}/[2(30 - \sqrt{p})]$; (b) $400 < p < 900$; (c) $-\$15{,}000$

## Chapter 4 Review Problem Set (p. 158)

1. 5120
3. $(9.5, \pm\sqrt{9.5})$
5. $x = 40/(4 + \pi) \approx 5.60$, $y = 20/(4 + \pi) \approx 2.80$
7. Length: $20\sqrt[3]{6}$ ft; width: $10\sqrt[3]{6}$ ft; depth: $20\sqrt[3]{6}$ ft
9. (a) 120; (b) \$66.73
11. (a) 200; (b) \$6200
13. \$600
15. About 775
17. 17 or 18 days from now.
19. About 61 mi/h
21. $y - 2 = -\frac{9}{13}(x - 1)$
23. (a) $160/3\pi \approx 16.98$ ft/min; (b) $160/27\pi \approx 1.89$ ft/min
25. $452.39 \pm 4.52$
27. (a) Elastic; (b) decreases; (c) $-\$500$
29. (a) $-2, 3$; (b) $f''(x) = 2 + 8x^{-3} > 0$ on $(0, \infty)$; (c) $3\sqrt[3]{4} \approx 4.762$; (d) 8.25; (e)

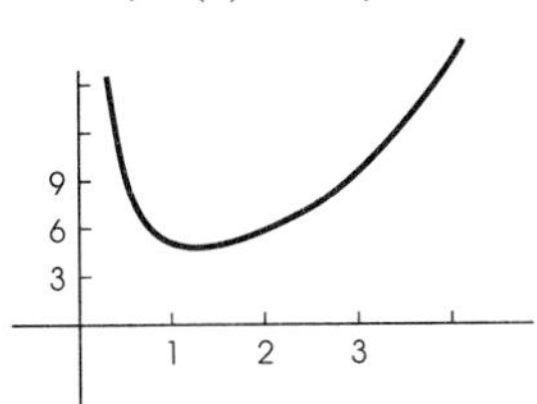


31.

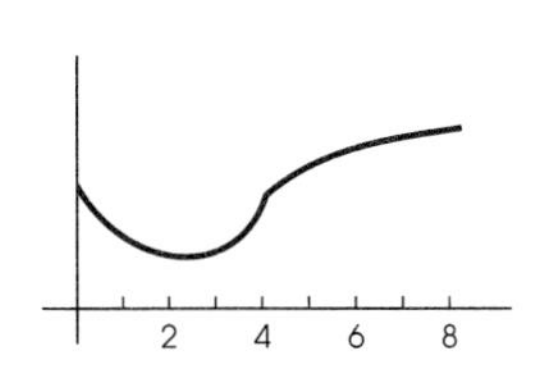


33. $(2\sqrt{2} - 3)/2\sqrt{2} = 1 - \frac{3}{4}\sqrt{2}$
35. 47,000

## Problem Set 5.1 (p. 167)

1. 64
3. 5
5. $\frac{1}{2}$
7. 3.388695
9. 0.0353581
11.

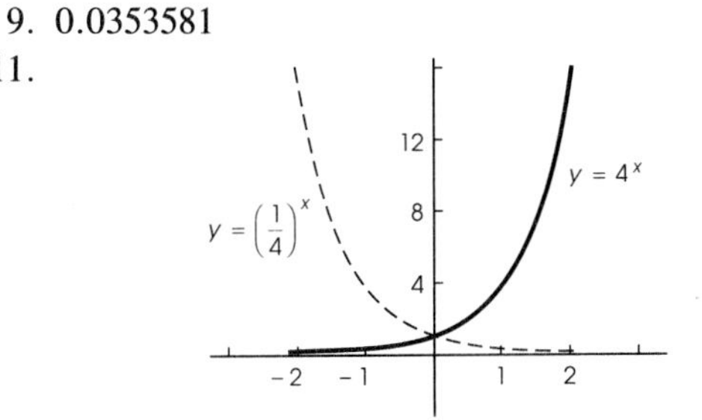


13.

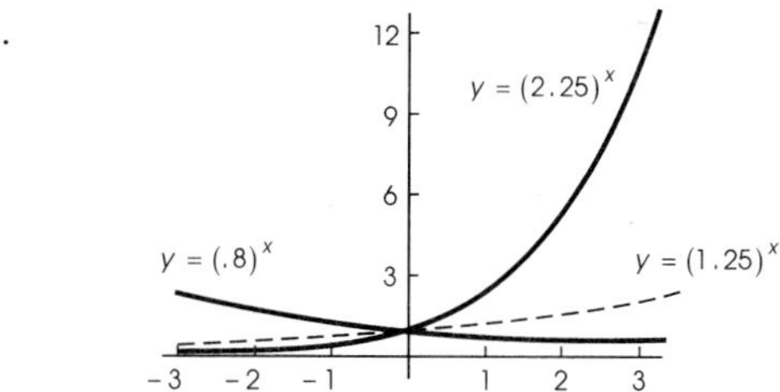


15. $b$
17. $1/c^2$
19. $b^3$
21. $1/\pi^8$
23. $(a + 1)^{1/6}$
25. 3
27. 5
29. $\frac{3}{2}$
31. $-\frac{3}{2}$
33. $-0.3644157$
35. $-0.089754$
37. 4.32
39. $2 \log_a 2 + \log_a 3 = 2.485$
41. $3 \log_a 3 - 2 \log_a 2 = 1.911$
43. $3 \log_a 2 + \log_a 5 - 1 = 2.688$
45. $(\log_a 5 + 2)/4 \log_a 3 = 0.821$
47. $\log_b(x/64)$
49. $\log_b(x^5/y^4)$
51. $\log_b(x^3 z^6/y^3)$
53.

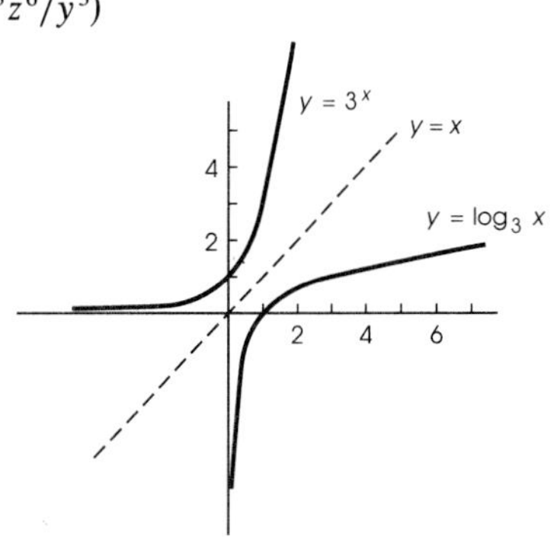

55.

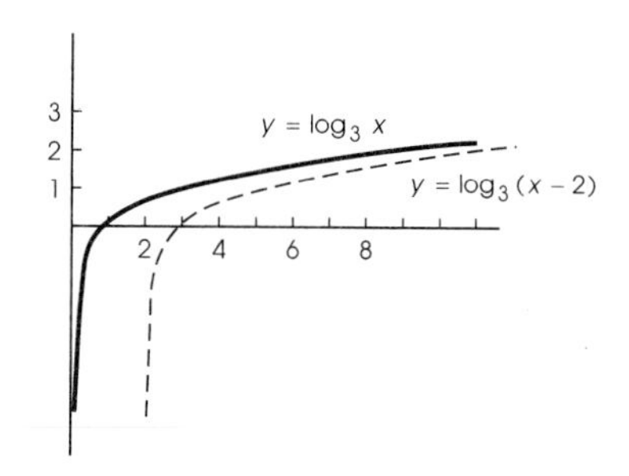


57. 4
59. 2
61. 9
63. 23
65. 2
67. $10^{1000}$
69. $13^{9000}$
71. $2^{28} - 1$ cents = \$2,684,354.55

## Problem Set 5.2 (p. 173)

1. 2.6810215
3. −4.6459922
5. 6.8532634
7. 1.3751734
9. 24.046754
11. 0.0432139
13. −4.1640482
15. 2.4649735
17. 1.7858507
19. 0.2905096
21. 2.7588016
23. 2.1037232, −1.1037232
25. 2, 1, $\frac{1}{3}$, $\frac{1}{100}$
27. $y - 3 = 3(x - e)/e$
29. $y - 4 = 4e(x - 1/e)$
31. (15, 12.124151)
33. $x + 2x \ln x$
35. $(6 \ln x)/x$
37. $-12/[x(\ln x)^4]$
39. $6(x^2 + 2 \ln x)^2(x + x^{-1})$
41. $(2x + 3)e^x$
43. $(-x^2e^x + 2xe^x + 6x)/(e^x + 3)^2$
45. $(e^x + x^4)/(5e^x + x^5)^{4/5}$
47. (a) Inc $(-1, \infty)$, dec $(-\infty, -1)$; (b) $f(-1) = -1/e$;
(c) up $(-2, \infty)$, down $(-\infty, -2)$;
(d)

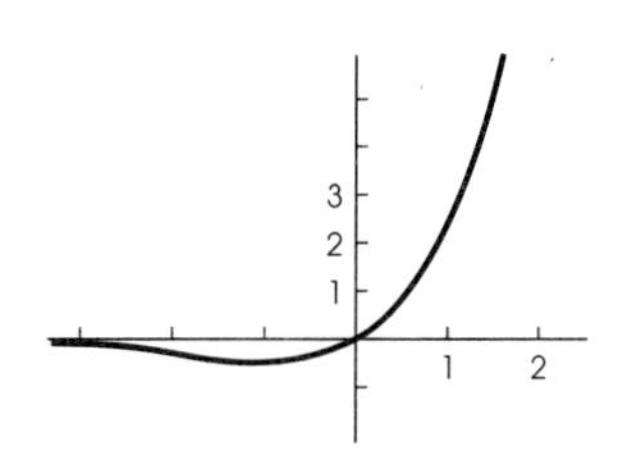


49. (a) 2.9611202; (b) 0.6714003
51. $\ln[9x(x + 1)]$
53. (a) 48; (b) $-\frac{5}{54}$
55. (a) Inc $(1, \infty)$, dec $(0, 1)$; (b) $f(1) = e$;
(c) up $(0, \infty)$
(d)

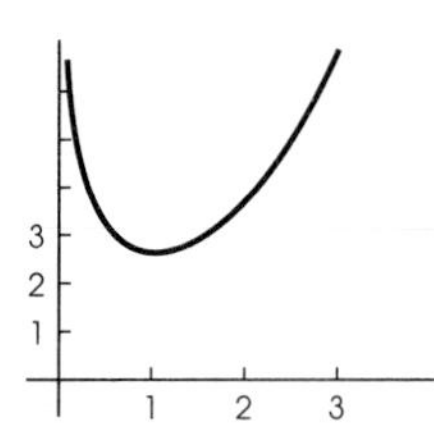


## Problem Set 5.3 (p. 179)

1. $12e^{3x}$
3. $2xe^{x^2}$
5. $8/(2x + 5)$
7. $(5x^2 - 1)/(5x^3 + x)$
9. $6e^{4\sqrt{x}}/\sqrt{x}$
11. $(-3x^2 - 2x + 6)/[(3x + 1)(x^2 + 2)]$
13. $[6/(x + 3)^2]e^{2x/(x+3)}$
15. $(18x^2 - 8x + 15)/[(2x^2 + 5)(3x - 2)]$
17. $(3x^2 + 2x)e^{3x}$
19. $-4xe^{-2x}$
21. $(2x + 6)/(x^2 + 4) - (3/x^2) \ln(x^2 + 4)$
23. $2e^{x^2}[1 + x^2 \ln(x^2)]/x$
25. $(2 + 5e^{5t})/(2t + e^{5t})$
27. $e^t(t - 1)^2/(t^2 + 1)^2$
29. $1/(t \ln t)$
31. $1/[t \ln t \ln(\ln t)]$
33. $18e^{2x}(3e^{2x} - 1)^2$
35. $8x[\ln(x^2 + 4)]^3/(x^2 + 4)$
37. $-8(e^x + \ln x)^{-3}(e^x + x^{-1})$
39. $x(e^x + 2)^2(3xe^x + 2e^x + 4)$
41. $4(2x + 1)e^{(2x+1)^2}$
43. $(3x + 12)[\ln(x + 4 \ln x)]^2/(x^2 + 4x \ln x)$
45. (a) $6e^{3t} + 4e^{-t}$; (b) $6t/(t^2 - 4)$;
(c) $t^2(e^t - t^{-1}) + 2t(e^t - \ln t)$;
(d) $e^{2t}(2t^2 - 2t - 10)/(t^2 - 5)^2$;
(e) $4(2t + 1)e^{(2t+1)^2}$;
(f) $2(4\sqrt{t} + 1)/[\sqrt{t}(2t + \sqrt{t})]$;
(g) $4[\ln(2t + \sqrt{t})]^3(4\sqrt{t} + 1)/[2\sqrt{t}(2t + \sqrt{t})]$;
(h) $6(4t - e^{2/t})^5(4 + 2t^{-2}e^{2/t})$
47. $e^{-x^2}(4x^4 - 10x^2 + 2)$
49. $y = 3x$
51. (a) \$16,404.98; (b) \$2213.64/yr; (c) about 7.7 yr
53. (a) About \$813,000; (b) \$2.04/dollar;
(c) An additional dollar of advertising will produce an additional profit of \$2.04 (\$1.04 after deducting the additional advertising cost).

## Problem Set 5.4 (p. 184)

1. $\ln 29/\ln 3 \approx 3.0650448$
3. $e^{2.63 \ln 4} \approx 38.319319$
5. $\ln 0.432/\ln 6 \approx -0.4684388$
7. $e^{2.4 \ln 3.2} \approx 16.306470$
9. $e^{t \ln 4.71} \approx e^{1.5496879t}$
11. $e^{-2t \ln 8.12} \approx e^{-4.1886603t}$
13. $3 \ln t/\ln 6 \approx 1.6743319 \ln t$
15. $1/(x \ln 5)$
17. $4^x \ln 4$
19. $2/[(2x + 1)\ln 3]$
21. $-6^{4-5x^2}(10x \ln 6)$
23. $18x/[(3x^2 + 1) \ln 4]$
25. $5^{4e^x+3}(4e^x \ln 5)$
27. $2^x 3^{-2x}(-2 \ln 3 + \ln 2) = 2^x 3^{-2x} \ln(\frac{2}{9})$
29. $(2x + 1)^x[2x/(2x + 1) + \ln(2x + 1)]$
31. $(x^2 + 5x)^{2x}[(2x)(2x + 5)/(x^2 + 5x) + 2 \ln(x^2 + 5x)]$
$= 2(x^2 + 5x)^{2x}[(2x + 5)/(x + 5) + \ln(x^2 + 5x)]$
33. $(5x)^{x^2-x^{-1}}[x - x^{-2} + (2x + x^{-2}) \ln(5x)]$
35. $(e^x + 4)^{\ln x}\left[\dfrac{e^x \ln x}{e^x + 4} + \dfrac{\ln(e^x + 4)}{x}\right]$
37. $x^{10}(2x - 1)^6\left(\dfrac{10}{x} + \dfrac{12}{2x - 1}\right)$
39. $\dfrac{(x + 1)^3(x + 2)^4}{(x + 3)^5}\left(\dfrac{3}{x + 1} + \dfrac{4}{x + 2} - \dfrac{5}{x + 3}\right)$
41. $(x^2 + 1)^8(2x + 5)^5\left(\dfrac{16x}{x^2 + 1} + \dfrac{10}{2x + 5}\right)$
43. $\dfrac{\sqrt{x^2 + 2}\,(x + 2)^4}{\sqrt[3]{3x - 5}}\left(\dfrac{x}{x^2 + 2} + \dfrac{4}{x + 2} - \dfrac{1}{3x - 5}\right)$
45. 6.6438562, 4.6051702, 2, 0.8691760
47. (a) 0; (b) $(2e^{2x} + 1)/[(e^{2x} + x) \ln 3]$;
(c) $6e^{2x}5^{3e^{2x}} \ln 5$;
(d) $2x(4 + 3x^2)^{x^2}[3x^2/(4 + 3x^2) + \ln(4 + 3x^2)]$
49. $-304$
51. $X + 2Y = -5$, $m = -\frac{1}{2}$, $b = -\frac{5}{2}$
53. (a) 3316; (b) $4.899h$
55. Let $L_w$ denote the loudness of a whisper in decibels. Then the ratio $I_c/I_w$ of the intensity of the sound to that of a whisper is $10^{0.4L_w}$.

## Chapter 5 Review Problem Set (p. 185)

1. (a) $\frac{1}{27}$; (b) $\frac{1}{5}$; (c) 3; (d) $\frac{1}{3}$; (e) $\frac{4}{3}$; (f) $\pi + 1$
3. (a) 118.9043 (b) 8.611396; (c) 7.550287;
(d) 139.0456; (e) 1.169610; (f) 2.200973
5.

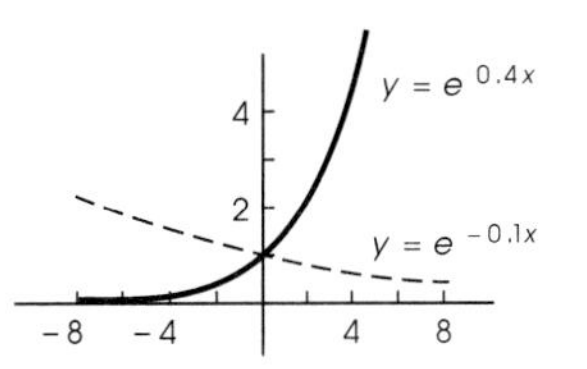


7. (a) 1.895; (b) $-0.377$; (c) $-0.304$
9. $\ln(125x^2/y^5)$
11. (a) $(e^4 - 1)/3 \approx 17.86605$;
(b) $-2/(3 - \ln \pi) \approx -1.078010$;
(c) $(1 \pm \sqrt{1 + 12e})/2 \approx -2.399111, 3.399111$;
(d) $\pm\sqrt{3 + \ln(\ln 140)} \approx \pm 2.144224$
13. (a) $4xe^{x^2-3}$; (b) $-15/[x(4x + 3)]$;
(c) $(2x + 1)^{-2}e^{x/(2x+1)}$;
(d) $4x(x^2 + 3)/[(x^2 + 3)^2 + 1]$;
(e) $2(x^2 + 4)/(2x + 1) + 2x \ln(2x + 1)$;
(f) $e^{2x}(2x + 3)/(x + 2)^2$;
(g) $8(3 + 2x \ln x)^3(1 + \ln x)$;
(h) $\frac{3}{2}e^{-2x}(1 - 2x)(xe^{-2x} - 1)^{1/2}$
15. (a) Inc $(1, \infty)$, dec $(-\infty, 1)$;
(b) up $(0, \infty)$, down $(-\infty, 0)$; (c) $f(1) = -e$;
(d)

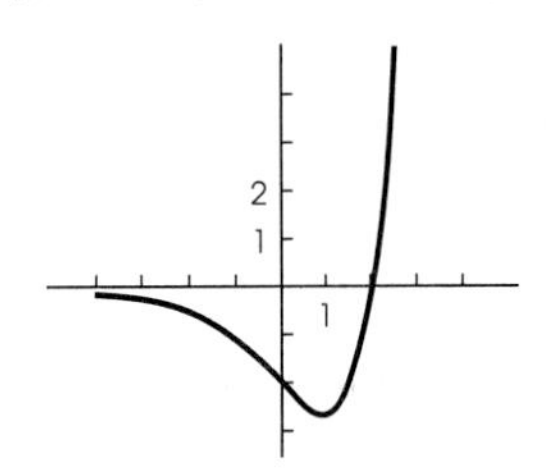


17. At $x = -1$ and $x = 1$
19. (a) About 50; (b) $-3.2$ points/mo; (c) 9 mo
21. $e^{t \ln(2\pi)} \approx e^{1.837877t}$
23. $(1/\ln 5) \ln t \approx 0.6213349 \ln t$
25. (a) $10^{4\sqrt{x}}(\ln 10)(2/\sqrt{x})$; (b) $36x/[(3x^2 + 5) \ln 4]$
27. (a) $2^2[2 + 3 \ln 2] \approx 16.317766$
(b) $155/3\sqrt{3} \approx 29.82976$
29. Let $Y = \ln y$, $X = x$; then $m = -3$ and $b = 4 + \ln 4$
31. $y - 1 = \frac{5}{4}(x - 2)$
33. $18e$
35. $\frac{2}{5} \ln 5$
37. Positive, positive
39. (a) $(0, 1) \cup (e, \infty)$; (b) $(\sqrt{e}, \infty)$; (c) $(0, e^{3/2})$

## Problem Set 6.1 (p. 194)

1. (a) $dy/dt = 5e^{0.03t}(0.03) = 0.03y$;
(b) similar to part (a)

3. $y = 16e^{0.8t}$; 9629.52
5. $y = 20e^{-0.05t}$; 13.4064
7. $y = 3.21e^{0.193t}$; 15.0333
9. (a) 800; (b) $100e^{0.0099021t}$; (c) 328
11. (a) $16{,}000e^{0.110146t}$; (b) 14.6118 h
13. About 35 yr
15. (a) $10.5e^{0.032965t}$; (b) about 21.7 million
17. $k = -0.153691$; 17.1507 g
19. About 0.00015 s
21. About 3841 yr
23. About 4.74%
25. 160,000
27. 11,140 yr
29. $y = 0.723379e^{0.875469t}$
31. 0.0669 g
33. 2.878 h

## Problem Set 6.2 (p. 200)

1. $y = 8 - 6e^{-0.09t}$; 4.50351
3. $y = 20 - 8e^{-0.45t}$; 18.6776
5. (a) 84%; (b) 12.81 h
7. (a) $dT/dt = k(T - T_m)$; (b) 114.5°F
9. (a) $W = 72 - 72e^{-0.6t}$; (b) 72 words/min; (c) 4.8 wk
11. 18.2 cm
13. 34,870
15. 21.4 days
17. About 3.5 yr
19. About 64 h
21. About 23,000
23. 690 lb

## Problem Set 6.3 (p. 206)

1. $1320
3. $1713
5. $3741.99
7. $1360.49
9. $3392.07
11. (a) $2367.36; (b) $2411.71; (c) $2435.19; (d) $2451.36; (e) $2459.33
13. $7389.59
15. $2459.60
17. $6946.51
19. 10.3813%
21. 9.3069%
23. 8.2% compounded semiannually
25. 9.6% compounded monthly
27. About 76 mo
29. 7.2963 yr
31. (a) $4243.20; (b) $5157.30; (c) $5172.56; (d) $5173.08
33. $1770.13
35. 5727 days
37. 8.3619%
39. 7.3241%
41. $4591.59

## Problem Set 6.4 (p. 212)

1. $4033.76
3. $1126.31
5. $1878.51
7. $2849.91
9. $3770.30
11. $6479.44
13. $3715.40
15. $3754.95
17. $5176.84
19. $11,481.70; 2.80193%
21. The true growth rates are 3.84615% and 3.63636%, so the first is better.
23. (a) $25,600; (b) $13,107.20; (c) $5368.71
25. 5.885 yr
27. $3135.19
29. $4651.24
31. (a) $6304.83; (b) $422.03
33. 9.5958%
35. 7.6 yr

## Chapter 6 Review Problem Set (p. 213)

1. $y = 80e^{-0.029t}$; 43.5116
3. $y = 12 - 9e^{-0.07t}$; 5.65781
5. $y = 20/(1 + 1.5e^{-t})$; 19.1333
7. (a) $y = 2400e^{0.13412t}$; (b) about 4700; (c) 368 bacteria/h; (d) 14.14 h
9. About 22,600 yr
11. About 44,200
13. About 7440
15. (a) $5400; (b) $6476.77; (c) $6658.92; (d) $6676.62
17. (a) About 8.33 yr; (b) about 8.15 yr
19. 9.8157%
21. $1775.12
23. 7.7593%
25. $3564.22, $7128.44
27. (a) $39,321.60; (b) about 3.106 yr
29. About 9480
31. About $52,641
33. $(d/dt)[L - (L - y_0)e^{-kt}] = (L - y_0)ke^{-kt} = k(L - y)$
35. $y' = (2A + B + 2Bt)e^{2t}$; $y'' = (4A + 4B + 4Bt)e^{2t}$; $y'' - 4y' + 4y = 0$
37. $dy/dt = kty$; faster
39. $d^2y/dt^2 = k(dy/dt)^2$

## Problem Set 7.1 (p. 222)

1. Antiderivative.
3. $f(x)$
5. $4x^3$
7. $\frac{1}{4}x^4 + C$

9. $-\frac{1}{2}x^{-2} + C$
11. $\frac{2}{7}x^{7/2} + C$
13. $x^{\pi+1}/(\pi + 1) + C$
15. $-\frac{1}{3}x^{-3} + C$
17. $\ln|x| + C$
19. $\frac{3}{2}x^2 + 2x + C$
21. $x^5 - 3x^2 + C$
23. $2x^{5/2} - 2x^{3/2} + C$
25. $2x^{3/2} + 5\ln|x| + C$
27. $-5x^{-1} + x^{-2} + C$
29. $5e^x - 4\ln|x| + C$
31. $\frac{9}{4}t^{4/3} + t^2 + C$
33. $\frac{1}{4}u^4 + 4u^{-1} + C$
35. $\frac{2}{3}x^3 + \frac{5}{2}x^2 - 3x + C$
37. $\frac{4}{3}t^3 + 6t^2 + 9t + C$
39. $x + 4\ln|x| - 2x^{-1} + C$
41. $2x^2 - 2x + 3$
43. $\frac{1}{3}x^3 - x^2 + 5x - 1$
45. $4e^x - 2$
47. $3\ln|x| + 2x^3 + 3$
49. (a) $x^2 - 4\ln|x| + C$; (b) $\frac{1}{3}x^3 + \frac{4}{5}x^{5/2} + \frac{1}{2}x^2 + C$; (c) $3x + 6\sqrt{x} + C$; (d) $3\ln|x| + 2x^{-2} + C$; (e) $6s + 4e^s + C$; (f) $e^t + t^{e+1}/(e + 1) + C$
51. \$3,487,500
53. $s = 2t^3 - 2t^2 + 2t + 12$
55. 26,800
57. (a) \$278,000; (b) \$10.86
63. $y = \ln|x^2 + 3x + 1| + 3$
65. (a) \$8495.26; (b) \$106.19
67. (a) $S(t) = 10(3 - 0.2t)^{5/2} - 50(3 - 0.2t)^{3/2} + 50$; (b) $S(15) - S(0) = 103.92$ ft

## Problem Set 7.2 (p. 228)

1. $\frac{32}{5}x^{15} + C$ both ways.
3. $\frac{1}{3}(x^2 + 3)^3 + C$; $\frac{1}{3}x^6 + 3x^4 + 9x^2 + K$; check that these answers are equivalent.
5. $\frac{1}{6}(x^3 - 4)^6 + C$
7. $\frac{1}{28}(x^4 + 5)^7 + C$
9. $\frac{1}{5}(x^2 + 6x + 4)^5 + C$
11. $\frac{1}{33}(x^3 + 3x - 2)^{11} + C$
13. $\frac{2}{5}(\sqrt{x} + 1)^5 + C$
15. $\frac{1}{3}e^{3x} + C$
17. $\frac{3}{2}e^{2t+1} + C$
19. $\frac{1}{4}e^{2x^2+1} + C$
21. $2e^{\sqrt{x}} + C$
23. $\ln|x^2 + 5x + 4| + C$
25. $\frac{1}{4}\ln|x^4 + 2x^2 - 2| + C$
27. $\frac{1}{3}\ln(3e^x + 1) + C$
29. $\frac{2}{5}(4 + \ln x)^{5/2} + C$
31. $\frac{1}{10}[\ln(2x + 1)]^5 + C$
33. $\frac{1}{4}[\ln(x^2 + 1)]^2 + C$
35. $\frac{1}{16}(x - 4)^{16} + \frac{4}{15}(x - 4)^{15} + C$
37. $\frac{1}{10}(2x + 1)^{10} - \frac{1}{9}(2x + 1)^9 + C$
39. $\frac{1}{5}x^5 + \frac{10}{3}x^3 + 25x + C$
41. $\frac{1}{2}x^2 + 4x^{3/2} + 9x + C$
43. $\frac{3}{2}x^2 + \ln|x| + 7x^{-1} + C$
45. $2x + 10\sqrt{x} + C$
47. $\frac{1}{2}x^2 + 2x + 3\ln|x - 1| + C$
49. $\frac{1}{2}x^2 + \frac{3}{2}\ln|x^2 - 1| + C$
51. $\frac{8}{7}u^7 + \frac{12}{5}u^5 + 2u^3 + u + C$
53. $\frac{1}{3}e^{x^3} + C$
55. $\frac{1}{12}(x + 1)^{12} - \frac{1}{11}(x + 1)^{11} + C$
57. $\frac{1}{3}\ln|t^3 + 3t - 4| + C$
59. $2\sqrt{t} + 2t + 3\ln|t| + C$
61. $\frac{1}{2}[\ln(e^t + 2)]^2 + C$

## Problem Set 7.3 (p. 235)

1. $\frac{1}{2}xe^{2x} - \frac{1}{4}e^{2x} + C$
3. $\frac{1}{2}x^2\ln x - \frac{1}{4}x^2 + C$
5. $-(1/x)(1 + \ln x) + C$
7. $\frac{1}{7}x(1 + x)^7 - \frac{1}{56}(1 + x)^8 + C$
9. $-\frac{1}{2}x^2e^{-2x} - \frac{1}{2}xe^{-2x} - \frac{1}{4}e^{-2x} + C$
11. $\frac{1}{2}x^2\ln(x + 2) - \frac{1}{4}x^2 + x - 2\ln(x + 2) + C$
13. $\dfrac{3}{x + 1} - \dfrac{3}{x + 2}$
15. $\dfrac{3}{x} + \dfrac{2}{x - 3}$
17. $\dfrac{9}{x + 1} - \dfrac{12}{x + 2} + \dfrac{3}{x + 3}$
19. $\dfrac{5}{x} + \dfrac{2}{x + 3} - \dfrac{4}{x - 2}$
21. $3\ln|x + 1| - 3\ln|x + 2| + C$
23. $3\ln|x| + 2\ln|x - 3| + C$
25. $9\ln|x + 1| - 12\ln|x + 2| + 3\ln|x + 3| + C$
27. $5\ln|x| + 2\ln|x + 3| - 4\ln|x - 2| + C$
29. $2e^{2x} + C$
31. $-\frac{1}{5}(x + 2)e^{-5x} - \frac{1}{25}e^{-5x} + C$
33. $\frac{1}{8}e^{4x^2} + C$
35. $2\sqrt{x}\ln x - 4\sqrt{x} + C$
37. $\ln|x^2 - x - 2| + C$
39. $\frac{3}{7}\ln|x + 5| + \frac{11}{7}\ln|x - 2| + C$
41. $4\ln|x + 1| + (x + 1)^{-1} + 3\ln|x - 4| + C$
43. $8e^{16} - \frac{1}{2}e^{16} + \frac{1}{2} \approx 66{,}645{,}829$ ft
45. 18.425 mg

## Problem Set 7.4 (p. 240)

1. $\frac{1}{3}y^3 = \frac{1}{2}y^2 + C$ or $2y^3 = 3y^2 + K$
3. $y - y^3 = \frac{1}{3}x^3 + C$ or $3y - 3y^3 = x^3 + K$
5. $-1/y = \frac{1}{2}\ln(1 + x^2) + C$ or $-1/y = \ln(K\sqrt{1 + x^2})$
7. $\frac{1}{2}\ln(1 + y^2) = \frac{1}{4}\ln(1 + x^4) + C$ or $1 + y^2 = K\sqrt{1 + x^4}$
9. $\ln|y| = -2\ln|x| + 2\ln|x - 1| + C$ or $y = K(x - 1)^2/x^2$

11. $ye^y - e^y = x^3 + C$
13. $y^2 = \frac{1}{2}x^2 + 14$
15. $y = 10/[(t^2 + 2)^5 - 22]$
17. $y = 4e^{0.02t}$
19. $y = 5e^{0.36(1-t)}$
21. $y = 12 - 4e^{-4t}$
23. $y = 60/(5 + e^{-0.36t})$
25. (a) $x = 80p^{-3/4}$; (b) $x = \frac{80}{27} \approx 2.96$
27. $y = y_0 e^{-4t}$
29. $y^2 = 2e^x + C$
31. $-1/y = \frac{1}{2}e^{x^2} + C$
33. $y = K(x - 2)^4/x^4$
37. 333
39. (a) $dy/dt = ky(5000 - y)$; (b) 2848
41. \$875.13

27. (a) Concave up nowhere: (b) $(-3, \infty)$; (c) $(2, \infty)$; (d) $y = 1$ is a horizontal asymptote.
(e)

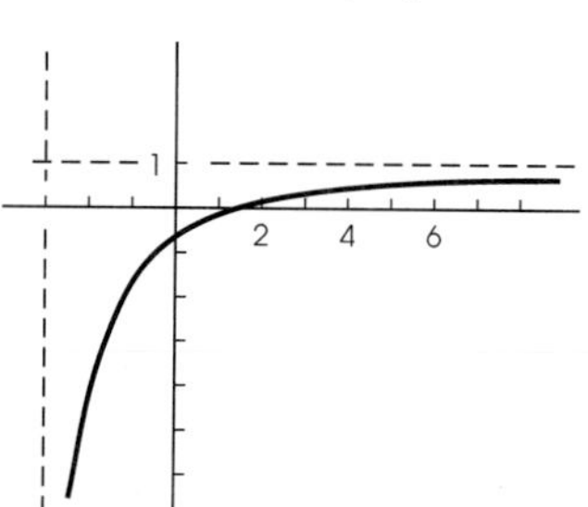


## Chapter 7 Review Problem Set (p. 242)

1. (a) $x^8 - x^4 + \pi x + C$; (b) $2x^{\pi} - 4x^{5/2} + C$; (c) $-4\ln|x| + \frac{1}{2}e^{4x} + C$; (d) $e^{x^3} - 2x + C$
3. (a) $x^3 - 6x^2 + 5x + C$; (b) $2x^{3/2} - 5x^{-1} + C$; (c) $2x^2 + 4x + \ln|x| + C$; (d) $\frac{4}{3}e^{3x} - 9x^{2/3} + C$
5. $F(x) = -4x^{-1} + 4x + \frac{1}{3}x^3 - \frac{17}{3}$
7. $f(t) = t^{-3} + 4t - 12$
9. $C(x) = 700 + 280x - 6x^2 + 0.2x^3$
11. 38,500
13. (a) $\frac{1}{5}(x^3 + 2x)^5 + C$; (b) $\frac{1}{8}(e^{2x} + 4)^4 + C$; (c) $\ln|x^2 + x - 7| + C$; (d) $-\frac{1}{8}(x^4 + 2x^2 - 7)^{-2} + C$; (e) $-1/\ln x + C$; (f) $\frac{1}{11}(x + 3)^{11} - \frac{3}{10}(x + 3)^{10} + C$; (g) $\frac{1}{2}x^2 + 4x - 3\ln|x - 3| + C$; (h) $\frac{1}{8}x^8 + \frac{4}{5}x^5 + 2x^2 + C$
15. (a) $\frac{1}{4}xe^{4x} - \frac{1}{16}e^{4x} + C$; (b) $-2x^{-1/2}\ln x - 4x^{-1/2} + C$
17. (a) $2\ln|x - 4| + \ln|x + 3| + C$; (b) $\ln|x| + 2\ln|x + 4| - \ln|x - 3| + C$
19. (a) $y = -1/(2x^4 + C)$; (b) $y = [\frac{1}{4}(\ln t)^2 + K]^2$
21. (a) $y^2 = 1/(10 - e^x)$; (b) $y = 3/[4(5 + x^2)^{3/2} - 102]$
23. $y^2 = 4(x^3 + 3)^{3/2} - \frac{7}{2}$.
25. (a) $dy/dt = ky(25{,}000 - y)$, $y_0 = 50$; (b) about 16,300 people.

## Problem Set 8.1 (p. 250)

1. 20
3. $6 + \frac{1}{2}\pi(\frac{5}{2})^2 \approx 15.82$
5. $9\sqrt{3} \approx 15.59$
7. $\frac{3}{2}(\sqrt{7} + \sqrt{27}) \approx 11.76$
9. $\frac{1}{2}\pi(5)^2 - \frac{1}{2}(6)(8) \approx 15.27$
11.

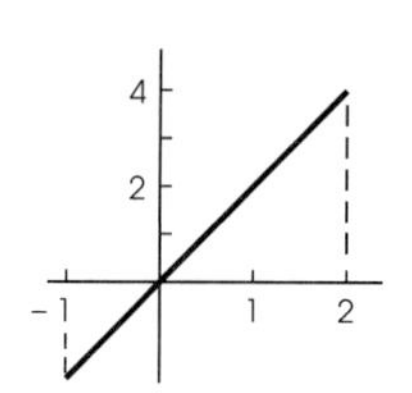


$\int_{-1}^{2} f(x)dx = 4 - 1 = 3$

13.

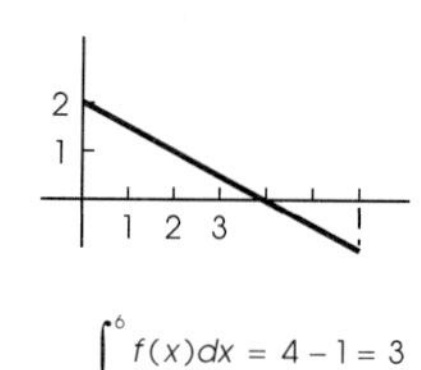


$\int_{0}^{6} f(x)dx = 4 - 1 = 3$

15.

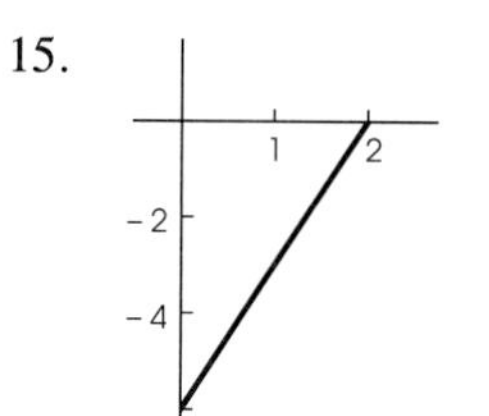


$\int_{0}^{2} f(x)dx = -6$

17.

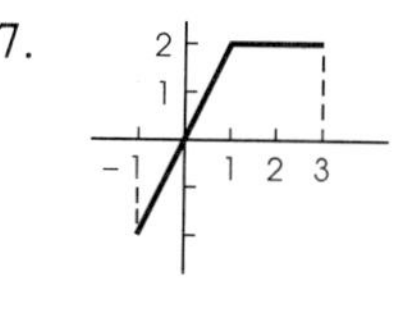


$\int_{-1}^{3} f(x)dx = 5 - 1 = 4$

19.

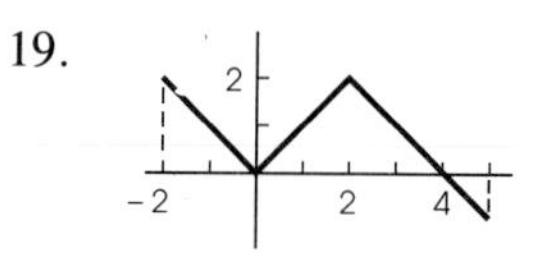


$\int_{-2}^{5} f(x)dx = 6 - \frac{1}{2} = \frac{11}{2}$

21.

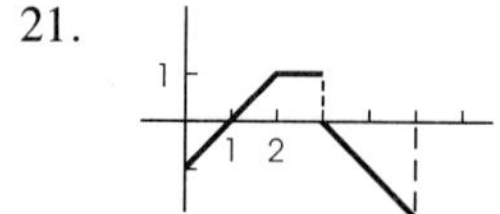


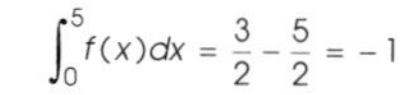

$\int_{0}^{5} f(x)dx = \frac{3}{2} - \frac{5}{2} = -1$

23. $2\pi - 10$
25. $\frac{9}{2}$
27. 9
29. $\frac{1}{5}\pi^5$
31. $\frac{16}{3}$
33. (a) Odd; (b) even; (c) neither; (d) odd.
35. 0
37. $2\int_0^1 x^4\,dx = \frac{2}{5}$
39. $2\int_0^2 x\,dx = 4$
41. 0
43. 4

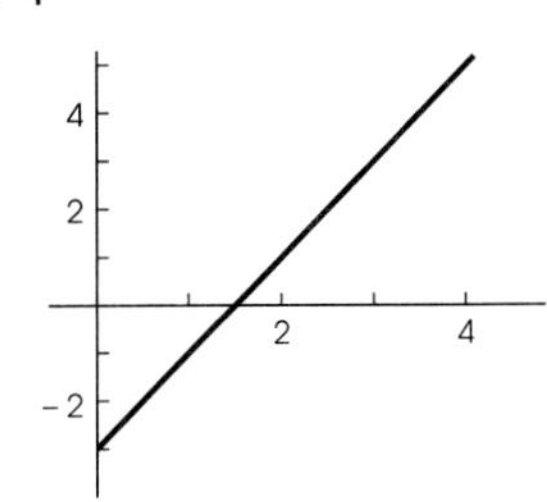


45. $\frac{5}{2}$ 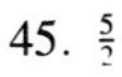

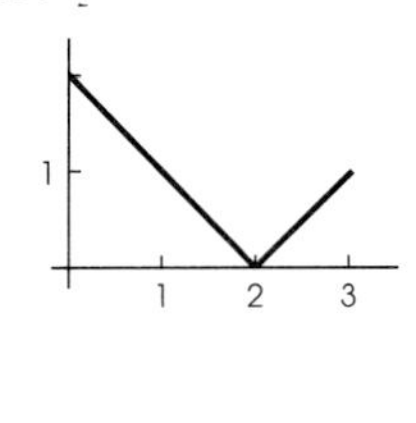


47. $\frac{9}{2}\pi$

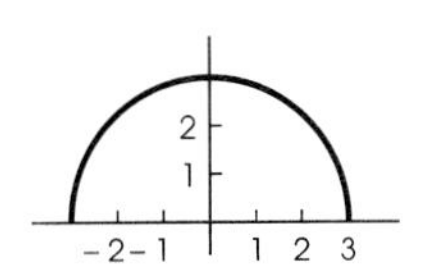


49. 6

51. $\frac{13}{2}$
53. $\frac{13}{2}$

## Problem Set 8.2 (p. 258)

1. 8
3. $-\frac{3}{2}$
5. $-\frac{9}{8}$
7. 70
9. 18
11. 0
13. $\frac{64}{3}$
15. 45
17. 1
19. $5e - 6$
21. $\frac{76}{3}$
23. $\frac{96}{5}$
25. 8
27. 16
29. 14
31. $-12$
33. 10
35. 7
37. $\frac{31}{6}$
39. $\frac{7}{3}$
41. 3
43. $\frac{9}{4}$
45. 3
47. 48
49. $e - 1 + (e + 1)^{-1}$
51. 5
53. $-3$
55. 1
57. $e - 1$
59. (a) $3(e^{10} - 1) \approx 66{,}076.4$ in.$^3$; (b) $10(e^3 - 1) \approx 190.9$ in.$^3$
61. (a) 73.92°F; (b) 72.44°F
63. 460 lb

## Problem Set 8.3 (p. 263)

1. $\frac{2}{3}$
3. $\frac{14}{3}$
5. $\frac{1}{5}$
7. $\frac{98}{3}$
9. $5 \ln 4$
11. $\frac{1}{4}\ln 9$
13. $\frac{3}{2}(e^2 - 1)$
15. $\frac{1}{2}(1 - e^{-4})$
17. $\frac{81}{8}$
19. 21
21. $\frac{1}{9}(\ln 10)^3$
23. $\frac{1}{2}(e^9 - 1)$
25. $\frac{3}{4}e^4 + \frac{1}{4}$
27. $\frac{6}{25}e^{5/2} + \frac{4}{25}$
29. $\frac{1}{30}$
31. $\frac{15}{2} - 8\ln 2$
33. $\frac{15}{2} - 8\ln 2$
35. $\frac{3}{2} - \frac{5}{4}\ln 7$
37. $\ln[(\sqrt{34} + 5)/9] \approx 0.1852$
39. $\frac{5}{2}(\sqrt{34} - 4) + \frac{9}{2}\ln[(\sqrt{34} + 5)/9] \approx 5.4107$
41. $e^2 + 1 \approx 8.3891$
43. $\frac{5}{4}e^6 + \frac{1}{4} \approx 504.5360$
45. $1/[15(\sqrt[3]{2} - 1)] \approx 0.2565$

## Problem Set 8.4 (p. 270)

1. $\frac{21}{5}$
3. 38
5. $\frac{3}{4}$
7. 31
9. $\sum_{i=1}^{10} i^3$
11. $\sum_{i=1}^{100} 3/2^i$
13. $\sum_{i=4}^{21} x_i$
15. $\sum_{i=1}^{n} g(t_i)\,\Delta t$
17. $\sum_{i=1}^{n} 3x_i\,\Delta x$
19. $\sum_{i=1}^{n} (x_i^3 + 4)\,\Delta x$
21. 120
23. 10
25. 50
27. 18
29. $\lim_{n\to\infty} \sum_{i=1}^{n} \left(4 - \frac{3i}{n}\right)\frac{3}{n} = \frac{15}{2}$
31. $\lim_{n\to\infty} \sum_{i=1}^{n} \left(2\frac{2i}{n} + 1\right)\frac{2}{n} = 6$
33. $\lim_{n\to\infty} \sum_{i=1}^{n} \left[\left(\frac{3i}{n}\right)^2 + 1\right]\frac{3}{n} = 12$
35. $\lim_{n\to\infty} \sum_{i=1}^{n} \left[\left(\frac{i}{n}\right)^2 - 2\frac{i}{n}\right]\frac{1}{n} = -\frac{2}{3}$
37. $\lim_{n\to\infty} \sum_{i=1}^{n} \left(1 + \frac{2i}{n} + 4\right)\frac{2}{n} = 12$
39. 124,750
41. 4905
43. 323,200
45. 323,400
47. $\int_0^4 (x^2 + 3x)\,dx$
49. $-\ln 101 \approx -4.615$
51. 6, 3.6

## Chapter 8 Review Problem Set (p. 271)

1. (a) $\frac{9}{2}\pi + 8$; (b) 10.5
3. (a) 2.5 (b) $2\pi$

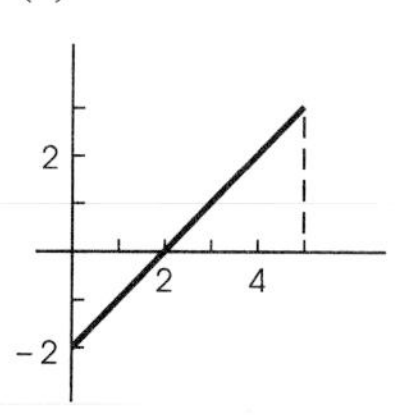


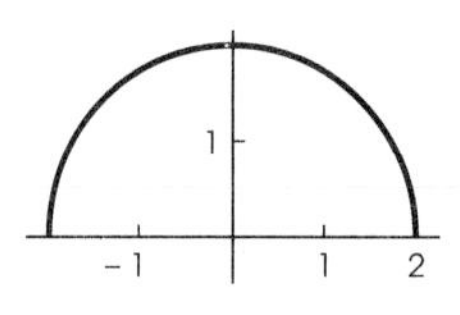


(c) $-2$

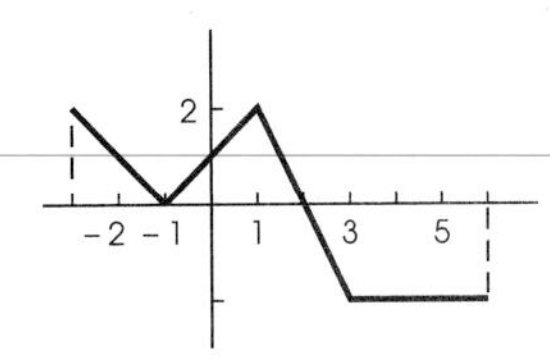


5. (a) $\frac{10}{7}$; (b) 0; (c) $-\frac{4}{5}$; (d) 0
7. 14.5 ft
9. 52
11. 12
13. $\frac{1}{2}\ln 5$
15. $e^4 - e^2$
17. 4
19. $\frac{1}{8}(\ln 5)^4$
21. 26
23. 1
25. $e^3 + 12$
27. $\frac{96}{7}$
29. $\sqrt{5}/18$
31. $\frac{59}{2}$
33. 94 in.
35. 100
37. (a) $\sum_{i=1}^{97} i(i+1)$; (b) $\sum_{i=1}^{n} (2x_i + 3)\Delta x$
39. 54
41. 76,527,700
43. 1
45. 14
47. $\int_0^4 \frac{x}{1+x^2}\,dx = \left[\frac{1}{2}\ln(1+x^2)\right]_0^4 = \ln\sqrt{17}$
49. (a)

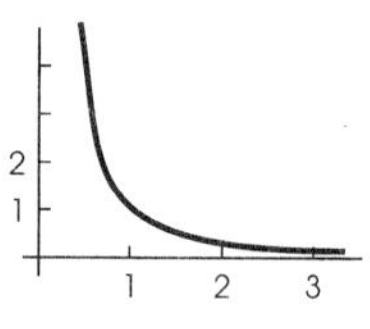
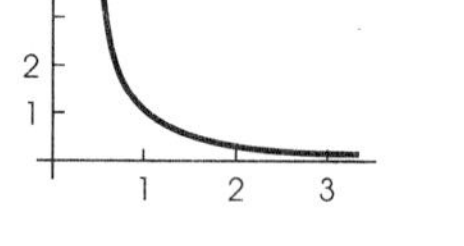


(b) $(2, \frac{1}{4})$; (c) $-0.005$; (d) 0.999 (e) 1

## Problem Set 9.1 (p. 281)

1. 21
3. $\frac{56}{3}$
5. $\frac{9}{4}$
7. $4e^2 - 4$
9. $\ln 3$
11. $\frac{1}{2}\ln 5$
13. $\frac{56}{3}$
15.

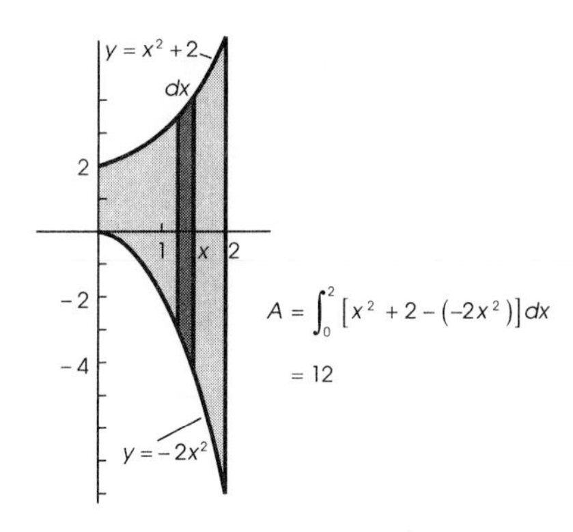


17.

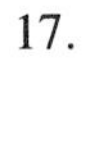

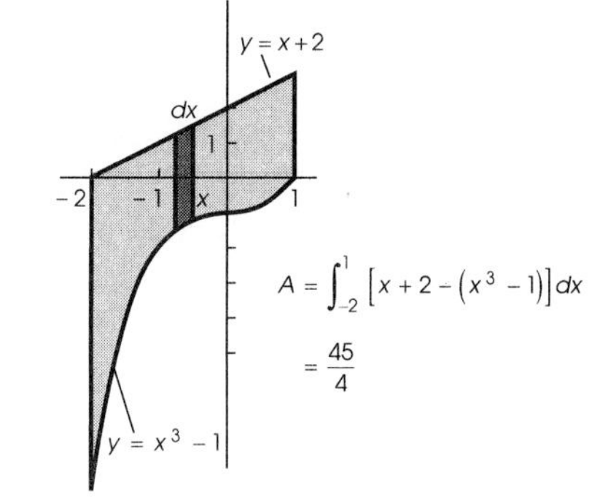


19.

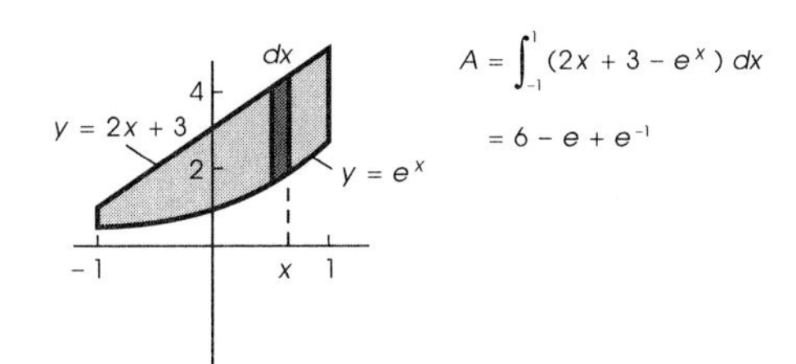


21.

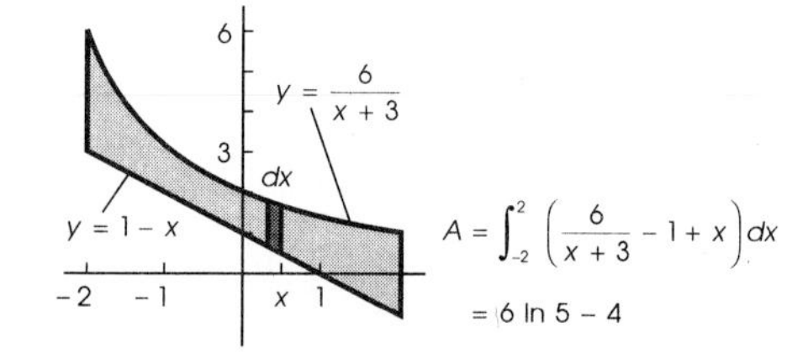

23. $\int_{-1}^{3} (2x + 5 - x^2 - 2)\,dx = \frac{32}{3}$

25. $\int_{-2}^{3} (5 - x^2 + x + 1)\,dx = \frac{125}{6}$

27. $\int_{1}^{3} (x - 1 - x^2 + 3x - 2)\,dx = \frac{4}{3}$

29. $\int_{1}^{4} \left(2\sqrt{x} - \frac{2}{3}x - \frac{4}{3}\right) dx = \frac{1}{3}$

31. $\int_{-3}^{3} (6 - x^2 - x^2 + 12)\,dx = 72$

33. $2\int_{0}^{2} (4x - x^3)\,dx = 8$

35. $\int_{-2/3}^{0} (2x^3 + x^3 - x^2 - 2x)\,dx + \int_{0}^{1} (-x^3 + x^2 + 2x - 2x^3)\,dx = \frac{253}{324} \approx 0.7809$

37. $2\int_{0}^{2} (4 - y^2)\,dy = \frac{32}{3}$

39. $\int_{0}^{3} (4y - y^2 - y)\,dy = \frac{9}{2}$

41. $\int_{1}^{4} (y - 1 - y^2 + 4y - 3)\,dy = \frac{9}{2}$

43. 36

45. $\frac{253}{12} \approx 21.0833$

47. ln 3

49. 22

51. $\ln 2 - \frac{1}{2}$

53. 20

## Problem Set 9.2 (p. 288)

1. $\pi\int_{0}^{3} (x^2)^2\,dx = \frac{243\pi}{5}$

3. $\pi\int_{0}^{3} (6x - x^2)^2\,dx = \frac{648\pi}{5}$

5. $\pi\int_{0}^{3} (6 - 2x)^2\,dx = 36\pi$

7.

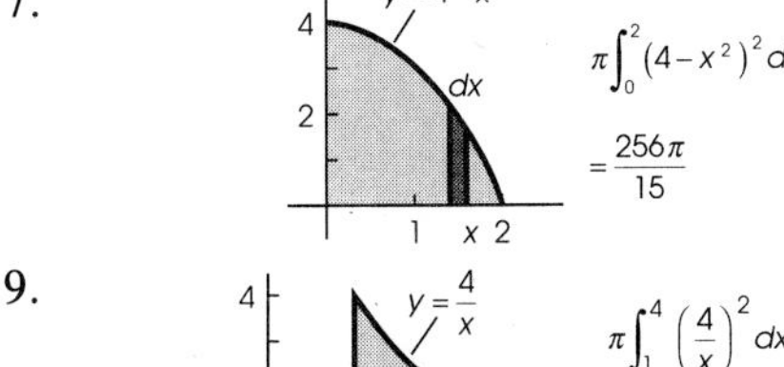


9.

$y = \frac{4}{x}$, dx

$\pi\int_1^4 \left(\frac{4}{x}\right)^2 dx = 12\pi$

11.

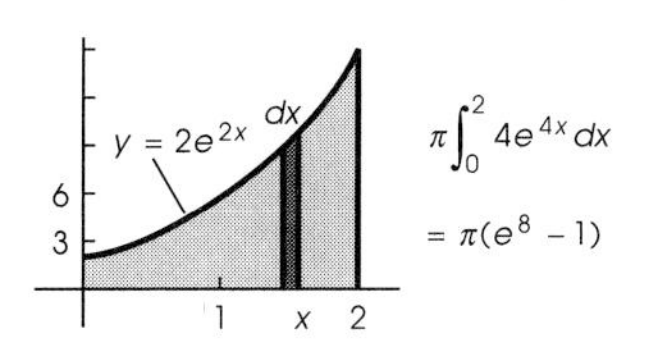


13.

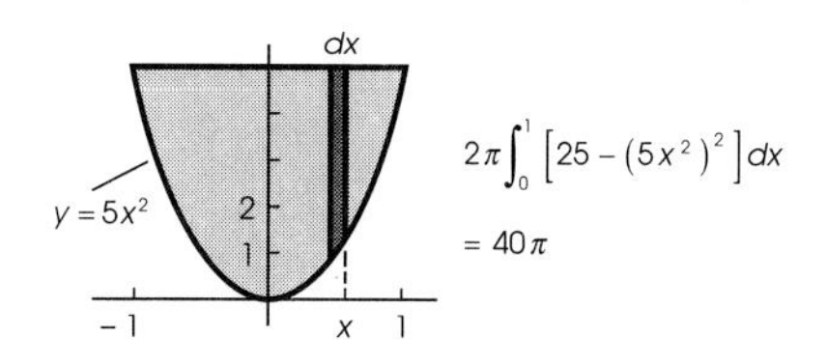


15.

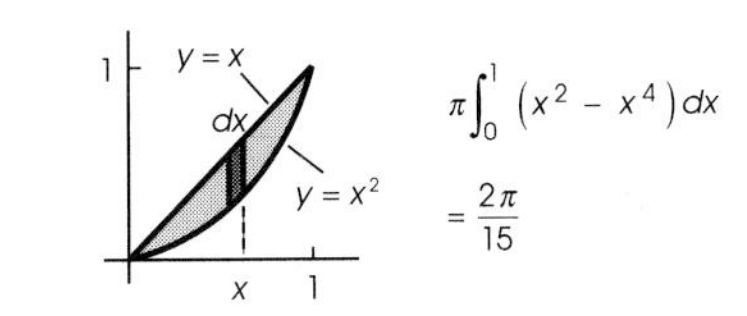


17.

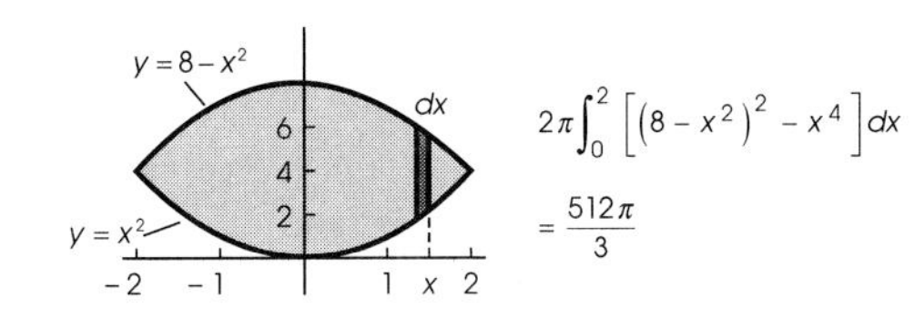


19.

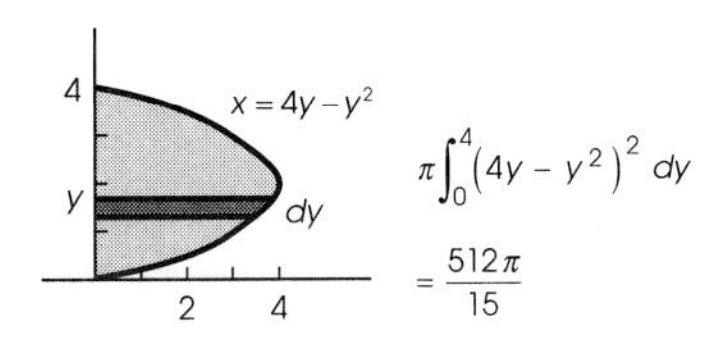


21.

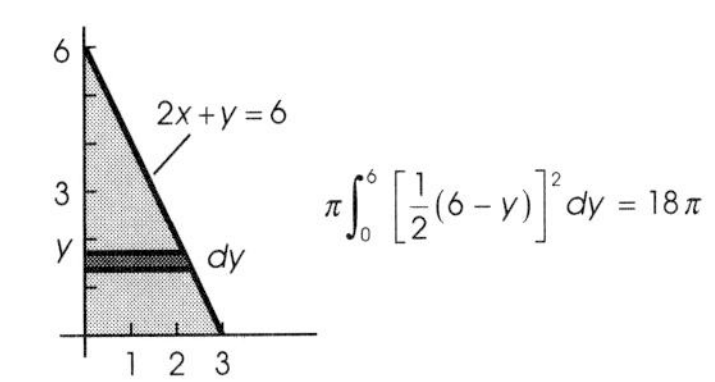


23.

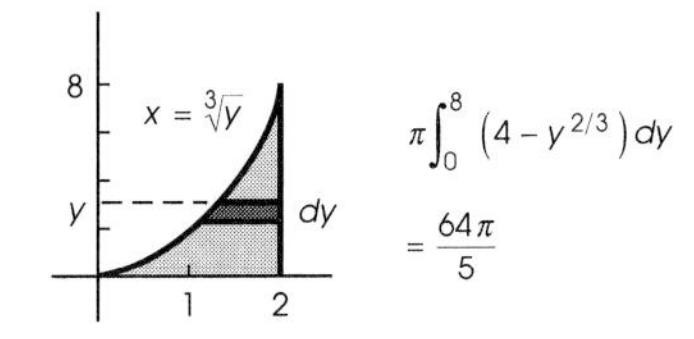

25.

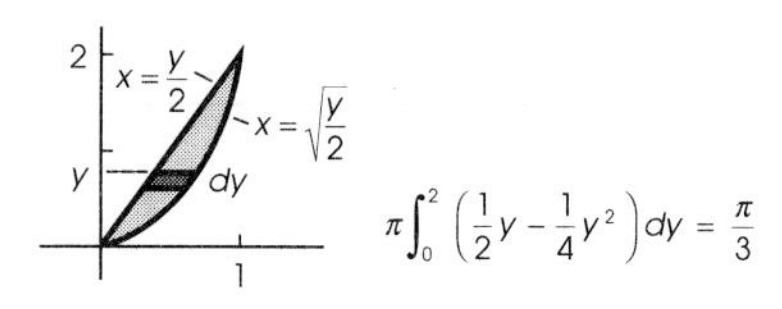


27. $423\pi/5 \approx 265.7787$
29. $972\pi/5 \approx 610.7256$
31. $72\pi$
33. $80\pi/3$
35. $7\pi/3$
37. $256\pi/3$
39. $3\pi e^4$

## Problem Set 9.3 (p. 294)

1. $\int_1^{16}(6t^2 + 30t^{1/4} + 100)\,dt = 10{,}434$
3. $50{,}000 + \int_0^8(450 + 35x^{4/3})\,dx = 55{,}520$
5. (a) $2000 + \int_0^{150}(30 - 0.12x)\,dx = \$6365$; (b) \$42.43
7. (a) $\int_0^{150}(50 + 0.36x - 0.003x^2)\,dx = \$8175$; (b) \$1810
9. 165
11. $p = \$4$
13. $p = \$100$
15. $p = \$100$
17.

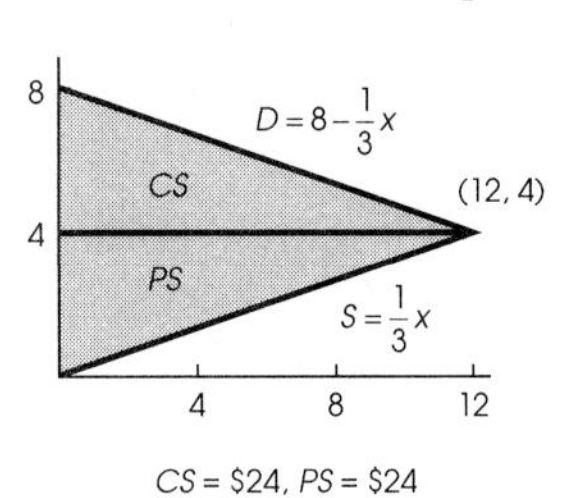


$CS = \$24$, $PS = \$24$

19.

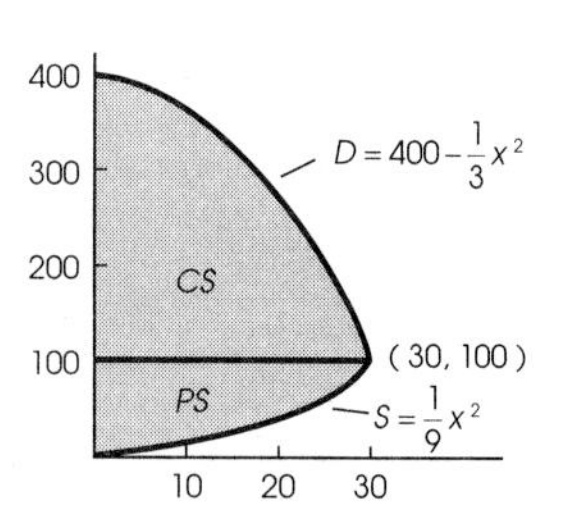


$CS = \$6000$, $PS = \$2000$

21.

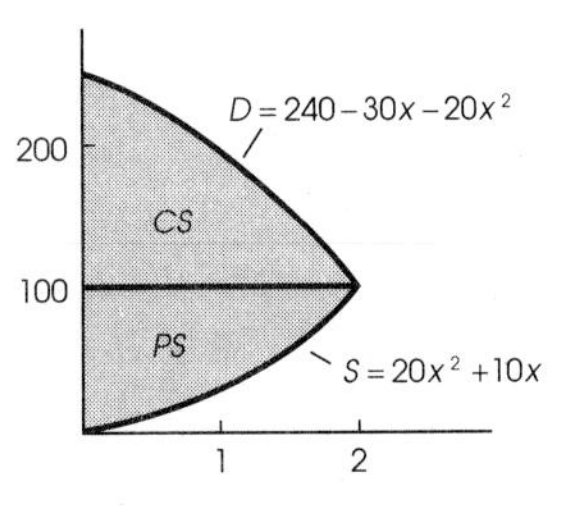


$CS = \$166.67$, $PS = \$126.67$

23. 144, \$3456
25. \$1272.59
27. $\frac{1}{2}$
29. $\frac{1}{6}$
31. $(3 - e)/(e - 1) \approx 0.1640$
33. \$34,995
35. 14
37. (a) \$18,000; (b) \$20,000

## Problem Set 9.4 (p. 300)

1. \$1481.64
3. \$5934.22
5. \$4393.24
7. \$161,383.94
9. \$170,918.48
11. $\frac{1}{2}$
13. $\frac{100}{9}$
15. $10000/81 \approx 123.4568$
17. \$88,888.89
19. \$101,234.57
21. \$9055.39
23. \$3441.69
25. \$6250
27. Diverges.
29. Converges, $\frac{1}{2}$
31. Diverges.
33. $\lim_{b\to\infty}\int_1^b x^{-p}\,dx = \lim_{b\to\infty}\left[x^{-p+1}/(-p+1)\right]_1^b = 1/(p-1)$
35. \$10,135,135.14

## Problem Set 9.5 (p. 306)

1. (a) $f(x) = \frac{1}{4}$ on $[1, 5]$; $f(x) = 0$ elsewhere; (b) 0.3
3. $\frac{1}{3}$
5. (a) $k = \frac{1}{12}$; (b) $\frac{3}{8}$
7. (a) $\frac{1}{2}\int_1^{e^2}(1/x)\,dx = \frac{1}{2}\ln(e^2) = 1$; (b) $\frac{1}{2}\ln 2 \approx 0.347$; (c) $a = e$
9. (a) $\frac{1}{144}\int_0^6(36 - x^2)\,dx = 1$;
   (b)

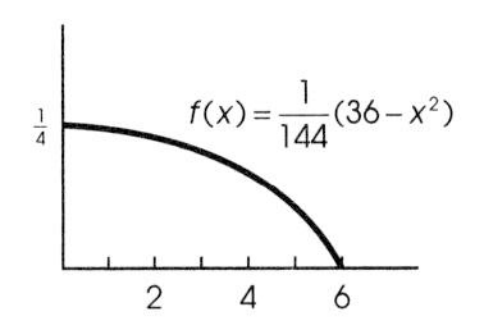


   (c) $325/432 \approx 0.752$; (d) $324/144 = 2.250$
11. 0.087
13. (a) 0.449; (b) 50 min
15. (a) $k = \frac{3}{4}$; (b) 0.061; (c) 1
17. (a) $k = 3$; (b) $\frac{3}{5}$
19. $1/\sqrt[3]{4}$
21. $\mu = 1/k$

## Chapter 9 Review Problem Set (p. 308)

1. $8 + 2\ln 3$
3. 8
5.

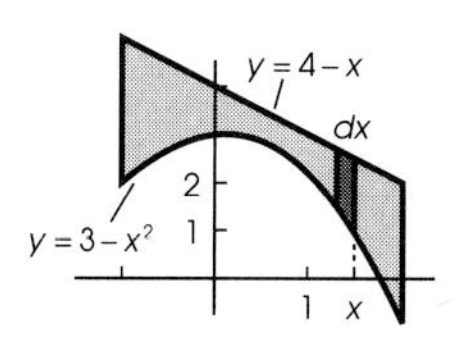


$A = \int_{-1}^{2} [4 - x - (3 - x^2)]\,dx = 4.5$

7.

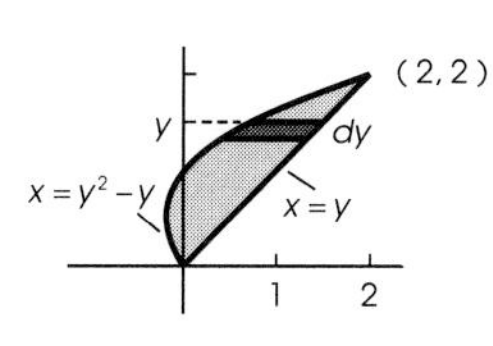


$A = \int_0^2 [y - (y^2 - y)]\,dy = \frac{4}{3}$

9. (a)

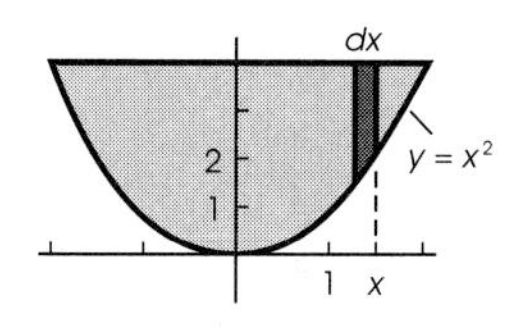


$V = 2\int_0^2 (\pi 4^2 - \pi x^4)\,dx = \frac{256\pi}{5}$

(b)

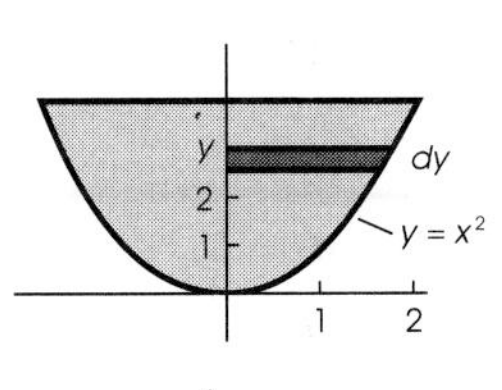


$V = \int_0^4 \pi y\,dy = 8\pi$

11. $\int_0^2 [\pi(4y - y^2)^2 - \pi y^4]\,dy = 32\pi/3$
13. (a) \$30,640; (b) \$14,360
15. \$16,000
17. $CS$ = \$37,944.34, $PS$ = \$54,000
19. $\frac{1}{24}$
21. (a) 10,990; (b) 110 people/yr
23. \$4416.10
25. \$657,894.74
27. (a) $\frac{1}{12}$; (b) $\frac{1}{16}$; (c) $\frac{32}{7}$
29. (a) $1/e$; (b) 8
31. (a) $\int_a^b |f(x)|\,dx$; (b) $\pi\int_a^b [f(x)]^2\,dx$
33. $a = 2$
35. $\int_a^b F(x)\,dx$
37. $\int_0^3 [1 + f(x)]^2\,dx$

## Problem Set 10.1 (p. 317)

1. (a) 13; (b) 4; (c) $3x^2$; (d) $2xh + h^2 - hy$
3. (a) $\frac{1}{256}$; (b) $\frac{1}{8}$; (c) $1/16x^3$; (d) $h/y^4$
5. (a) 4; (b) $7e^{-2}$; (c) $(3x + 1)e^{2x-4}$; (d) $3he^{y-4}$
7. (a) $-3$; (b) $-30$; (c) $3z^2$; (d) $2yz + xy - 3xz$
9. (a) 1; (b) $-1$; (c) $-2(z + 1)/(z + 4)$; (d) $(yz - 2x)/(xy - 2z)$
11. (a) 6; (b) 1176; (c) 0; (d) $(z + 2y)(y + 2x)(x + 2z)$
13. (a) $f(x, y) = 2x + 2y$, $g(x, y) = xy$; (b) 16, 12; (c) $y = 2x/(x - 2)$
15. (a) $V(x, y) = x^2y$; $S(x, y) = 2x^2 + 4xy$; (b) it is doubled; (c) $C(x, y) = 0.6x^2 + 0.72xy$
17. (a) 22,680; 10,080; (b) multiply by $\sqrt[4]{2}$; multiply by $\sqrt[4]{8}$; double.
19. (a) $(4, 1, -8)$, $(2, -6, 12)$; (b) $3x + 2y = 6$, $3x + z = 6$, $2y + z = 6$; (c)

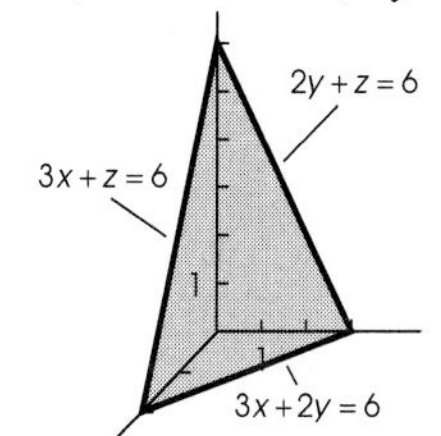


21.

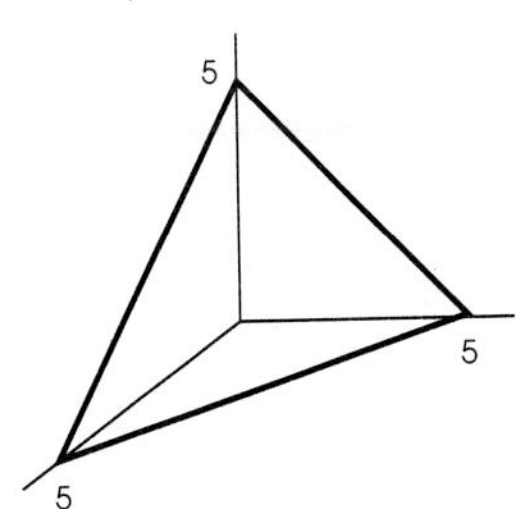


23.

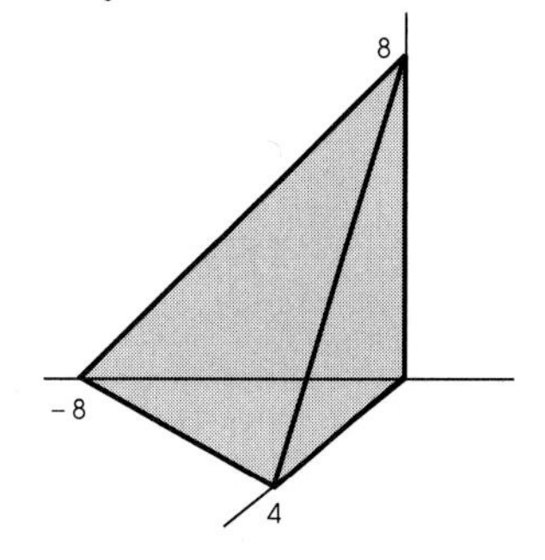

25.

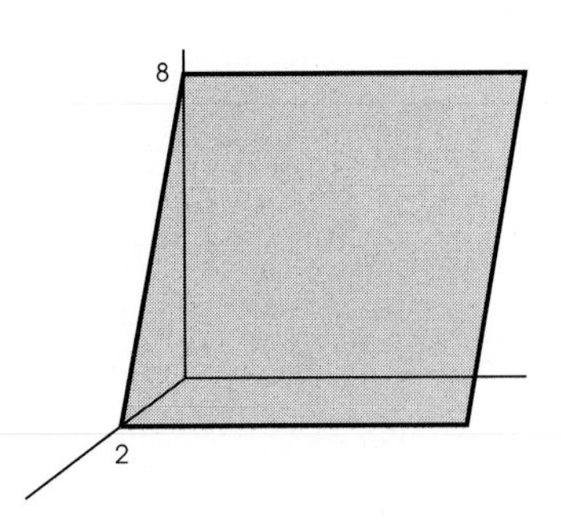


27.

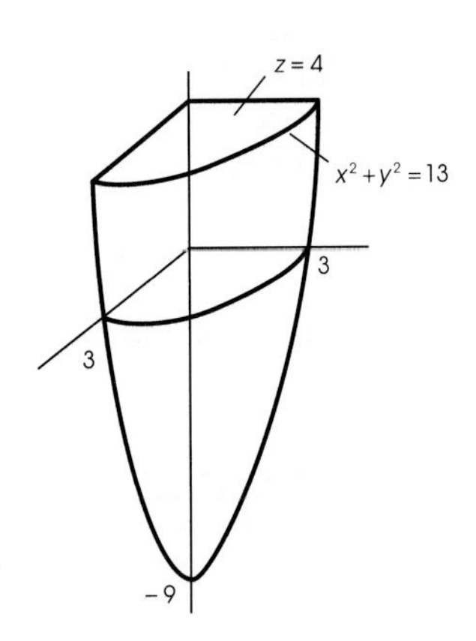


29.

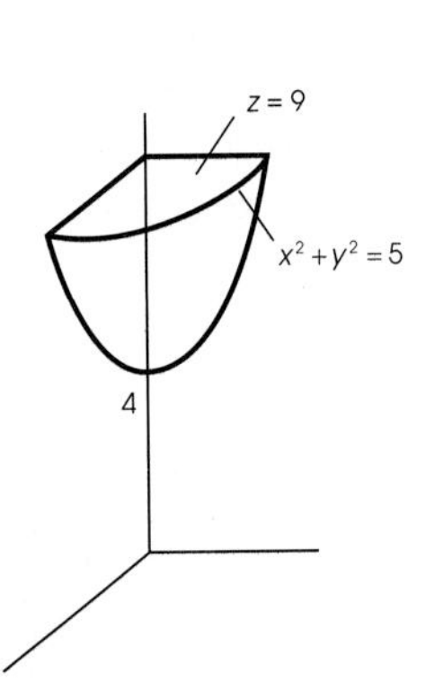


31.

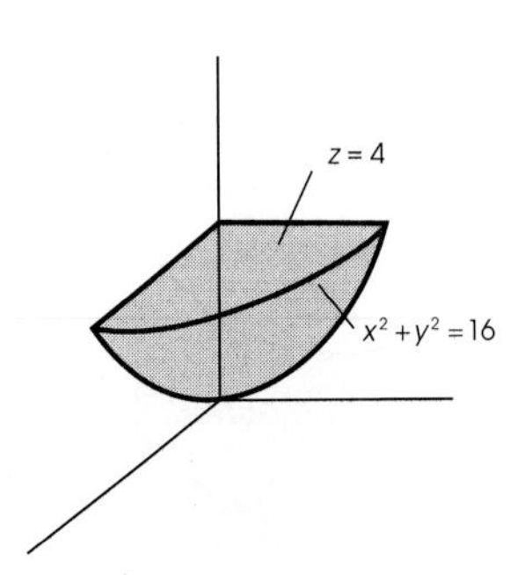


33.

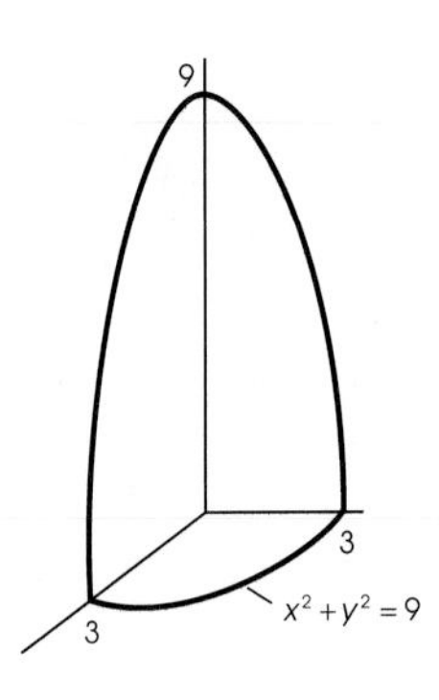


35.

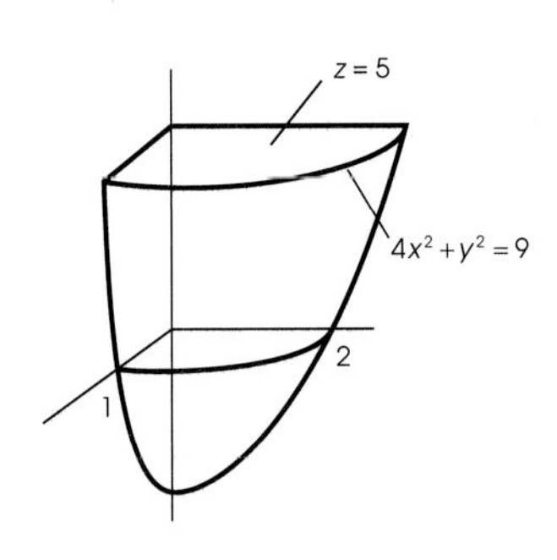


37. $P(x, y) = 30x + 30y - 2000$

39. $S(w, r) = 0.00003wr^2$; $S(2400, 60) = 259.2$ ft

41. (a)

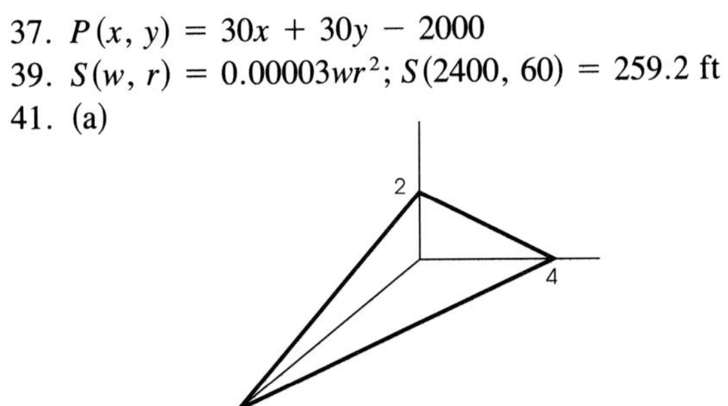


(b)

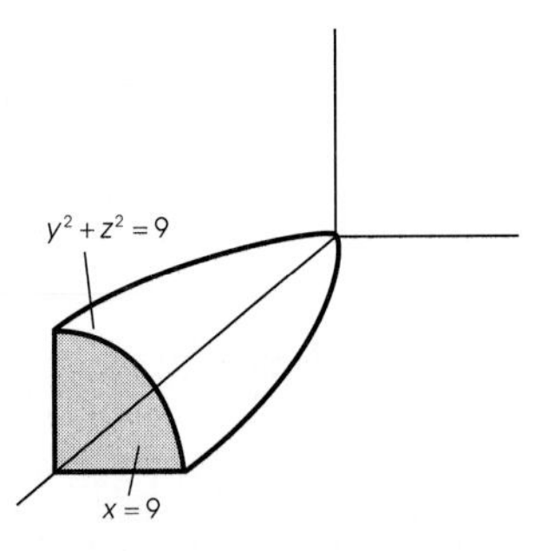

(c)

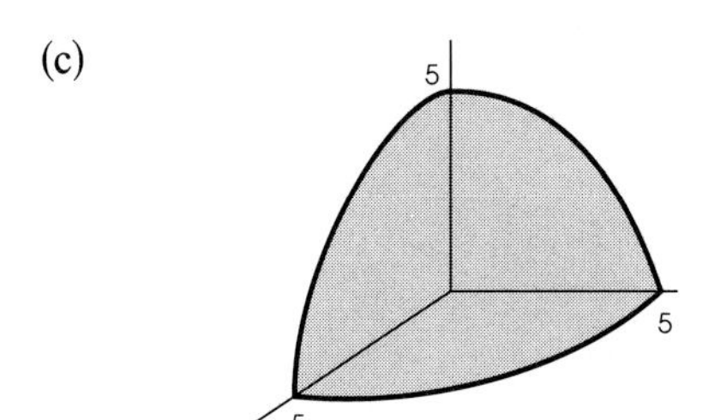


## Problem Set 10.2 (p. 324)

1. $2xy^3$, $3x^2y^2$
3. $y^2 - 8x$, $2xy + 3$
5. $2\sqrt{y}$, $xy^{-1/2} - 8y^{-3}$
7. $-4x^{-2}y^{-2}$, $-8x^{-1}y^{-3}$
9. $2xye^{2x}(x + 1)$, $x^2e^{2x}$
11. $6x(x^2 + 2y)^2 \ln y$, $(x^2 + 2y)^3y^{-1} + 6(x^2 + 2y)^2 \ln y$
13. $ye^{2y}(2xy + y^2)^{-1/2}$, $2(2xy + y^2)^{1/2}e^{2y} + (x + y)(2xy + y^2)^{-1/2}e^{2y}$
15. $3y/[x(2x + y)]$, $(4x - y)/[y(2x + y)]$
17. $2xy + 6yz$
19. $xye^{z+2} + xy^{-1}$
21. 6, −9
23. $\frac{2}{7}$, 3
25. 1, 2
27. −4
29. $-\frac{3}{4}$
31. About 0.00136%
33. About 208.14%
35. (a) $32\pi$; (b) $12\pi$
37. (a) $294H^{0.73}/W^{0.58}$, $511W^{0.42}/H^{0.27}$; (b) 1108 cm$^2$; (c) 749 cm$^2$
39. (a) $-4y$; (b) $6y^2 - 4x$; (c) $6y^2 - 4x$
41. (a) $y^4e^{xy^2}$; (b) $2ye^{xy^2}(xy^2 + 1)$; (c) $2ye^{xy^2}(xy^2 + 1)$
43. (a) $24x^2(x^2 + 2y) + 6(x^2 + 2y)^2$; (b) $24x(x^2 + 2y)$; (c) $24x(x^2 + 2y)$
45. Complementary.
47. Competitive.
49. $R_x(20, 30) = 180$. Increasing research from \$20 to \$21 million while keeping advertising at \$30 million will increase revenue by about \$180 million. $R_y(20, 30) = 40$. Increasing advertising from \$30 to \$31 million while keeping research at \$20 million will increase revenue by about \$40 million.
51. $T_x(4, 2) = -4$. As you move from (4, 2) to $(4 + \Delta x, 2)$, the temperature drops about 4 $\Delta x$ degrees. $T_y(4, 2) = -6$. As you move from (4, 2) to $(4, 2 + \Delta y)$, the temperature drops by about 6 $\Delta y$ degrees.
53. (a) $A(x, y, z) = (6x + 9y + 12z)/(x + y + z)$; (b) $A_x(10, 20, 10) = -\frac{3}{40} = -7.5$¢; (c) Increasing the first group of workers from 10 to 11 while keeping the second and third groups at 20 and 10 decreases the average hourly wage by about 7.5¢.
55. 34.5
57. 691

## Problem Set 10.3 (p. 330)

1. (2, −3): local min.
3. (−3, 5): local min.
5. (2, 2): local max.
7. (2, 2): local max.
9. (2, −2): saddle point; (−2, 2): saddle point.
11. (0, 0): saddle point; (4, 4): local min.
13. (0, 2): saddle point.
15. Length: 8 ft; width: 8 ft; depth: 4 ft
17. Length: 28 in., width: 14 in., height: 14 in.
19. $p = \$53.65$, $q = \$67.67$
21. $x = 30$, $y = 25$
23. (0, 0): saddle point; (8, 2): local min; (−8, −2): local min.
25. $x = \frac{20}{3}$, $y = 6$
27. $R(5, \frac{1}{2}) = 50/e \approx 18.39$

## Problem Set 10.4 (p. 337)

1. $f(8, 2) = 76$
3. $f(10, 5, 11) = 689$
5. Max value: $f(4, 2) = f(-4, -2) = 18$; min value: $f(-4, 2) = f(4, -2) = 2$
7. Length: 75 m; width: 37.5 m
9. $f(2, 1, 3) = 14$
11. $f(-4, -2, -8) = -42$
13. Length: 2 ft; width: 2 ft; height: 3 ft
15. $x = 300$, $y = 120$; 0.368
17. 9.75 in. by 9.75 in.
19. $\sqrt{3.5} \approx 1.87$
21. $x = 12$, $y = 14$, $z = 20$; \$740
23. $(5\sqrt{2}/2, \frac{5}{2}, \frac{5}{2})$ and $(-5\sqrt{2}/2, -\frac{5}{2}, -\frac{5}{2})$

## Problem Set 10.5 (p. 342)

1. $y = -\frac{7}{26}x + \frac{42}{13} = -0.269x + 3.231$
3. $y = 0.32x + 1.44$
5. $y = \frac{39}{35}x - \frac{13}{35} = 1.114x - 0.371$
7. $y = 0.8x + 1$
9. $y = 0.1469x + 1.381$; \$4,025; \$7,257
11. $y = 3.1x + 180.5$; \$258,000
13. $x = -135.3p + 2514$; 890
15. $y = 1.58e^{0.68x}$

## Problem Set 10.6 (p. 348)

1. $12x^3$
3. $48 + 8y$
5. $2x^2 \ln(\frac{3}{2})$
7. $\frac{1}{2}y^2(e^{14} - 1)$
9. $(e - 1)/x$
11. $x^{7/2} - x^{5/2}$
13. $-\frac{1}{4}x^3 + \frac{1}{2}x^2$
15. 208

17. $\frac{27}{2}$ 19. $\frac{52}{3}$

21. $-\frac{11}{70}$ 23. 3

25. $\displaystyle\int_0^1 \left[\int_{x^2}^{x} 1\, dy\right] dx = \frac{1}{6}$

27. $\displaystyle\int_0^3 \left[\int_{x}^{4x-x^2} 1\, dy\right] dx = \frac{9}{2}$

29.

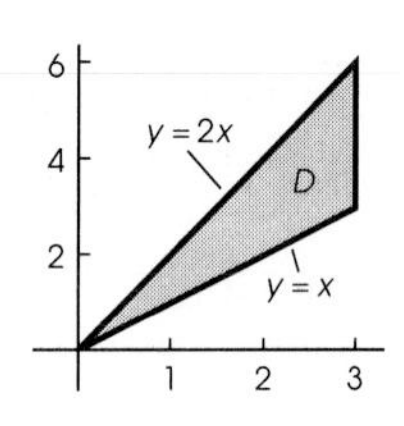


$\displaystyle\int_0^3 \left[\int_x^{2x} 1\, dy\right] dx$

31.

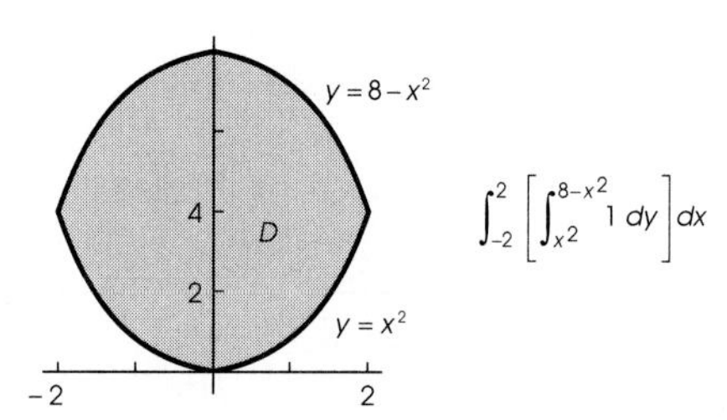


$\displaystyle\int_{-2}^{2} \left[\int_{x^2}^{8-x^2} 1\, dy\right] dx$

33.

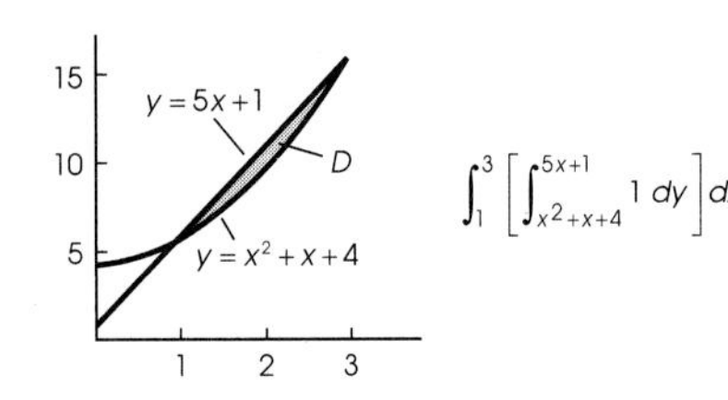


$\displaystyle\int_1^3 \left[\int_{x^2+x+4}^{5x+1} 1\, dy\right] dx$

35. $\frac{3}{35}$ 37. $\frac{1107}{20} = 55.35$

39. 28 41. $e^3 + e - 2$

43. 0.1714 45. 12

47. 36 49. $-\frac{7}{10}$

51. 0 53. \$141,400

## Chapter 10 Review Problem Set (p. 351)

1. (a) $-144$; (b) $x^9/2$; (c) 2
3. $f(r, h) = 4\pi r^2 + 4\pi(r + h)^2$; $85\pi$
5. $g(a, b) = \sqrt{3}\, ab^2/4$

7. (a)

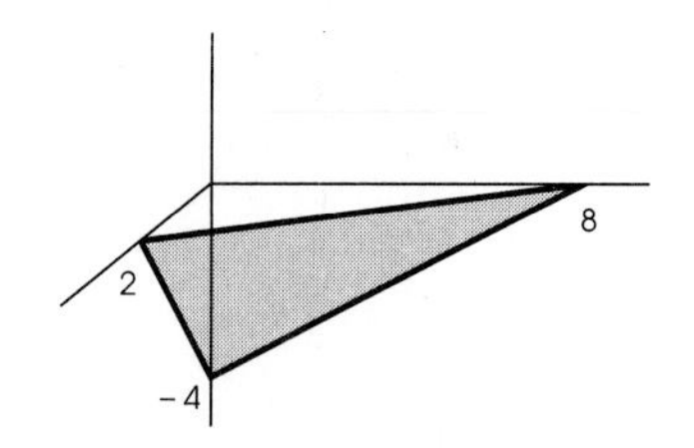


(b)

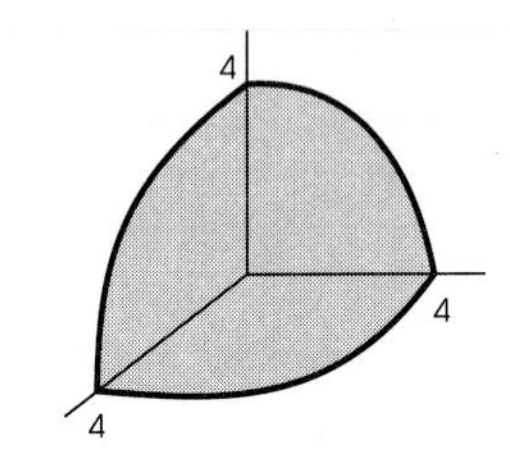


(c)

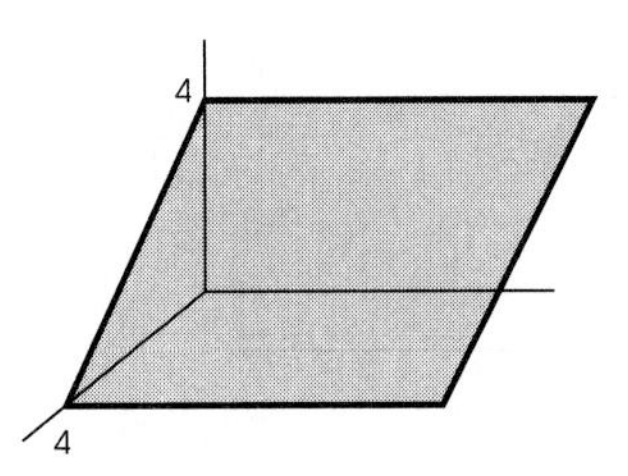


(d)

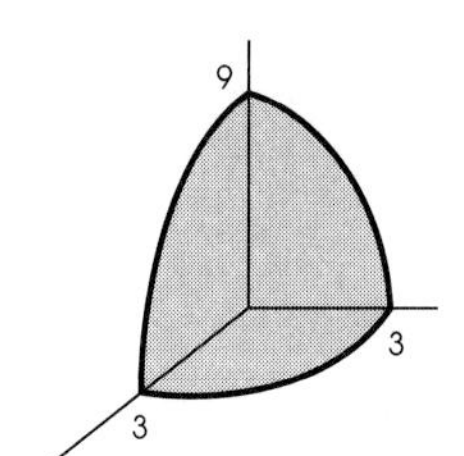


9. (a) $4x - 3y$; $-3$;
   (b) $2x^2/(2x + 3y) + 2x \ln(2x + 3y)$; $\frac{3}{2}$;
   (c) $2xye^{x^2y}$; 1;
   (d) $4(3x - 2y/x)^3(3 + 2y/x^2)$; $-216$
11. $-6$ 13. $-\frac{4}{9}$
15. $f_P(2000, 0.08) \approx 4.66$. Increasing $P$ by \$1 increases the amount by about \$4.66. $f_r(2000, 0.08) \approx 172{,}628$. Increasing $r$ by $\Delta r$ (that is, from 0.08 to $0.08 + \Delta r$) increases the amount by about \$172,628 $\Delta r$.
17. (0, 0): saddle point; (18, 6): local min.
19. $x = \frac{1}{3}$, $y = \frac{2}{3}$, $z = \frac{2}{3}$ 21. $p = 800$, $q = 1000$
23. $f(3/\sqrt{2}, \sqrt{2}) = 3$
25. (a) $y = 3.686x + 27.6$; (b) \$6446
27. (a) 18; (b) $\frac{19}{60}$ 29. $\displaystyle\int_{-2}^{2} \left[\int_{x^2-2}^{6-x^2} 1\, dy\right] dx$
31. $\frac{9}{2} \ln 3 - 2$

33. $f_x(a, b) = \lim_{h \to 0} \dfrac{f(a + h, b) - f(a, b)}{h}$

35. (a) $k = 14{,}000$;
 (b) it decreases the safe load by about 70 lb.

37. $\int_0^2 \left[ \int_0^{2y} e^{y^2}\, dx \right] dy = e^4 - 1$

## Basic Algebra Review Problem Set (p. 364)

1. $22$
2. $8$
3. $9$
4. $80$
5. $169$
6. $100$
7. $0.41\overline{6}$, $0.417$
8. $0.458\overline{3}$, $0.458$
9. $0.\overline{714285}$, $0.714$
10. $0.\overline{384615}$, $0.385$
11. $\frac{8}{15}$, $0.5333$
12. $-\frac{1}{9}$, $-0.1111$
13. $-\frac{2}{7}$, $-0.2857$
14. $-\frac{1}{19}$, $-0.0526$
15. $\frac{7}{8}$, $0.8750$
16. $\frac{101}{8}$, $12.6250$
17. $-\frac{7}{6}$, $-1.1667$
18. $\frac{41}{3}$, $13.6667$
19. $\frac{19}{59}$, $\frac{13}{40}$, $0.3\overline{3}$, $0.334$
20. $\frac{41}{79}$, $\frac{13}{25}$, $0.\overline{52}$, $0.5\overline{3}$
21. $0.5958$
22. $-13.4259$
23. $-8.9645$
24. $1.6753$
25. $0.0281$
26. $4.4586$
27. $6$
28. $1125$
29. $x^{5/3}/y^{1/4}$
30. $a^4b^5c^6$
31. $a^{8/3}b^{1/2}$
32. $a$
33. $9x^2y^2 - 12xy + 4$
34. $1 - 8xy + 16x^2y^2$
35. $9x^2y^2 - 4$
36. $1 - 16x^2y^2$
37. $2x^2 + 5xy - 12y^2$
38. $20x^2 - 14x + 2$
39. $a - 2a^{1/2}b^{1/2} + b$
40. $a^{2/3} + 2a + a^{4/3}$
41. $a - 2 + 1/a$
42. $x^4 - y^4$
43. $(5a - 3)(5a + 3)$
44. $(5a - 1)^2$
45. $(5a + 3)(a - 1)$
46. $(a - 1)(a + 1)(a^2 + 1)$
47. $(3xy - z)(xy + 2z)$
48. $(x + 2 - y)(x + 2 + y)$
49. $-5/(x - y)$
50. $(3b - 2a + b^2)/(a^2b^3)$
51. $18/(a^2 - 9)$
52. $(5x - 2)/(x^2 + x - 6)$
53. $(x - 2)/[3x(x + 4)]$
54. $2y(x + 4y)/[x(x + 2y)]$
55. $-(x + y)/y^2$
56. $x^2 + y^2$
57. $a - b^2$
58. $x - 4\sqrt{x}\sqrt{y} + 4y$
59. $2h/\sqrt{x + h}$
60. $(2 + h)/(x + 2 + h)$
61. $\sqrt{4x^4 + 9}$
62. $2x^2 + 3$
63. $-\frac{6}{5}$
64. $\frac{8}{11}$
65. $-5, 0$
66. $-9, 4$
67. $-4, 2$
68. $\frac{3}{2}$
69. $-1, 1$
70. $-2, 0, 3$
71. $-\frac{1}{2}, \frac{2}{3}$
72. $\frac{3}{4}, 4$
73. $\pm\sqrt{3}$
74. $-3, 1$
75. $1 \pm \sqrt{3}$
76. $-2 \pm 2\sqrt{2}$
77. $(5 \pm \sqrt{17})/4$
78. $(-7 \pm \sqrt{13})/6$
79. $x \geq \frac{7}{2}$ 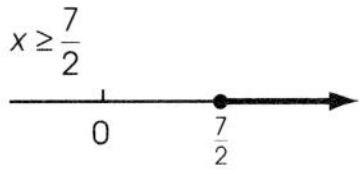
80. $x > -\frac{8}{3}$ 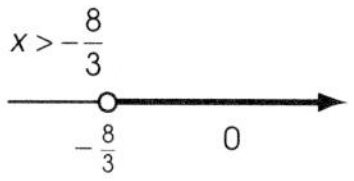
81. $-3 < x < 2$
82. $x \leq -6$ or $x \geq 3$ 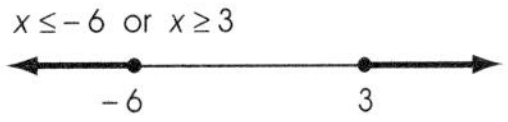
83. $x < -1$ or $x > \frac{3}{2}$ 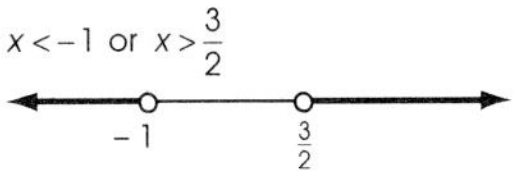
84. $-\frac{2}{3} \leq x \leq 2$ 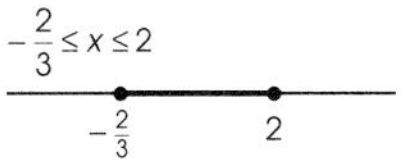
85. $x \leq -2$ or $-1 \leq x \leq 2$ 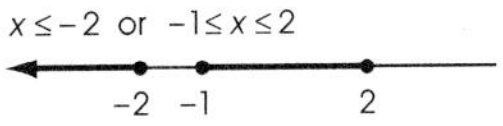
86. $-2 < x < 0$ or $x > 4$

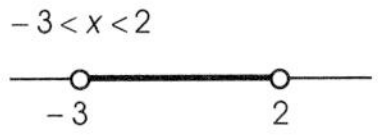

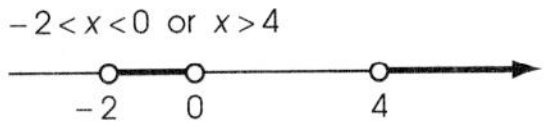

# Index

**E**



**F**

# GEOMETRY

## Triangles

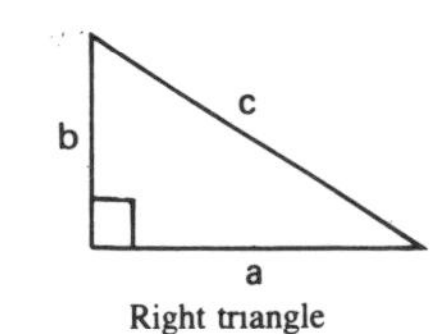

Right triangle

Pythagorean Theorem

$a^2 + b^2 = c^2$

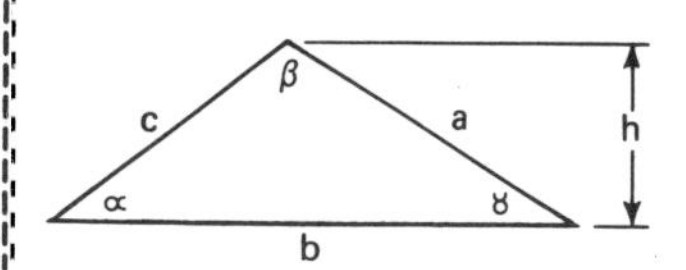

Any triangle

Angles $\alpha + \beta + \gamma = 180°$

Area $A = \frac{1}{2}bh$

## Circles

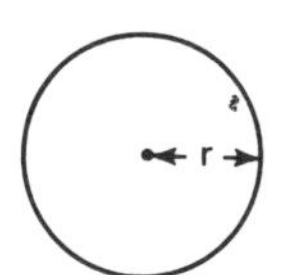


Circumference $C = 2\pi r$

Area $A = \pi r^2$

## Cylinders

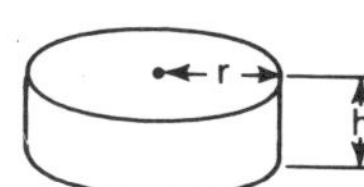


Surface area $S = 2\pi r^2 + 2\pi rh$

Volume $V = \pi r^2 h$

## Cones

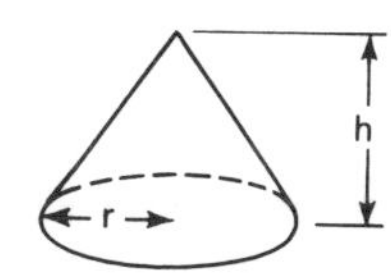


Surface area $S = \pi r^2 + \pi r\sqrt{r^2 + h^2}$

Volume $V = \dfrac{1}{3}\pi r^2 h$

## Spheres

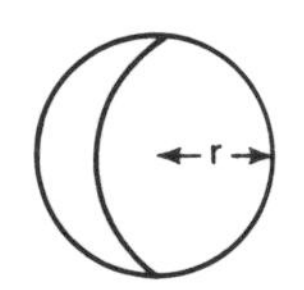


Surface area $S = 4\pi r^2$

Volume $V = \dfrac{4}{3}\pi r^3$

## Conversions

1 inch = 2.54 centimeters
1 liter = 1000 cubic centimeters
1 kilogram = 2.20 pounds
1 kilometer = .62 miles
1 liter = 1.057 quarts
1 pound = 453.6 grams
$\pi$ radians = 180 degrees

**FORMULA CARD**
to accompany
**Applied Calculus for Management, Social, and Life Sciences**
**Dale Varberg and Walter Fleming**
PRENTICE HALL, Englewood Cliffs, N. J. 07632

# ALGEBRA

## Exponents

$a^m a^n = a^{m+n}$

$(a^m)^n = a^{mn}$

$\dfrac{a^m}{a^n} = a^{m-n}$

$(ab)^n = a^n b^n$

$\left(\dfrac{a}{b}\right)^n = \dfrac{a^n}{b^n}$

## Radicals

$(\sqrt[n]{a})^n = a$

$\sqrt[n]{a^n} = a$ if $a \geq 0$

$\sqrt[n]{ab} = \sqrt[n]{a}\sqrt[n]{b}$

$\sqrt[n]{\dfrac{a}{b}} = \dfrac{\sqrt[n]{a}}{\sqrt[n]{b}}$

## Logarithms

$\log_a MN = \log_a M + \log_a N \qquad \log_a (M/N) = \log_a M - \log_a N$

$\log_a (N^P) = P \log_a N$

## Quadratic Formula

Solutions to $ax^2 + bx + c = 0$ are $x = \dfrac{-b \pm \sqrt{b^2 - 4ac}}{2a}$

## Factoring Formulas

$x^2 - y^2 = (x - y)(x + y)$

$x^2 + 2xy + y^2 = (x + y)^2$

$x^2 - 2xy + y^2 = (x - y)^2$

$x^3 - y^3 = (x - y)(x^2 + xy + y^2)$

$x^3 + y^3 = (x + y)(x^2 - xy + y^2)$

$x^3 + 3x^2y + 3xy^2 + y^3 = (x + y)^3$

## Binomial Formula

$(x + y)^n = {}_nC_0 x^n y^0 + {}_nC_1 x^{n-1} y^1 + \cdots + {}_nC_{n-1} x^1 y^{n-1} + {}_nC_n x^0 y^n$

${}_nC_r = \dfrac{n!}{(n - r)!\, r!} = \dfrac{n(n - 1)\cdots(n - r + 1)}{r(r - 1)\cdots 3 \cdot 2 \cdot 1}$

${}_nC_0 = {}_nC_n = 1$

# PARABOLAS

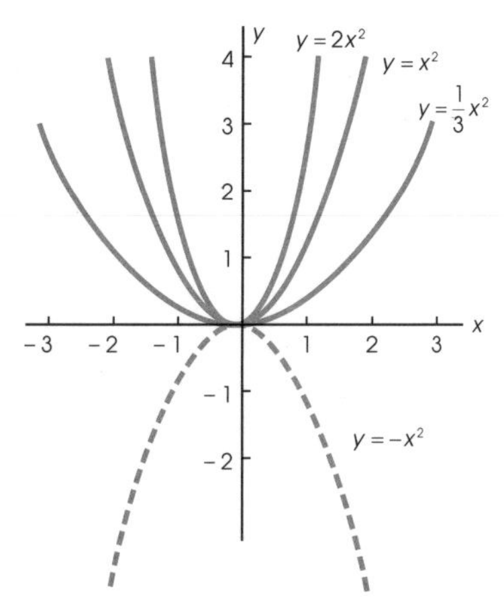


Multiplying by a constant

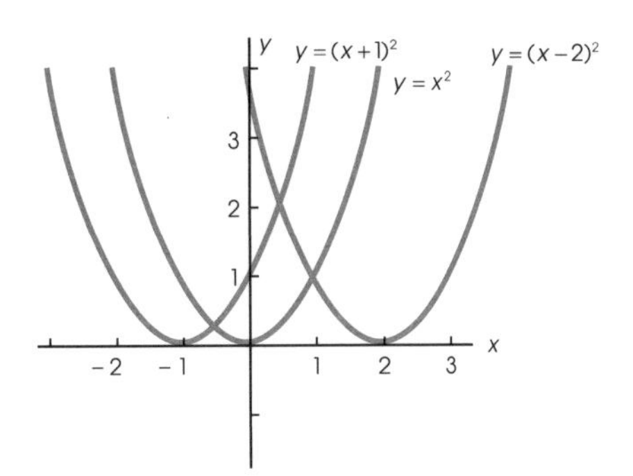


Replacing $x$ by $x - c$

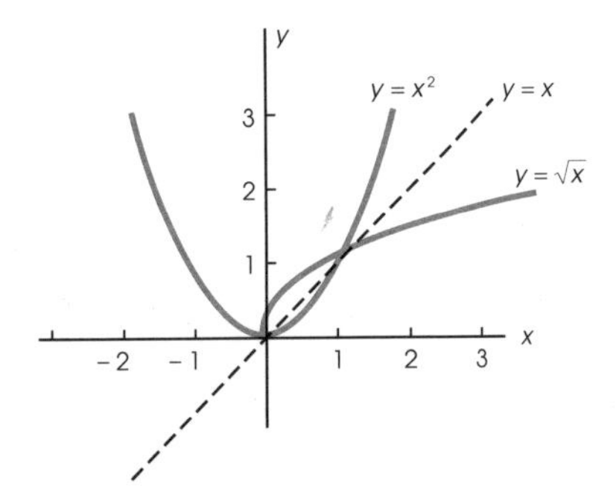


Powers and roots

# CALCULUS

## Derivatives

$$\frac{d}{dx} x^r = rx^{r-1}$$

$$\frac{d}{dx} \ln x = \frac{1}{x}, \quad x > 0$$

$$\frac{d}{dx} \log_a x = \frac{1}{x \ln a}, \quad x > 0$$

$$\frac{d}{dx} e^x = e^x$$

$$\frac{d}{dx} a^x = a^x \ln a$$

$$\frac{dy}{dx} = \frac{dy}{du}\frac{du}{dx}$$

## Integrals

$$\int u^r \, du = \frac{u^{r+1}}{r+1} + C, \quad r \neq -1$$

$$\int \frac{1}{u} \, du = \ln |u| + C$$

$$\int e^u \, du = e^u + C$$

$$\int a^u \, du = \frac{a^u}{\ln a} + C$$

$$\int [g(x)]^r g'(x) \, dx = \frac{[g(x)]^{r+1}}{r+1} + C, \quad r \neq -1$$

$$\int u \, dv = uv - \int v \, du$$

## Sums

$$1 + 2 + 3 + \cdots + n = \frac{n(n+1)}{2}$$

$$1^2 + 2^2 + 3^2 + \cdots + n^2 = \frac{n(n+1)(2n+1)}{6}$$

$$1^3 + 2^3 + 3^3 + \cdots + n^3 = \left[\frac{n(n+1)}{2}\right]^2$$

## Two Important Constants

$$\pi = 3.14159265358979\ldots$$

$$e = \lim_{h \to 0} (1 + h)^{1/h} = 2.718281828\ldots$$

# Table of Integrals

## ELEMENTARY FORMS

1. $\int u\,dv = uv - \int v\,du$
2. $\int u^n\,du = \frac{1}{n+1}u^{n+1} + C \quad \text{if } n \neq -1$
3. $\int \frac{du}{u} = \ln|u| + C$
4. $\int e^u\,du = e^u + C$
5. $\int a^u\,du = \frac{a^u}{\ln a} + C$
6. $\int \sin u\,du = -\cos u + C$
7. $\int \cos u\,du = \sin u + C$
8. $\int \sec^2 u\,du = \tan u + C$
9. $\int \csc^2 u\,du = -\cot u + C$
10. $\int \sec u \tan u\,du = \sec u + C$
11. $\int \csc u \cot u\,du = -\csc u + C$
12. $\int \tan u\,du = \ln|\sec u| + C$
13. $\int \cot u\,du = \ln|\sin u| + C$
14. $\int \sec u\,du = \ln|\sec u + \tan u| + C$
15. $\int \csc u\,du = \ln|\csc u - \cot u| + C$
16. $\int \frac{du}{\sqrt{a^2 - u^2}} = \sin^{-1}\frac{u}{a} + C$
17. $\int \frac{du}{a^2 + u^2} = \frac{1}{a}\tan^{-1}\frac{u}{a} + C$
18. $\int \frac{du}{a^2 - u^2} = \frac{1}{2a}\ln\left|\frac{u+a}{u-a}\right| + C$
19. $\int \frac{du}{u\sqrt{u^2 - a^2}} = \frac{1}{a}\sec^{-1}\left|\frac{u}{a}\right| + C$

## TRIGONOMETRIC FORMS

20. $\int \sin^2 u\,du = \frac{1}{2}u - \frac{1}{4}\sin 2u + C$
21. $\int \cos^2 u\,du = \frac{1}{2}u + \frac{1}{4}\sin 2u + C$
22. $\int \tan^2 u\,du = \tan u - u + C$
23. $\int \cot^2 u\,du = -\cot u - u + C$
24. $\int \sin^3 u\,du = -\frac{1}{3}(2 + \sin^2 u)\cos u + C$
25. $\int \cos^3 u\,du = \frac{1}{3}(2 + \cos^2 u)\sin u + C$
26. $\int \tan^3 u\,du = \frac{1}{2}\tan^2 u + \ln|\cos u| + C$
27. $\int \cot^3 u\,du = -\frac{1}{2}\cot^2 u - \ln|\sin u| + C$
28. $\int \sec^3 u\,du = \frac{1}{2}\sec u \tan u + \frac{1}{2}\ln|\sec u + \tan u| + C$
29. $\int \csc^3 u\,du = -\frac{1}{2}\csc u \cot u + \frac{1}{2}\ln|\csc u - \cot u| + C$
30. $\int \sin au \sin bu\,du = \frac{\sin(a-b)u}{2(a-b)} - \frac{\sin(a+b)u}{2(a+b)} + C \quad \text{if } a^2 \neq b^2$
31. $\int \cos au \cos bu\,du = \frac{\sin(a-b)u}{2(a-b)} + \frac{\sin(a+b)u}{2(a+b)} + C \quad \text{if } a^2 \neq b^2$
32. $\int \sin au \cos bu\,du = -\frac{\cos(a-b)u}{2(a-b)} - \frac{\cos(a+b)u}{2(a+b)} + C \quad \text{if } a^2 \neq b^2$
33. $\int \sin^n u\,du = -\frac{1}{n}\sin^{n-1} u \cos u + \frac{n-1}{n}\int \sin^{n-2} u\,du$
34. $\int \cos^n u\,du = \frac{1}{n}\cos^{n-1} u \sin u + \frac{n-1}{n}\int \cos^{n-2} u\,du$
35. $\int \tan^n u\,du = \frac{1}{n-1}\tan^{n-1} u - \int \tan^{n-2} u\,du \quad \text{if } n \neq 1$
36. $\int \cot^n u\,du = \frac{-1}{n-1}\cot^{n-1} u - \int \cot^{n-2} u\,du \quad \text{if } n \neq 1$
37. $\int \sec^n u\,du = \frac{1}{n-1}\sec^{n-2} u \tan u + \frac{n-2}{n-1}\int \sec^{n-2} u\,du \quad \text{if } n \neq 1$
38. $\int \csc^n u\,du = \frac{-1}{n-1}\csc^{n-2} u \cot u + \frac{n-2}{n-1}\int \csc^{n-2} u\,du \quad \text{if } n \neq 1$

39a. $\int \sin^n u \cos^m u\,du = -\frac{\sin^{n-1} u \cos^{m+1} u}{n+m} + \frac{n-1}{n+m}\int \sin^{n-2} u \cos^m u\,du \quad \text{if } n \neq -m$

39b. $\int \sin^n u \cos^m u\,du = \frac{\sin^{n+1} u \cos^{m-1} u}{n+m} + \frac{m-1}{n+m}\int \sin^n u \cos^{m-2} u\,du \quad \text{if } m \neq -n$

40. $\int u \sin u\,du = \sin u - u \cos u + C$
41. $\int u \cos u\,du = \cos u + u \sin u + C$
42. $\int u^n \sin u\,du = -u^n \cos u + n\int u^{n-1} \cos u\,du$
43. $\int u^n \cos u\,du = u^n \sin u - n\int u^{n-1} \sin u\,du$

## FORMS INVOLVING $\sqrt{u^2 \pm a^2}$

**44** $$\int \sqrt{u^2 \pm a^2}\, du = \frac{u}{2}\sqrt{u^2 \pm a^2} \pm \frac{a^2}{2}\ln|u + \sqrt{u^2 \pm a^2}| + C$$

**45** $$\int \frac{du}{\sqrt{u^2 \pm a^2}} = \ln|u + \sqrt{u^2 \pm a^2}| + C$$

**46** $$\int \frac{\sqrt{u^2 + a^2}}{u}\, du = \sqrt{u^2 + a^2} - a\ln\left(\frac{a + \sqrt{u^2 + a^2}}{u}\right) + C$$

**47** $$\int \frac{\sqrt{u^2 - a^2}}{u}\, du = \sqrt{u^2 - a^2} - a\sec^{-1}\frac{u}{a} + C$$

**48** $$\int u^2\sqrt{u^2 \pm a^2}\, du = \frac{u}{8}(2u^2 \pm a^2)\sqrt{u^2 \pm a^2} - \frac{a^4}{8}\ln|u + \sqrt{u^2 \pm a^2}| + C$$

**49** $$\int \frac{u^2\, du}{\sqrt{u^2 \pm a^2}} = \frac{u}{2}\sqrt{u^2 \pm a^2} \mp \frac{a^2}{2}\ln|u + \sqrt{u^2 \pm a^2}| + C$$

**50** $$\int \frac{du}{u^2\sqrt{u^2 \pm a^2}} = \mp\frac{\sqrt{u^2 \pm a^2}}{a^2 u} + C$$

**51** $$\int \frac{\sqrt{u^2 \pm a^2}}{u^2}\, du = -\frac{\sqrt{u^2 \pm a^2}}{u} + \ln|u + \sqrt{u^2 \pm a^2}| + C$$

**52** $$\int \frac{du}{(u^2 \pm a^2)^{3/2}} = \frac{\pm u}{a^2\sqrt{u^2 \pm a^2}} + C$$

**53** $$\int (u^2 \pm a^2)^{3/2}\, du = \frac{u}{8}(2u^2 \pm 5a^2)\sqrt{u^2 \pm a^2} + \frac{3a^4}{8}\ln|u + \sqrt{u^2 \pm a^2}| + C$$

## FORMS INVOLVING $\sqrt{a^2 - u^2}$

**54** $$\int \sqrt{a^2 - u^2}\, du = \frac{u}{2}\sqrt{a^2 - u^2} + \frac{a^2}{2}\sin^{-1}\frac{u}{a} + C$$

**55** $$\int \frac{\sqrt{a^2 - u^2}}{u}\, du = \sqrt{a^2 - u^2} - a\ln\left|\frac{a + \sqrt{a^2 - u^2}}{u}\right| + C$$

**56** $$\int \frac{u^2\, du}{\sqrt{a^2 - u^2}} = -\frac{u}{2}\sqrt{a^2 - u^2} + \frac{a^2}{2}\sin^{-1}\frac{u}{a} + C$$

**57** $$\int u^2\sqrt{a^2 - u^2}\, du = \frac{u}{8}(2u^2 - a^2)\sqrt{a^2 - u^2} + \frac{a^4}{8}\sin^{-1}\frac{u}{a} + C$$

**58** $$\int \frac{du}{u^2\sqrt{a^2 - u^2}} = -\frac{\sqrt{a^2 - u^2}}{a^2 u} + C$$

**59** $$\int \frac{\sqrt{a^2 - u^2}}{u^2}\, du = -\frac{\sqrt{a^2 - u^2}}{u} - \sin^{-1}\frac{u}{a} + C$$

**60** $$\int \frac{du}{u\sqrt{a^2 - u^2}} = -\frac{1}{a}\ln\left|\frac{a + \sqrt{a^2 - u^2}}{u}\right| + C$$

**61** $$\int \frac{du}{(a^2 - u^2)^{3/2}} = \frac{u}{a^2\sqrt{a^2 - u^2}} + C$$

**62** $$\int (a^2 - u^2)^{3/2}\, du = \frac{u}{8}(5a^2 - 2u^2)\sqrt{a^2 - u^2} + \frac{3a^4}{8}\sin^{-1}\frac{u}{a} + C$$

## EXPONENTIAL AND LOGARITHMIC FORMS

**63** $$\int ue^u\, du = (u - 1)e^u + C$$

**64** $$\int u^n e^u\, du = u^n e^u - n\int u^{n-1}e^u\, du$$

**65** $$\int \ln u\, du = u\ln u - u + C$$

**66** $$\int u^n \ln u\, du = \frac{u^{n+1}}{n+1}\ln u - \frac{u^{n+1}}{(n+1)^2} + C$$

**67** $$\int e^{au}\sin bu\, du = \frac{e^{au}}{a^2 + b^2}(a\sin bu - b\cos bu) + C$$

**68** $$\int e^{au}\cos bu\, du = \frac{e^{au}}{a^2 + b^2}(a\cos bu + b\sin bu) + C$$

## INVERSE TRIGONOMETRIC FORMS

**69** $$\int \sin^{-1} u\, du = u\sin^{-1} u + \sqrt{1 - u^2} + C$$

**70** $$\int \tan^{-1} u\, du = u\tan^{-1} u - \frac{1}{2}\ln(1 + u^2) + C$$

**71** $$\int \sec^{-1} u\, du = u\sec^{-1} u - \ln|u + \sqrt{u^2 - 1}| + C$$

**72** $$\int u\sin^{-1} u\, du = \frac{1}{4}(2u^2 - 1)\sin^{-1} u + \frac{u}{4}\sqrt{1 - u^2} + C$$

**73** $$\int u\tan^{-1} u\, du = \frac{1}{2}(u^2 + 1)\tan^{-1} u - \frac{u}{2} + C$$

**74** $$\int u\sec^{-1} u\, du = \frac{u^2}{2}\sec^{-1} u - \frac{1}{2}\sqrt{u^2 - 1} + C$$

**75** $$\int u^n\sin^{-1} u\, du = \frac{u^{n+1}}{n+1}\sin^{-1} u - \frac{1}{n+1}\int \frac{u^{n+1}}{\sqrt{1 - u^2}}\, du + C \quad \text{if } n \neq -1$$

**76** $$\int u^n\tan^{-1} u\, du = \frac{u^{n+1}}{n+1}\tan^{-1} u - \frac{1}{n+1}\int \frac{u^{n+1}}{1 + u^2}\, du + C \quad \text{if } n \neq -1$$

**77** $$\int u^n\sec^{-1} u\, du = \frac{u^{n+1}}{n+1}\sec^{-1} u - \frac{1}{n+1}\int \frac{u^n}{\sqrt{u^2 - 1}}\, du + C \quad \text{if } n \neq -1$$